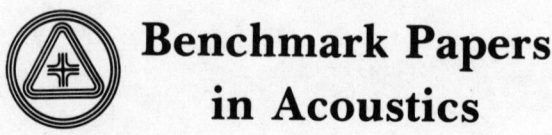

Benchmark Papers in Acoustics

Series Editor: R. Bruce Lindsay
Brown University

Published Volumes and Volumes in Preparation

UNDERWATER SOUND
  Vernon M. Albers
ACOUSTICS: Historical and Philosophical Development
  R. Bruce Lindsay
SPEECH SYNTHESIS
  James L. Flanagan and L. R. Rabiner
PHYSICAL ACOUSTICS
  R. Bruce Lindsay
ARCHITECTURAL ACOUSTICS
  Thomas D. Northwood
MUSICAL ACOUSTICS: The Violin Family
  Carleen Hutchins
MUSICAL ACOUSTICS: Piano and Wind Instruments
  Earl L. Kent
PSYCHOLOGICAL ACOUSTICS
  Arnold M. Small, Jr.
PHYSIOLOGICAL ACOUSTICS
  Arnold M. Small, Jr., and Joel S. Wernick
LIGHT AND SOUND INTERACTION
  Osman K. Mawardi
VIBRATION PROBLEMS
  Arturs Kalnins
ACOUSTICAL INSTRUMENTATION
  Benjamin B. Bauer
BIOACOUSTICS
  Floyd Dunn

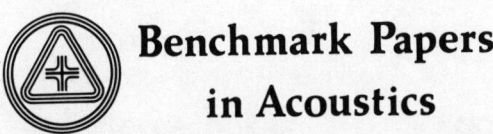

Benchmark Papers in Acoustics

———— A *BENCHMARK*® Books Series ————

PHYSICAL ACOUSTICS

Edited by
R. BRUCE LINDSAY
Brown University

Dowden, Hutchinson & Ross, Inc.
Stroudsburg, Pennsylvania

Copyright © 1974 by **Dowden, Hutchinson & Ross, Inc.**
Benchmark Papers in Acoustics, Volume 4
Library of Congress Card Number: 73-12619
ISBN: 87933-040-6

All rights reserved. No part of this book covered by the copyrights hereon may be reproduced or transmitted in any form or by any means—graphic, electronic, or mechanical, including photocopying, recording, taping, or information storage and retrieval systems—without written permission of the publisher.

Library of Congress Cataloging in Publication Data

```
Lindsay, Robert Bruce, 1900-      comp.
    Physical acoustics.

    (Benchmark papers in acoustics, v. 4)
    Includes bibliographical references.
    1. Sound.  2. Sound-waves.  I. Title.
QC225.7.L56         534         73-12619
ISBN 87933-040-6
```

Manufactured in the United States of America.

Exclusive distributor outside the United States and Canada: John Wiley & Sons, Inc.

Acknowledgments and Permissions

ACKNOWLEDGMENTS

American Physical Society—*Physical Review*
   "Acoustic Wave Filters"
   "Vortices and Streams Caused by Sound Waves"
   "Dispersion and Absorption of High-Frequency Sound Waves"

National Academy of Sciences—*Proceedings of the National Academy of Sciences*
   "Acoustical Impedance, and the Theory of Horns and of the Phonograph"
   "On the Scattering of Light by Supersonic Waves"

Royal Society, London—*Philosophical Transactions of the Royal Society, London*
   "On the Influence of Heat Conduction in a Gas on Sound Propagation"

PERMISSIONS

The following papers are reprinted with the permission of the authors and copyright owners.

Acoustical Society of America—*Journal of the Acoustical Society of America*
   "The Theory of Acoustic Filtration in Solid Rods"
   "Scattering of Sound by Sound"
   "Experimental Investigation of an End-Fire Array"
   "The Effect of Humidity upon the Absorption of Sound in a Room, and a Determination of the Coefficients of Absorption of Sound in Air"
   "The Absorption of Sound in Air, in Oxygen, and in Nitrogen—Effects of Humidity and Temperature"
   "Attenuation and Scattering of High-Frequency Sound Waves in Metals and Glasses"

Acoustical Society of America—*Sound*
   "Strange Sounds in the Atmosphere: Part I"
   "Strange Sounds in the Atmosphere: Part II"

American Academy of Arts and Sciences—*Proceedings of the American Academy of Arts and Sciences*
   "Piezoelectric Crystal Oscillators Applied to the Precision Measurement of the Velocity of Sound in Air and CO_2 at High Frequencies"

American Physical Society—*Physical Review*
   "Third and Fourth Sound in Liquid Helium II"
   "Experimental Determination of the Fourth Sound Veolcity in Helium II"
   "Ultrasonic Attenuation Due to Lattice-Electron Interaction in Normal Conducting Metals"
   "Saturation of Nuclear Quadrupole Energy Levels by Ultrasonic Excitation"

American Physical Society—*Physical Review Letters*
   "Excitation of Very-High-Frequency Sound in Quartz"
   "Fourth Sound in Helium II"

American Physical Society—*Reviews of Modern Physics*
   "Acoustic Radiation Pressure of Plane Compressional Waves"
   "Absorption of Sound in Fluids"

American Telephone and Telegraph Company—*The Bell System Technical Journal*
"A Study of the Regular Combination of Acoustic Elements, with Applications to Recurrent Acoustic Filters, Tapered Acoustic Filters, and Horns"

Cambridge University Press—*Proceedings of the Cambridge Philosophical Society*
"The Energy Exchanges Between Molecules"

Dover Publications, Inc.—*Scientific Papers of Lord Rayleigh*
"On Waves Propagated Along the Plane Surface of an Elastic Solid"
"Aerial Plane Waves of Finite Amplitude"
"On the Pressure of Vibrations"
"The Explanation of Certain Acoustical Phenomena"

Institute of Electrical and Electronics Engineers, Inc.—*Proceedings of the IRE*
"The Piezo-electric Resonator"

Pergamon Press Ltd.—*Men of Physics—L. D. Landau*
"The Theory of Superfluidity of Helium II"

Taylor & Francis Ltd.—*Philosophical Magazine*
"The Physical and Biological Effects of High-Frequency Sound Waves of Great Intensity"

Series Editor's Preface

The "Benchmark Papers in Acoustics" constitute a series of volumes that make available to the reader in carefully organized form important papers in all branches of acoustics. The literature of acoustics is vast in extent and much of it, particularly the earlier part, is inaccessible to the average acoustical scientist and engineer. These volumes aim to provide a practical introduction to this literature, since each volume offers an expert's selection of the seminal papers in a given branch of the subject, that is, those papers which have significantly influenced the development of that branch in a certain direction and introduced concepts and methods that possess basic utility in modern acoustics as a whole. Each volume provides a convenient and economical summary of results as well as a foundation for further study for both the person familiar with the field and the person who wishes to become acquainted with it.

Each volume has been organized and edited by an authority in the area to which it pertains. In each volume there is provided an editorial introduction summarizing the technical significance of the field being covered. Each article is accompanied by editorial commentary, with necessary explanatory notes, and an adequate index is provided for ready reference. Articles in languages other than English are either translated or abstracted in English. It is the hope of the publisher and editor that these volumes will constitute a working library of the most important technical literature in acoustics of value to students and research workers.

The present volume, *Physical Acoustics,* has been edited by the series editor. Through its 40 seminal articles it is intended to stress the physical principles that form the basis of acoustics and those physical properties of matter with which acoustical radiation is most closely concerned. With but four exceptions all the included material is from the post-1900 period. More than half the papers were originally published during the last half century. A principal aim is to show that, in its recent developments, a large and vital part of acoustics is still in the mainstream of physical research. Further details about the nature of the book will be found in the Introduction.

R. Bruce Lindsay

Editor's Acknowledgment

I am deeply indebted to Patricia Galkowski and her colleagues in the Sciences Library of Brown University for assistance in the location of source material. I am grateful to Susan Desilets Proto of the Department of Physics at Brown University for typing the translated material.

A great debt is owed to colleagues and friends who have graciously permitted the inclusion of their published articles in this volume.

R. B. L.

Contents

Acknowledgments and Permissions	v
Series Editor's Preface	vii
Editor's Acknowledgment	viii
Contents by Author	xiii

Introduction. The Nature of Physical Acoustics ... 1

I. ACOUSTIC RADIATION AND SCATTERING

Editor's Comments on Paper 1 ... 5
1. Kirchhoff, J. G.: "On the Influence of Heat Conduction in a Gas on Sound Propagation" ... 7
   Translated from *Ann. Physik Chem.* (Fifth Ser.), **134**, 177–193 (1868)

Editor's Comments on Paper 2 ... 20
2. Stokes, G. G.: "On the Communication of Vibrations from a Vibrating Body to a Surrounding Gas" ... 22
   Phil. Trans. Roy. Soc., London, **158**, 447–456 (1868)

Editor's Comments on Paper 3 ... 32
3. Frank, P. G., P. G. Bergmann, and A. Yaspan: "Ray Acoustics" ... 34
   Summary Tech. Rept. Division 6, NDRC, **8**, 41–49, 51–54, 59–68 (n.d.)

Editor's Comments on Paper 4 ... 57
4. Webster, A. G.: "Acoustical Impedance, and the Theory of Horns and of the Phonograph" ... 58
   Proc. Natl. Acad. Sci., **5**, 275–282 (1919)

Editor's Comments on Papers 5a and 5b ... 66
5a. Cook, R. K.: "Strange Sounds in the Atmosphere. Part I" ... 67
   Sound, **1**, 12–16 (1962)
5b. Cook, R. K., and J. M. Young: "Strange Sounds in the Atmosphere. Part II" ... 72
   Sound, **1**, 25–33 (1962)

Editor's Comments on Papers 6, 7, and 8 ... 81
6. Stewart, G. W.: "Acoustic Wave Filters" ... 83
   Phys. Rev., **20**, 528–540 (1922)
7. Mason, W. P.: "A Study of the Regular Combination of Acoustic Elements, with Applications to Recurrent Acoustic Filters, Tapered Acoustic Filters, and Horns" ... 96
   Bell System Tech. J., **6**, 258–275 (1927)
8. Lindsay, R. B., and F. E. White: "The Theory of Acoustic Filtration in Solid Rods" ... 114
   J. Acoust. Soc. Amer., **4**, 155–165 (1932)

Editor's Comments on Paper 9 ... 125
9. Strutt, J. W., (Lord Rayleigh): "On Waves Propagated Along the Plane Surface of An Elastic Solid" ... 126
   Scientific Papers of Lord Rayleigh, Vol. 2, 441–447 (1964)

II. MACROSONICS, NONLINEAR ACOUSTICS, AND SHOCK WAVES

Editor's Comments on Paper 10 — 135
10. Lord Rayleigh: "Aerial Plane Waves of Finite Amplitude" — 136
    Scientific Papers of Lord Rayleigh, Vol. 5, 573–610 (1964)

Editor's Comments on Paper 11 — 174
11. Eckart, C.: "Vortices and Streams Caused by Sound Waves" — 175
    Phys. Rev., **73**, 68–76 (1948)

Editor's Comments on Papers 12 and 13 — 184
12. Lord Rayleigh: "On the Pressure of Vibrations" — 185
    Scientific Papers of Lord Rayleigh, Vol. 5, 41–48 (1964)
13. Borgnis, F. E.: "Acoustic Radiation Pressure of Plane Compressional Waves" — 193
    Rev. Mod. Phys., **25**, 653–664 (1953)

Editor's Comments on Papers 14 and 15 — 205
14. Westervelt, P. J.: "Scattering of Sound by Sound" — 206
    J. Acoust. Soc. Amer., **29**, 934–935 (1957)
15. Bellin, J. L. S., and R. T. Beyer: "Experimental Investigation of an End-Fire Array" — 208
    J. Acoust. Soc. Amer., **34**, 1051–1054 (1962)

Editor's Comments on Paper 16 — 212
16. Lord Rayleigh: "The Explanation of Certain Acoustical Phenomena" — 213
    Scientific Papers of Lord Rayleigh, Vol. 1, 348–354 (1964)

III. ULTRASONICS. INTERACTION OF SOUND WITH THE MEDIUM

Editor's Comments on Paper 17 — 222
17. Cady, W. G.: "The Piezo-electric Resonator" — 223
    Proc. IRE, **10**, 83–95, 106–109 (1922)

Editor's Comments on Paper 18 — 239
18. Wood, R. W., and A. L. Loomis: "The Physical and Biological Effects of High-Frequency Sound Waves of Great Intensity" — 240
    Phil. Mag., Seventh Series, **4**, 417–436 (1927)

Editor's Comments on Paper 19 — 267
19. Einstein, A.: "Sound Propagation in Partially Dissociated Gases" — 268
    Translated from *Sitzber. Preussischem Akad. Wiss., Berlin*, **24**, 380–385 (1920)

Editor's Comments on Papers 20 Through 26 — 273
20. Pierce, G. W.: "Piezoelectric Crystal Oscillators Applied to the Precision Measurement of the Velocity of Sound in Air and CO_2 at High Frequencies" — 277
    Proc. Amer. Acad. Arts Sci., **60**, 271–291 (1925)
21. Herzfeld, K. F., and F. O. Rice: "Dispersion and Absorption of High-Frequency Sound Waves" — 298
    Phys. Rev., **31**, 691–695 (1928)
22. Kneser, H. O.: "The Dispersion Theory of Sound" — 303
    Ann. Physik (Folge 5), **11**, 761–776 (1931)
23. Henry, P. S. H.: "The Energy Exchanges Between Molecules" — 316
    Proc. Cambridge Phil. Soc., **28**, 249–255 (1932)
24. Knudsen, V. O.: "The Effect of Humidity upon the Absorption of Sound in a Room, and a Determination of the Coefficients of Absorption of Sound in Air" — 323
    J. Acoust. Soc. Amer., **3**, 126–138 (1931)
25. Knudsen, V. O.: "The Absorption of Sound in Air, in Oxygen, and in Nitrogen—Effects of Humidity and Temperature" — 336
    J. Acoust. Soc. Amer., **5**, 112–121 (1933)

26. Markham, J. J., R. T. Beyer, and R. B. Lindsay: "Absorption of Sound in Fluids" **346**
    Rev. Mod. Phys., **23**, 353–411 (1951)
Editor's Comments on Paper 27 **405**
27. Bömmell, H. E., and K. Dransfeld: "Excitation of Very-High-Frequency Sound in Quartz" **406**
    Phys. Rev. Letters, **1**, 234–235 (1958)
Editor's Comments on Papers 28 and 29 **408**
28. Debye, P., and F. W. Sears: "On the Scattering of Light by Supersonic Waves" **410**
    Proc. Natl. Acad. Sci., **18**, 409–414 (1932)
29. Brillouin, L.: "Diffusion of Light by a Transparent Homogeneous Body" **416**
    Translated from *Compt. Rend. Acad. Sci., Paris*, **158**, 1331–1334 (1914)
Editor's Comments on Papers 30 Through 34 **419**
30. Tisza, L.: "On the Thermal Superconductivity of Liquid Helium II and the Bose-Einstein Statistics" **421**
    Translated from *Compt. Rend. Acad. Sci., Paris*, **207**, 1035–1037 (1938)
31. Landau, L. D.: "The Theory of Superfluidity of Helium II" **424**
    Men of Physics—L. D. Landau, Vol. 1, 54–63, 80–87 (1965)
32. Atkins, K. R.: "Third and Fourth Sound in Helium II" **442**
    Phys. Rev., **113**, 962–965 (1959)
33. Rudnick, I., and K. A. Shapiro: "Fourth Sound in Helium II" **446**
    Phys. Rev. Letters, **9**, 191–193 (1962)
34. Shapiro, K. A., and I. Rudnick: "Experimental Determination of the Fourth Sound Velocity in Helium II" **449**
    Phys. Rev., **137**, A1383–A1391 (1965)
Editor's Comments on Papers 35 and 36 **458**
35. Mason, W. P., and H. J. McSkimin: "Attenuation and Scattering of High Frequency Sound Waves in Metals and Glasses" **459**
    J. Acoust. Soc. Amer., **19**, 464–473 (1947)
36. Mason, W. P.: "Ultrasonic Attenuation Due to Lattice–Electron Interaction in Normal Conducting Metals" **469**
    Phys. Rev., **97**, 557–558 (1955)
Editor's Comments on Paper 37 **471**
37. Proctor, W. G., and W. H. Tanttila: "Saturation of Nuclear Electric Quadrupole Energy Levels by Ultrasonic Excitation" **472**
    Phys. Rev., **98**, 1854 (1955)

Author Citation Index **473**
Subject Index **477**

Contents by Author

Atkins, K. R., 442
Bellin, J. L. S., 208
Bergmann, P. G., 34
Beyer, R. T., 208, 346
Bömmel, H. E., 406
Borgnis, F. E., 193
Brillouin, L., 416
Cady, W. G., 223
Cook, R. K., 67, 72
Debye, P., 410
Dransfeld, K., 406
Eckart, C., 175
Einstein, A., 268
Frank, P. G., 34
Henry, P. S. H., 316
Herzfeld, K. F., 298
Kirchhoff, J. G., 7
Kneser, H. O., 303
Knudsen, V. O., 323, 336
Landau, L. D., 424
Lindsay, R. B., 114, 346
Loomis, A. L., 240

McSkimin, H. J., 459
Markham, J. J., 346
Mason, W. P., 96, 459, 469
Pierce, G. W., 277
Proctor, W. G., 472
Rice, F. O., 298
Rudnick, I., 446, 449
Sears, F. W., 410
Shapiro, K. A., 446, 449
Stewart, G. W., 83
Stokes, G. G., 22
Strutt, J. W. (Lord Rayleigh), 126, 136, 185, 213
Tanttila, W. H., 472
Tisza, L., 421
Webster, A. G., 58
Westervelt, P. J., 206
White, F. E., 114
Wood, R. W., 240
Yaspan, A., 34
Young, J. M., 72

Introduction
The Nature of Physical Acoustics

The term *physical acoustics* may be interpreted in both a broad and a narrow sense. In its broad definition it is the study of all phenomena relating to the production, propagation, and reception of sounds of every kind, together with the theoretical analysis that purports to describe and explain these phenomena. This is the point of view which insists that all acoustics is a branch of physics and subject to the fundamental principles of physical science. The various volumes of the Benchmark Series in Acoustics exemplify this view in its application to the many subfields of acoustics.

In the present volume we shall adopt a narrower point of view. The articles reprinted here are classified in three main groups: (1) those relating to the fundamental theory of mechanical wave propagation in continuous material media, including the attenuation, scattering, and filtration of sound, and the connection between wave and ray acoustics; (2) those relating to nonlinear acoustical radiation (i.e., radiation of such high intensity that the normal linear wave equation does not apply); and (3) those relating primarily to the interaction between sound radiation and the properties of the medium through which it passes.

Some of the basic papers in the first category have already been included in the volume *Acoustics: Historical and Philosophical Development* of this series. However, important new developments have occurred in the twentieth century and the corresponding papers are included in the present volume. Similarly, it is true that the foundations of high-intensity, nonlinear acoustics, now known as macrosonics, were laid in the nineteenth century; here again, the full development and exploitation of the field had to await the application of electrical techniques to the production of sound radiation of high intensity over a very wide range of frequencies, a special feature of twentieth-century acoustics. Finally, the third category is largely concerned with the properties of high-frequency, or ultrasonic, radiation, that have been found to depend on the molecular constitution of the medium. This has led to the development of a wholly new branch of the subject called *molecular acoustics*. Most of the interesting new acoustical research discoveries are connected with this field. Not only do high-frequency sound waves, which have been produced up to a frequency of 10^{10} Hz, interact with the medium to affect vitally the attenuation and dispersion of the radia-

The Nature of Physical Acoustics

tion, but intense ultrasound can have decided effects on the properties of the medium, such as cavitation in liquids and emulsification of mixtures not normally miscible. Intense ultrasonic radiation can also affect chemical reactions, as well as bring about biological changes, through either heating or mechanical action. The propagation of sound is markedly influenced by other physical phenomena: light, for example. Very intense sound waves can become shock waves, leading to the propagation of what are very close to discontinuities in density and pressure throughout the medium.

Many important papers in physical acoustics will be found in other volumes of this series because of their relevance to particular domains such as underwater sound, bioacoustics, acoustical instrumentation, and the interaction of sound and light.

Each article is prefaced by a brief editorial commentary explaining its significance and providing biographical information about the author or authors. Where the article is not reproduced in full, editorial notes describe briefly the deleted contents.

Acoustic Radiation and Scattering

I

Editor's Comments on Paper 1

1 **Kirchhoff:** *On the Influence of Heat Conduction in a Gas on Sound Propagation*

In his famous 1845 paper (*Trans. Cambridge Phil. Soc.*, **8**, 287, 1845), G. G. Stokes in the course of his investigation of fluid viscosity deduced the effect of this viscosity on the propagation of sound through the fluid (see the reproduction of that paper in *Acoustics: Historical and Philosophical Development*, R. B. Lindsay, ed., Dowden, Hutchinson & Ross, Stroudsburg, Pa., 1973, p. 261). In Stokes' time, methods for the experimental measurements of sound velocity and attenuation in fluids were not precise enough to test his theoretical results. Some twenty-three years after the work by Stokes, Helmholtz investigated the effect of viscosity on the propagation of sound through a fluid contained in a very narrow tube. His results, as far as sound velocity is concerned, do not agree with the experimental values for that velocity obtained by Kundt at about the same time. This led Kirchhoff to consider the problem in order to follow up on Kundt's suggestion that heat conduction in the fluid should not be neglected in connection with sound propagation. Kirchhoff proceeded, in very general fashion, to study the propagation of plane and spherical sound waves in an unbounded medium, taking into account both viscosity and heat conduction. He thus generalized Stokes' work, confirming the latter's results with respect to viscosity, but showing that for a gaseous fluid like air, the effect of heat conduction on the attenuation of sound is of the same order as that of viscosity. He also showed that the combined effect of both on the velocity of sound is of the second order compared with the effect on attenuation.

Kirchhoff proceeded to apply his theory to the case of sound propagation in narrow tubes and obtained a result agreeing in theoretical form, to a certain extent, but not in all details, with Kundt's experimental results. This has proved to be a complicated problem.

Kirchhoff's work on propagation in unbounded media constituted a seminal contribution to physical acoustics. It served to stimulate further research on the nature of sound propagation in fluids, especially when it was found that even the combination of the effects of viscosity and heat conduction is inadequate to account for observed values of the attenuation of sound in a fluid. Kirchhoff's classic paper is presented here in its entirety in English translation.

Editor's Comments on Paper 1

Gustav Kirchhoff (1824–1887) was one of the most notable of nineteenth-century German physicists. He was, for many years, a professor of physics in Heidelberg and Berlin. Best known for his pioneer research in radiation, particularly the spectral analysis of light, he also made important contributions to mechanics and electric currents.

1

On the Influence of Heat Conduction in a Gas on Sound Propagation

G. KIRCHHOFF

Translated expressly for this Benchmark volume by R. Bruce Lindsay from "Über den Einfluss der Wärmeleitung in einem Gase auf die Schallbewegung," Ann. Physik Chem. (Fifth Ser.), 134, 177–193 (1868)

In his beautiful investigation of the velocity of sound in air in tubes, Kundt[1] has shown that the sound velocity in narrow tubes is the smaller the narrower the tube and the lower the frequency. Helmholtz[2] has investigated theoretically the propagation of sound in a cylindrical tube in its relation to friction. He has derived a formula for the sound velocity which agrees with Kundt's results to the extent that the velocity becomes smaller as the radius of the tube and the sound frequency decrease. However, Kundt has shown that the values for the velocity which he has observed for his narrower tubes are very much smaller than those given by the Helmholtz formula. From this he concludes that friction or viscosity is not sufficient to explain the phenomena observed by him. He suggests that heat exchange between the air through which the sound is propagated and the tube wall is the essential cause of his result. The heat conductivity of the air which must bring about such a heat exchange is, according to the newer gas theory, connected with the viscosity in such a way that in a motion of a gas in which temperature changes take place that may not be neglected, the influence of heat conduction must be at least of the same order as that of viscosity. It is, therefore, suggestive to investigate whether Kundt's experimental results cannot be theoretically explained more completely by taking heat conduction into account than by neglecting it.

If u, v, and w are the infinitely small components of the fluid velocity and if ρ is the density at time t at the point x, y, and z, we have to begin with

$$\frac{1}{\rho}\frac{\partial \rho}{\partial t} + \frac{\partial u}{\partial x} + \frac{\partial v}{\partial y} + \frac{\partial w}{\partial z} = 0$$

---

[1] *Monatsber. Berliner Akad.*, Dec. 19, 1867.
[2] *Verhandlungen des natur-historisch-medizinischen Vereins zu Heidelberg von Jahre* 1868, Bd. III, 8, 16.

If p is the pressure and μ' and μ'' are two constants depending on the friction, we have further

$$\frac{\partial u}{\partial t} + \frac{1}{\rho}\frac{\partial p}{\partial x} = \mu' \nabla^2 u + \frac{\mu''}{\rho}\frac{\partial^2 \rho}{\partial x \partial t}$$

$$\frac{\partial v}{\partial t} + \frac{1}{\rho}\frac{\partial p}{\partial y} = \mu' \nabla^2 v + \frac{\mu''}{\rho}\frac{\partial^2 \rho}{\partial y \partial t}$$

$$\frac{\partial w}{\partial t} + \frac{1}{\rho}\frac{p}{z} = \mu' \nabla^2 w + \frac{\mu''}{\rho}\frac{\partial^2 \rho}{\partial z \partial t}$$

where

$$\nabla^2 F = \frac{\partial^2 F}{\partial x^2} + \frac{\partial^2 F}{\partial y^2} + \frac{\partial^2 F}{\partial z^2}$$

We still have to add a fifth equation to the four above. If we neglect heat conduction this equation is

$$c\rho\, dp - c'p\, d\rho = 0$$

in which c denotes the specific heat of the gas at constant volume and c' its specific heat at constant pressure. If we take heat conduction into account, this equation is replaced by a more complicated one. In order to derive it we introduce p_0 and ρ_0 to designate the pressure and density, respectively, in the state of rest. We also introduce θ as the temperature reckoned from the temperature in the state of rest as base. We call α the coefficient of expansion of the gas, so that

$$\frac{p}{\rho} = \frac{p_0}{\rho_0}(1 + \alpha\theta)$$

Suppose that in a quantity of gas of mass M the pressure is increased by amount dp and at the same time the density is increased by $d\rho$. This means that a quantity of heat dW must be given to this mass of gas, where

$$dW = P\, dp + R\, d\rho$$

where, further,

$$P = \frac{Mc\rho_0}{\alpha p_0 \rho}$$

$$R = -\frac{Mc'p\rho_0}{\alpha p_0 \rho^2}$$

Now suppose the quantity of gas in question is that which occupies a rectangular parallelepiped whose corner is the point x, y, z and whose sides are parallel, respectively, to the coordinate axes and have lengths dx, dy, dz, respectively. In the time interval dt a quantity of heat is communicated to this, equal to

$$k\, dx\, dy\, dz\, dt\, \nabla^2 \theta$$

If we remember that u, v, w are very small, it follows that the above expression also equals

$$\left(P\frac{\partial p}{\partial t} + R\frac{\partial \rho}{\partial t}\right)dt$$

if we take

$$M = \rho \, dx \, dy \, dz$$

If we further assume that ρ differs only infinitesimally from ρ_0, we then get

$$k\nabla^2\theta = \frac{1}{\alpha\rho_0}\left(c\rho\frac{\partial p}{\partial t} - c'p\frac{\partial \rho}{\partial t}\right)$$

We now use the equation

$$dp = \frac{p_0}{\rho_0}d\rho + \alpha p_0 \, d\theta$$

to eliminate p from the above equations. Further, we set

$$\rho = \rho_0(1+\sigma)$$

$$\theta = \frac{c'-c}{\alpha c}\Theta$$

$$k = \nu c \rho_0$$

$$\frac{p_0}{\rho_0}\frac{c'}{c} = a^2$$

$$\frac{p_0}{\rho_0} = b^2$$

Here a represents what the sound velocity would be if there were no influence of friction and heat conduction. Similarly, b represents what the sound velocity would be if there were no temperature change as a result of the density change. The five differential equations then take the following form:

$$\frac{\partial \sigma}{\partial t} + \frac{\partial u}{\partial x} + \frac{\partial v}{\partial y} + \frac{\partial w}{\partial z} = 0$$

$$\frac{\partial u}{\partial t} + b^2\frac{\partial \sigma}{\partial x} + (a^2-b^2)\frac{\partial \Theta}{\partial x} = \mu'\nabla^2 u - \mu''\frac{\partial^2 \sigma}{\partial x \partial t}$$

$$\frac{\partial v}{\partial t} + b^2\frac{\partial \sigma}{\partial y} + (a^2-b^2)\frac{\partial \Theta}{\partial y} = \mu'\nabla^2 v - \mu''\frac{\partial^2 \sigma}{\partial y \partial t}$$

$$\frac{\partial w}{\partial t} + b^2\frac{\partial \sigma}{\partial z} + (a^2-b^2)\frac{\partial \Theta}{\partial z} = \mu'\nabla^2 w - \mu''\frac{\partial^2 \sigma}{\partial z \partial t}$$

$$\frac{\partial \Theta}{\partial t} - \frac{\partial \sigma}{\partial t} = \nu\nabla^2\Theta$$

Of the three constants μ', μ'', ν occurring here, only μ' up to now has been experimentally determined and then, indeed, only for the case in which changes in density can be neglected. According to Meyer[1] for atmospheric air at temperature around 20°C and a pressure of 1 atmosphere, if we use the second as the unit of time, $\sqrt{\mu'} = 4.86$ mm. According to the theory of Stokes,[2] $\mu'' = \frac{1}{3}\mu'$. The same relation results from the theory of Maxwell.[3] Moreover, according to Maxwell's theory, $\nu = \frac{5}{2}\mu'$ if the gas molecules are considered as material points. Moreover, μ', μ'', and ν are inversely proportional to the pressure and directly proportional to the square of the absolute temperature. The value of ν is, however, perhaps higher than it should be from Maxwell's theory, since the latter takes no account of heat radiation, and the radiation increases the heat conduction without changing the form of the law, if one is permitted to assume that the heat radiation emitted by the gas particles is completely absorbed after passing through very small distances.

The equations that have just been set up will now be developed further under the assumption that the unknown functions u, v, w, σ, and Θ contain the factor e^{ht} but otherwise do not depend on t. Here h denotes a constant which will later be taken as imaginary. If we now denote the functions of x, y, z which result from separation of the factor e^{ht} from u, v, w, σ, and Θ by the same symbols, we obtain

$$\frac{\partial u}{\partial x} + \frac{\partial v}{\partial y} + \frac{\partial w}{\partial z} + h\sigma = 0$$

$$hu - \mu' \nabla^2 u = -\frac{\partial P}{\partial x}$$

$$hv - \mu' \nabla^2 v = -\frac{\partial P}{\partial y}$$

$$hw - \mu' \nabla^2 w = -\frac{\partial P}{\partial z}$$

$$P = (b^2 + h\mu'')\sigma + (a^2 - b^2)\Theta$$

$$\sigma = \Theta - \frac{\nu}{h}\nabla^2\Theta$$

By the use of the last of these equations the one next-to-last becomes

$$P = (a^2 + h\mu'')\Theta - \frac{\nu}{h}(b^2 + \mu''h)\nabla^2\Theta$$

and the first one becomes

$$\frac{\partial u}{\partial x} + \frac{\partial v}{\partial y} + \frac{\partial w}{\partial z} + h\Theta - \nu\nabla^2\Theta = 0$$

---

[1] *Pogg. Ann.*, **125**, 527 and 549.
[2] *Cambridge Phil. Trans.*, **8**, 297 (1845).
[3] *London Phil. Trans.*, **157** (part 1), 49 (1867).

We then obtain from the three remaining equations, by differentiating them with respect to x, y, z, respectively, and adding the results,

$$h^2 \Theta - [a^2 + h(\mu' + \mu'' + \nu)] \nabla^2 \Theta + \frac{\nu}{h}[b^2 + h(\mu' + \mu'')] \nabla^4 \Theta = 0$$

One can find a solution of this equation for Θ by calling λ_1 and λ_2 the roots of the quadratic equation

$$h^2 - [a^2 + h(\mu' + \mu'' + \nu)]\lambda + \frac{\nu}{h}[b^2 + h(\mu' + \mu'' + \nu)]\lambda^2 = 0$$

and then determine two functions Q_1 and Q_2 from

$$\nabla^2 Q_1 = \lambda_1 Q_1$$
$$\nabla^2 Q_2 = \lambda_2 Q_2$$

and set

$$\Theta = A_1 Q_1 + A_2 Q_2$$

where A_1 and A_2 are arbitrary constants.

For this value of Θ we obtain particular solutions of the second, third, and fourth of our equations by equating u, v, w, respectively, to the derivatives with respect to x, y, z of

$$B_1 Q_1 + B_2 Q_2$$

and obtain B_1 and B_2 suitably from

$$B_1 = -A_1 \left(\frac{h}{\lambda_1} - \nu\right)$$

$$B_2 = -A_2 \left(\frac{h}{\lambda_2} - \nu\right)$$

We obtain more general solutions of these same equations by adding to the functions already found the new ones u', v', w', which satisfy

$$\nabla^2 u' = \frac{h}{\mu'} u'$$

$$\nabla^2 v' = \frac{h}{\mu'} v'$$

$$\nabla^2 w' = \frac{h}{\mu'} w'$$

From this it follows that

$$u = u' + B_1 \frac{\partial Q_1}{\partial x} + B_2 \frac{\partial Q_2}{\partial x}$$

$$v = v' + B_1 \frac{\partial Q_1}{\partial y} + B_2 \frac{\partial Q_2}{\partial y}$$

Influence of Heat Conduction in a Gas on Sound Propagation

$$w = w' + B_1 \frac{\partial Q_1}{\partial z} + B_2 \frac{\partial Q_2}{\partial z}$$

in which the above values of B_1 and B_2 are to be inserted.

If we substitute these expressions for u, v, w in our revised form of the first of the above fundamental equations, we find that the relation connecting u', v', w', is still

$$\frac{\partial u'}{\partial x} + \frac{\partial v'}{\partial y} + \frac{\partial w'}{\partial z} = 0$$

The results that have just been derived will now be applied to the case of plane waves which are propagated in unbounded space in the positive x direction. We then set $v' = w' = 0$ and assume that u', Q_1, and Q_2 are independent of y and z. Then u' must satisfy

$$\frac{d^2 u'}{dx^2} = \frac{h}{\mu} u'$$

$$\frac{du'}{dx} = 0$$

From this it results that we also have

$$u' = 0$$

The equations for Q_1 and Q_2 then become

$$\frac{d^2 Q_1}{dx^2} = \lambda_1 Q_1$$

$$\frac{d^2 Q_2}{dx^2} = \lambda_2 Q_2$$

From these we can set

$$Q_1 = \exp(-x\sqrt{\lambda_1})$$

$$Q_2 = \exp(-x\sqrt{\lambda_2})$$

where the signs of the square roots ($\sqrt{\lambda_1}$ and $\sqrt{\lambda_2}$) are to be so chosen that their real parts are positive so that Q_1 and Q_2 do not become infinitely large at infinity. We, therefore, get

$$u = A_1 \sqrt{\lambda_1} \left(\frac{h}{\lambda_1} - \nu \right) \exp(-x\sqrt{\lambda_1}) + A_2 \sqrt{\lambda_2} \left(\frac{h}{\lambda_2} - \nu \right) \exp(-x\sqrt{\lambda_2})$$

$$\Theta = A_1 \exp(-x\sqrt{\lambda_1}) + A_2 \exp(-x\sqrt{\lambda_2})$$

If when $x = 0$, we take

$$u = u_0 \quad \text{and} \quad \Theta = \Theta_0$$

the quantities A_1 and A_2 are determined through the equations

$$u_0 = A_1 \sqrt{\lambda_1} \left(\frac{h}{\lambda_1} - \nu \right) + A_2 \sqrt{\lambda_2} \left(\frac{h}{\lambda_2} - \nu \right)$$

$$\Theta_0 = A_1 + A_2$$

It follows from the quadratic equation whose roots are λ_1 and λ_2 that

$$\frac{1}{\lambda_1} + \frac{1}{\lambda_2} = \frac{a^2}{h^2} + \frac{\mu' + \mu'' + \nu}{h}$$

$$\frac{1}{\lambda_1 \lambda_2} = \frac{\nu}{h^3}[b^2 + h(\mu' + \mu'')]$$

If we now designate μ', μ'', and ν as infinitely small quantities of the first order, one of the two quantities $1/\lambda_1$ and $1/\lambda_2$ must also be infinitely small of the first order, while the other remains finite. Let us take λ_1 as the finite root and λ_2 as the infinitely great one. From the equations for A_1 and A_2 it follows that these quantities are finite if u_0 and Θ_0 are taken as finite, and further that the second term in the expressions for u must be infinitely small for $x = 0$ and, indeed, of the order of $1/\sqrt{\lambda_2}$. For finite, nonvanishing x, it must however be infinitely small, of the order of $(1/\sqrt{\lambda_2})\exp(-x\sqrt{\lambda_2})$. If we neglect infinitely small quantities of higher order we get for finite, nonvanishing x,

$$u = A_1 \sqrt{\lambda_1}\left(\frac{h}{\lambda_1} - \nu\right)\exp(-x\sqrt{\lambda_1})$$

If in the formation of the value of λ_1 we consider only infinitely small quantities of the lowest order, we find

$$\sqrt{\lambda_1} \doteq \left\{\frac{h}{a} - \frac{h^2}{2a^3}\left[\mu' + \mu'' + \nu\left(1 - \frac{b^2}{a^2}\right)\right]\right\}$$

or if we set $h = 2\pi n i$, where $i = \sqrt{-1}$ and n is the frequency of the sound, we get

$$\sqrt{\lambda_1} = \frac{2\pi^2 n^2}{a^3}\left[\mu' + \mu'' + \nu\left(1 - \frac{b^2}{a^2}\right) + \frac{2\pi n i}{a}\right]$$

If we set

$$\frac{2\pi^2 n^2}{a^3}\left[\mu' + \mu'' + \nu\left(1 - \frac{b^2}{a^2}\right)\right] = m'$$

and restore in the expression for u the time-dependent factor, we finally get

$$u = C\exp\left[-m'x + 2\pi n\left(t - \frac{x}{a}\right)i\right]$$

where C denotes a new constant.

If we now join to this expression for u that which results from replacing i by $-i$ and changing the constant, we obtain

$$u = e^{-m'x}\left[D\sin 2\pi n\left(t - \frac{x}{a}\right) + E\cos 2\pi n\left(t - \frac{x}{a}\right)\right]$$

where D and E are two real, arbitrary constants.

We see that the quantity m' controls the decrease experienced by the amplitude of the oscillations in their progress through the medium. The sound velocity is here unchanged by the friction and heat conduction. Strictly speaking, this quantity also suffers a change[1] but one which is of the order of the square of the quantities μ', μ'', and ν. We find this change by retaining quantities of this order in the expansion of $\sqrt{\lambda_1}$.

Spherical waves can be handled in the same fashion as plane waves. We set

$$x^2 + y^2 + z^2 = r^2$$

$$u' = s'x, \qquad v' = s'y, \qquad w' = s'z$$

and take s', Q_1, and Q_2 as functions of r. The four differential equations satisfied by u', v', w' then give

$$\frac{d^2 s'}{dr^2} + \frac{1}{r}\frac{ds'}{dr} = \frac{h}{\mu'} s'$$

$$r\frac{ds'}{dr} + 3s' = 0$$

From these it follows that $s' = 0$. The equations for Q_1 and Q_2 become

$$\frac{d^2 Q_1}{dr^2} + \frac{2}{r}\frac{dQ_1}{dr} = \lambda_1 Q_1$$

$$\frac{d^2 Q_2}{dr^2} + \frac{2}{r}\frac{dQ_2}{dr} = \lambda_2 Q_2$$

from which we get

$$Q_1 = \frac{1}{r}\exp(-r\sqrt{\lambda_1})$$

$$Q_2 = \frac{1}{r}\exp(-r\sqrt{\lambda_2})$$

in which the radicals must be given the sign such that their real parts are positive in order that the motion may not become infinitely great at $r = \infty$. Therefore, we have

$$u = \frac{sx}{r}, \qquad v = \frac{sy}{r}, \qquad w = \frac{sz}{r}$$

$$s = -\frac{d}{dr}\left[A_1\left(\frac{h}{\lambda_1} - \nu\right)\frac{1}{r}\exp(-r\sqrt{\lambda_1}) + A_2\left(\frac{h}{\lambda_2} - \nu\right)\frac{1}{r}\exp(-r\sqrt{\lambda_2})\right]$$

$$\Theta = A_1 \frac{1}{r}\exp(-r\sqrt{\lambda_1}) + A_2 \frac{1}{r}\exp(-r\sqrt{\lambda_2})$$

---

[1] See Stefan, *Sitzber. Wiener Akad.*, **54**, 529 (1866).

The quantities A_1 and A_2 are determined if s and Θ are given for some value of r. For every somewhat greater value of r the term in A_2 loses its significance, since λ_2 is infinitely great. Hence we get by a method similar to that employed for the plane-wave case,

$$s = \frac{d}{dr}\frac{1}{r} e^{-m'r}\left[D \sin 2\pi n\left(t - \frac{r}{a}\right) + E \cos 2\pi n\left(t - \frac{r}{a}\right)\right]$$

where m' has the same value as above.

It will now be assumed that the mass of air under consideration is confined in a circular tube with circular cross section. It is assumed that the motion is symmetrical with respect to the axis of the tube, which is taken as the x axis. We let

$$y^2 + z^2 = r^2$$

$$v = \frac{sy}{r}, \quad w = \frac{sz}{r}$$

$$v' = \frac{s'y}{r}, \quad w' = \frac{s'z}{r}$$

and u, u', s, s', Q_1, and Q_2 are to be functions of x and r. It will now be assumed that all these functions have the factor e^{mx}, where m is a constant independent of x. For the functions of r which remain after separating out the factor e^{mx}, we have

$$\frac{d^2Q_1}{dr^2} + \frac{1}{r}\frac{dQ_1}{dr} = (\lambda_1 - m^2)Q_1$$

$$\frac{d^2Q_2}{dr^2} + \frac{1}{r}\frac{dQ_2}{dr} = (\lambda_2 - m^2)Q_2$$

For s' and u' we get the three differential equations

$$\frac{d^2u'}{dr^2} + \frac{1}{r}\frac{du'}{dr} = \left(\frac{h}{\mu'} - m^2\right)u'$$

$$\frac{d^2s'}{dr^2} + \frac{1}{r}\frac{ds'}{dr} - \frac{s'}{r^2} = \left(\frac{h}{\mu'} - m^2\right)s'$$

$$mu' + \frac{ds'}{dr} + \frac{s'}{r} = 0$$

These three equations will be satisfied if we determine u' from the first and set

$$s' = -\frac{m}{\frac{h}{\mu'} - m^2}\frac{du'}{dr}$$

a relation that results from differentiating the third equation with respect to r and subtracting from the second. We now set

$$u' = AQ$$

where A is a constant and Q is a function of r satisfying the equation

$$\frac{d^2Q}{dr^2} + \frac{1}{r}\frac{dQ}{dr} = \left(\frac{h}{\mu'} - m^2\right)Q$$

We then have

$$u = AQ - A_1 m\left(\frac{h}{\lambda_1} - \nu\right)Q_1 - A_2 m\left(\frac{h}{\lambda_2} - \nu\right)Q_2$$

$$s = -A\frac{m}{\frac{h}{\mu'} - m^2}\frac{dQ}{dr} - A_1\left(\frac{h}{\lambda_1} - \nu\right)\frac{dQ_1}{dr} - A_2\left(\frac{h}{\lambda_2} - \nu\right)\frac{dQ_2}{dr}$$

$$\Theta = A_1 Q_1 + A_2 Q_2$$

The quantities u, s must satisfy certain conditions at the tube wall. We here follow the hypothesis that the air particles that touch the wall adhere to it and possess the same constant temperature as the wall. The expressions for u, s, and Θ must therefore vanish if r is set equal to the radius of the tube. This demands that for this value of r the determinant of the coefficient A, A_1, and A_2 in the expressions for u, s, and Θ must vanish. Hence

$$\frac{m^2 h}{\frac{h}{\mu'} - m^2}\left(\frac{1}{\lambda_1} - \frac{1}{\lambda_2}\right)\frac{d\lg Q}{dr} + \left(\frac{h}{\lambda_1} - \nu\right)\frac{d\lg Q_1}{dr} - \left(\frac{h}{\lambda_2} - \nu\right)\frac{d\lg Q_2}{dr} = 0$$

The functions Q, Q_1, and Q_2 must have the property of remaining finite for $r = 0$. Consequently, they will be completely determined save for a multiplicative arbitrary constant. All three can be expressed by the function

$$1 + \frac{q^2}{1^2} + \frac{q^4}{(1.2)^2} + \frac{q^6}{(1.2.3)^2} + \cdots$$

which may be denoted by $\mathcal{J}(q)$ and satisfies the differential equation

$$\frac{d^2\mathcal{J}}{dq^2} + \frac{1}{q}\frac{d\mathcal{J}}{dq} = 4\mathcal{J}$$

We can therefore set

$$Q = \mathcal{J}\left(\frac{r}{2}\sqrt{\frac{h}{\mu'} - m^2}\right)$$

$$Q_1 = \mathcal{J}\left(\frac{r}{2}\sqrt{\lambda_1 - m^2}\right)$$

$$Q_2 = \mathcal{J}\left(\frac{r}{2}\sqrt{\lambda_2 - m^2}\right)$$

If now we assume that μ', μ'', and ν are all equal to zero, we have

$$m^2 = \frac{h^2}{a^2} \quad \text{and} \quad \lambda_1 = \frac{h^2}{a^2}$$

whence $\lambda_1 = m^2$. From this it follows that if μ', μ'', and ν are taken as infinitely small, then $\lambda_1 - m^2$ is also infinitely small. Since then

$$\frac{h}{\mu'} - m^2 \quad \text{and} \quad \lambda_1 - m^2$$

are infinitely great, one has to consider the function J only for infinitely small and infinitely large values of its argument.

For a very small value of q, we have

$$J = 1 + q^2$$

For a very large value of q, however,

$$J = \frac{1}{\sqrt{2\pi}} \frac{e^{2q}}{\sqrt{q}}$$

provided that the real part of q is positive and infinitely large and that the sign of $\sqrt{q}$ is chosen so that the real part is positive. The importance of these conditions can be demonstrated easily by beginning with the expression

$$J = \frac{1}{2\pi} \int_{-\pi}^{+\pi} e^{2q \cos x} dx$$

If we then take account of terms of the highest order only, we obtain

$$\frac{d\lg Q}{dr} = \sqrt{\frac{h}{\mu'}}$$

$$\frac{d\lg Q_1}{dr} = \frac{r}{2} (\lambda_1 - m^2)$$

$$\frac{d\lg Q_2}{dr} = \sqrt{\lambda_2}$$

If we now set here, as well as in the coefficients in the equation for m,

$$\lambda_1 = \frac{h^2}{a^2}, \quad \lambda_2 = \frac{ha^2}{\nu b^2}$$

and write h^2/a^2 for m^2 in the associated term with the factor $\sqrt{\mu'}$, we have

$$m^2 = \frac{h^2}{a^2} \left(1 + \frac{2\gamma}{r\sqrt{h}}\right)$$

where

$$\gamma = \sqrt{\mu'} + \left(\frac{a}{b} - \frac{b}{a}\right)\sqrt{\nu}$$

The sign of $\sqrt{h}$ is to be taken so that its real part is positive.

If we now again set

$$h = 2\pi n i \quad \text{and} \quad \sqrt{h} = \sqrt{n\pi}(1 + i)$$

we get

$$m = \pm(m' + im') \qquad \text{where } m' = \gamma\sqrt{\pi n}/ar$$

$$m'' = \frac{2\pi n}{a} + \frac{\gamma\sqrt{\pi n}}{ar}$$

If we now reintroduce the factors depending on x and t, we have

$$u = BRe^{ht+mx}$$

$$s = BR'e^{ht+mx}$$

$$\Theta = BR''e^{ht+mx}$$

where B is an arbitrary constant, and R, R', and R'' are certain functions of r which vanish when r is put equal to the radius of the tube. For points that lie at a finite distance from the tube wall, these functions take on the values $R = 1$, $R' = 0$, $R'' = -(1/a)$ if we neglect infinitely small quantities of higher order. If we now form the expression that represents the velocity u for points at finite distance from the tube wall for both signs of i and m and put the four resulting expressions together with different arbitrary multiplicative constants, we get

$$u = C_1 e^{m'x} \sin(2\pi nt + m''x + \delta_1) + C_2 e^{-m'x} \sin(2\pi nt - m''x + \delta_2)$$

where C_1, C_2, δ_1, and δ_2 denote four real arbitrary constants. Here m' controls the decrease which the oscillations experience in their propagation. Similarly, m'' refers to the velocity of propagation of the wave. This velocity is indeed

$$\frac{2\pi n}{m''}$$

which is equal to

$$a\left(1 - \frac{2r\sqrt{n\pi}}{\gamma}\right)$$

This expression agrees with that of Helmholtz, except that here γ has a different meaning.

If the tube is closed at one end, say where $x = 0$, then here $u = 0$ and hence

$$C_2 = -C_1 \qquad \text{and} \qquad \delta_2 = \delta_1$$

or

$$u = C_1\sqrt{e^{2m'x} + e^{-2m'x} - 2\cos 2m''x} \sin(2\pi nt + \delta)$$

where δ depends to a certain extent on x. The radical in u does not vanish for any other value of x save $x = 0$, but it has a series of minima which differ from zero by only a very small amount. These minima correspond to the nodes of the wave. The maxima and minima of the radical are determined by the equation

$$m'(e^{2m'x} - e^{-2m'x}) + 2m'' \sin 2m''x = 0$$

or, if we neglect quantities of the order γ^2, through the equation $\sin 2m''x = 0$. Therefore,

At a great distance from the sphere the function $f_n(r)$ becomes ultimately equal to 1, and we have

$$\varphi = -\frac{c^2}{r} e^{im(at-r+c)} \Sigma \frac{U_n}{F_n(c)}. \quad\quad\quad\quad\quad\quad (12)$$

It appears from (3) that the component of the velocity along the radius vector is of the order r^{-1}, and that in any direction perpendicular to the radius vector of the order r^{-2}, so that the lateral motion may be disregarded except in the neighbourhood of the sphere.

In order to examine the influence of the lateral motion in the neighbourhood of the sphere, let us compare the actual disturbance at a great distance with what it would have been if all lateral motion had been prevented, suppose by infinitely thin conical partitions dividing the fluid into elementary canals, each bounded by a conical surface having its vertex at the centre.

On this supposition the motion in any canal would evidently be the same as it would be in all directions if the sphere vibrated by contraction and expansion of the surface, the same all round, and such that the normal velocity of the surface was the same as it is at the particular point at which the canal in question abuts on the surface. Now if U were constant the expansion of U would be reduced to its first term U_0, and seeing that $f_0(r)=1$ we should have from (11)

$$\varphi = -\frac{c^2}{r} e^{im(at-r+c)} \frac{U_0}{F_0(c)}.$$

This expression will apply to any particular canal if we take U_0 to denote the normal velocity at the sphere's surface for that particular canal; and therefore to obtain an expression applicable at once to all the canals we have merely to write U for U_0. To facilitate a comparison with (11) and (12) I shall, however, write ΣU_n for U. We have then

$$\varphi = -\frac{c^2}{r} e^{im(at-r+c)} \frac{\Sigma U_n}{F_0(c)}. \quad\quad\quad\quad\quad\quad (13)$$

It must be remembered that this is merely an expression applicable at once to all the canals, the motion in each of which takes place wholly along the radius vector, and accordingly the expression is not to be differentiated with respect to θ or ω with the view of applying the formulæ (3).

On comparing (13) with the expression for the function φ in the actual motion at a great distance from the sphere (12), we see that the two are identical with the exception that U_n is divided by two different constants, namely $F_0(c)$ in the former case and $F_n(c)$ in the latter. The same will be true of the leading terms (or those of the order r^{-1}) in the expressions for the condensation and velocity*. Hence if the mode of vibration of

---

* Of course it would be true if the *complete* differential coefficients with respect to r of the right-hand members of (12) and (13) were taken, but then the former does not give the velocity u' except as to its leading term, since (12) has been deduced from the exact expression (11) by reducing $f_n(r)$ to its first term 1; nor again is it true, except as to terms of the order r^{-1}, of the actual motion of the unimpeded fluid that the whole velocity is in the direction of the radius vector.

the sphere is such that the normal velocity of its surface is expressed by a LAPLACE's Function of any one order, the disturbance at a great distance from the sphere will vary from one direction to another according to the same law as if lateral motion had been prevented, the amplitude of excursion at a given distance from the centre varying in both cases as the amplitude of excursion, in a normal direction, of the surface of the sphere itself. The only difference is that expressed by the symbolic ratio $F_n(c) : F_0(c)$. If we suppose $F_n(c)$ reduced to the form $\mu_n(\cos \alpha_n + \sqrt{-1} \sin \alpha_n)$, the amplitude of vibration in the actual case will be to that in the supposed case as μ_0 to μ_n, and the phases in the two cases will differ by $\alpha_0 - \alpha_n$.

If the normal velocity of the surface of the sphere be not expressible by a single LAPLACE's Function, but only by a series, finite or infinite, of such functions, the disturbance at a given great distance from the centre will no longer vary from one direction to another according to the same law as the normal velocity of the surface of the sphere, since the modulus μ_n and likewise the amplitude α_n of the imaginary quantity $F_n(c)$ vary with the order of the function.

Let us now suppose the disturbance expressed by a LAPLACE's Function of some one order, and seek the numerical value of the alteration of intensity at a distance, produced by the lateral motion which actually exists.

The intensity will be measured by the *vis viva* produced in a given time, and consequently will vary as the density multiplied by the velocity of propagation multiplied by the square of the amplitude of vibration. It is the last factor alone that is different from what it would have been if there had been no lateral motion. The amplitude is altered in the proportion of μ_0 to μ_n, so that if

$$\frac{\mu_n^2}{\mu_0^2} = I_n,$$

I_n is the quantity by which the intensity which would have existed if the fluid had been hindered from lateral motion has to be divided.

For the first five orders the values of the function $F_n(c)$ are as follows:—

$$F_0(c) = imc + 1,$$

$$F_1(c) = imc + 2 + \frac{2}{imc},$$

$$F_2(c) = imc + 4 + \frac{9}{imc} + \frac{9}{(imc)^2},$$

$$F_3(c) = imc + 7 + \frac{27}{imc} + \frac{60}{(imc)^2} + \frac{60}{(imc)^3},$$

$$F_4(c) = imc + 11 + \frac{65}{imc} + \frac{240}{(imc)^2} + \frac{525}{(imc)^3} + \frac{525}{(imc)^4}.$$

If λ be the length of the sound-wave corresponding to the period of the vibration, $m = \frac{2\pi}{\lambda}$, so that mc is the ratio of the circumference of the sphere to the length of a wave. If we suppose the gas to be air and λ to be 2 feet, which would correspond to about 550 vibrations in a second, and the circumference $2\pi c$ to be 1 foot (a size and pitch which would correspond with the case of a common house bell), we shall have $mc = \frac{1}{2}$. The following Table gives the values of the square of the modulus and of the ratio I_n for the functions $F_n(c)$ of the first five orders, for each of the values 4, 2, 1, $\frac{1}{2}$, and $\frac{1}{4}$ of mc. It will presently appear why the Table has been extended further in the direction of values greater than $\frac{1}{2}$ than it has in the opposite direction. Five significant figures at least are retained.

mc.	$n=0$.	$n=1$.	$n=2$.	$n=3$.	$n=4$.	
4	17	16·25	14·879	13·848	20·177	
2	5	5	9·3125	80	1495·8	Values of μ_n^2.
1	2	5	89	3965	300137	
0·5	1·25	16·25	1330·2	236191	72086371	
0·25	1·0625	64·062	20878	14837899	18160×10^6	
4	1	0·95588	0·87523	0·81459	1·1869	
2	1	1	1·8625	16	299·16	Values of I_n.
1	1	2·5	44·5	1982·5	150068	
0·5	1	13	1064·2	188953	57669097	
0·25	1	60·294	19650	13965×10^3	17092×10^6	

When $mc = \infty$ we get from the analytical expressions $I_n = 1$. We see from the Table that when mc is somewhat large I_n is liable to be *a little* less than 1, and consequently the sound to be *a little* more intense than if lateral motion had been prevented. The possibility of this is explained by considering that the waves of condensation spreading from those compartments of the sphere which at a given moment are vibrating positively, *i.e.* outwards, after the lapse of a half period may have spread over the neighbouring compartments, which are now in their turn vibrating positively, so that these latter compartments in their outward motion work against a somewhat greater pressure than if each compartment had opposite to it only the vibration of the gas which it had itself occasioned; and the same explanation applies *mutatis mutandis* to the waves of rarefaction. However, the increase of sound thus occasioned by the existence of lateral motion is but small in any case, whereas when mc is somewhat small I_n increases enormously, and the sound becomes a mere nothing compared with what it would have been had lateral motion been prevented.

The higher be the order of the function, the greater will be the number of compartments, alternately positive and negative as to their mode of vibration at a given moment, into which the surface of the sphere will be divided. We see from the Table that for a given periodic time as well as radius the value of I_n becomes considerable when n is somewhat high. However practically vibrations of this kind are produced when the elastic sphere executes, not its principal, but one of its subordinate vibrations, the pitch corresponding to which rises with the order of the vibration, so that m increases with that order. It was for this reason that the Table was extended from

$mc=0.5$ further in the direction of high pitch than low pitch, namely, to three octaves higher and only one octave lower.

When the sphere vibrates symmetrically about the centre, *i. e.* so that any two opposite points of the surface are at a given moment moving with equal velocities in opposite directions, or more generally when the mode of vibration is such that there is no change of position of the centre of gravity of the volume, there is no term of the order 1. For a sphere vibrating in the manner of a bell the principal vibration is that expressed by a term of the order 2, to which I shall now more particularly attend.

Putting, for shortness, $m^2c^2 = q$, we have

$$\mu_0^2 = q+1, \quad \mu_2^2 = (q^{\frac{1}{2}} - 9q^{-\frac{1}{2}})^2 + \left(4 - \frac{9}{q}\right)^2 = q - 2 + \frac{9}{q} + \frac{81}{q^2},$$

$$I_2 = \frac{q^3 - 2q^2 + 9q + 81}{q^2(q+1)}.$$

The minimum value of I_2 is determined by

$$q^3 - 6q^2 - 84q - 54 = 0,$$

giving approximately

$$q = 12.859, \quad mc = 3.586, \quad \mu_0^2 = 13.859, \quad \mu_2^2 = 12.049, \quad I_2 = .86941;$$

so that the utmost increase of sound produced by lateral motion amounts to about 15 per cent.

I come now more particularly to Leslie's experiments. Nothing is stated as to the form, size, or pitch of his bell; and even if these had been accurately described, there would have been a good deal of guesswork in fixing on the size of the sphere which should be considered the best representative of the bell. Hence all we can do is to choose such values for m and c as are comparable with the probable conditions of the experiment.

I possess a bell, belonging to an old bell-in-air apparatus, which may probably be somewhat similar to that used by Leslie. It is nearly hemispherical, the diameter is 1·96 inch, and the pitch an octave above the middle C of a piano. Taking the number of vibrations 1056 per second, and the velocity of sound in air 1100 feet per second, we have $\lambda = 12.5$ inches. To represent the bell by a sphere of the same radius would be very greatly to underrate the influence of local circulation, since near the mouth the gas has but a little way to get round from the outside to the inside, or the reverse. To represent it by a sphere of half the radius would still apparently be to underrate the effect. Nevertheless for the sake of rather underestimating than exaggerating the influence of the cause here investigated, I will make these two suppositions successively, giving respectively $c = .98$ and $c = .49$, $mc = .4926$, and $mc = .2463$ for air.

If it were not for lateral motion the intensity would vary from gas to gas in the proportion of the density into the velocity of propagation, and therefore as the pressure into the square root of the density under a standard pressure, if we take the factor depending on the development of heat as sensibly the same for the gases and gaseous mixtures with which we have to deal. In the following Table the first column gives the gas, the second the pressure p, in atmospheres, the third the density D under the pressure p, referred to the density of air at the atmospheric pressure as unity, the fourth, Q_r, what would have been the intensity had the motion been wholly radial, referred to the intensity in air at atmospheric pressure as unity, or, in other words, a quantity varying as $p \times$ (the density at pressure 1)$^{\frac{1}{2}}$. Then follow the values of q, I_2, and Q, the last being the actual intensity referring to air as before.

Gas.	$p.$	D.	$Q_r.$	$c=\cdot98.$			$c=\cdot49.$		
				$q.$	$I_2.$	Q.	$q.$	$I_2.$	Q.
Air........................	1	1	1	·2427	1136	1	·06067	20890	1
Hydrogen	1	·0690	·2627	·01674	284700	·001048	·004186	4604000	·001191
Air, rarefied	·01	·01	·01	·2427	1136	·01	·06067	20890	·01
The same filled with H	1	·0783	·2798	·01900	220600	·001440	·004751	3572000	·001637
Air of same density ...	·0783	·0783	·0783	·2427	1136	·0783	·06067	20890	·0783
Air rarefied ½	·5	·5	·5	·2427	1136	·5	·06067	20890	·5
The same filled with H	1	·5345	·7311	·1297	4322	·1921	·0324	74890	·2039

An inspection of the numbers contained in the columns headed Q will show that the cause here investigated is amply sufficient to account for the facts mentioned by LESLIE.

It may be noticed that while q is 4 times smaller, and I_2 is 16 or 18 times larger, for $c=\cdot49$ than for $c=\cdot98$, there is no great difference in the values of Q in the two cases for hydrogen and mixtures of hydrogen and air in given proportions. This arises from the circumstance that q is sufficiently small to make the last terms in μ_0^2 and μ_2^2, namely, 1 and $81q^{-2}$, the most important, so that I_n does does not greatly differ from $81q^{-2}$. If this result had been exact instead of approximate, the intensity in different gases, supposed for simplicity to be at a common pressure, would have varied as $D^{\frac{5}{2}}$; and it will be found that for the cases in which $p=1$ the values of Q in the above Table, especially those in the last column, do not greatly deviate from this proportion. But the simplicity of this result depends on two things. First, the vibration must be expressed by a LAPLACE's function of the order 2; for a different order the power of D would have been different; and this is just one of the points respecting which we cannot infer what would be true of a bell of the ordinary shape from what we have proved for a sphere. Secondly, the radius must be sufficiently small, or the pitch sufficiently low, to make q small; at the other extremity of the scale, in which c is supposed to be very large, or λ very small, Q varies nearly as $D^{\frac{1}{2}}$ instead of $D^{\frac{5}{2}}$, whatever be the order of the LAPLACE's function. Hence no simple relation can be expected between the numbers furnished by experiment and the numerical constants of the gas in such experiments as those of M. PEROLLE*, in which the same bell was rung in succession in different gases.

[*Editor's Note:* The remainder of the paper, not reproduced here, deals with radiation from a vibrating cylinder.]

* Mémoires de l'Académie des Sciences de Turin, t. iii. (1786–7); Mém. des Correspondans, p. 1.

Editor's Comments on Paper 3

3 Frank, Bergmann, and Yaspan: *Ray Acoustics*

    The wave theory of sound is very old, and historically, most problems in acoustics have been handled in terms of this theory. However, just as in the case of light, where the concept of light rays as being normal to the advancing wave fronts has proved to be very useful in the field of geometrical optics, sound rays may be conveniently used in certain problems in acoustics. It is, of course, true that the relatively long wavelength of audible sound, promoting easy diffraction around obstacles, renders the use of sound rays more questionable than their employment in light, except at very high frequencies.

    The important problem is the establishment of the analytical conditions under which one may solve problems in wave propagation by means of rays. Historically the problem goes back to Sir William Rowan Hamilton, who put geometrical optics on a firm analytical foundation in his famous theory of systems of rays. (See *Mathematical Papers of Sir William Rowan Hamilton*, Volume 1, edited for the Royal Irish Academy by A. W. Conway and J. L. Synge, Cambridge University Press, New York, 1931, pp. 1–164.) Here Hamilton introduced his characteristic function W, later generalized by H. Bruns (*Abhandl. Königliche Sächsische Ges. Wiss. Math. Phys. Kl.*, **21**, 323, 1895) and renamed the *eikonal*. It does not appear that either Hamilton or Bruns investigated the analytical justifications for the use of rays in place of the strict propagation of wave fronts. This was certainly known to L. de Broglie, who, in connection with his development of wave mechanics, leaned heavily on Hamilton's earlier work. (See L. de Broglie, *Introduction à l'étude de la mécanique ondulatoire*, Hermann & Cie, Paris, 1930, Chap. 4.) A simple presentation of the problem will be found in R. B. Lindsay (*Mechanical Radiation*, McGraw-Hill, New York, 1960), pp. 41–47.

    The paper presented here is probably the best exposition of the fundamental nature of ray acoustics and some of its applications. It is Chapter Three in Volume One (Transmission) of *Physics of Sound in the Sea* (originally issued as a Summary Technical Report, Volume 8 of Division Six of the National Defense Research Committee, reprinted and distributed by the Research Analysis Group of the Committee on Undersea Warfare of the National Research Council, Washington, D.C., pp. 41–68). In our presentation certain pages of the original have been omitted as being of less interest from the standpoint of fundamentals.

Philipp G. Frank (1884–1966) was an Austrian physicist, who, after having served as professor of physics in Prague, succeeding Einstein in that post, came to the United States in 1941 and spent the rest of his life in this country, lecturing at Harvard and other American universities. He was an authority on classical theoretical physics, relativity, and the philosophy of science.

Peter G. Bergmann (1915–   ) was born in Berlin and received his higher education in Europe. He has been a naturalized American citizen since 1942 and for many years has been a professor of physics in Syracuse University. As a theoretical physicist he has worked in relativity and quantum theory.

Arthur Yaspan is an American mathematician who has taught at the Polytechnic Institute of Brooklyn.

Chapter 3

RAY ACOUSTICS

P. G. Frank, P. G. Bergmann, and A. Yaspan

CHAPTER 2 was devoted to the rigorous computation of the acoustic pressure p as a function of position in the fluid and of time. In situations where the acoustic pressure could be determined the sound intensity at an arbitrary spot and at an arbitrary time could be calculated. However, it was noted that, in most situations involving initial and boundary conditions similar to those met with in actual sound transmission in the ocean, this computation was at best very laborious and at worst completely impossible to carry out. Ray acoustics provides a more convenient though less rigorous approach.

In the study of sound the ray concept has not played so great a role as in optics. The reason for this is that the wavelengths of most audible sounds are not small compared to the obstacles in the path of the sound. Consequently, sounds audible to the ear do not travel straight-line or nearly straight-line paths; they bend around corners and fill almost all of any space into which they are directed. However, for the short wavelengths used in supersonic sound, the ray methods have an important, if limited, application. This chapter elaborates the theory of sound rays, describes the computation of sound intensity from the ray pattern, and finally, examines the conditions under which ray intensities may be expected to approximate the intensities calculated according to the rigorous methods of the second chapter.

3.1 WAVE FRONTS AND RAYS

3.1.1 Spherical Waves

The wave equation was solved explicitly for p in one very important case: an impulse sent out by a point source into a homogeneous medium under the assumption of spherical symmetry. The pressure as a function of time and space was found to be

$$p = \frac{1}{r} f\left(t - \frac{r}{c}\right). \quad (1)$$

In this expression, r is the distance from the source, and the function $f(t - r/c)$ is determined by the output of the source. Specifically, the source can be characterized by the statement that, while the pressure itself at the source is infinite, the product rp in the immediate vicinity of the source has a finite value at every instant, namely $f(t)$.

Obviously, this function $f(t - r/c)$ determines when a pulse emitted by the source at a particular instant will arrive at a given point of space. If the pulse should, for instance, have started in time at some instant $t = \epsilon$ so that for all values of the argument less than ϵ the function $f(t)$ vanishes, then the onset of the disturbance at a distance r from the source will be observed at the time

$$t = \epsilon + \frac{r}{c}.$$

Likewise, we find that the front of the pulse will have reached at the time t a distance r, given by

$$r = c(t - \epsilon). \quad (2)$$

What has just been stated about the front of the pulse might just as well have been said about any other specified part of the pulse; only, ϵ would in that event characterize the time at which the specified part of the pulse was radiated by the source. What has been called, vaguely, part of the pulse, is often more concisely called *phase*, particularly in connection with harmonic pulses. The term ϵ then characterizes the phase of the pulse considered and is ordinarily referred to as a *phase constant*.

The surface defined by equation (2) is, of course, a sphere of radius $c(t - \epsilon)$ at the time t. As the time increases, the radius of the sphere increases at the rate of c units per second. The surface of this sphere of constant phase is called the *wave front*. The energy represented by the disturbance of equilibrium conditions clearly spreads out radially from the source. We may focus attention on the direction of energy flow by mentally drawing an infinite number of radii from the source to the wave front. These radii may be regarded as representing the paths of energy flow and may be called *sound rays*, in analogy with light rays. Sound energy may be regarded as traveling out along these rays with the speed c. The wave fronts assume in this description the secondary role of surfaces everywhere normal to the rays.

An individual sound ray cannot exist as a physical phenomenon. An isolated sound ray would mean a state of the fluid where the condensation was confined to the immediate neighborhood of a particular straight line. Beams of narrow cross section can be produced by directing a wave front onto a very narrow slit; but if the size of the slit becomes comparable to the wavelength of the sound, the sound leaving the slit is not a narrow beam, but a cone. This phenomenon is called *diffraction*, and will be discussed in Section 3.7. It is mentioned here only to indicate that the concept of a sound ray refers not to the propagation of a narrow beam with sharp edges, but merely to the direction of propagation of actual wave fronts.

Spherical wave fronts represent only one particular case of sound propagation, even in an infinite fluid. First, the wave front can be spherical only if the initial disturbance has no preferential direction (a vibrating bubble satisfies this condition). Second, the expanding surface remains a sphere concentric with the origin only if the sound velocity c is either constant throughout the fluid, or has spherical symmetry about the sound source.

The wave factor $f(t - r/c)$ in equation (1) is responsible for the conclusion that the disturbance is propagated with the velocity c. The remaining factor, $1/r$, is called the amplitude factor since it is responsible for the decrease in sound intensity as the distance from the source increases. The rate at which sound intensity is weakened with distance can be easily computed by using the concept of rays as carriers of sound energy, provided we assume that energy is generated only at the source and then flows through space without gain or loss. For reasons of symmetry, the energy flow from the source must take place along the radial sound rays. There will be a definite number of rays inside a unit solid angle. These rays will intercept an area of 1 sq ft on a sphere of radius 1 ft whose center is at the source, an area of 4 sq ft on a sphere of radius 2 ft, and generally r^2 sq ft on a sphere of radius r. Since the total energy flow is the same for all these spherical surfaces, the energy flow per unit area, or sound intensity, must be inversely proportional to the square of the distance of the unit area from the source.

3.1.2 General Waves

The frontal attack on the wave equation was the solution of the boundary problem by the method of normal modes. This method was found to be too complicated. It was shown in Section 3.1.1 that the method of sound rays gave a simple and plausible account of sound propagation for the case of spherical symmetry. A natural approach to the general problem would be to generalize the definition of sound rays, and see if light could thereby be thrown on the general case of variable sound velocity and arbitrary initial distributions of p.

First we must generalize the definition of wave fronts. In what follows we shall restrict ourselves to harmonic sound waves, that is, sound waves which have been produced by a sound source which undergoes single-frequency harmonic vibrations. In accordance with Section 2.4.3, the pressure at any point inside the fluid can be represented as the real part of an expression having the form

$$p = A(x,y,z) e^{i\theta(x,y,z,t)} \qquad (3)$$

in which the angle θ at each point in space increases linearly with time,

$$\theta = 2\pi f[t - \epsilon(x,y,z)]. \qquad (4)$$

We shall now call a wave front all those points at which the phase angle θ has a specified value, say θ_0. At any time t, this surface is defined by the equation

$$\epsilon(x,y,z) = t - \frac{\theta_0}{2\pi f}. \qquad (5)$$

For later convenience, we shall replace $\epsilon(x,y,z)$ by an expression $W(x,y,z)/c_0$, in which c_0 is the velocity of sound under certain designated standard conditions. Equation (3) then takes the form

$$p = A(x,y,z) e^{2\pi i f\left[t - \frac{W(x,y,z)}{c_0}\right]}, \qquad (6)$$

both A and W being real functions of the space coordinates. The defining equation (5) of an individual wave front assumes the form

$$W(x,y,z) = c_0(t - t_0) \qquad (7)$$

where

$$t_0 = \frac{\theta_0}{2\pi f}.$$

The term t_0 has different values for different wave fronts, but is constant both in space and in time for a given wave front. The function W clearly has the dimension of a length.

In order to make use of the concept of sound rays to describe the energy propagated by such generalized wave fronts, we must also generalize the definition of a ray. We can no longer assume that the rays are straight lines since we concede the possibility of refraction and reflection. We shall, however, retain the property that the rays are everywhere perpendicular to the wave fronts. It is, of course, by no means obvious that the results of this new approach will agree with results from a direct solution of the wave equation plus initial and boundary conditions. A comparison between the results from the ray pattern approach and the results from a rigorous treatment of the wave equation will be carried out in Section 3.6 once the ray method has been fully described. It will be found that in many practical situations these two approaches lead to similar results.

Geometrically, the rays and successive wave fronts can be constructed as in Figure 1. The wave front at

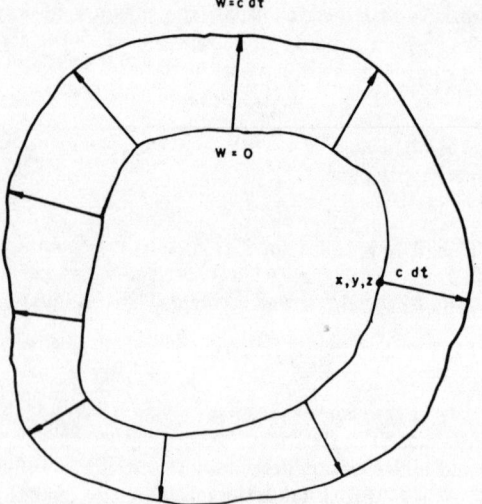

FIGURE 1. Huyghens' method for constructing successive wave fronts.

time $t = 0$ (whose equation is given by $W = -c_0 t_0$) is first drawn. In order to determine the wave front at the time dt, the small ray elements are drawn as straight-line segments perpendicular to the initial wave front, as at (x_1,y_1,z_1). In the time dt, the end point of the ray starting at (x_1,y_1,z_1) will have progressed to a point a distance $c\,dt$ from the initial wave front, where c is the velocity at the point (x_1,y_1,z_1). If this process is carried through for all the points on the initial wave surface, the end points of all the small ray elements will determine a second surface, which may be regarded as the wave front at

the time dt. By performing this process many times, the wave front can be obtained at any time t. This method of determining wave fronts by gradually widening an initial wave front was first suggested by the Dutch physicist, Huyghens, in the seventeenth century, for the solution of problems in optics.

3.2 FUNDAMENTAL EQUATIONS

3.2.1 Differential Equation of the Wave Fronts

Since the construction of wave fronts described in the preceding section is purely geometrical, it must be reformulated in mathematical terms for use in an algebraic analysis of the sort we are carrying out.

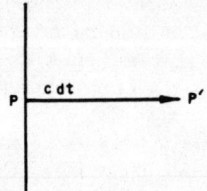

FIGURE 2. Differential ray path.

Let P in Figure 2 be any point on the wave front at time t. The equation of the wave front is given by equation (7). Let the coordinates of P be (x,y,z); let PP' be the ray element emanating from P at the end of a time interval dt; and let α,β,γ be the direction cosines of PP'. Then the coordinates of P' are $(x + \alpha c dt, y + \beta c dt, z + \gamma c dt)$. Further, the wave front at the time $t + dt$ is given by the equation

$$W(x + \alpha c dt, y + \beta c dt, z + \gamma c dt) = c_0(t - t_0 + dt). \quad (8)$$

If $c dt$ is assumed to be very small, the left-hand side of equation (3) is very nearly equal to

$$W(x,y,z) + \left(\alpha \frac{\partial W}{\partial x} + \beta \frac{\partial W}{\partial y} + \gamma \frac{\partial W}{\partial z}\right) c dt.$$

If we substitute this expression into equation (8), and use equation (7), equation (8) reduces to

$$\alpha \frac{\partial W}{\partial x} + \beta \frac{\partial W}{\partial y} + \gamma \frac{\partial W}{\partial z} = \frac{c_0}{c}. \quad (9)$$

The direction cosines α,β,γ will next be eliminated from equation (9). It is a well-known theorem[1] of analytical geometry that the direction cosines of the normal to the surface $W = $ constant at the point (x,y,z) satisfy the proportion

$$\alpha : \beta : \gamma = \frac{\partial W}{\partial x} \cdot \frac{\partial W}{\partial y} \cdot \frac{\partial W}{\partial z}.$$

Because the sum of the squares of the direction cosines equals unity,

$$\alpha^2 + \beta^2 + \gamma^2 = 1,$$

the constant of proportionality in the multiple proportion above can be determined, and we obtain

$$\alpha = \left[\left(\frac{\partial W}{\partial x}\right)^2 + \left(\frac{\partial W}{\partial y}\right)^2 + \left(\frac{\partial W}{\partial z}\right)^2\right]^{-\frac{1}{2}} \frac{\partial W}{\partial x} \quad (10)$$

$$\beta = \left[\left(\frac{\partial W}{\partial x}\right)^2 + \left(\frac{\partial W}{\partial y}\right)^2 + \left(\frac{\partial W}{\partial z}\right)^2\right]^{-\frac{1}{2}} \frac{\partial W}{\partial y}, \text{ etc.}$$

By substituting these values of α, β, γ into equation (9) and squaring both sides,

$$\left(\frac{\partial W}{\partial x}\right)^2 + \left(\frac{\partial W}{\partial y}\right)^2 + \left(\frac{\partial W}{\partial z}\right)^2 = \frac{c_0^2}{c^2(x,y,z)}. \quad (11)$$

If we define n, the *index of refraction*, by

$$n(x,y,z) = \frac{c_0}{c(x,y,z)}, \quad (12)$$

equation (11) becomes

$$\left(\frac{\partial W}{\partial x}\right)^2 + \left(\frac{\partial W}{\partial y}\right)^2 + \left(\frac{\partial W}{\partial z}\right)^2 = n^2(x,y,z). \quad (13)$$

Equation (13), often called the *eikonal equation*, is the fundamental equation of ray acoustics. It is a partial differential equation satisfied by all functions W which can define wave fronts according to equation (7). Initial conditions for equation (13) are usually of the form that W has the value zero for all points (x,y,z) on a particular surface.

Once the solution W of equation (13) has been found, the ray pattern can easily be drawn. The direction cosines of the rays at every point of space can be computed from equation (10); more simply, if equations (10) and (13) are combined,

$$\alpha = \frac{1}{n}\frac{\partial W}{\partial x}; \quad \beta = \frac{1}{n}\frac{\partial W}{\partial y}; \quad \gamma = \frac{1}{n}\frac{\partial W}{\partial z}. \quad (14)$$

Later we shall eliminate the function W from equation (14), and derive a set of *ordinary* differential equations, which together determine the course of each individual ray. First, however, we shall give a simple example illustrating how the ray pattern may be calculated from the partial differential equation (13) for the wave fronts.

Let us consider the special case where the sound velocity c depends only on the vertical depth coordinate y. Thus, the sound velocity is assumed constant everywhere on a particular horizontal plane. We shall examine only the ray pattern in one vertical plane, which we can take as the xy plane. Then equation (13) reduces to

$$\left(\frac{\partial W}{\partial x}\right)^2 + \left(\frac{\partial W}{\partial y}\right)^2 = n^2(y). \quad (15)$$

To find a simple solution of equation (15), we assume that $W(x,y)$ is the sum of a function of x and a function of y.

$$W(x,y) = W_1(x) + W_2(y).$$

Substituting this expression into equation (15), we obtain

$$\left(\frac{dW_1}{dx}\right)^2 + \left(\frac{dW_2}{dy}\right)^2 = n^2(y).$$

To obtain a family of solutions, we put $dW_1/dx = k$, where k is an arbitrary constant. Then, the differential equation will be satisfied if

$$\frac{dW_2}{dy} = \sqrt{n^2(y) - k^2}; \quad W_2 = \int_0^y \sqrt{n^2(y) - k^2}\, dy.$$

Therefore, in view of the assumed nature of W, the equation (15) will be satisfied by all functions W defined by

$$W(x,y) = kx + \int_0^y \sqrt{n^2(y) - k^2}\, dy \quad (16)$$

where k is any constant. A particular choice of k corresponds to a particular solution $W(x,y)$ and therefore to a particular set of wave fronts (7).

The direction cosines of the rays, corresponding to this choice of k, can be calculated from equations (14) and (16).

$$\alpha = \frac{k}{n}; \quad \beta = \sqrt{1 - \frac{k^2}{n^2}}.$$

We use these expressions to obtain the equations $y = y(x)$ of the sound rays. If we denote by dy/dx the slope of the direction of the ray at the point (x,y), then

$$\frac{dy}{dx} = \frac{\beta}{\alpha} = \sqrt{\frac{n^2(y)}{k^2} - 1}.$$

This equation integrates immediately to

$$x = \int_0^y \frac{dy}{\sqrt{\frac{n^2(y)}{k^2} - 1}} + x_0 \quad (17)$$

where x_0 is an arbitrary constant. Regard the k as fixed, and the x_0 as variable; then equation (17) gives an infinite set of curves which satisfy the definition of rays.

If n is a constant, that is, if the sound velocity is independent of depth, the rays (17) are clearly straight lines.

3.2.2 Differential Equations of Rays

It may be argued that the replacement of the wave equation by the ray treatment as represented by the differential equation (13) has little to recommend itself. It appears that one difficult partial differential equation has merely been replaced by another, which might resist attempts at solution as effectively as the first one.

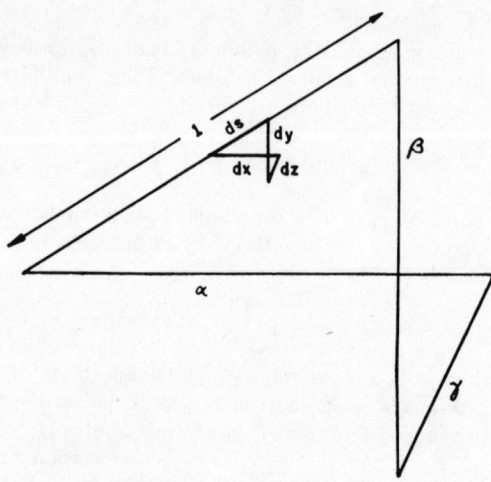

FIGURE 3. Specification of direction of ray element ds by direction cosines.

Further examination shows, however, that the new equation (13) has two properties which tend to simplify its solution. First, equation (13) contains no time derivatives. This means that it describes the propagation of a disturbance in terms independent of the frequencies which make up this disturbance. Second, it is possible to set up *ordinary* differential equations that describe the path of individual rays; the latter equations will be derived in this section.

We start with the equations (14), from which we proceed to eliminate W. This can easily be done by use of the formulas $\partial^2 W/\partial x \partial y = \partial^2 W/\partial y \partial x$, etc. By differentiating the first equation of (14) with respect to y, the second with respect to x, and by equating the results, a relation between α, β, and n is obtained. Proceeding similarly with the other equations, we obtain the following relationships which must hold at any point of the ray pattern:

$$\frac{\partial(n\alpha)}{\partial y} = \frac{\partial(n\beta)}{\partial x} ; \frac{\partial(n\alpha)}{\partial z} = \frac{\partial(n\gamma)}{\partial x} ; \frac{\partial(n\beta)}{\partial z} = \frac{\partial(n\gamma)}{\partial y}. \tag{18}$$

These equations can be developed further to yield the changes of α, β, and γ along the path of an individual ray. If the arc length along the ray path from a given starting point is denoted by s, we have

$$\frac{d(n\alpha)}{ds} = \frac{\partial(n\alpha)}{\partial x}\frac{dx}{ds} + \frac{\partial(n\alpha)}{\partial y}\frac{dy}{ds} + \frac{\partial(n\alpha)}{\partial z}\frac{dz}{ds}. \tag{19}$$

We see from Figure 3 that

$$\frac{dx}{ds} = \alpha ; \frac{dy}{ds} = \beta ; \frac{dz}{ds} = \gamma. \tag{20}$$

Thus equation (19) turns into

$$\frac{d(n\alpha)}{ds} = \alpha\frac{\partial(n\alpha)}{\partial x} + \beta\frac{\partial(n\alpha)}{\partial y} + \gamma\frac{\partial(n\alpha)}{\partial z},$$

which, upon using the relations (18), becomes

$$\frac{d(n\alpha)}{ds} = \alpha\frac{\partial(n\alpha)}{\partial x} + \beta\frac{\partial(n\alpha)}{\partial x} + \gamma\frac{\partial(n\alpha)}{\partial x}$$

$$= (\alpha^2 + \beta^2 + \gamma^2)\frac{\partial n}{\partial x}$$

$$+ \left(\alpha\frac{\partial \alpha}{\partial x} + \beta\frac{\partial \beta}{\partial x} + \gamma\frac{\partial \gamma}{\partial x}\right)n. \tag{21}$$

The first parenthesis equals unity, because it is the sum of squares of direction cosines; while the second parenthesis, which is equal to one-half times the derivative of the first one, vanishes. Thus equation (21) simplifies to

$$\frac{d(n\alpha)}{ds} = \frac{\partial n}{\partial x}.$$

After similar calculations are carried out for $d(n\beta)/ds$ and $d(n\gamma)/ds$, we get the following set of three ordinary differential equations:

$$\frac{d(n\alpha)}{ds} = \frac{\partial n}{\partial x} ; \frac{d(n\beta)}{ds} = \frac{\partial n}{\partial y} ; \frac{d(n\gamma)}{ds} = \frac{\partial n}{\partial z}. \tag{22}$$

It is understood that n, the index of refraction, is a given function of x,y,z.

We now deduce an important result for the special case where the sound velocity is a function of the vertical depth coordinate y alone. We shall show that for this case the entire path of an individual ray lies in a plane determined by the vertical line through the projector and the initial direction of the ray.

Let the origin of coordinates be taken at the projector, and let the direction cosines of a ray leaving the projector be $\alpha_0, \beta_0, \gamma_0$, as in Figure 4. Since n depends only on y, equations (22) simplify to

$$\frac{d(n\alpha)}{ds} = 0 \; ; \; \frac{d(n\beta)}{ds} = \frac{dn}{dy} \; ; \; \frac{d(n\gamma)}{ds} = 0. \quad (23)$$

Thus, along any individual ray we have $n\alpha = $ constant, $n\gamma = $ constant, which in turn implies

$$\frac{\gamma}{\alpha} = \text{constant} = \kappa$$

along the ray. Then, the initial direction of the ray is $(\alpha_0, \beta_0, \kappa\alpha_0)$.

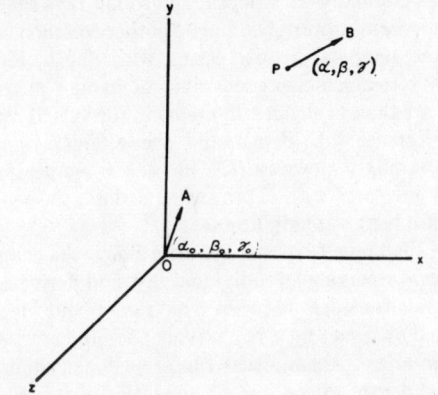

FIGURE 4. Change of ray direction between point $(0, 0, 0)$ and point (x, y, z).

The direction at a general point P along the ray will be characterized by the direction cosines $\alpha, \beta, \kappa\alpha$ because of the equations (23). It can easily be shown by the methods of analytical geometry that the normal to the plane determined by OY (direction cosines $0,1,0$) and OA (direction cosines $\alpha_0, \beta_0, \kappa\alpha_0$) has the direction cosines $\kappa/\sqrt{\kappa^2 + 1}, 0, -1/\sqrt{\kappa^2 + 1}$. The direction of the ray at P is characterized by the direction cosines $\alpha, \beta, \kappa\alpha$; thus the ray direction at P is perpendicular to the normal to the plane AOY; hence the segment PB lies in that plane. Since P was any point on the ray, the entire ray must lie in the plane AOY.

3.3 RAY PATHS FOR VERTICAL VELOCITY GRADIENTS

3.3.1 Derivation of the Equations of Ray Paths

We now solve the equations (22) for the special case where the sound velocity depends only on the vertical depth coordinate y and discuss this solution in detail. It is intuitively obvious that if we carry through the solution for the xy plane, then the ray pattern in any other plane through the vertical (y) axis will be identical in size and shape.

Since the water depth increases in the downward direction, we shall take the y axis positive downward. We shall denote the angle which a direction in the xy plane makes with the positive x direction by θ, as in Figure 5. To avoid ambiguity, we must specify care-

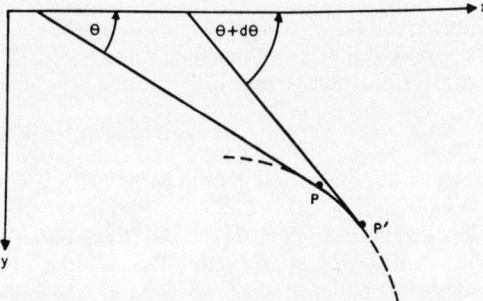

FIGURE 5. Change in ray direction over ray element PP'.

fully the sign of the angle θ. We shall be concerned only with rays moving in the direction of increasing x, in other words, to the right in the figure. If the ray is gaining depth with increasing range, we give the angle θ a positive sign; while if the ray is losing depth with increasing range, we give θ a negative sign. These conventions, illustrated in Figure 6, enable us to use

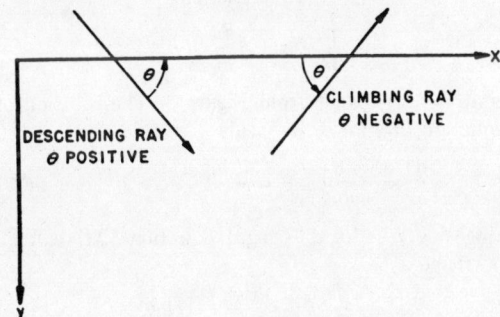

FIGURE 6. Conventions fixing sign of θ.

the following relations both for climbing and descending rays:

$$\alpha = \cos\theta \; ; \; \beta = \sin\theta \; ; \; \gamma = 0. \quad (24)$$

Since the sound velocity is assumed to depend only on y, we have

$$\frac{\partial n}{\partial x} = \frac{\partial n}{\partial z} = 0,$$

and by reason of relations (24) the equations (22) reduce to

$$\frac{d(n \cos \theta)}{ds} = 0; \quad \frac{d(n \sin \theta)}{ds} = \frac{dn}{dy}. \quad (25)$$

From the first equations it follows that $n \cos \theta$ has a constant value along a particular single ray. That is, if P and P' are two points on the ray, then

$$\frac{c_0}{c} \cos \theta = \frac{c_0}{c'} \cos \theta'.$$

If, in particular, P is located at the depth where $c(y) = c_0$, and if θ_0 is the direction of the ray at this point, this equation becomes

$$\frac{\cos \theta}{\cos \theta_0} = \frac{c}{c_0} \equiv \frac{1}{n}. \quad (26)$$

Equation (26) is identical in form with Snell's law in optics.

The second equation in (24) is used to compute the *curvature* of the ray at any point. The *curvature* of a curve at a point on it is defined as $d\theta/ds$, the angle through which the tangent turns as one travels along the curve for unit distance. Because of our conventions for the sign of the direction angle θ, upward bending is always associated with negative curvature, and downward bending with positive curvature.

From the relations (25), we have

$$\begin{aligned}\frac{dn}{dy} &= n\frac{d(\sin \theta)}{ds} + \sin \theta \frac{dn}{ds} \\ &= n\frac{d(\sin \theta)}{d\theta}\frac{d\theta}{ds} + \sin \theta \frac{dn}{dy}\frac{dy}{ds} \quad (27)\\ &= n \cos \theta \frac{d\theta}{ds} + \sin^2 \theta \frac{dn}{dy}\end{aligned}$$

since $dy/ds = \sin \theta$, from Figure 5. The solution of equation (27) for $d\theta/ds$ yields

$$\frac{d\theta}{ds} = \frac{1}{n}\frac{dn}{dy} \cos \theta = \frac{d(\log n)}{dy} \cos \theta. \quad (28)$$

Since $\log n = \log c_0 - \log c$, equation (28) can be rewritten as

$$\frac{d\theta}{ds} = -\frac{d(\log c)}{dy} \cos \theta. \quad (29)$$

We can use equation (29) to describe, qualitatively, what happens when a ray travels to a layer just above it ($dy < 0$) of different sound velocity. If the new layer has higher sound velocity, the curvature $d\theta/ds$ has a positive sign, and the ray is bent downward. If the layer just above has lower sound velocity, the curvature $d\theta/ds$ is negative, and the ray is bent upward. We get the opposite result if the ray is traveling to a layer just below it ($dy > 0$) of different sound velocity. Thus we can say, in general, that a ray entering a layer of higher sound velocity is bent away from the layer, and a ray entering a layer of lower sound velocity is bent into the layer.

In the open ocean the vertical velocity gradient usually falls into one of two types, depending on the temperature-depth variation. If the temperature does not depend on the depth, the velocity is determined by the pressure, which increases with depth; therefore, in such isothermal water the sound velocity increases gradually with depth, and sound rays should possess slight upward bending. Another common case has the temperature decreasing with depth. Since velocity is much more sensitive to changes in temperature than to changes in pressure, the velocity will also decrease with depth, and the sound rays will bend strongly downward. The water temperature rarely increases with depth; when it does, the sound rays are bent strongly upward.

We shall now examine, quantitatively, the change of curvature along an individual ray, and derive certain relationships between the range and depth reached at time t by a ray leaving the projector at a certain angle. Assume that the projector is situated at the depth where $c = c_0$; thus the ray may be characterized by its initial angle θ_0 at the projector. Because of equation (26), equation (29) becomes

$$\frac{d\theta}{ds} = -\frac{dc}{dy}\frac{\cos \theta_0}{c_0}. \quad (30)$$

The advantage of the representation (30) is that it gives the curvature along a single ray as a function of dc/dy only, since θ_0 is constant for that particular ray.

We consider, in particular, the case where the velocity gradient has the constant value a; that is,

$$c = c_0 + ay, \quad (31)$$

if the origin of coordinates is taken at the projector. At all points on the ray, in view of equation (30),

$$\frac{d\theta}{ds} = -\frac{a \cos \theta_0}{c_0}. \quad (32)$$

We see from equation (32) that the curvature is constant along the ray; this means that the ray must be an arc of a circle. As the radius of curvature is the reciprocal of the curvature $d\theta/ds$, the radius r of this circle must be given by

$$r = \left|\frac{c_0}{a \cos \theta_0}\right|. \quad (33)$$

If a is positive, the curvature (32) is negative, and

the circular arc bends upward; but if a is negative, the circular arc bends downward.

We can determine the center of the circle defining the ray by a simple geometrical construction. Figure 7 shows the path of a ray leaving the projector at the

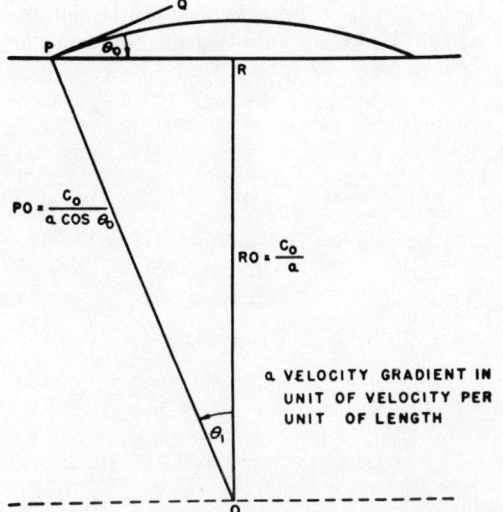

FIGURE 7. Geometrical construction of ray path.

angle θ_0 into a medium of constant negative velocity gradient. The center of the circle is obtained by following the perpendicular to PQ down through the medium a distance $c_0/a \cos \theta_0$. It is a simple consequence of the geometry of the situation that this center will lie on the horizontal line a distance c_0/a below the projector. For, from the illustration, $RO = (c_0/a)(\cos \theta_1/\cos \theta_0)$, and θ_1 clearly equals θ_0. Similarly, if the constant velocity gradient is positive so that the rays are bent upward, the centers of the defining circles lie on a horizontal line a distance c_0/a *above* the projector. It is easily shown that the dashed horizontal "line of centers" in Figure 7 is at the depth where the velocity c would equal zero if the assumed linear gradient extended indefinitely.

An approximate solution in the general case where c is an arbitrary function of y can be obtained by repeated use of the solution for constant gradient. Even a complicated velocity-depth curve can be closely approximated, as in Figure 8, by dividing the depth interval into a relatively small number of segments in each of which the velocity is assumed to change linearly with depth. Within each layer the ray path is an arc of a circle; and the total ray path is a consecutive series of such arcs.

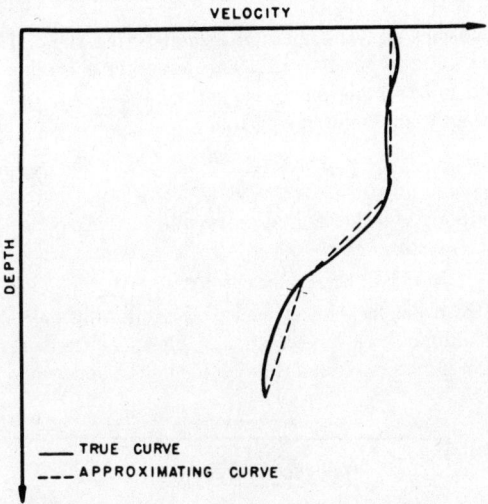

FIGURE 8. Approximating velocity-depth curve by a succession of linear gradients.

In practice, the path of the ray cannot be conveniently plotted as a sum of circular arcs because the horizontal ranges are much greater than the depths of interest, and therefore their scale must be contracted. Instead, the ray is usually traced by calculating the angles θ_1 and θ_2 at which it enters and leaves a given layer, and the horizontal distance it travels in the layer. This calculation is illustrated in Figure 9,

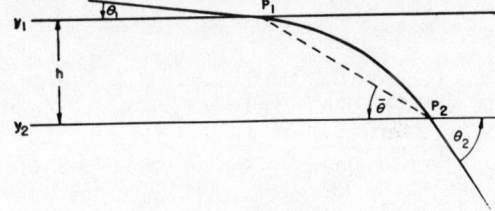

FIGURE 9. Ray path in layer of linear gradient.

where the top of the layer is at depth y_1, the bottom is at depth y_2, and the thickness of the layer is h.

The ray leaves the projector at an angle θ_0, enters the layer at the angle θ_1, and leaves the layer at the angle θ_2. Then, by equation (26),

$$\theta_1 = \arccos\left[\frac{c(y_1) \cos \theta_0}{c_0}\right]$$
$$\theta_2 = \arccos\left[\frac{c(y_2) \cos \theta_0}{c_0}\right] \quad (34)$$

where $c(y_1)$ and $c(y_2)$ are calculated from equation (31).

Consider now the chord P_1P_2 converting the end points of the circular arc. The direction $\bar{\theta}$ of this chord is by simple plane geometry $\frac{1}{2}(\theta_1 + \theta_2)$; and its length is therefore given by

$$P_1P_2 = \frac{h}{\sin \frac{1}{2}(\theta_1 + \theta_2)}. \quad (35)$$

The increase in horizontal range due to the passage of the ray through the layer is $P_1P_2 \cos \bar{\theta}$, or

$$\text{Range in layer} = h \cot \tfrac{1}{2}(\theta_1 + \theta_2). \quad (36)$$

This result may be applied to the following problem. Suppose we have a sum of layers of the sort shown in Figure 10; and we wish to find the horizontal

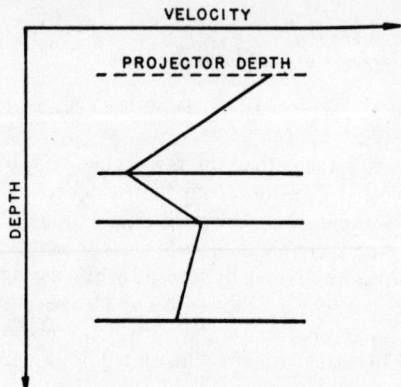

FIGURE 10. Succession of linear gradients.

range attained by the time the ray reaches the depth H below the projector. We let the bottom layer extend just to the depth H; suppose this is the third layer below the projector. We know θ_0 and we calculate $\theta_1, \theta_2, \theta_3$ by the relations (34). Then the horizontal range to the depth H will be the sum of terms of the form (36):

Horizontal range to $H = h_1 \cot \tfrac{1}{2}(\theta_0 + \theta_1) +$
$\quad h_2 \cot \tfrac{1}{2}(\theta_1 + \theta_2) + h_3 \cot \tfrac{1}{2}(\theta_2 + \theta_3). \quad (37)$

The inverse problem is a little more complicated. Suppose we wish to find the depth reached by a ray of initial direction θ_0 by the time it has traveled a horizontal distance R in a stratified medium that consists of layers of thickness h_1, h_2, h_3, etc. We calculate the range R_1 in the first layer, R_2 in the second layer, and so on, until the sum of these partial ranges is greater than R:

$$R_1 + R_2 + R_3 < R$$
$$R_1 + R_2 + R_3 + R_4 > R.$$

Then the depth the ray reaches at range R will be greater than $h_1 + h_2 + h_3$ and less than $h_1 + h_2 + h_3 + h_4$. Its value may be obtained with sufficient accuracy by interpolation.

The ray-tracing methods described in this section are too cumbersome to use in practice. A number of devices have been developed to facilitate the plotting of rays bent by known velocity gradients; these devices will be discussed in Section 3.5.1.

* * * * * * *

the point to which the range is measured, and on refraction conditions, or more specifically on the temperature-depth variation indicated by the bathythermograph.

The intensity contour diagram is a set of lines drawn on a ray diagram indicating the intensity loss. On each contour the intensity loss has a constant value, in a fashion similar to the curves of constant barometric pressure on a weather map. The contours are obtained from a ray diagram by using one of the methods discussed in Sections 3.4 and 3.5. On each ray, or for each pair of adjacent rays, the intensity, or transmission anomaly, is computed at suitably chosen intervals. Then one finds, by interpolation, the points where the intensity loss is 55 db, 60 db, 65 db, and so on. After this process is carried through for all the rays, intensity contours can be drawn by joining the points of equal transmission loss on all the rays.

Sample intensity contour diagrams for the oceanographic situations treated in Section 3.5.2 are given in Figure 25. The contour diagram for isothermal water is shown for comparison since it indicates optimum sound-ranging conditions, that is, the intensity losses which would be observed if the water had no temperature gradients, and if there were no attenuation losses; for this situation, the intensity loss out to the range R is given by the inverse square law and amounts to 20 log R. The contour diagrams for the split-beam cases are identical with that for the isothermal case at depths near the sea surface and at short to moderate ranges; at depths below the thermocline, however, the predicted spreading loss is much increased; the amount of increase depends on the depth to the thermocline and the sharpness of the thermocline gradient. In the case of downward refraction, the intensity contours which denote large values of the intensity loss are piled together in the vicinity of the predicted shadow boundary.

A more detailed discussion of intensity contours with a derivation of some of the basic equations derived at the beginning of this chapter is given in a report by UCDWR.[6] Sample theoretical intensity contours for different temperature patterns are also discussed in this reference. A comparison of these predicted intensities with sound intensities found from explosive pulses is given in Chapter 9. The encyclopedia of ray diagrams in reference 5 includes intensity contours on most of the diagrams and thus may be used to find the type of predicted sound field for many different varieties of temperature-depth patterns.

It will be seen in Chapter 5 that the intensity predictions of the contour diagram are not, in general, sufficiently accurate to be trusted for the prediction of maximum echo ranges. However, they are useful for various special purposes, such as indicating how sound intensities should vary with depth at a fixed range.

3.6 VALIDITY OF RAY ACOUSTICS

In Sections 3.1 to 3.5 of this chapter the method of ray acoustics has been presented as an independent theory without much connection with the rigorous treatment of wave propagation presented in Chapter 2. We first noted in Section 3.1.1 that the important features of the propagation of spherical waves could be derived equally well by using the concept of wave fronts connecting points which have equal phase of condensation, and the concept of energy transported by rays perpendicular to these wave fronts. Then we generalized the definition of wave fronts and rays, derived differential equations for the ray paths from these definitions, and solved these differential equations for the ray paths and the resulting sound intensity.

It is important to remember, however, that the method of wave fronts for the general case placed no requirement on the wave front, except for stipulating that it be of the form (7) for some function $W(x,y,z)$. To make the idea of wave fronts intuitively significant, it was implied that the wave front should always join points of constant phase of condensation; but this implication was never used. The ray paths depended only on the form of the function W and the variation of c; the intensity calculations depended, in addition, on the assumption that energy is transported out along the rays. In this section, where we try to find a connection between ray acoustics and wave acoustics, we must assume a physical significance for the wave fronts. Accordingly, we shall make the explicit assumption that the wave fronts join points of equal phase of condensation since we already know that the assumption brings ray acoustics and wave acoustics into agreement for the case of spherical waves.

In this section, we shall examine whether wave acoustics and ray acoustics with this definition of wave fronts are equivalent in general or only under some special conditions. Since sound field calculations are much simpler by the ray method than by a rigorous solution of the wave equation, it will be extremely valuable to know when the ray theory can be applied without much error and when it will lead to definitely wrong results.

3.6.1 Eikonal Wave Fronts versus General Wave Fronts

It will be remembered that the entire method of rays was based on the eikonal equation (13), which in turn was based on the assumption that the wave fronts (7) "grow" perpendicularly to themselves. That is, the eikonal equation was derived by assuming that the wave front at time $t + dt$ is found from the wave front at time t by moving each point on the latter a distance cdt along the outward normal. We shall now show that wave fronts ordinarily do not obey this law of propagation rigorously, but that the assumption often provides a good approximation.

It is intuitively apparent that wave fronts, defined purely as surfaces of constant phase without reference to the way they grow, exist in the exact case, at least when the dependence on time is harmonic. We shall define these wave fronts in the rigorous case by

$$V(x,y,z) = c_0(t - t_0) \qquad (91)$$

reserving the expressions W for those cases where the wave fronts grow perpendicularly to themselves, and where W therefore satisfies the eikonal equation. We shall call surfaces (91) *general wave fronts*, and surfaces defined by similar equations, with V replaced by W, *eikonal wave fronts*.

We know that in instances where the sound source vibrates harmonically with a single frequency f the solution of the wave equation can be expressed as the real part of the complex expression

$$p = A(x,y,z)e^{2\pi i f\{t - [V(x,y,z)]/c_0\}}. \qquad (92)$$

This expression is identical with equation (6), except that we assume that the expression (92) with the function $V(x,y,z)$ is a rigorous solution of the wave equation, while the expression (6) with the function $W(x,y,z)$ was obtained by means of a Huyghens construction so that $W(x,y,z)$ would satisfy the eikonal equation.

We now shall see under what conditions the expression (92) can satisfy the wave equation and, simultaneously, $V(x,y,z)$ satisfy the eikonal equation. Suppose p satisfies equation (27) of Chapter 2, and simultaneously V satisfies the eikonal equation (13). The latter condition is

$$\left(\frac{\partial V}{\partial x}\right)^2 + \left(\frac{\partial V}{\partial y}\right)^2 + \left(\frac{\partial V}{\partial z}\right)^2 - n^2 = 0. \qquad (93)$$

The former condition may be simply calculated by noting that equation (92) may be written as

$$p = e^{\log A - 2\pi i f(V/c_0)} e^{2\pi i f t}. \qquad (94)$$

Substitution of the expression (94) into the wave equation, performance of the indicated differentiations, and collection of terms is a straightforward calculation which will not be reproduced here. The real and imaginary parts must vanish separately; these parts are

$$\left(\frac{\partial V}{\partial x}\right)^2 + \left(\frac{\partial V}{\partial y}\right)^2 + \left(\frac{\partial V}{\partial z}\right)^2 - n^2 - \frac{\lambda_0^2}{4\pi}\left\{\frac{\partial^2(\log A)}{\partial x^2}\right.$$
$$+ \frac{\partial^2(\log A)}{\partial y^2} + \frac{\partial^2(\log A)}{\partial z^2} + \left[\frac{\partial(\log A)}{\partial x}\right]^2$$
$$\left. + \left[\frac{\partial(\log A)}{\partial y}\right]^2 + \left[\frac{\partial(\log A)}{\partial z}\right]^2\right\} = 0. \qquad (95)$$

and

$$\frac{\partial^2 V}{\partial x^2} + \frac{\partial^2 V}{\partial y^2} + \frac{\partial^2 V}{\partial z^2} + 2\left[\frac{\partial V}{\partial x}\frac{\partial(\log A)}{\partial x}\right.$$
$$\left. + \frac{\partial V}{\partial y}\frac{\partial(\log A)}{\partial y} + \frac{\partial V}{\partial z}\frac{\partial(\log A)}{\partial z}\right] = 0. \qquad (96)$$

Clearly, V will satisfy condition (93) only if

$$\lambda_0^2\left\{\frac{\partial^2(\log A)}{\partial x^2} + \frac{\partial^2(\log A)}{\partial y^2} + \frac{\partial^2(\log A)}{\partial z^2}\right.$$
$$\left. + \left[\frac{\partial(\log A)}{\partial x}\right]^2 + \left[\frac{\partial(\log A)}{\partial y}\right]^2 + \left[\frac{\partial(\log A)}{\partial z}\right]^2\right\} = 0. \qquad (97)$$

This can happen if λ_0 is zero, or if

$$B \equiv \frac{1}{A}\left(\frac{\partial^2 A}{\partial x^2} + \frac{\partial^2 A}{\partial y^2} + \frac{\partial^2 A}{\partial z^2}\right) = 0, \qquad (98)$$

since the expression in braces in (97) easily reduces to the above. This condition (98) is usually not satisfied. While it happens to be satisfied by the pressure wave of a point source in a homogeneous medium, it does not hold, for instance, for the radiation of a double source. In general, equations (93) and (95) will be rigorously equivalent only if the wavelength λ_0 vanishes.

3.6.2 Conditions for Nearly Eikonal Wave Fronts

We derived in Section 3.6.1 the conditions under which wave fronts, defined as expanding surfaces of constant phase of condensation, expand perpendicularly to themselves. It is more useful to know how large the frequency must be, relative to the other parameters of the problem, before the function $V(x,y,z)$ of equation (92) very nearly satisfies the eikonal equation; we will then know under what conditions the wave fronts are very nearly perpendicularly expanding.

Clearly the expression B of equation (98), the remainder term will be negligible compared with the other terms if

$$\lambda_0 (\log A)' \ll V' \qquad (99)$$
$$\lambda_0^2 (\log A)'' \ll (V')^2, \qquad (100)$$

where the prime denotes *any* spatial derivative, and $\ll$ means "is negligible compared with." If V even approximately satisfies the eikonal equation (13), then

$$V' \sim n, \qquad (101)$$

where the symbol $\sim$ signifies "is of the same order of magnitude as."

Another useful relation is obtained from equation (96). The functions A and V must satisfy equation (96) as long as the surface (91) has the significance of a general wave front. But equation (96) implies that

$$V'' \sim V' (\log A)', \qquad (102)$$

which in turn implies that

$$\lambda_0 V'' \sim V' \lambda_0 (\log A)' \ll V' V' \qquad (103)$$

because of equation (99). Combining equations (103) and (101),

$$\lambda_0 V'' \ll n^2. \qquad (104)$$

In the ocean the index of refraction n is of the order of magnitude of unity. Then, the relation (104) may be stated in the following words. The first spatial derivative of V must not change much over a spatial distance of one wavelength. The first spatial derivatives of V give the direction of the rays; while the second derivatives, yielding the rate of change of ray direction, give the curvature of the rays. Therefore, the condition (104) becomes the following. The direction of the ray must not change much over a distance of one wavelength. In regions where the ray curves very strongly, ray acoustics cannot be applied safely.

Differentiating the eikonal equation (13), we get $V'V'' \sim nn'$ or $V'' \sim n'$ because of equation (101). In view of equation (104), this means that

$$\lambda_0 n' \ll n^2 \sim 1. \qquad (105)$$

In other words, the index of refraction must not change much over a distance of one wavelength.

We derive one more restriction — this time on the amplitude function A. From equations (102) and (104), we also have

$$\lambda_0 (\log A)' < 1. \qquad (106)$$

The relation (106) means that $\log A$ must not change much over one wavelength. Since this change is very nearly $\lambda_0 A'/A$, this means that the percentage change in A over one wavelength must be very small.

We can summarize our conclusions as follows. The eikonal equation usually will not lead to a good approximation (1) if the radius of curvature of the rays is anywhere of the order of, or smaller than, one wavelength, or (2) if the velocity of sound changes appreciably over the distance of one wavelength, or (3) if the percentage change in the amplitude function A is not small over the distance of one wavelength.

3.6.3 Comparison of Ray Intensities and Rigorous Intensities

It follows from the results of Section 2.7.3 that if the general wave fronts are defined by equation (91), and the instantaneous acoustic pressure by equation (92), then the rigorous intensity is given by

$$I = \frac{a^2}{2\rho c_0} \sqrt{\left(\frac{\partial V}{\partial x}\right)^2 + \left(\frac{\partial V}{\partial y}\right)^2 + \left(\frac{\partial V}{\partial z}\right)^2} \qquad (107)$$

and, further, that the direction of energy flow is characterized by the direction numbers $\partial V/\partial x : \partial V/\partial y : \partial V/\partial z$. The latter direction is perpendicular to the general wave front; thus, if the wave fronts are eikonal wave fronts, the energy flows along the rays in the rigorous case. If the wave fronts are approximately eikonal wave fronts, then the directions perpendicular to these wave fronts represent very nearly the true direction of energy flow.

Thus, if the conditions for eikonal wave fronts derived in Section 3.6.2 are satisfied, the energy emanating from the source into all solid angles will remain within the tubular confines assumed in deriving the ray intensity. We can therefore say, intuitively, that if the wave fronts are very nearly eikonal wave fronts, the ray intensity will be very close to the rigorous intensity. Further, we can say that in both cases the intensity will be given by

$$I = \frac{a^2}{2\rho c}, \qquad (108)$$

since $\left[\left(\frac{\partial V}{\partial x}\right)^2 + \left(\frac{\partial V}{\partial y}\right)^2 + \left(\frac{\partial V}{\partial z}\right)^2\right]^{\frac{1}{2}} \approx n = \frac{c_0}{c}.$

3.7 SHADOW ZONE AND DIFFRACTION

When the velocity decreases from the surface downwards, the ray theory predicts a sharp shadow boundary across which no sound ray penetrates; a typical ray diagram for such an instance is shown in Figure 24. At the shadow boundary the ray theory predicts a discontinuous drop of intensity from a finite value on one side to a zero value on the other. It was shown in Section 3.6 that the ray theory cannot be trusted whenever it predicts such a rapid change of intensity in a distance of only a few wavelengths. Thus, it is necessary to use the wave equation directly to compute the intensity of sound which penetrates the so-called shadow zone.

The simplest case of a shadow zone is that produced by a screen in front of a light source. As shown in Figure 26, the ray theory predicts that no light

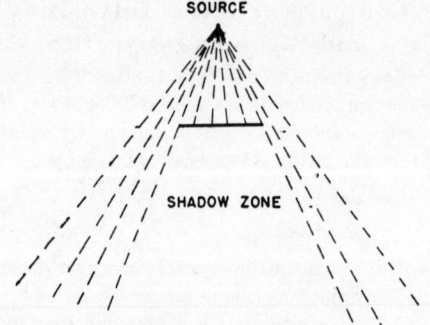

FIGURE 26. Optical shadow zone produced by screen.

can reach the shadow zone behind the screen. When the rays carrying the energy are curved, as in Figure 24, it is the surface of the ocean that intercepts the curved rays and "casts a shadow." In either case, however, some energy actually appears inside the predicted shadow zone, and the wave is said to be "diffracted."

The computation of diffracted sound in the shadow zone is a rather complicated problem in the general case. To indicate the type of analysis required, and to show the general nature of the results, a simplified problem will be considered here. As shown in Figure 27, a sound projector is assumed to be placed against a vertical wall, which extends down to great depths. The introduction of the wall simplifies the problem without changing the final results essentially. The water is assumed to be so deep that bottom-reflected sound may be neglected. The projector face is assumed to be so wide that the horizontal spreading of the sound beam may be neglected; thus, only the two-dimensional problems need be considered. The sound velocity c is assumed to vary according to the law

$$c^2 = \frac{c_0^2}{1 + By} \qquad (109)$$

where B is a constant, and y represents depth below the surface. Since B is in practice very small, this gradient is indistinguishable from a linear gradient at depths of interest. The exact velocity gradient at the depth y is given by

$$\frac{dc}{dy} = -\frac{c_0^2}{2c} \frac{B}{(1 + By)^2}. \qquad (110)$$

Thus, at the surface, where $y = 0$, the velocity gradient $-b$ is given by

$$-b = \frac{Bc_0}{2} = -\frac{dc}{dy}\bigg|_{y=0}. \qquad (111)$$

The gradient (109) is chosen instead of a simple linear gradient not for physical reasons, but because it simplifies the following computations.

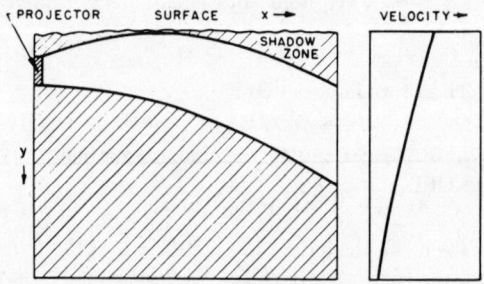

FIGURE 27. Sound shadow cast by sea surface.

To solve the wave equation under these conditions, it is necessary to use the method of normal modes developed in Chapter 2. In particular, we must find a solution to the wave equation (27) in Chapter 2 which satisfies the boundary conditions we shall impose. As in Section 2.7.2, we look for a solution which is the product of three functions, one dependent only on the time t, another dependent on the depth y, and the third, a function of the horizontal distance x. The coordinate z need not be considered in the two-dimensional case under discussion.

Following the analysis of Section 2.7.2, we therefore write

$$p(x,y,z,t) = e^{2\pi i f t} F(y) G(x). \qquad (112)$$

By substitution of equation (112) into the wave equation (27) of Chapter 2, and by dividing through by C^2, it is found that F and G satisfy an equation of the form

$$G\frac{d^2F}{dy^2} + F\frac{d^2G}{dx^2} + \frac{4\pi^2 f^2}{c^2} FG = 0. \qquad (113)$$

If equation (113) is divided through by FG, and equation (109) used for c,

$$\left(\frac{1}{G}\frac{d^2G}{dx^2}\right) + \left[\frac{1}{F}\frac{d^2F}{dy^2} + \frac{4\pi^2 f^2}{c_0^2}(1 + By)\right] = 0. \quad (114)$$

Since the first bracket depends only on x and the second only on y, equation (115) can be satisfied only if each bracket is constant. If we denote the first bracket by $-\mu^2$, the second bracket must be $+\mu^2$, and we have

$$\frac{d^2F}{dy^2} + \left[\frac{4\pi^2 f^2}{c_0^2}(1 + By) - \mu^2\right]F = 0 \quad (115)$$

$$\frac{d^2G}{dx^2} + \mu^2 G = 0. \quad (116)$$

The basic problem is to find solutions of equations (115) and (116) which satisfy the boundary conditions. First, we have the boundary conditions for equation (115). In the analysis in Section 2.7.2, these boundary conditions were that the pressure vanished both at the surface and at the bottom. Here, also, the pressure must vanish at the surface. However, the water is so deep that the condition at the bottom disappears. Instead, there is simply the condition that at some distance below the projector no sound is coming upwards; that is, any sound present at these depths is coming down from shallower depths. Although this boundary condition is somewhat complicated to formulate exactly, the general result is the same as that found in the solution of equation (161) of Chapter 2. In this earlier instance it was found that $\sin 2\pi y/\lambda_y$, corresponding to $F(y)$ in equation (112), when B is zero, satisfied the two boundary conditions only if λ_y had one of a number of fixed values. Similarly, the function $F(y)$ can satisfy the two boundary conditions only if μ has one of a certain number of values. These values, which are called characteristic values of μ, may be denoted by μ_1, μ_2, μ_3, and so on, or more generally by μ_j, where j can be any integral number. For each of the characteristic values μ_j, equation (115) has a particular solution $F_j(y)$ which satisfies the boundary conditions.

Once a value of μ_j has been chosen, the solution of equation (116) is very simple. For each value of μ_j,

$$G = A_j e^{-i\mu_j x} \quad (117)$$

where A_j is an arbitrary constant.[a] Thus the wave equation is satisfied by any product of the type

$$p_j = A_j e^{2\pi i f t} F_j(y) e^{-i\mu_j x}. \quad (118)$$

Equation (118) satisfies the boundary conditions at the surface and at great depth since $F_j(y)$ satisfies these conditions. However, the boundary conditions at the vertical plane $x = 0$, the assumed vertical wall, must also be satisfied. These conditions are that the particle velocity at the sound projector must be $v_0 \cos 2\pi f t$, and that the particle velocity at all other points in the plane $x = 0$ must be zero.

To satisfy this boundary condition at the plane $x = 0$ requires a combination of an infinite number of possible solutions of the form (118). Each A_j must be chosen in such a way that the sum has the required properties. Methods for doing this have been developed, but are beyond the scope of this discussion. However, the final result is that the pressure p is the sum of many terms of the type (118) with $e^{2\pi i f t}$ the only common factor.

Within the direct sound field a large number of these terms are important, and an exact computation is necessary to find p. In the shadow zone, on the other hand, one term dominates, and the other terms may be neglected. This is because all the μ_j are partly real, partly imaginary, with the result that the absolute value of $\exp(i\mu_j x)$ decreases exponentially for sufficiently great values of x. It can be shown that the range at which only one term dominates is approximately the range to the shadow boundary computed from the ray theory. This dominant term is the one for which μ_j has the smallest imaginary part. Thus, the theory predicts that in the shadow zone the sound intensity falls off exponentially with increasing range, or, in other words, that the predicted transmission anomaly in the shadow zone increases linearly with increasing range.

Although the exact determination of the different characteristic values μ_j is somewhat involved, it is relatively simple to show how these values depend on the frequency f, the velocity gradient, and the sound velocity c_0 at the surface. This is useful since it indicates how the attenuation into the shadow zone may be expected to vary under different conditions. In order to investigate this dependence of μ_j on the other variables, we rewrite equation (115) in a simplified dimensionless form. Let

$$n = \frac{4\pi^2 f^2}{c_0^2} - \mu^2 \quad (119)$$

and

$$y = \frac{u}{D} \quad (120)$$

---

[a] The negative sign must be taken in the exponent so that p_j in equation (118) will correspond to a wave moving away from the projector; that is, p_j must be a function of $2\pi f t - \mu_j x$, where μ_j is positive.

where D is an arbitrary constant to be determined later. Then equation (115) becomes, on dividing through by D^2,

$$\frac{d^2F}{du^2} + \left(\frac{n}{D^2} + \frac{4\pi^2 f^2 Bu}{c_0^2 D^3}\right) F = 0. \qquad (121)$$

If D is chosen so that

$$D^3 = \frac{4\pi^2 f^2 B}{c_0^2},$$

then equation (115) becomes

$$\frac{d^2F}{du^2} + (K + u)F = 0 \qquad (122)$$

where

$$K = \frac{nc_0^2}{(\pi^2 f^2 B c_0)^{\frac{2}{3}}}. \qquad (123)$$

Equation (122) has solutions of the type desired only for certain characteristic values of K, denoted by the symbol K_i. The different values of K_i are determined only by the nature of the differential equation (122) and by the two boundary conditions, namely that the sound pressure is zero at the surface and that no sound is coming up from below the projector. Thus the values of K_i are independent of the frequency, sound velocity, and velocity gradient.

Once these characteristic values of K have been found, the corresponding values of μ to be used in equations (115) and (116) can be found directly. By substitution in equations (123) and (119), we find

$$\mu_i^2 = \frac{4\pi^2 f^2}{c_0^2} - \frac{(\pi^2 f^2 B c_0)^{\frac{2}{3}}}{c_0^2} K_i. \qquad (124)$$

The second term in equation (124) is always very much less than the first in cases of practical importance. Even for a temperature gradient as large as 1 F per ft of depth increase, and for a frequency of only 100 c, the second term is less than 1 per cent of the first for K_i less than 10, the region of practical interest. Thus we may take the square root of equation (124), expand in a series, and retain only the first two terms. This process gives

$$\mu_i = \frac{2\pi}{\lambda}\left[1 - \frac{K_i}{2}\left(\frac{Bc_0}{8\pi f}\right)^{\frac{2}{3}}\right]$$

$$= \frac{2\pi}{\lambda} - \frac{K_i}{4}\left(\frac{\pi f B^2}{c_0}\right)^{\frac{1}{3}}. \qquad (125)$$

Let K_1 be the characteristic value of K with the smallest imaginary part, and let this imaginary part be denoted by iK_1'. Let the theoretical sound pressure associated with the characteristic value K_1 be p_1. In the shadow zone the intensity is proportional to the square of p_1 since the sound pressures associated with the other characteristic values K_j may be neglected.

The intensity level found from equation (119) is

$$L = 20 \log p_1 = C - 20(\log_{10} e)\frac{K_1'}{4}\left(\frac{\pi f B^2}{c_0}\right)^{\frac{1}{3}} x \qquad (126)$$

where C includes A_1 and the other variables taken over from equation (118). While C changes gradually with position, it is nearly constant along the shadow boundary. Multiplying out terms in equation (126), and using equation (111) for B, we get, finally,

$$L = C - \frac{5.05 K_1' f^{\frac{1}{3}}(-dc/dy)^{\frac{2}{3}} x}{c_0}. \qquad (127)$$

It should be emphasized that equations (126) and (127) apply only in the shadow zone. In the main beam other terms corresponding to other values of K_i must be considered.

The analysis in a report by Columbia University Division of War Research[7] considers the radiation in three dimensions sent out by a point source and is thus more general than the simple analysis presented here. However, the final result for the sound in the shadow zone is nearly identical with equation (127); the only difference is that the term $5.05 K_1'$ becomes 25.7 in the exact computation of reference 7. With this substitution, we have the following formula for a, the attenuation coefficient beyond the shadow boundary in decibels per unit distance.

$$a = \frac{25.7 f^{\frac{1}{3}}(-dc/dy)}{c_0}. \qquad (128)$$

In this equation f is the sound frequency in cycles per second, and dc/dy is the velocity gradient in feet per second per foot. If c_0 is in feet per second, formula (128) gives the attenuation in decibels per foot; if c_0 is in yards per second, the result is the attenuation in decibels per yard.

Since inverse-square spreading is quite negligible compared to the intensity drop at the shadow boundary, equation (128) gives the slope of the transmission anomaly at points beyond the shadow boundary. However, this equation cannot be used at shorter ranges and must therefore be regarded as an expression for the local attenuation coefficient in the shadow zone.

Equation (128) is compared with observational data under *Attenuation Coefficient at Shadow Boundary* in Section 5.4.1, where it is shown that the observed local attenuation coefficients beyond the shadow boundary are not more than about half the predicted values. In other words, in practice much more sound appears in the shadow zone than is predicted by equation (128).

Editor's Comments on Paper 4

4 Webster: *Acoustical Impedance, and the Theory of Horns and of the Phonograph*

The concept of impedance was introduced into the theory of electric circuits by Oliver Heaviside eighty-five years ago and it has proved to be a notation of great importance in electrical problems. A. G. Webster saw the value of extending the significance of the concept to the case of mechanical oscillations and acoustic wave propagation. As far as is known, he was the first to do this in his paper "Acoustical Impedance, and the Theory of Horns and of the Phonograph," which is reproduced here in full. The concept of acoustic impedance has been found to be of great value, not only in all problems in physical acoustics, but also in all applications of acoustics, as will be clear from examination of the other volumes of the Benchmark Series in Acoustics.

Arthur Gordon Webster (1863–1923) was an American physicist who received his doctorate from the University of Berlin in 1890 and spent practically the whole of his professional career as a professor of physics at Clark University, Worcester, Massachusetts. His research included theoretical and experimental work in mechanics, acoustics, electricity and magnetism, and heat.

ACOUSTICAL IMPEDANCE, AND THE THEORY OF HORNS AND OF THE PHONOGRAPH

By Arthur Gordon Webster

Department of Physics, Clark University

Communicated, May 8, 1919*

The introduction more than thirty years ago of the term 'impedance' by Mr. Oliver Heaviside has been productive of very great convenience in the theory of alternating currents of electricity. Unfortunately, engineers have not seemed to notice that the idea may be made as useful in mechanics and acoustics as in electricity. In fact, in such apparatus as the telephone one may combine the notions of electrical and mechanical impedance with great advantage. Whenever we have permanent vibrations of a single given frequency, which is here denoted, as usual, by $n/2\pi$, the notion of impedance is valuable in replacing all the quantities involved in the reactions of the system by a single complex number. If we follow the convenient practice of denoting an oscillating quantity by e^{int} and taking its real part (as introduced by Cauchy) all the derivatives of e^{int} are obtained by multiplication by powers of in, or graphically by advancing the representative vector by the proper number of right angles.

If we have any oscillating system into which a volume of air X periodically enters under an excess pressure p, I propose to define the impedance by the *complex* ratio $Z = p/X$. If we call $dX/dt = I$ the current as in electricity, if we followed electrical analogy we should write $Z = pI$ so that the definition as given above makes our impedance lead by a right angle the usual definition. I believe this to be more convenient for our purposes than the usual definition and it need cause no confusion.

If we have a vibrating piston of area S as in the phonometer, we shall refer its motion to the volume $S\xi$ it carries with it and the force acting on it to the pressure, so that $F = Sp$. The differential equation of the motion is

$$m\frac{d^2\xi}{dt^2} + \kappa \frac{d\xi}{dt} + f\xi = F = Sp, \qquad X = S\xi, \qquad (1)$$

we have

$$Z_1 = (f - mn^2 + i\kappa n)/S^2, \qquad (2)$$

where m is the mass, κ the damping, f the stiffness. The real part of S^2Z, $f - mn^2$, is the uncompensated stiffness, which is positive in a system tuned too high, when the displacement lags behind the force, by an angle between zero and one right angle, negative when the system is tuned too low, when the

---

* This article was read in December 1914 at the meeting of the American Physical Society at Philadelphia, and has been held back because of the continual development of the experimental apparatus described in a previous paper in these Proceedings.

lag is between one and two right angles, as shown in figure 1. If we force air into a chamber of volume V, the compression $s = X/V$ will be related to the excess pressure p by the relation $p = es$, where e is the modulus of elasticity of the air $e = \rho a^2$, ρ being the density and a the velocity of sound. Consequently we have

$$Z_0 = \frac{e}{V} = \frac{\rho a^2}{V}, \qquad (3)$$

and the analogy is to a condenser. If we have air passing through an orifice or short tube of conductivity c its inertia gives an apparent mass ρ/c, and if it

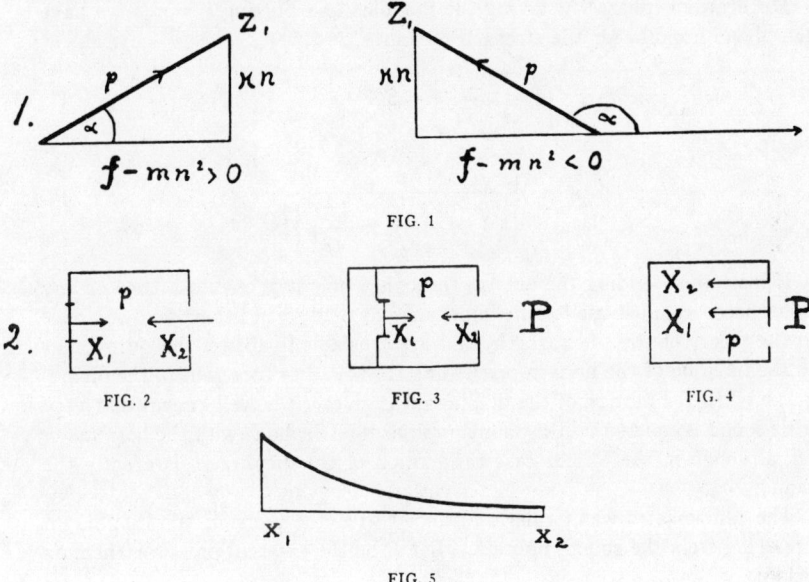

FIG. 1

FIG. 2   FIG. 3   FIG. 4

FIG. 5

escapes from a circular hole in an infinite plane it dissipates energy so that the whole impedance credited to the hole is

$$Z_2 = -\frac{\rho n^2}{c} + \frac{\rho n^3}{2\pi a} i = ek^2 \left\{ \frac{k}{2V} i - \frac{1}{c} \right\}, \text{ where } k = \frac{n}{a}. \qquad (4)$$

These three typical impedances will be at constant use in acoustics. It is to be remembered that systems in series have their impedances added and in parallel have the reciprocals of impedance added. Also that the free vibrations of a system are obtained by equating the impedances to zero.

As a simple example consider the phone described in the previous article, figure 3.

Let $X_1 = S\xi$ be the volume introduced by the piston X_2 that entering by the hole. Then

$$p = Z_0(X_1 + X_2) = -Z_2 X_2, \qquad (5)$$

$$X_2 = \frac{-Z_0}{Z_0 + Z_2} X_1,$$

and inserting values,

$$X_2 = -\frac{S\xi}{1 + Vk^2\left(\dfrac{ki}{2\pi} - \dfrac{1}{c}\right)} \qquad (6)$$

Disregarding phase by taking the modulus and putting $k = n/a$ we have the phone formula for the strength of source.

$$A = \left|\frac{dX}{dt}\right| = S\,\Psi\,|\xi|, \qquad (7)$$

where

$$\Psi = \frac{n}{\left\{\left(1 - \dfrac{Vk^2}{c}\right)^2 + \dfrac{V^2 k^6}{4\pi^2}\right\}^{\frac{1}{2}}} \qquad (8)$$

If instead of sending the air out through a hole it goes into a cone or any other horn, we must use for the impedance Z_2 that given below, and we arrive at the theory of the phonograph, and are thus able to answer the question as to the function of the horn in persuading the sound to come out of the phonograph when the motion of the diaphragm is given (it is well known that very little sound is emitted by the phonograph or the telephone with the horn taken off, although in the former case the motion of the diaphragm is exactly the same).

The phonometer was formerly arranged with the back of the diaphragm protected from the sound, figure 3. Let P be the external pressure, then, as before,

$$p = Z_0(X_1 + X_2) \qquad (9)$$

and in addition,

$$-p = Z_1 X_1,$$
$$P - p = Z_2 X_2, \qquad (10)$$

from which

$$X_1 = \frac{-PZ_0}{Z_0 Z_1 + Z_1 Z_2 + Z_2 Z_0}, \qquad (11)$$

giving the formula for the measurement of the pressure,

$$P = \varphi\,\xi/S \qquad (12)$$

$$\varphi = \frac{\gamma}{[\{uv - (\alpha\beta + \gamma^2)\}^2 + \{\beta u + \alpha v\}^2]^{\frac{1}{2}}} \qquad (13)$$

and φ may be termed the sensitiveness of the phonometer. Where

$$\gamma = S^2 Z_0 = S^2 \rho a^2/V, \qquad \alpha = \kappa n, \qquad \beta = S^2 \rho n^3/2\pi a,$$
$$u = f - n^2 m + S^2 \rho a^2/V, \qquad v = S^2 \{\rho a^2/V - \rho u^2/c\} = S^2 Z_0 (1 - k^2 V/c) \quad (14)$$

As described in my recent article the back of the piston is exposed to the sound, figure 4. Then

$$P - p = Z_1 X_1 = Z_2 X_2$$
$$p = Z_0 (X_1 + X_2) \tag{15}$$

from which

$$X_1 = \frac{Z_2 P}{Z_0 Z_1 + Z_1 Z_2 + Z_2 Z_0} \tag{16}$$

$$\varphi = \left[\frac{(v-\gamma)^2 + \beta^2}{\{uv - (\alpha\beta + \gamma^2)\}^2 + (\alpha v + \beta u)^2} \right]^{\frac{1}{2}}, \tag{17}$$

Tubes and Horns.—Beside the above described phone and phonometer, the theory of which assumed a resonator so small that the pressure is supposed to be the same at every internal point, I have made use of many arrangements employing tubes or cones, in which we must take account of wave-motion. The familiar theory of cylindrical pipes may be included in the following generalized theory, which I have found experimentally to serve well.

Let us consider a tube of infinitesimal cross section σ varying as a function of the distance x from the end of the tube. Then if q is the displacement of the air, p the pressure, s the compression, we have the fundamental equations

$$p = es = \rho a^2 s = -e \operatorname{div} q = -\frac{e}{\sigma} \frac{d(q\sigma)}{dx}, \tag{18}$$

$$\frac{d^2 p}{dt^2} = a^2 \Delta p = a^2 \operatorname{div} \operatorname{grad} p = a^2 \left\{ \frac{1}{\sigma} \frac{d}{dx} \left(\sigma \frac{dp}{dx} \right) \right\} \tag{19}$$

$$\frac{d^2 q}{dt^2} = -\frac{1}{\rho} \frac{dp}{dx} = a^2 \frac{d}{dx} \left\{ \frac{1}{\sigma} \frac{d}{dx} (q\sigma) \right\}. \tag{20}$$

For a simple periodic motion we put p, q proportional to e^{int}, and obtain

$$\frac{d^2 p}{dx^2} + \frac{d \log \sigma}{dx} \frac{dp}{dx} + k^2 p = 0, \quad \frac{d^2 q}{dt^2} + \frac{d \log \sigma}{dx} \frac{dq}{dx} + \frac{d^2 \log \sigma}{dx^2} q + k^2 q = 0. \tag{21}$$

Both these linear equations may be solved by means of series, and if we call $u(kx)$, $v(kx)$ two independent solutions we have

$$p = Au + Bv, \quad \beta q = Au' + Bv', \quad \beta = \rho a^2 k,$$

where the accents signify differentiation according to kx. If we denote values at one end $x = x_1$ and at the other end $x = x_2$ by suffixes 1, 2, respectively, and form the determinants

$$D_1 = \begin{vmatrix} u_1, v_1 \\ u_1', v_1' \end{vmatrix}, \quad D_2 = \begin{vmatrix} u_2, v_2 \\ u_2', v_2' \end{vmatrix}, \quad D_3 = \begin{vmatrix} u_1, v_1 \\ u_2, v_2 \end{vmatrix},$$
$$D_4 = \begin{vmatrix} u_2, v_2 \\ u_1', v_1' \end{vmatrix}, \quad D_5 = \begin{vmatrix} u_1, v_1 \\ u_2', v_2' \end{vmatrix}, \quad D_6 = \begin{vmatrix} u_1', v_1' \\ u_2, v_2 \end{vmatrix},$$
(22)

which satisfy the relation,

$$D_1 D_2 = D_3 D_4 + D_5 D_6$$

we may determine the constants A, B in terms of any two out of p_1, q_1, p_2, q_2, so that we obtain

$$p_2 = (p_1 D_4 + \beta q_1 D_5)/D_1, \qquad \beta q_2 = (-p_1 D_6 + \beta q_1 D_3)/D_1,$$
$$p_1 = (p_2 D_3 - \beta q_2 D_5)/D_2, \qquad \beta q_1 = (p_2 D_6 + \beta q_2 D_4)/D_2.$$
(23)

As it is more convenient to deal with the volumes $X_1 = \sigma_1 q_1$, $X_2 = \sigma_2 q_2$ we shall have in general

$$p_2 = a p_1 + b X_1, \quad X_2 = c p_1 + d X_1,$$
(24)

where

$$a = \frac{D_4}{D_1}, \quad b = \frac{\beta}{\sigma_1} \frac{D_5}{D_1}, \quad c = -\frac{\sigma_2}{\beta} \frac{D_6}{D_1}, \quad d = \frac{\sigma_2}{\sigma_1} \frac{D_3}{D_1}, \quad ad - bc = \frac{\sigma_2}{\sigma_1} \frac{D_2}{D_1},$$

and for the impedances belonging to the ends of the tube

$$Z_2 = \frac{aZ_1 + b}{cZ_1 + d}, \qquad Z_1 = \frac{dZ_2 - b}{-cZ_2 + a},$$
(25)

so that the impedance at either end of the tube is a linear fractional function of the other. According to the apparatus attached to an end the impedance attached to that end is known. A tube for which a, b, c, d are given may be replaced by any other tube having the same constants.

Examples.—Cylindrical tube, σ constant. Put $x_2 - x_1 = l$,

$$\frac{d^2 p}{dx^2} + k^2 p = 0,$$

$$u = \cos kx, \quad v = \sin kx, \quad u' = -\sin kx, \quad v' = \cos kx, \quad (26)$$

$$D_1 = D_2 = 1, \quad D_3 = D_4 = \cos kl, \quad D_5 = D_6 = \sin kl, \quad (27)$$

$$a = d = \cos kl, \quad b = \frac{\beta}{\sigma} \sin kl, \quad c = -\frac{\sigma}{\beta} \sin kl,$$

$$Z_2 = \frac{\beta}{\sigma} \frac{Z_1 \cos kl + \frac{\beta}{\sigma} \sin kl}{-Z_1 \sin kl + \frac{\beta}{\sigma} \cos kl}, \quad Z_1 = \frac{\beta}{\sigma} \frac{Z_2 \cos kl - \frac{\beta}{\sigma} \sin kl}{Z_2 \sin kl + \frac{\beta}{\sigma} \cos kl} \quad (28)$$

Conical tube, $\sigma = \sigma_0 x^2$

$$\frac{d^2p}{dx^2} + \frac{2}{x}\frac{dp}{dx} + k^2 x = 0 \tag{29}$$

$$u = \frac{\cos kx}{kx}, \quad v = \frac{\sin kx}{kx}, \quad u' = -\left(\frac{\sin kx}{kx} + \frac{\cos kx}{k^2 x^2}\right), \quad v' = \frac{\cos kx}{kx} - \frac{\sin kx}{k^2 x^2}$$

$$D_1 = \frac{1}{k^2 x_1^2}, \quad D_2 = \frac{1}{k^2 x_2^2}, \quad D_2 = \frac{\cos kl}{k^2 x_1 x_2},$$

$$D_4 = \frac{\cos kl}{k^2 x_1 x_2} + \frac{\sin kl}{k^3 x_1 x_2^2}, \quad D_5 = \frac{\sin kl}{k^2 x_1 x_2}$$

and if we introduce two lengths ϵ_1, ϵ_2, defined by the equations

$$\tan k\epsilon_1 = kx_1, \quad \tan k\epsilon_2 = kx_2,$$

we easily get

$$a = \frac{x_1}{x_2}\frac{\sin k(l + \epsilon_1)}{\sin k\epsilon_1}, \quad b = \frac{\beta}{\sigma_1}\frac{x_1}{x_2}\sin kl, \tag{30}$$

$$c = -\frac{\sigma_2}{\beta}\frac{x_1}{x_2}\frac{\sin k(l + \epsilon_1 - \epsilon_2)}{\sin k\epsilon_1 \sin k\epsilon_2}, \quad d = \frac{\sigma_2}{\sigma_1}\frac{x_1}{x_2}\frac{\sin k(l - \epsilon_2)}{\sin k\epsilon_2}, \tag{31}$$

$$Z_2 = -\frac{\beta}{\sigma_2}\frac{Z_1\dfrac{\sin k(l + \epsilon_1)}{\sin k\epsilon_1} + \dfrac{\beta}{\sigma_1}\sin kl}{Z_1\dfrac{\sin k(l + \epsilon_1 - \epsilon_2)}{\sin k\epsilon_1 \sin k\epsilon_2} + \dfrac{\beta}{\sigma_1}\dfrac{\sin k(l - \epsilon_2)}{\sin k\epsilon_2}}$$

$$Z_1 = -\frac{\beta}{\sigma_1}\frac{Z_2\dfrac{\sin k(l - \epsilon_1)}{\sin k\epsilon_2} + \dfrac{\beta}{\sigma_2}\sin kl}{Z_2\dfrac{\sin k(l + \epsilon_1 - \epsilon_2)}{\sin k\epsilon_1 \sin k\epsilon_2} + \dfrac{\beta}{\sigma_2}\dfrac{\sin k(l + \epsilon_1)}{\sin k\epsilon_2}} \tag{32}$$

The formulae (31), (32) were used by Professor G. W. Stewart in designing horns to be used during the war.

It is not true, as is frequently stated in books on musical instruments, that the brass instruments of the orchestra are hyperbolic in profile, but I have found for all practical purposes the bell of every instrument may be represented by one of the three formulae

$$\sigma = \sigma_0 x^n, \quad \sigma = \sigma_0 e^{-mx}, \quad \sigma = \sigma_0 e^{-mx^2}$$

Even if an equation cannot be given to the profile the differential equation may be easily integrated graphically, or the length may be divided up into sections and different values of n used for different sections, as is customary in the theory of ballistics.

Case 1. $\sigma = \sigma_0 x^n$. (Change units so that $k = 1$)

$$\frac{d^2 p}{dx^2} + \frac{n}{x}\frac{dp}{dx} + p = 0, \qquad \frac{d^2 X}{dx^2} - \frac{n}{x}\frac{dX}{dx} + X = 0 \qquad (34)$$

We have

$$p = J_{\pm\frac{n-1}{2}}(x)/x^{\frac{n-1}{2}}, \qquad X = J_{\pm\frac{n+1}{2}}(x) x^{\frac{n+1}{2}}, \qquad (35)$$

Examples.

$$n = 0, \qquad\qquad n = 2, \qquad\qquad n = -2$$

$$J_{\frac{1}{2}}(x) = \sin x/\sqrt{x}, \qquad J_{\frac{3}{2}}(x) = \sin x/x^{\frac{3}{2}} - \cos/x\sqrt{x},$$

$$J_{-\frac{1}{2}}(x) = \cos x/\sqrt{x}, \qquad J_{-\frac{3}{2}}(x) = -\sin x/\sqrt{x} - \cos x/x^{\frac{3}{2}}$$

These include the straight cylinder, the straight cone, and the purely hyperbolic horn. In the latter case we have figure 5, where x_1, is the bell. If the horn is closed at x_2 we have

$$Z_2 = \infty$$

$$Z_1 = ck^2 \left\{ \frac{1}{c} - \frac{k}{2\pi} i \right\} = -\frac{d}{c} = \frac{(\sin kl + kx_1 \cos kl)\beta x_1}{\sigma_0 k \sin kl_1}$$

and if we put $\xi = kl$

$$\operatorname{ctn} \xi = \frac{\sigma_1}{lc}\xi - \frac{l}{x_1}\frac{1}{\xi}, \qquad (36)$$

which may be easily discussed graphically.

On the other hand if the horn is open at x_2 we have

$$\tan \xi/\xi = \left(1 - \frac{\sigma x_1 x_2}{cl^3}\xi^2\right) \Big/ \left(1 + \xi^2 \left\{\frac{x_1 x_2}{l^2} - \frac{\sigma_1 x_1}{cl^2}\right\}\right). \qquad (37)$$

These formulae were confirmed experimentally by my then assistant Dr. H. K. Stimson in 1915 on a coach-horn, a trombone, and a phonograph horn, with the following results:

		CALCULATED	OBSERVED
For the coach-horn	Closed	177	181
	Open	254	202
For the trombone	Closed	286	305
	Open	418	432
For the phonograph	Closed	311	304
	Open	329	415

These results give a fair agreement considering that we have used for the conductivity of the mouth the simple formula $c = 0.6\,R$ which is true only for cross-sections infinitesimal compared with the wave-length, whereas in the case of the wooden phonograph horn, the actual radius is nearly one-fourth of the wave-length.

A paper on the subject of the impedance of such an end will shortly appear.

In the case of an exponential section we have

$$\sigma = \sigma_0 e^{-mx}$$

$$\frac{d^2p}{dx^2} + m\frac{dp}{dx} + p = 0, \qquad \frac{d^2X}{dx^2} - m\frac{dX}{dx} + X = 0,$$

$$p = e^{-\sqrt{4-k^2}\,x}\{A\cos kx + B\sin kx\},$$

$$X = e^{-\sqrt{4-k^2}\,x}\{C\cos kx + D\sin kx\}.$$

and it is noticeable that the pressure vanishes at the same cross-section as for a straight tube.

Finally, in the case

$$\sigma = \sigma_0 e^{-mx^2}$$

we may solve the equation by means of the confluent hyper-geometric function.

It is to be noticed that in none of these cases, except the straight tube, are the different overtones harmonic. Thus, the characteristic tone of the "brass" is not due to the substance, but is entirely a matter of geometry as is shown by the heavy casting in plaster of Paris of a trombone bell used by the writer, the tone of which cannot be distinguished from that of the brass bell. I believe this phenomenon is well known.

Inasmuch as all musical instruments are composed either of resonators combined with strings, bars, plates, and horns, I feel that the above theory, while merely an approximation as to accuracy, will go far toward enabling us to complete the theory of musical instruments. Of course, the actual tones emitted by a brass instrument will depend upon the dynamics of the lips which is reserved for a future paper.

Editor's Comments on Papers 5a and 5b

5a **Cook:** *Strange Sounds in the Atmosphere. Part I*

5b **Cook and Young:** *Strange Sounds in the Atmosphere. Part II*

It will be recalled that the average frequency limits for audible sound run from about 20 Hz to about 20 kHz. Inaudible sound of frequencies above 20 kHz is termed *ultrasonic*, whereas inaudible sound of frequencies below 20 Hz is called *infrasound*. A pendulum swinging in air with a period of 1 second generates infrasound; our ears are unable to detect it. Suitably sensitive instrumentation can detect infrasound, however, and it has been found that there are many sources of it. The senior author of the two papers reprinted here has been engaged for many years in the study of infrasound in the atmosphere, arising from a variety of sources. These papers constitute an interesting review of early pioneer work in a field that has become an important branch of geophysical investigation.

Dr. Cook (1910–   ) is an American acoustical physicist, long associated with the National Bureau of Standards. Dr. Young is also a staff member of the Bureau at its Boulder, Colorado, Laboratory.

Strange Sounds in the Atmosphere

Part I

Richard K. Cook

NATIONAL BUREAU OF STANDARDS
Washington 25, D. C.

THE atmosphere occasionally seems very quiet to the ear. But, at infrasonic frequencies too low to be heard, sounds of substantial intensity are always present. These natural sounds have many causes, including distant tornadoes, volcanoes, earthquakes, and magnetic storms. Many of the sounds, including the ubiquitous and mysterious microbaroms, are of unknown origin.

We are concerned with infrasound in the atmosphere—sound waves whose frequencies of oscillation are less than the lowest frequency, about 15 cps, which can be heard. Of particular interest are those waves whose frequencies are less than 1 cps, because such waves can be propagated thousands of kilometers without substantial loss of energy. We see later why this is true.

Early Observations

A tremendous explosion of the volcano Krakatoa occurred in the East Indies in 1883. Infrasound from this explosion traveled around the world several times, a distance of over 100 000 km. Even the audible sound was heard on Rodrigues Island, 5000 km away in the western part of the Indian Ocean. At that time, electroacoustic equipment suitable for measurement of infrasound did not exist. But the inaudible sound waves from this disturbance had sound pressures so great that readable deflections were produced on barographs (recording barometers) all over the world. The waves were still detectable after traveling several times around the earth because of the very low absorption of infrasound.

Similar infrasonic waves, but of lesser intensity, were generated by the impact of the great Siberian meteor in 1908.

About the time of World War I, another aspect of the propagation of sound through the atmosphere was noticed. It became apparent that the sound of cannon fire could be heard within a radius of about 100 km from the source and often beyond 200 km, but not at distances between 100 and 200 km. The anomalous audibility was particularly evident when large explosions occurred. The zones of audibility for the explosion which occurred at Oppau on the Rhine River in Germany on September 21, 1921, are shown in Fig. 1. The black dots indicate locations where the sound was heard, and the open circles where observers did not hear the explosion. The zone of audibility for the direct sound reached out to a distance of about 100 km. Then, a zone of relative inaudibility extended from 100 to 200 km. The sound was again audible at distances greater than 200 km from Oppau. This remarkable phenomenon—where the sound again became audible at large distances in spite of inaudibility nearer to the source—was observed for other explosions as well.

The phenomenon was correctly ascribed to the reflection of sound waves from a layer of air in the upper atmosphere in which the speed of sound was greater than at sea level. Such an interpretation re-

---

Richard K. Cook is Chief of the Sound Section of the National Bureau of Standards. He received his Ph.D. in physics at the University of Illinois, Urbana, in 1935 and has worked as a physicist at the National Bureau of Standards since that time. He is a Fellow of the Acoustical Society of America and served as its president in 1957–58. Dr. Cook is an authority on many aspects of physical acoustics.

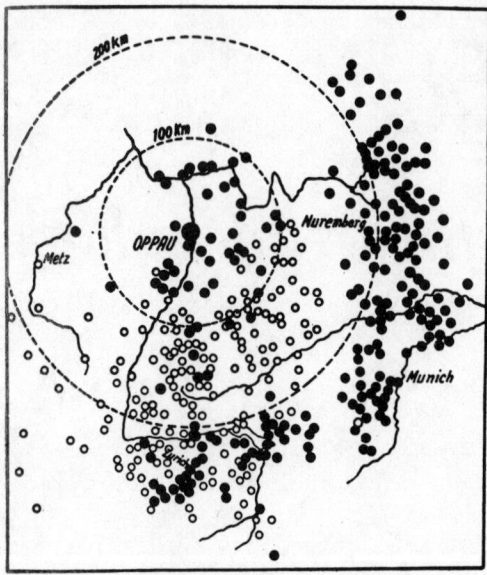

Fig. 1. Zones of audibility and of silence for the explosion at Oppau, Germany, on September 21, 1921. The black dots represent places where the sound was heard. The open circles represent the positions of observers who did not hear the explosion. Towards east and south, there is a zone of audibility of roughly 100 km extent. The zone 100 to 200 km is a zone of silence. Beyond 200 km, there is again a zone of audibility. (After Quervain.)

quired that the reflecting layer be at a higher temperature than the air at ground level.

This early hypothesis suggested quantitative measurement of the total transit time for propagation from the source to the point of observation, and measurement of the angle of descent of the sound from the upper atmosphere. The conclusion from such measurements was that the temperature of the atmosphere rises with increasing altitude in the region above about 30 km. The earlier supposition had been that the stratosphere was isothermal at $220°K$ ($-53°C$) up to altitudes above 50 km.

The rise in temperature, first suggested by the anomalous audibility of sounds from explosions, was subsequently confirmed by observations on the heights of appearance and disappearance of meteor trails. From the meteor data, it was concluded that the atmospheric temperature must be at least $300°K$ ($27°C$) at altitudes near 50 km. Within the last fifteen years or so, direct observations of the upper atmosphere with instrumental rockets have yielded more detailed information, and densities and temperatures have been deduced from orbital changes in satellites above 160 km.

It has been established that the relatively high temperature near altitudes of 50 km is due to the strong absorption of ultraviolet energy in sunlight by ozone in the atmosphere. The ozone is distributed in a layer between about 10 and 50 km above the earth's surface, its center of mass being at about 25 km. But, the absorption of sunlight by ozone is so great that practically all of the heating effect is confined to the uppermost part of the ozone layer which, therefore, serves as a vast heat reservoir maintained at a temperature well above the temperature of the stratosphere.

Propagation of Infrasound through the Atmosphere

Before embarking on a study of the various natural sources of infrasound, we first need to know a few facts about the sound waves themselves, about the influence of the atmosphere on their propagation, and about instruments for measuring them.

Infrasonic waves have the same speed c as audible sound, about 344 m/sec at $20°C$. It is convenient to describe an infrasonic wave by its period of oscillation $T = 1/f$, where $f =$ the frequency of oscillation. For frequencies $f < 1$ c/s, the corresponding periods $T = 1/f > 1$ sec. The wavelength $\lambda = cT$, and so waves of $T > 1$ sec have $\lambda > 344$ m. Several natural sources have periods of about 10 sec, and hence wavelengths of about 3.4 km.

A sound wave causes pressure oscillations as it traverses the atmosphere. For infrasound of natural origin, the amplitude p of the sound pressure is often in the range 1 to 10 d/cm^2, and infrasonic microphones are usually designed to respond to such pressures.

microphone which is in a thermally insulated can. The nects the microphone to the noise-reducing pipe line.

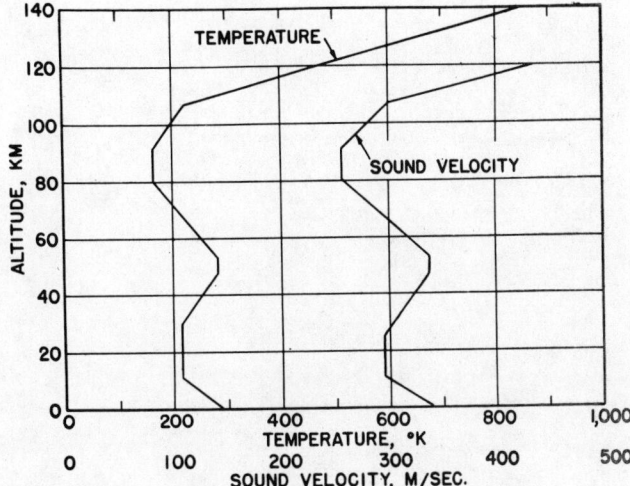

Fig. 2. Temperature and sound velocity in the model atmosphere as developed by the Air Research and Development Command in 1959. Details of the real atmosphere vary with location on the earth's surface and with the seasons. Wind speed also influences sound velocity.

The absorption of infrasound in the atmosphere, due to viscosity and heat conduction, is considerably less than the absorption for audible sounds because of the low frequency of oscillation. The absorption coefficient α, defined by the spatial variation of p, $p(x) = p_0 \exp(-\alpha x)$, is about $1.6 \times 10^{-4}/T^2 B$ db/m, where B is the barometric pressure in d/cm². For a plane wave of sound in the lower atmosphere at $T = 10$ sec, the absorption is, therefore, less than 2×10^{-9} db/km. Hence, the loss due to this absorption mechanism is totally insignificant, even for propagation over distances of thousands of kilometers. The absorption in the upper atmosphere is substantially greater because of the lower barometric pressure. At an altitude of 90 km, where the barometric pressure ≈ 1 d/cm², the absorption $\approx 2 \times 10^{-3}$ db/km for waves of 10-sec periods.

Up to altitudes of about 10 km in the troposphere, the absorption due to water vapor should be considered. The exact variation of this absorption with barometric pressure is not accurately known for infrasonic frequencies. We estimate that at sea level (altitude = 0 km) the absorption coefficient might be as large as $5 \times 10^{-9}/T^2$ db/m, which is about 30 times greater than the absorption for viscosity and heat conduction, as indicated previously. But the absorption due to water vapor is still insignificant for infrasound at $T = 10$ sec, being only 5×10^{-8} db/km. This corresponds to an energy loss of less than one percent after propagation half-way around the earth, a distance of 20 000 km.

At very low frequencies, there is an absorption due to relaxation of the thermal energy stored in vibrations of the diatomic molecules in air. We estimate the absorption coefficient, based on unpublished measurements made by Prof. M. C. Henderson at The Catholic University of America, to be almost 1000 times greater than that of the viscosity-heat-conduction loss. Therefore, waves in the lower atmosphere at $T = 10$ sec have $\alpha \approx 10^{-6}$ db/km. Again, this is an insignificant loss.

The atmosphere has inhomogeneities in temperature and density arising, e.g., from solar heating of the ground. Inhomogeneities in density and motion are associated with turbulence in the wind as it passes over trees, buildings, hills, etc.; furthermore, sound waves are scattered by such obstacles. All of these effects cause absorption of sound-wave energy. But, the absorption is estimated to be quite small when the wavelength is greater than about 1 km.

The net result is that the total absorption for infrasound in the atmosphere is small enough so that propagation can occur over thousands of kilometers without substantial loss of energy. An example of this is the sound from the tremendous explosion, mentioned above, of the volcano Krakatoa. The absorption of infrasound from the explosion was low enough so that the waves were still detectable after traveling around the earth several times.

The variation of sound speed with altitude is a gross feature of the atmosphere which has a substantial effect on the propagation of sound to great distances. The data on the atmosphere obtained from sound propagation, rockets, and satellites show that the temperature and, hence, the velocity of sound depend on position on the earth's surface, and vary with time. The average properties have been incorporated into various "standard atmospheres." The temperature distribution with altitude is of particular interest for the propagation

Calibration of an infrasonic microphone (in front) which is connected by a hose to the calibrating barrel. On top is an oscillating piston which produces accurately known sound pressures inside the barrel at various low frequencies. The metal barrel is padded on the outside to reduce thermal effects.

of infrasound to large distances. The distribution for the standard atmosphere developed by the Air Research and Development Command is shown in Fig. 2. Also shown is the distribution of sound velocity with altitude, the speed of sound being proportional to the square root of the absolute temperature. The curves should be regarded as averages over all seasons of the year for northern temperate latitudes.

Let us give detailed attention to the relationship between temperature and sound velocity in the atmosphere. The square of the speed c^2 for sound in a gas is the ratio of its modulus of elasticity to its density; $c^2 = \gamma B/\rho$, where B (as above) is the barometric pressure in d/cm², and ρ is the density in g/cc. For air, the adiabatic gas constant γ (which is dimensionless) is $\gamma = 1.402$. γB is the adiabatic modulus of elasticity for the atmosphere. But the equation of state for air is $B = \rho R K$, wherein K is the absolute temperature and R is a constant independent of B, ρ, and K. Therefore, $c^2 = \gamma B/\rho = \gamma \rho R K/\rho = \gamma R K$, and so finally

$$c = \text{constant} \times \sqrt{K}. \qquad (1)$$

Formula (1) shows that the speed of sound is independent of the density of the atmosphere, but the speed is directly proportional to the square root of the absolute temperature. For air at a temperature of $20°C = 293°$ Absolute, the speed $c = 344$ m/sec. From this, the velocity can be found at other temperatures by means of Formula (1), which was used to obtain the sound-velocity curve (Fig. 2) from the temperature curve. The vector velocity of the wind must be added to the sound velocity of a wave at any point in the atmosphere in order to arrive at the total sound velocity relative to the ground. Formula (1) is applicable for all sound waves from the low infrasonic frequency of $f = 0.01$ cps ($\lambda = 34$ km) through audible frequencies ($f \approx 1000$ cps) to ultrasonic frequencies, $f > 20\,000$ cps.

A detailed mathematical analysis for the propagation of sound waves through the atmosphere shows that the speed minimum in the stratosphere results in the waves being "channeled" between the ground and the layer of relatively high sound speed at 50-km altitude. Loosely speaking, the layer serves as a reflector, albeit a poor one. For the shorter waves, $T < 15$ sec (approximately), sound-ray trajectories are useful for studying propagation. In general, the rays from a source at ground level are alternately reflected between the

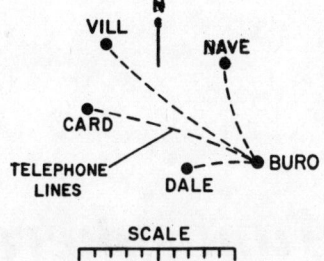

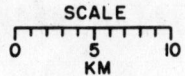

Fig. 3. Location of microphones at the infrasonics station of the National Bureau of Standards in Washington, D. C. Recordings are made at the BURO location.

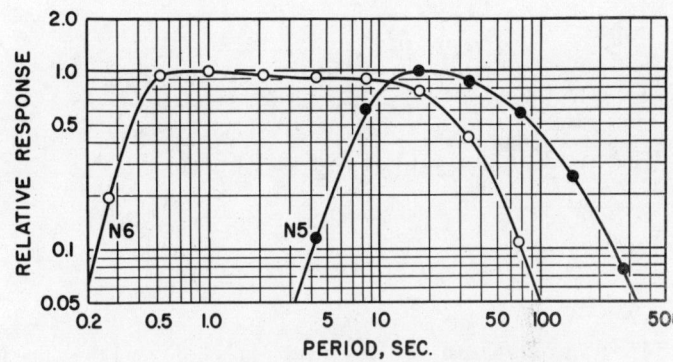

Fig. 4. Response curves of some filters used with an infrasonic system.

Over-all Response of N5 and N6 Channels

layer at 50-km altitude and the surface of the ground.

Instruments for Measurement

We next describe the system which is now used at the National Bureau of Standards for detection and measurement of infrasonic waves. The system is designed for determination of three characteristics. The first is the amount of incident sound pressure, the second is the direction of approach of the incident wave, and the third is its speed across the earth's surface (usually called the "trace velocity" or "horizontal phase velocity").

The system consists of four microphones located at ground level, approximately in the same plane, and about 7 km apart in the vicinity of Washington, D. C. (see photograph of location of microphone and Fig. 3). Effects on each microphone of pressure fluctuations due to local turbulent wind conditions are minimized by noise-reducing lines of pipe which are about 300 m long, have capillary inlets, and are connected to the inlet to the microphone.[1] For sound waves of periods greater than about 10 sec, this antenna is essentially nondirectional and does not attenuate the sound pressure appreciably. However, noise due to random pressure fluctuations in the period range from 1 to 30 sec, such as that caused by wind turbulence, is reduced considerably.

The microphones are of the electrostatic condenser type, and produce frequency-modulated voltages proportional to the incident sound pressures. These voltages are transmitted by telephone wires to a central location where they are demodulated, amplified, and recorded as ink-on-paper traces. When a sound wave of sufficient magnitude is present, about 1 d/cm² or greater, similar traces are produced on each of the four paper records. The direction of approach of the wave and the trace velocity are obtained by comparing the different times of appearance at the four microphones. Wind pressures are also recorded, but such effects are purely local to each microphone installation. Band-pass filters can be introduced into the amplifiers when a higher signal-to-noise ratio is desired for the sound under study. Earthquake waves, for example, are best studied with a band-pass filter passing sounds having periods between 0.4 and 20 sec, as in Fig. 4.

Calibration of each microphone is accomplished by connecting its inlet through a short hose to a calibrating barrel. An oscillating piston on top of the barrel produces accurately known sinusoidal sound pressures in the barrel at various low frequencies, as is shown in the photograph.

The next installment of this article will describe the results of observations on a few of the many natural sounds occurring at infrasonic frequencies in the atmosphere.

[1] Noise-reducing line microphone for frequencies below 1 cps, by Fred B. Daniels, J. Acoust. Soc. Am. **31**, 529 (1959).

Strange Sounds in the Atmosphere

Part II

Richard K. Cook and Jessie M. Young

NATIONAL BUREAU OF STANDARDS
Washington 25, D. C.

THE first installment of this article [1] explained how infrasonic waves, particularly sound waves having periods of vibration greater than about one second, can be propagated through the atmosphere. There were two main conclusions. First, there is very little absorption for waves of such long periods, and they can, therefore, be propagated for distances of thousands of kilometers without substantial loss of energy. Second, the atmosphere has a layer of relatively high sound speed at an altitude of about 50 km, and sound waves being propagated to great distances are, therefore, "channeled" between the ground and the layer at 50 km. A description of instruments suitable for measuring infrasonic waves was given.

We proceed to study a few of the many natural sounds occurring at infrasonic frequencies in the atmosphere. But first it is necessary to distinguish between pressure variations due to sound waves

In this second installment, Dr. Cook is joined by Dr. Young in the presentation of observations on infrasound in the atmosphere. Dr. Young joined the staff of the Bureau about two years ago, following work at the Dupont Corporation. He received his degree in physics from the University of Notre Dame in 1955. Earlier he had worked on underwater-sound measurement problems for the U. S. Navy.

Dr. Cook is Chief of the Sound Section of the National Bureau of Standards. He received his Ph.D. in physics at the University of Illinois in 1935, and has worked as a physicist at the National Bureau of Standards since that time. He is a Fellow of the Acoustical Society of America and served as its President in 1957-58. Dr. Cook is an authority on many aspects of physical acoustics.

and those due to the turbulent passage of the wind over obstructions on the landscape, such as buildings, trees, and hills. The turbulent motion can cause tremendous pressure fluctuations; these cause the familiar rattling of windows and doors on a windy day. The amount of the pressure variation can be estimated from the Bernoulli principle which tells us that the sum $p + (1/2)\rho v^2 = $ a constant, p being the pressure and ρ the atmospheric density in a wind moving with speed v. Hence, the magnitude of the pressure fluctuation is $\Delta p \approx \rho v \Delta v$ when the wind-speed fluctuation is Δv. Thus, a wind whose speed varies irregularly from 15 to 25 mph has a corresponding random pressure fluctuation of about 500 d/cm². Such pressures are much greater than those due to natural sounds at infrasonic frequencies. The latter are seldom more than 50 d/cm², and could be as small as 0.1 d/cm². Smaller sound pressures are rarely observable, because there are few times when the irregular motions of the wind are sufficiently small. The wind pressure is, therefore, a noise which interferes with observations on infrasonic waves. Its influence on the microphones of a measuring system is considerably reduced by the noise-reducing lines of pipe described under Instruments in the first installment, but wind noise remains as the greatest single nuisance.

The noise pressure is convected through the atmosphere at approximately the mean wind speed. Hence, it can be distinguished from sound waves which have the same period of oscillation by the lack of correlation between fluctuating wind pressures at points several kilometers apart.

Microbaroms

About twenty years ago, Benioff and Gutenberg [2] observed very small variations in the barometric pressure at their Seismological Laboratory near Pasadena in southern California. They found the pressure variations to "have predominant periods of 4 sec and 20 sec to several hundred seconds. . . ." Their further studies [3] showed that the pressure variations having periods near 5 sec were very similar in waveform to microseisms, and so it was natural to call them "microbaroms." Furthermore, the waves traveled at approximately the speed of sound in the atmosphere.

Several years later, Saxer [4, 5] began to observe and measure microbaroms at the University of Fribourg, Switzerland. By means of apparatus similar to that described earlier under <u>Instruments</u>, he measured the strengths of the waves and found their sound pressures were sometimes greater than 1 d/cm². Some of the recordings which he obtained are shown in Fig. 1. Dessauer and his associates [6] measured also the direction of propagation, and they found that the waves passed over Fribourg going from a northwest to a southeast direction with a trace velocity of about 400 m/sec.

Most interesting of all, the strengths of the microbaroms at Fribourg were correlated both with the heights of water waves during storms in the North Atlantic Ocean and with the strength of microseisms at Strasbourg, about 200 km north of Fribourg. When the water-wave heights increased during an oceanic storm, the microbaroms gained in strength, and they waned as the storm and water waves subsided (Fig. 2). Similarly, the waxing and waning of microbaroms corresponded to changes in microseisms at Strasbourg, as Fig. 3 clearly indicates.

A thought comes to mind that the gravity waves on the surface of a body of water can radiate sound waves into the atmosphere, but detailed analysis apparently does not support this. The equations for sound propagation above a moving sinusoidal surface show that there is indeed a sound pressure created above the surface (Fig. 4). But the pressure decreases exponentially with height, because the water waves travel with a speed much less than

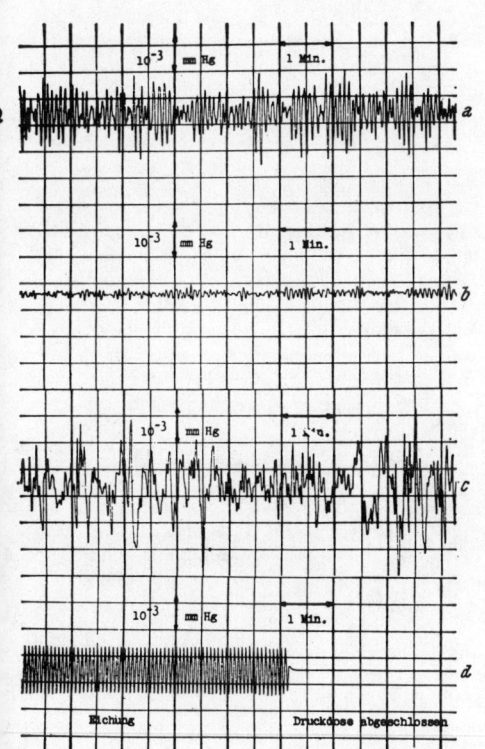

Fig. 1. Microbaroms observed at Fribourg, Switzerland. a = large microbaroms of 5-sec periods. b = small microbaroms. c = wind noise. d = calibration with sinusoidal pressure of 4.5-sec period. (From Saxer.[5])

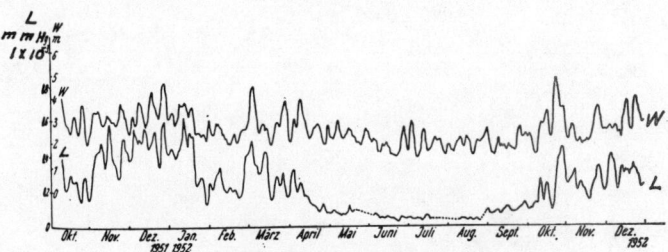

Fig. 2. Correlation between microbaroms and ocean waves. W = daily average heights of water waves in the North Atlantic Ocean. L = daily average sound pressure of microbaroms at Fribourg. (From Saxer.[5])

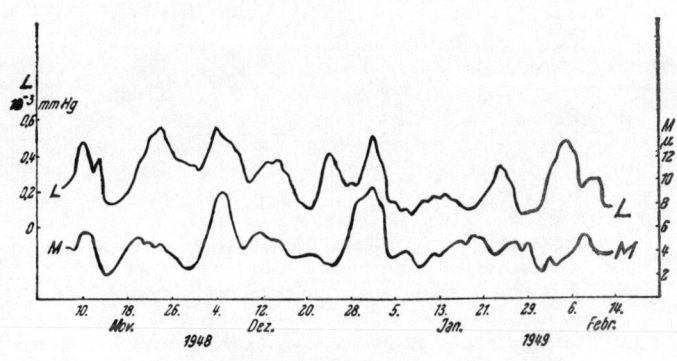

Fig. 3. Correlation between microbaroms and microseisms. L = daily average sound pressure of microbaroms at Fribourg. M = daily average amplitudes of microseisms at Strasbourg. (From Saxer.[5])

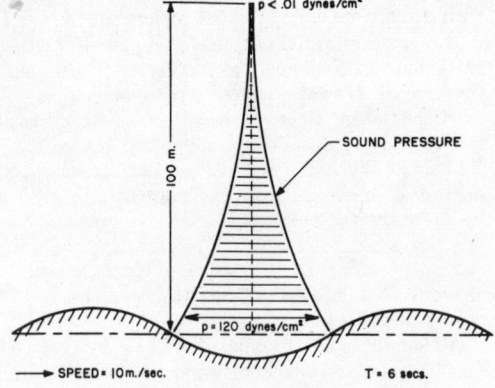

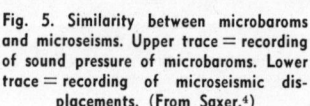

Fig. 4. Pseudosound pressure in the air above a slowly moving sinusoidal surface, such as traveling gravity waves on a body of water.

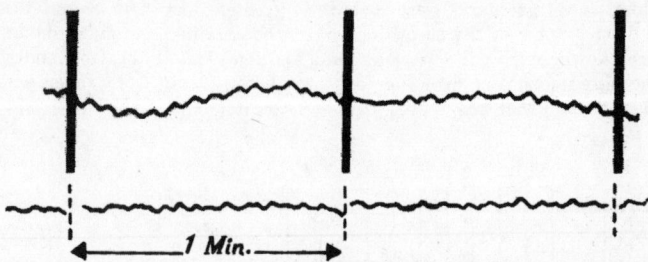

Fig. 5. Similarity between microbaroms and microseisms. Upper trace = recording of sound pressure of microbaroms. Lower trace = recording of microseismic displacements. (From Saxer.[4])

the speed of sound in the atmosphere. No sound power is radiated. We are forced to conclude that traveling waves on the surfaces of the oceans apparently do not cause microbaroms by direct radiation of sound into the atmosphere.

There still remains the question of the correlation between the strengths of microbaroms and microseisms as noted by Saxer.

Let us digress briefly to summarize some of the known facts about microseisms. These are very small oscillations of the surface of the earth. They are recorded on the seismographs of seismological observatories all over the world, and at almost all times. The waves have periods of 2–10 sec, more often 4–6 sec, and average amplitudes of 1–5 μ [the micron (μ) is one millionth of a meter]. Occasionally, during a so-called microseismic storm, the amplitudes may gradually increase to as much as 100 μ, and then gradually subside. The waves travel across continental surfaces with the speed of Rayleigh waves, about 3 km/sec. At one time, the waves were presumed to originate from the impact of oceanic surf on the shores of the continents. But the presence of microseisms at locations near the centers of the continents, with the waves coming from directions where there is no appreciable coastal surf, seems to make this hypothesis untenable. A more recent theory suggests that standing gravity water waves, produced by storms, o period T sec on the surface of the ocean will pr(duce an alternating pressure of period $T/2$ sec on the bottom of the ocean. This theory is supported by the observation that microseism activity usually increases near the eastern seaboard of the United States of America during the passage of cyclonic storms and hurricanes across the North Atlantic Ocean. This puzzling subject has resulted in hundreds of research papers, but neither the theories cited above nor several other theories have, as yet, been able to explain quantitatively all the observations.

Let us return now to microbaroms. Their similarity in waveform to microseisms is seen in Fig. 5. One is immediately tempted to think either that microbaroms are sound waves radiated into the atmosphere by microseismic oscillations of the earth's surface, or, conversely, that the sound pressure of microbaroms causes the observed microseismic deformation of the ground. But neither is the case, as we can see from the following analysis.

We can estimate the atmospheric sound pressure produced by microseisms from the sound-radiation formula $p = \rho c v$. Here, p is the radiated sound pressure caused by a surface vibratory velocity v, ρ being as above the density of the atmosphere and c the speed of sound. For microseisms of 10μ ampli-

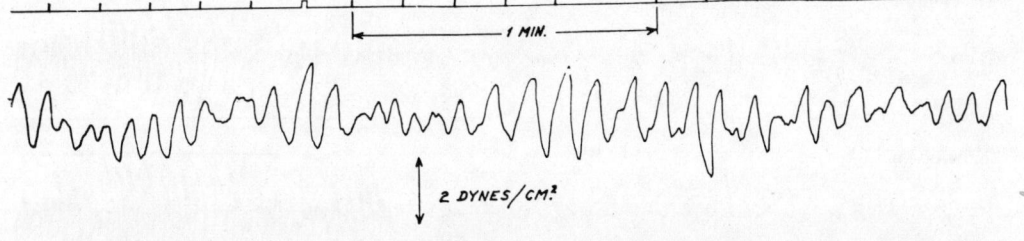

Fig. 6. Microbaroms of 5.5-sec periods recorded at Washington, D. C., on March 11, 1961.

tude at $T = 6$ sec, $v \approx 10^{-3}$ cm/sec. Since $\rho c \approx 40$ (cgs units), the radiated sound pressure is $p = 0.04$ d/cm². But the sound pressure of microbaroms is often about 1.0 d/cm² and occasionally as much as 6 d/cm². Evidently the microseisms are too weak to produce the observed strength of microbaroms.

Conversely, on looking into the elastic deformations of the earth's surface, produced by the strongest microbaroms of 6 d/cm², we find them to be much less than 1 μ. Furthermore, since the horizontal phase velocity of microbaroms (about 0.3 km/sec) is much less than the phase velocity of microseisms (about 3 km/sec), no power can be transferred from the atmospheric waves to the ground waves.

Before closing this subject, we note that Gutenberg and Benioff had failed to find any notable correlation between microbaroms and microseisms in Pasadena. But they did observe that microbaroms came mainly from the direction of the Pacific Ocean.

Microbaroms are frequently received in the Washington, D. C., area at the National Bureau of Standards. A typical recording is shown in Fig. 6. A program for studying them has been started. Preliminary results indicate that the waves travel parallel to the earth's surface with the speed of sound, and come mainly from an easterly direction, possibly from Chesapeake Bay or the Atlantic Ocean.

In summary, microbarom waves probably have a worldwide distribution. They appear to originate over or near the shores of the oceans, and are propagated inland over the continents. They might be generated by gravity waves on bodies of water, but the mechanism of this has not yet been elucidated.

Earthquake Waves

After a strong earthquake, the waves from it spread over the earth's surface and radiate sound into the atmosphere. The surface may be likened to the diaphragm of a loudspeaker, with vertical motions giving rise to sound radiations. There are several different types of waves in the earth produced by the earthquake, and they all travel with speeds much greater than the velocity of sound in air. Therefore, the radiations are propagated upward in a direction almost perpendicular to the earth's surface. The same radiation formula, $p = \rho c v$, can be used for computing the relationship between radiated sound pressure and surface motion as was used above for microseisms.

We showed (above) that microseismic tremors of the earth's surface are too weak to produce measurable sound pressure. But the amplitude of an earthquake wave can be hundreds or thousands of times greater than the amplitudes of microseisms. The sound radiation is often easily measurable even at distances thousands of kilometers away from the epicenter of the earthquake.

Let us look into the characteristics of a few of the waves which spread out from the focus of an earthquake. The focus is the location on or near the surface at which the earthquake occurs (Fig. 7). The epicenter is the point on the surface where a

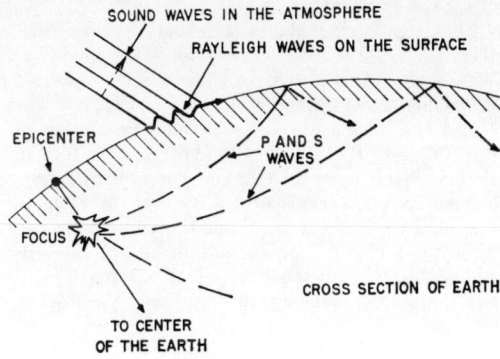

Fig. 7. Seismic waves from the focus of an earthquake, and sound radiated into the atmosphere by the seismic waves.

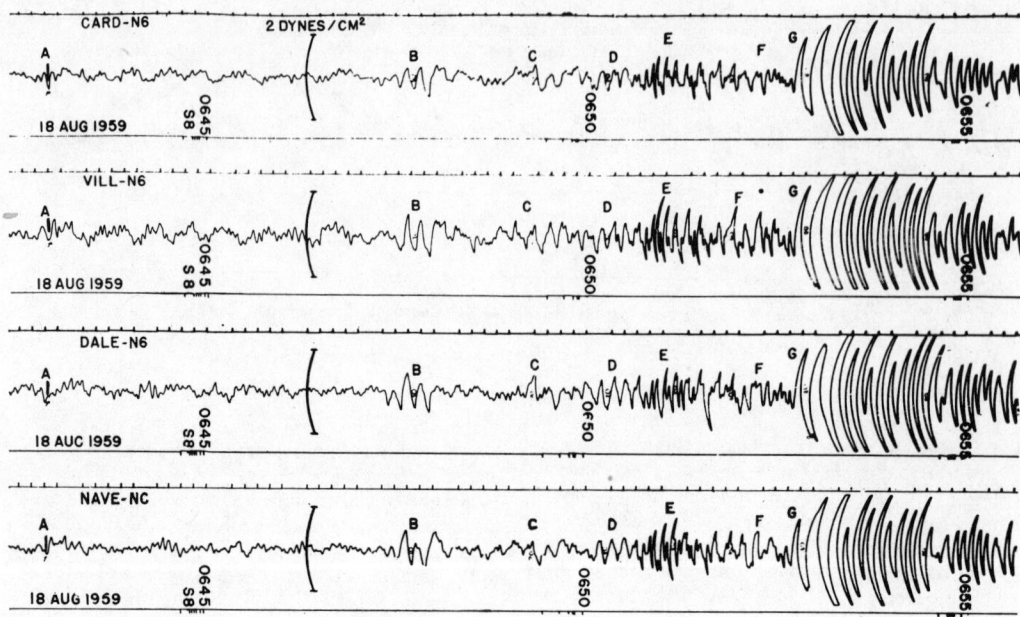

Fig. 8. Sound waves radiated by earthquake waves passing through the Washington, D. C., area. See text for explanation of waves at A, B, C, etc.

radius vector terminates on passing from the center of the earth through the focus. There are three waves of principal interest to our discussion. The first wave to arrive at a distant location is a longitudinal wave which has passed through the body of the earth; this is called a P wave. The second to arrive is a more-slowly-traveling transverse or shear wave, designated as an S wave. The third wave, which travels entirely on the surface, is the Rayleigh wave. These three are accompanied by many others, for example, P and S waves reflected from the boundary between the mantle and core 2900 km below the surface. From measurements of arrival times of the waves received at several seismological stations, the epicenter and focal depth of an earthquake can be determined very accurately.

The great earthquake in Montana on August 18, 1959, produced seismic waves strong enough in the Washington, D. C., area to cause easily measured infrasonic waves in the atmosphere (Fig. 8). The recordings were made at the infrasonics station of the National Bureau of Standards, described in Part I. Wind noise on that day, as well as microbaroms having periods of 4.3 sec, can be seen on the portions to the left. From the recordings can be deduced (1) the vertical amplitudes of the seismic waves, (2) their direction of approach, and (3) their trace velocity through the Washington area.

The epicenter in Montana was 2860 km away from the Washington station on a great-circle bearing 66° west of north. The earthquake occurred at 0637.27 UT on August 18, 1959. Incidentally, all times indicated are in "universal time" (UT) which is five hours ahead of Eastern Standard Time. Thus, 0637.27 UT = 1:37.27 A.M., EST. Point A in Fig. 8 indicates the time at which the P wave from the earthquake arrived at the Coast and Geodetic Survey seismological station in downtown Washington. The sound radiated by the P wave is obscured by wind noise and microbaroms, and cannot be distinguished with certainty.

The arrival of the S waves caused the sound waves indicated at B in Fig. 8. The shear waves came from 44° west of north with a period of 11 sec and a trace velocity of 6.0 km/sec. The earth's displacement in Washington deduced from a peak-to-peak sound pressure of 0.8 d/cm² was 0.34 mm.

The waves at C, D, E, and F came mainly from directions north of the great-circle direction to the epicenter, as did the S waves at B. Note that the latter came from a direction 22° north of the great-circle bearing.

We look next at the very strong Rayleigh waves at G. These had periods initially about 15 sec, which shortened to about 8 sec after a minute or so. This change came about because the group velocity of the 8-sec waves was less than for the 15-sec waves,

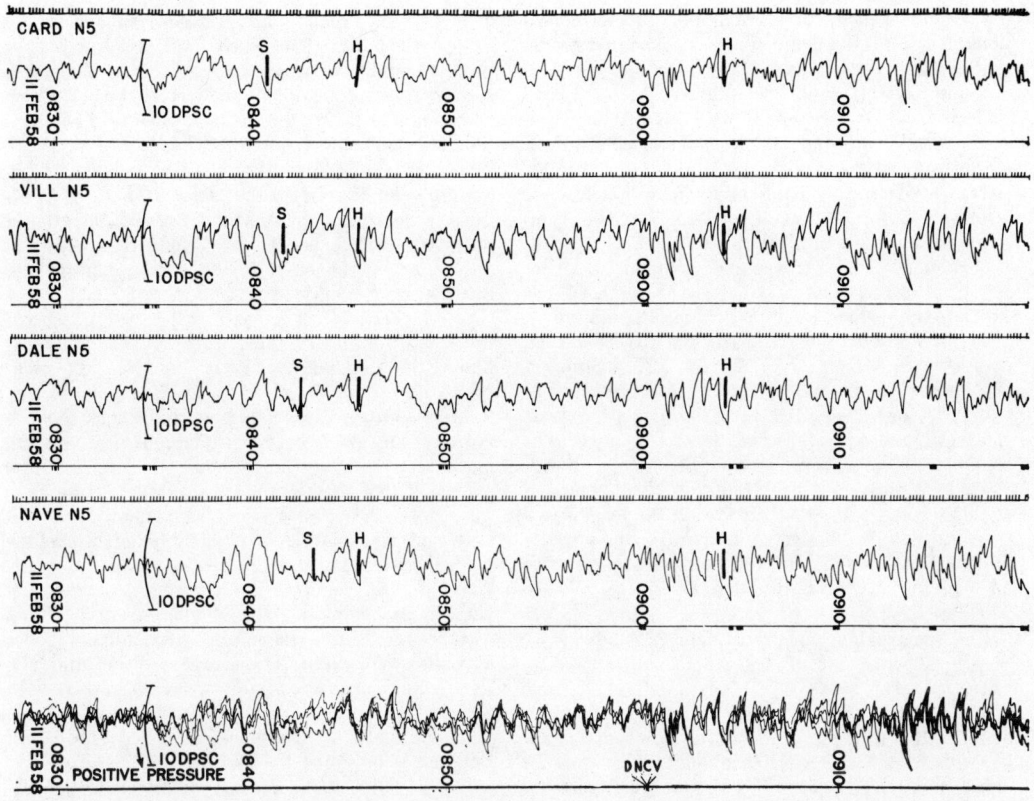

Fig. 9. Infrasonic waves at Washington, D. C., during the magnetic storm of February 11, 1958. The bottom trace is the superposition of the top four, and shows the correlation between sound pressures at the four microphone sites.

and so it took longer for the 8-sec waves to travel across the country to Washington. The earth's displacement, calculated from the large peak-to-peak sound pressure of 5 d/cm², was 3.0 mm. The average trace velocity of the Rayleigh waves was 3.8 km/sec. They came mainly from the great-circle direction of the epicenter, with the later waves coming from a slightly more northerly direction.

We can only conjecture about the reason for the arrival of the seismic waves from directions mostly north of the great-circle bearing. Refraction on passing from the Appalachian Mountains onto the Piedmont Plateau might be the cause.

An interesting property of these acoustical measurements on earthquake waves is that the sound pressure at any point is a measure of the average displacement of the earth over a considerable area in the vicinity of the point. A seismometer measurement, on the other hand, gives the earth's displacement only at the point at which the instrument is located.

Magnetic Storms

Another class of infrasonic waves is found in the atmosphere during magnetic storms. Before discussing these waves, we digress and present a short description of a magnetic storm and related phenomena.

With the advent of a solar flare or a sun storm, electromagnetic radiation reaches the earth almost immediately. An ionized-gas cloud sometimes appears one or two days later. This plasma cloud perturbs the magnetic field of the earth. Midlatitude observatories see a rise in the horizontal component of the magnetic field, followed by a larger decrease, and a recovery lasting several days. The strong and erratic variations that result are known as magnetic storms, magnetic activity, or disturbance variations. A measure of this solar-particle radiation effect is furnished by the planetary magnetic index K_p which is derived from data from a number of participating magnetic observatories. One of a series of numbers from 0 to 9 is given to each three-hour interval of each day, a larger number indicating a greater departure from undisturbed conditions. During large magnetic storms, magnetic fluctuation with periods from a few seconds to several minutes occurs, radio communications are disturbed, x-rays are observed with instruments carried in balloons, and the aurora is observed in midlatitudes.

Waves recorded during the magnetic storm of February 11, 1958, exhibited a more-or-less typical behavior pattern. The storm began on February 11 at 0126 UT. It was accompanied by an intense red aurora visible in Washington, D. C. The first distinguishable sound waves arrived about 0642 UT from a north-northwest direction, and had a trace velocity of 775 msec. A portion of these waves is shown in Fig. 9. Measurements at 0905 UT indicate a direction slightly more from the west, and a trace velocity of 750 msec. The sound waves decreased in amplitude and disappeared between 1100 and 1200 UT. The comparison of the large trace velocity with the local speed of sound is often enough to distinguish these waves from other infrasound. There are variations with time in the apparent direction and trace velocity of the waves.

The remarkably consistent changes in direction of arrival with time of day are shown in Fig. 10. Direction changes from the northeast in the evening, through north about midnight, then northwest in the morning, and to the northeast again somewhat suddenly after local noon. The data in this figure were restricted to signals with trace velocities above 390 msec to help prevent possible confusion with sound from other sources.

Sound waves usually arrive at Washington, D. C., within 5 or 6 hr of rise of K_p to a value of 6 or higher. Predominant periods range from 20 to several hundred seconds. The pressure amplitude is usually less than 3 d/cm², but is sometimes 7 or greater. Durations range from 1 or 2 hr to more than 24 hr, with a mean of about 6 hr. Additional details are given by Chrzanowski et al.[8]

A very simple hypothesis seems capable of explaining qualitatively the experimental observations concerning these sound waves. Imagine a somewhat diffuse source fixed in geomagnetic coordinates on the side of the earth opposite the sun. Let its magnetic latitude be that of the auroral zone, or about 66°. The earth will turn underneath the source once each day. This qualitatively explains the diurnal change of direction of infrasound that is observed at Washington, D. C. The relative absence of short periods and the large trace velocity suggest a high-altitude source where the mean free path of the molecules is long and the modes excited in the atmospheric wave guide have wave normals with a vertical component. It is realized that this picture is oversimplified. Since the aurora moves south with increasing geomagnetic activity, it is possible that the sound source may vary in geomagnetic latitude with strength of the disturbance. Fluctuations in longitude of the source may also take place. The source position and motion could be studied more advantageously with the aid of another infrasonic station, such as the one planned for the National Bureau of Standards' Boulder Laboratory in Colorado.

On the basis of the above hypothesis and the observations of duration and amplitude at Washington, D. C., it seems reasonable to assume that perhaps one quarter of the earth's surface is simultaneously bathed in acoustic radiation with an average pressure of about 1 d/cm². This suggests an acoustic source of roughly 10⁹ w during a typical magnetic storm. A detailed study of these acoustical radiations, together with associated magnetic, electromagnetic, and other phenomena during magnetic storms, should be a great help in understanding the dynamic behavior of our atmospheric environment.

Tornadic Storms

We are all familiar with some of the natural sounds produced by weather phenomena. Who has not experienced some of the audible and subaudible effects of thunder that follow a lightning flash? Severe tornadic storms also apparently emit infrasonic waves. The most violent tornadoes are commonest in the central and southern areas of the U.S.S.R., in southern Australia, and in the central United States of America. They occur most frequently in the afternoon and evening of the spring and summer months.

A comparison of infrasonic waves observed in Washington, D. C., in May 1960, with U. S.

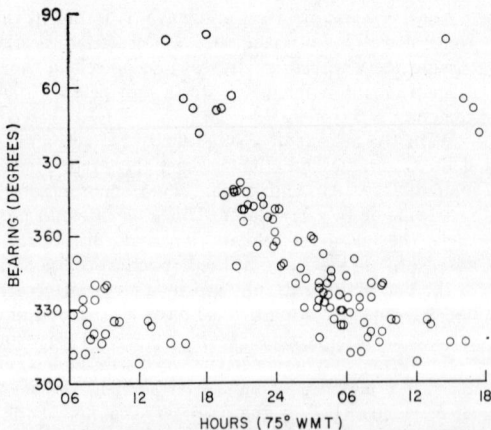

Fig. 10. Azimuth of infrasound from magnetic storms as a function of local time in Washington, D. C. Bearing (= azimuth) = angle, in degrees east of north, to the direction from which the sound comes.

Weather Bureau reports of tornadoes and funnel clouds is shown in Table I. During the eight tabulated time intervals, there is good correlation between the two phenomena. The infrasound had predominant periods between 12 and 50 sec, amplitudes up to about 1 d/cm^2, and horizontal phase velocities near the local acoustic velocity in air. Figure 11 illustrates the situation on May 5, 1960, when the sound arrived from between the two indicated directions. The sound must have been produced by some of the 20 tornadoes sighted in Oklahoma, northern Texas, and Kansas. An example of a portion of the signal on May 20 is shown in Fig. 12, and this signal could well have originated from a tornado in Meriden, Kansas. Time-lapse radar photographs showed a tornadic cloud with a hook in the vicinity of Meriden (see reference 9).

Sound waves having frequencies and amplitudes similar to those mentioned above were found during nine other intervals in May 1960. Three of them are thought to be associated with magnetic storms, one may have originated from the May 21 Chilean earthquake, and the remaining five are from unknown sources.

Summary and Comment

We have described sound waves at infrasonic frequencies produced by earthquakes, magnetic storms, and tornadoes, and we have looked into the origins of microbaroms. But these are only a few of the many natural sounds in the atmosphere. Volcanoes and meteors can serve as at least two additional sources of sound waves which have been observed in Washington, D. C. There have been numerous waves received which have not as yet been identified with any natural phenomena.

Waves of very long periods could be studied profitably. Their long wavelengths can be comparable to the height of the atmosphere, and, therefore, the propagation is strongly influenced by the density and temperature distribution with altitude.

TABLE I. Correlation between acoustic signals and tornado reports. The reader should note that the infrasound arrives in Washington, D. C., late in the afternoon and evening hours. This is partly because the interference from wind noise in Washington is less during these hours, and partly because the tornadoes occur in the afternoon, the sound waves needing about two hours to travel the distance.

	Acoustic Signals		Number of Tornado Reports Within	
Date. May 1960	Time interval (EST)	Azimuth[a] (degrees)	Time interval of col. 2	Time and Azimuth interval ($\pm 5°$) of cols. 2 & 3
3–4	2230–0300	255–262	2	2
4–5	2200–0215	259–268	20	19
5	2320–2400	263	1	1
16	1805–2200	259–269	3	1
17	1540–2100	214	6	3
19	1540–2400	260–272	27	22
24	1820–2150	266–274	7	4
25	1745–2300	238–270	15	6

[a] Azimuth = angle, in degrees east of north, to the direction from which the sound comes.

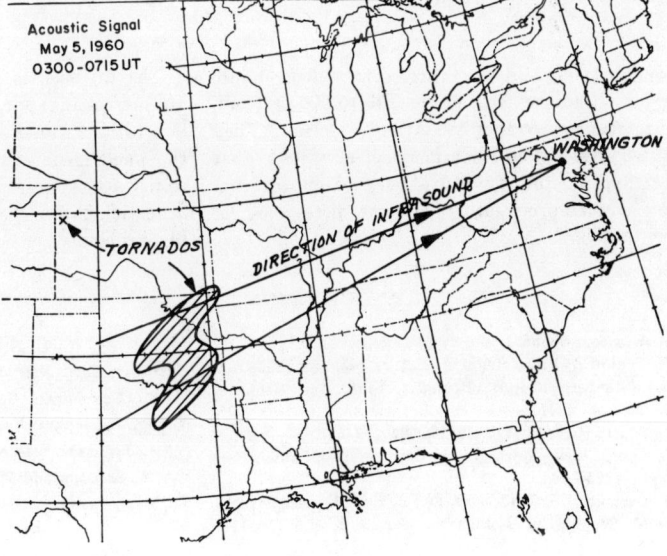

Fig. 11. Directions from which infrasonic waves arrived at Washington, D. C., on May 5, 1960, and location of tornadoes which could have produced such waves.

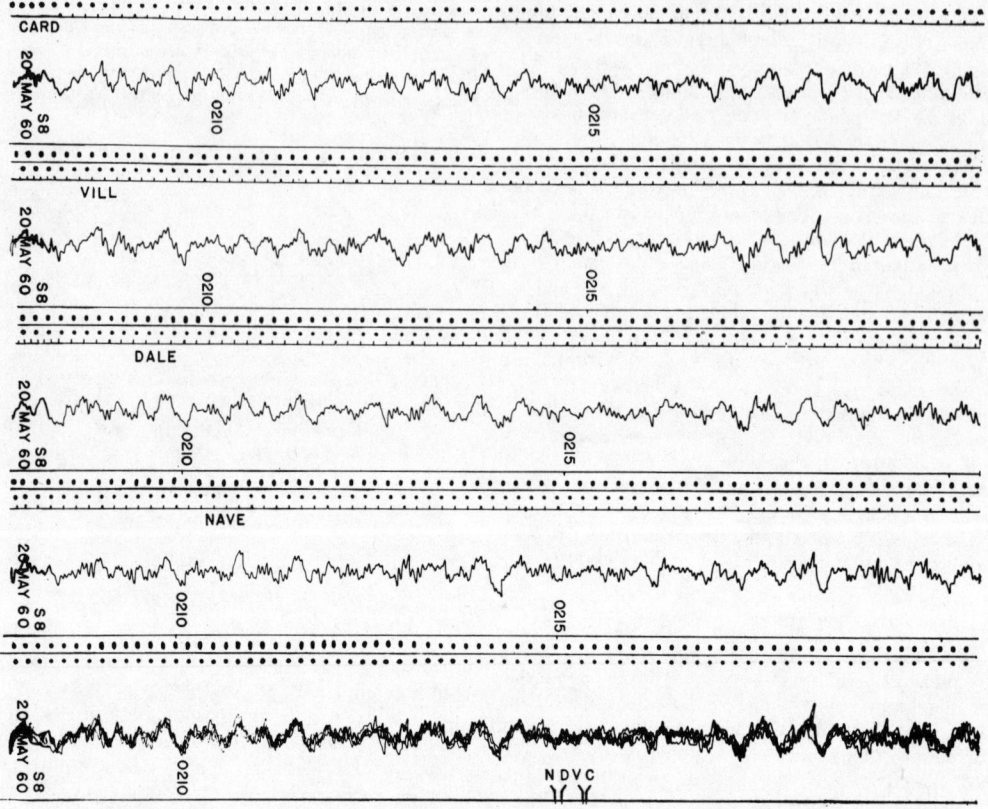

Fig. 12. Acoustic signal in Washington, D. C., from tornadoes in the midwest United States of America, May 20, 1960. The bottom trace is the superposition of the top four, and shows the correlation between sound pressures at the four microphone sites.

Recently, Donn and McGuinness have looked into long-period atmospheric waves, 240 to 600 sec, coupled to gravity waves on the Atlantic Ocean.[10] They find, with respect to the atmospheric waves, "that oscillations of this type are very often internal surface waves propagating on some meteorological discontinuity."

The infrasonics area of the science of sound is an interesting one, with many puzzling phenomena hitherto unexplained. The global character of the phenomena requires close cooperation between upper-atmosphere physicists, meteorologists, and acousticians for the solution of the many outstanding problems.

---

[1] R. K. Cook, Sound 1, No. 2, 12 (1962).
[2] H. Benioff and B. Gutenberg, Nature 144, 478 (1939).
[3] B. Gutenberg and H. Benioff, Trans. Am. Geophys. Union 22, 424 (1941).
[4] L. Saxer, Helv. Phys. Acta 18, 527 (1945).
[5] L. Saxer, Arch. Meteorolo. Geophys. u. Bioklimatol. Ser. A 6, 451 (1953–54).
[6] F. Dessauer, W. Graffunder, and J. Schaffhauser, Archiv Meteorol. Geophys. u. Bioklimatol. Ser. A 3, 453 (1951).
[7] See, e.g., L. L. Beranek, Acoustics (McGraw-Hill Book Company, Inc., New York, 1954), p. 35.
[8] P. Chrzanowski, G. Greene, K. T. Lemmon, and J. M. Young, J. Geophys. Research 66, 3727 (1961).
[9] A. Sadowski, Science 132, 736 (1960).
[10] W. L. Donn and W. T. McGuinness, J. Meteorol. 17, 515 (1960).

Editor's Comments on Papers 6, 7, and 8

6 Stewart: *Acoustic Wave Filters*

7 Mason: *A Study of the Regular Combination of Acoustic Elements, with Applications to Recurrent Acoustic Filters, Tapered Acoustic Filters, and Horns*

8 Lindsay and White: *The Theory of Acoustic Filtration in Solid Rods*

By 1920 the concept of electric wave filters had been well established and it was perhaps natural that the idea of an acoustic analogy would occur to someone working in acoustics. The scientist in this case was G. W. Stewart of the University of Iowa. His development of the concept of acoustic filtration was also stimulated by his early experiments on acoustic transmission through tubes with expansions, or constrictions, or side holes. In all these cases, transmission varies with frequency, with large transmission for certain frequency ranges and small transmission for others. This of course immediately suggests the idea of filtration. Stewart proceeded to develop a theory of acoustic filtration in air confined in tubes, using freely the concept of acoustic impedance. His theory, as set forth in his paper "Acoustic Wave Filters," the first part of which is reproduced here, is called a lumped impedance theory, since he assumed that the acoustic elements of inertia, resistance, and stiffness, characterizing the air in the different parts of the acoustic transmission line, are concentrated at definite locations and not distributed. Even with this limitation, his theory proved to be in rather good agreement with experimental results. His pioneer work in the field of acoustic filtration led to a large amount of subsequent research on wave transmission through iterative structures.

The remaining part of the article, which is not reprinted here, discusses further applications of the lumped impedance theory.

As has been made clear in the preceding paper, G. W. Stewart was the first to suggest in detail the possibility of constructing acoustic filters and to devise an adequate theory of acoustic filtration. He used the lumped impedance theory for acoustic elements with some success. It seemed desirable, however, to supplement this with a theory treating the actual transmission of acoustic radiation through an acoustic line and the effect on this of various iterated changes in the structure of the line. This theory was first developed by Warren P. Mason of the Bell Telephone Laboratories. We reproduce here, in part, his pioneer paper "A Study of the Regular Combination of Acoustic Elements, with Applications to Recurrent Acoustic Filters, Tapered Acoustic Filters, and Horns." The part of the paper not reprinted refers to specific applications not necessary for an understanding of the author's method.

Editor's Comments on Papers 6, 7, and 8

Warren Perry Mason (1900–), American physicist, a native of Colorado, received his Ph.D. from Columbia University in 1925. His professional career was spent almost entirely at the Bell Telephone Laboratories, from which he retired in 1965. Since that time, he has held professorships at Columbia and George Washington universities. His research has covered most of the important fields of physical acoustics, including acoustic filtration, piezoelectric transducers, and solid-state acoustics, in which he has become an internationally recognized authority.

As has been mentioned earlier in this volume, the lumped impedance theory of acoustic filtration in air was developed by G. W. Stewart and successfully applied by him to the construction of both low- and high-pass acoustic filters. The inadequacy of that theory was recognized by W. P. Mason, who developed a wave transmission theory. These two points of view are represented in the papers just discussed. A convenient modification of Mason's theory was developed by Stewart and Lindsay (See *Acoustics, a Text on Theory and Applications*, Van Nostrand Reinhold, New York, 1930, p. 334) and applied by Lindsay and White to solid acoustic filters in the paper "The Theory of Acoustic Filtration in Solid Rods," reproduced here in part. (The last part of the paper, providing further illustrations of the general method, is not reprinted.) It led to a considerable amount of additional research on acoustic filtration in solids and other acoustic structures. A review of this work will be found in "The Filtration of Sound," by R. B. Lindsay (*J. Appl. Phys.*, **9**, 612–622, 1938; and *ibid.*, **10**, 680–687, 1939).

Frederick Elmer White (1909–), originally a student under R. B. Lindsay at Brown University, has been for many years a professor of physics at Boston College.

ACOUSTIC WAVE FILTERS.

By G. W. Stewart.

Synopsis.

Acoustic Wave Filters Composed of a Series of Like Sections.—(1) *Theory.* Taking the impedance of any part of an acoustic circuit to be equal to the complex ratio of the applied pressure difference to the rate of change of volume displacement, it is shown that, neglecting dissipative forces, it is possible to construct a filter having limiting frequency values of no attenuation determined by the formulæ $Z_1/Z_2 = 0$ and $Z_1/Z_2 = -4$, where Z_1 is the impedance of the transmitting conduit circuit and Z_2 of each branch of each section. The impedance of any section depends on the inertance M of dimensions mass per unit area squared, and the capacitance C which has the dimensions of stiffness per unit area squared. If M and C are in parallel, $Z = iM\omega/(1 - MC\omega^2)$, whereas if they are in series, $Z = i(M\omega - 1/C\omega)$. For instance, in the case of a closed chamber or resonator, M and C are in series and are equal to ρ/c and $V/\rho a^2$ respectively where ρ is the density of the medium, c is the conductivity of the mouth, a the velocity of sound and V the volume. Formulæ are derived for various assumed cases. On account of the uncertainty as to whether a tube may be considered sa having the equivalent inertness and capacitance connected in parallel or in series, the application of these formulæ to actual cases is somewhat empirical. (2) *Construction and test of filters of three types.* Low-frequency-pass filters were made, for example, by two concentric cylinders joined by walls equally spaced and perpendicular to the axis. Each chamber thus formed had a row of apertures in the inner cylinder which served as the transmission tube. In one case the volume of each chamber was 6.5 cm.^{3}, the radius of the inner tube 1.2 cm. the length between apertures, 1.6 cm. A chamber and one such length of the inner tube is called a section. Four such sections were found to transmit 90 per cent. of the sound from zero to approximately 3,200 d.v. where the attenuation became very high, resuiting in zero transmission up to about 4,600 d.v. where transmission again appeared, Other similar filters of different dimensions attenuated through wider or narrower ranges. The lower limit of attenuation was found to correspond within 8 per cent. with the formula: $f = (1/\pi)(M_1C_2 + 4M_2C_2)^{-1/2}$. The upper limit was not predicted theoretically. High-frequency pass filters were made with a straight tube for transmission and short side tubes, for example, 0.5 cm. long and 0.28 cm. in diameter, opening through a hole with conductivity 0.08 into a tube 10 cm. long and 1 cm. in diameter. Six sections of such a filter would transmit about 90 per cent. of sounds above 800 but would refuse transmission to sounds of lower frequency. As would be expected, the cut off is not sharp. Filters with other dimensions were found to have an upper limit of attenuation varying from 450 to 2,300 d.v., agreeing with the formula $f = (1/2\pi)(1/4M_2C_1 + 1/M_1C_2)^{1/2}$, within about 13 per cent., on the average. The single-band filters made were a combination of the other two types, having side tubes leading to chambers of considerable volume. For instance, three sections each 5 cm. long and 0.5 cm. in diameter, with side tubes of the same size and 2.2 cm. long leading to a volume of 28 cm.^{2}, transmitted between 270 and 370 d.v. The frequencies of the edges of the band of small attenuation are determined by the following formulae,

$$2\pi f = [M_2C_2(1 + M_2'/M_2)]^{-\frac{1}{2}} \text{ and}$$

$$2\pi f = [M_2C_2(1 + 4M_2'/M_1)^{-1}(1 + M_2'/M_2 + 4M_2'/M_1)]^{-\frac{1}{2}}. \text{ Such filters}$$

exhibit the same variations from theoretical performance as would be expected from a combination of the other two types. However, the agreement of each type with theory is sufficient to enable filters to be designed to fulfill set conditions. The *attenuation secured* with only four sections is very great, the transmission being certainly less than 10^{-7} in the attenuated region, while it may rise to 90 per cent. in unattenuated regions. *Possible applications* of these simple filters in laboratory work and in connection with specking devices, are briefly suggested.

I. General Introduction.

THE selective transmission of an acoustic wave of a given frequency is well known. A Helmholtz resonator with a small ear opening is such a filter. Cylindrical tubes such as shown in Fig. 1 will also serve as filters, (*a*) transmitting chiefly the resonating frequencies of *cd* and (*b*) giving poor transmission especially for the resonating frequencies of *cd*. The acoustic wave filters which this article describes

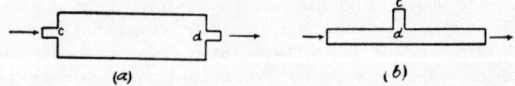

Fig. 1.

are different in principle and in performance in that they do not depend upon resonance itself, but upon the interaction between recurrent similar sections of a transmission "line," these sections containing the elements upon which free vibrations are to depend, and having over-all dimensions that are small in comparison with a wave-length of the sound.

These new filters are remarkable in that selected groups of frequencies, extending over a large range, can be eliminated in the transmission. Up to the present time, three kinds have been constructed and tested. The low frequency pass filter will give approximately zero transmission at all frequencies above, and a fairly good transmission below a certain predetermined frequency. The high-frequency filter will transmit above a minimum frequency. The single-band filter will transmit a group of frequencies. In all cases, the frequency limits are ascertained approximately by calculation from the dimensions so that the filters may be designed to fit specifications. In these filters, the cutoffs are not sharp and the performances are not exactly as just stated, but, as will be shown, the explanation is found in the fact that the experimental conditions only approximate the theoretical.

II. Theory.

A. General.

An exact theoretical discussion of the acoustic wave filter may be possible, but certainly there is much to be gained in securing a theory which, though only approximate, will aid in understanding the operation of the filters and in developing new designs. The well-known electric wave filters suggested to the writer the possibility of analogous effects in acoustics. Naturally, the theory[1] of the electric filters has been of much assistance.

Certain limitations will be imposed in our discussion. We will assume that the length of any selected section of an acoustic "line" or conductor is so small in comparison with a wave-length that no change in phase occurs therein. We will consider only sinusoidal waves. The term "acoustic impedance" will be used. Its absolute value is the ratio of the maximum pressure difference applied to the maximum rate of change of volume displacement. When complex notation is used for simplicity, as in alternating electric current theory, the acoustic impedance is the complex ratio of the pressure difference applied and the rate of change of volume displacement. The mathematical procedure will be based upon three hypotheses which are obviously reasonable approximations. They are: (1) the rate of volume displacement in any selected portion of the line, with a harmonically varying applied difference of pressure, can always be expressed by $Ie^{i\omega t}$ where I is complex; (2) the product of acoustic impedance and rate of volume displacement in any selected portion of the line equals the difference of pressure applied; (3) the algebraic sum of the volume displacements at any junction of lines is zero. By acoustic "line" is meant a bounded region of fluid or solid, forming a tube or channel and capable of transmitting sound waves in the direction of the tube or channel only.

We will consider an acoustic wave filter consisting of equal acoustic impedances in series, divided into sections by acoustic impedances in what might be termed shunt branches. In Fig. 2, let a sound wave of a

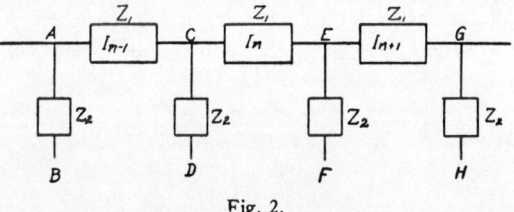

Fig. 2.

frequency $\omega/2\pi$ be transmitted through the line AG, a portion of an infinite line, containing a series of equal impedances, Z_1. Let each branch line AB, CD, EF, GH, etc., contain an impedance Z_2 and terminate in a

[1] U. S. Patent 1,227,113 by George A. Campbell. Chapter XVI. of Pierce's "Electrical Oscillations and Electric Waves," 1920.

volume of gas otherwise at rest, so that there is a common constant pressure at or near these termini. Let $I_n e^{i\omega t}$, etc., represent the rates of change of volume displacement in the corresponding lines, I_n being a complex quantity, and let the positive direction be from A to G. From the three conditions stated in the foregoing paragraph the following equation may be secured:

$$Z_2(I_{n-1} - I_n) = Z_2(I_n - I_{n+1}) + Z_1 I_n$$

or

$$I_{n+1} - \left(2 + \frac{Z_1}{Z_2}\right) I_n + I_{n-1} = 0. \tag{1}$$

This equation does not require an infinite line, but we will for convenience impose that condition. We may then write, using ΔP as the complex pressure difference over a branch,

$$\Delta P_{CD} = I_n(Z_1 + Z_\infty)$$

and

$$\Delta P_{EF} = I_{n+1}(Z_1 + Z_\infty),$$

wherein Z_∞ is the impedance of the infinite network to the right in the figure and has the same value in both equations.[1] Substituting the values,

$$\Delta P_{CD} = Z_2(I_{n-1} - I_n),$$
$$\Delta P_{EF} = Z_2(I_n - I_{n+1})$$

and dividing we have,

$$\frac{I_{n+1}}{I_n} = \frac{I_n - I_{n+1}}{I_{n-1} - I_n}$$

or

$$\frac{I_{n+1}}{I_n} = \frac{I_n}{I_{n-1}}. \tag{2}$$

Let us call the ratio of these successive I's, e^Y, where Y is unknown but in general is complex. Substitute in (1), and we have,

$$e^Y + e^{-Y} = 2 + \frac{Z_1}{Z_2}$$

or

$$\cosh Y = 1 + \tfrac{1}{2}\frac{Z_1}{Z_2}. \tag{3}$$

[1] The filter through its branches terminates in the undisturbed medium. The pressure at E can be expressed by $I_n Z_\infty$. From the equation of motion of a portion of a vibrating medium having a sinusoidal impressed force, and possessing mass, stiffness and dissipation, it can readily be shown that the impedance Z is a function of mass, stiffness, the dissipative factor and frequency only. In as much as these factors are the same for our Z_∞, whether taken from E or from G or any junction point, the assumption of identity of the Z_∞'s is justified.

If Y is a pure imaginary the rates of volume displacement in two adjacent sections differ only in phase. If Y is not a pure imaginary the rate of volume displacement is diminished in transmission, obviously the diminution occurring away from the input end. If Y is a pure imaginary, since $\cosh ix = \cos x$

$$+1 > \left(1 + \tfrac{1}{2}\frac{Z_1}{Z_2}\right) > -1 \tag{4}$$

and hence the limiting values of no attenuation are determined by the following:

$$\frac{Z_1}{Z_2} = 0, \tag{5}$$

$$\frac{Z_1}{Z_2} = -4. \tag{6}$$

We can find the limiting values of no attenuation in a filter by utilizing the actual values of Z_1 and Z_2. It would therefore appear that an acoustic wave filter can be constructed, the only uncertainty being the manner of constructing Z_1 and Z_2. In order to determine upon the practical development of the filter, idealized conditions will first be discussed.

B. Inertance and Capacitance in Parallel.

Consider two idealized diaphragms, a and b, supposed to move as a whole, the former having mass, m_a, and not stiffness, and the latter stiffness, f_b, and not mass. Let them have areas S_a and S_b and displacements ξ_a and ξ_b and let them be connected in parallel as branches of an acoustic line so that they are subject to the same fluid pressure differences, $Pe^{i\omega t}$. Then,

$$m_a \frac{d^2\xi}{dt^2} = S_a P e^{i\omega t} \quad \text{and} \quad f_b \xi_b = S_b P e^{i\omega t},$$

the latter being merely the definition of f_b.

If we are concerned with the total volume displacement of the gas at the diaphragms, X_a and X_b, these equations become:

$$\frac{m_a}{S_a^2}\frac{d^2 X_a}{dt^2} = Pe^{i\omega t} \quad \text{and} \quad \frac{f_b X_b}{S_b^2} = Pe^{i\omega t}. \tag{7}$$

We wish to obtain an expression for dX/dt or the rate of change of volume displacement in the main line at the junction of these two branches. By integration we find,

$$\frac{m_a}{S_a^2}\frac{dX_a}{dt} = \frac{P}{i\omega}e^{i\omega t} \tag{8}$$

the constant of integration vanishing since we are dealing with a variation in dX/dt caused by P and vanishing with P. By differentiation we obtain,

$$\frac{f_b \dfrac{dX_b}{dt}}{S_b^2} = i\omega P e^{i\omega t}. \tag{9}$$

But we know that at the junction of the branches the algebraic sum of the rate of volume displacements must be zero, or

$$\frac{dX_a}{dt} + \frac{dX_b}{dt} = \frac{dX}{dt}. \tag{10}$$

Substituting the values of dX_a/dt and dX_b/dt just found, we obtain,

$$\frac{dX}{dt} = Pe^{i\omega t}\left(\frac{1 - MC\omega^2}{iM\omega}\right), \tag{11}$$

wherein

$$M = \frac{m_a}{S_a^2} \quad \text{and} \quad \frac{1}{C} = \frac{f_b}{S_b^2}. \tag{12}$$

We will call M the *inertance*, C the *capacitance*, both defined by (12) and (7). We thus have for the impedance of the combination of the two circuits in parallel,

$$Z = \frac{iM\omega}{1 - MC\omega^2}. \tag{13}$$

Let us now assume that our impedances Z_1 and Z_2 are each composed of two such branches in parallel. The branch having mass, m_a, will vanish when $m_a = \infty$ or $M = \infty$ and the branch having stiffness, f_b, will vanish when $f_b = \infty$ or $C = 0$.

If we apply condition (5) to Z_1 and Z_2 as just determined in (13) and we have as a limiting frequency, f_1:

$$\omega^2 = \frac{1}{M_2 C_2} \quad \text{or} \quad f_1 = \frac{1}{2\pi}\sqrt{\frac{1}{M_2 C_2}}. \tag{14}$$

If we apply condition (6), we have:

$$\omega^2 = \frac{M_1 + 4M_2}{M_1 M_2(4C_1 + C_2)} \quad \text{or} \quad f_2 = \frac{1}{2\pi}\sqrt{\frac{M_1 + 4M_2}{M_1 M_2(4C_1 + C_2)}}. \tag{15}$$

In (14) and (15) the subscripts of M and C correspond to those adopted for the Z's. These frequencies f_1 and f_2 are those that limit the range in which there is no attenuation. It is to be observed that the range of no attenuation is one in which $\left(1 + \tfrac{1}{2}\dfrac{Z_1}{Z_2}\right)$ can change from $+1$ to -1.

But, from (13), Z_1/Z_2 is a continuous function of ω or $2\pi f$ and hence there is a continuous frequency range wherein there is no attenuation. A filter in which Z_1 and Z_2 each consists of an inertance and capacitance in parallel is thus possible.

C. Inertance and Capacitance in Series.

Consider each Z_1 and Z_2 to be an inertance and capacitance in series. The current must be the same throughout Z_1 or Z_2. Let P_1 be the pressure difference over the inertance, M_1, and P_2, over the capacitance, C_1. Then, from (7) and (8)

$$\frac{P_1}{i\omega M_1} = I = \frac{P_2}{\frac{1}{C_1 i\omega}}.$$

Hence

$$\frac{P_1 + P_2}{i\omega M_1 + \frac{1}{iC_1\omega}} = I$$

or

$$Z_1 = i\left(M_1\omega - \frac{1}{C_1\omega}\right). \tag{16}$$

If we now ascertain the limits of no attenuation, we have from (5),

$$f_1 = \frac{1}{2\pi}\sqrt{\frac{1}{M_1 C_1}} \tag{17}$$

and from (6)

$$f_2 = \frac{1}{2\pi}\sqrt{\frac{C_2 + 4C_1}{C_1 C_2 (M_1 + 4M_2)}}. \tag{18}$$

Again we have the possibility of a filter.

A special and yet a common case of inertance and capacity in series is one wherein M_1 is the inertance of an orifice entering a chamber and C_1 is the capacitance of the chamber. The condition for resonance, or $Z = 0$, occurs as shown by (16) when,

$$M_1\omega = \frac{1}{C_1\omega} \quad \text{or} \quad f = \frac{1}{2\pi}\sqrt{\frac{1}{M_1 C_1}}. \tag{19}$$

But in orifices which are short compared with their diameters we cannot use for M_1 the mass divided by the square of the area, for this expression neglects the end effects of the channel. The well-known formula for the vibration of such a system is,[1] neglecting dissipation[2]

---

[1] Rayleigh, Theory of Sound, Vol. II., p. 195.

[2] Insertion of dissipation seems to lead to difficulties which are avoided by this approximation.

$$\frac{\rho}{c}\ddot{X} + \frac{\rho a^2}{V}X = Pe^{i\omega t} \tag{20}$$

from the solution of which we get,

$$\dot{X} = \frac{Pe^{i\omega t}}{i\rho\left(\dfrac{\omega}{c} - \dfrac{a^2}{V\omega}\right)}. \tag{21}$$

In these equations ρ is the density, c is the "conductivity" of the channel, a the velocity of sound and V the volume of the chamber.

It is evident that the new value of Z_1 viz.,

$$Z = i\rho\left(\frac{\omega}{c} - \frac{a^2}{V\omega}\right), \tag{22}$$

differs from (16) only by the apparent substitution of ρ/C for M_1 and $V/\rho a^2$ for C_1.

D. *Inertance and Capacitance of a Tube.*

In order to assist in constructing a filter, we will now discuss the nature of a column of gas, which we must use as Z_1 in our conducting line.[1]

Assume an acoustic plane simple harmonic wave passing along a tube in the direction of x. Let ξ be the displacement of a particle from its mean position; ρ the density, a the velocity of sound, and p the excess pressure over the mean.

Then

$$\rho \frac{\partial^2 \xi}{\partial t^2} = -\frac{\partial p}{\partial x} \tag{23}$$

(an exact equation) and for small vibrations,[2]

$$\frac{\partial^2 \xi}{\partial t^2} = a^2 \frac{\partial^2 \xi}{\partial x^2}. \tag{24}$$

From (23) and (24)

$$\rho a^2 \frac{\partial^2 \xi}{\partial x^2} = -\frac{\partial p}{\partial x}. \tag{25}$$

The integration of (25), since

$$\frac{\partial \xi}{\partial x} = 0$$

at all times when $p = 0$, gives

$$\rho a^2 \frac{\partial \xi}{\partial x} = -p. \tag{26}$$

---

[1] Similar discussions occur in Drysdale, Jl. of Inst. Elec. Engs., July, 1920, Vol. 58, p. 591, and Kennelly and Kurowaka, Proc. Am. Acad., Feb., 1921, Vol. 56, No. 1, p. 29.

[2] Lamb's Hydrodynamics, 1916, p. 474.

We may write (26) as follows:

$$\rho a^2 \frac{\partial \xi}{\partial x} \frac{\Delta x}{\Delta x} = -p$$

and if we represent the volume in a length Δx by ∇V, we have,

$$\rho a^2 \frac{\Delta V}{V} = -p$$

or $\rho a^2 = -E$ where E is the modulus of elasticity of volume. Since the compression is due to p and not to changes of pressure along the line, the stiffness can be considered as analogous to stiffness in the walls of the tube upon which the pressure difference p acts. Our tube possesses inertance and capacitance as we know, but the above shows that these are not the equivalent of inertance and capacitance connected either in parallel or in series. Indeed, the capacitance can be thought of as between the inside and outside of the tube instead of along the tube.

Inasmuch as we must use tubes in a practical construction it is essential that we consider whether or not a tube can be used for either Z_1, or Z_2, arranged as in the preceding theory. Consideration shows that Z_1 cannot accurately be composed of a tube, for Z_1 is wholly in the line and not between the line and the outside. Hence we must ascertain whether or not such a substitution would be a sufficient approximation. In short, can we consider a tube as having the equivalent of inertance and capacitance connected in parallel or in series, or may we consider it as having inertance only or as capacitance only?

Assume that any tube we may use will be short as compared with a wave-length. Consider the gas to move as a whole. Then the δp is due to inertance. Or consider the gas to be stationary acting as a cushion. Then δp is due to capacitance. In both cases we have neglected the change of phase along the tube. Whichever case is the better approximation will depend upon the service the tube is rendering, or, in other words, will depend upon the adjacent construction or the composition of the filter. It might appear that the tube can be approximated by lumping the capacitance. This cannot be done at the center of the section for the general theory does not permit of a side branch other than Z_2. It cannot be lumped at Z_2 as can be shown by comparison of the resulting theory with experiment. The assumption of M_1 and C_1 in series does not appear reasonable for the total pressure over the section certainly acts on M_1. We are therefore forced to the policy of using M_1 and C_1 in parallel in the line with the understanding that experiment will determine empirically whether M_1 or C_1 or both shall be used in the computations. As will subsequently be shown, experiments thus far seem to demand the introduction of C_1 in only one of the three types of filters. Thus the somewhat arbitrary manner of introducing C_1 in parallel is minimized. It is fair to say that, in this one case, the assumption seems not without a theoretical justification. It is thus observed that at the outset we are driven to an approximation which demands an empirical selection of formulæ.

III. Construction of Filters.

A. General Limitations.

If we are to construct an acoustic filter that will have good transmission we must avoid any changes in the nature of the medium and in the diameter of the transmitting conduit or conductor. If the filter is to have good attenuation, similar conditions would hold for the branch lines. For first experiments, then, a single medium and a transmitting line of constant cross section are chosen. These selections suggest, from the standpoint of convenience, the use of air and the use of a cylindrical tube as the boundary of the transmission line.

The only limitation to our application is that there be a series of like sections with Z_1 in the main line and Z_2 in the side branch. With no limitation placed upon the constitution of Z_1 or of Z_2, an infinite number of designs may be possible. But, as already stated, considerations lead to the selection of a cylindrical tube containing air as the transmission line. The impedance of a short section of such a line cannot be accurately expressed by an equivalent M_1 and C_1 connected either in series or parallel. We will assume M_1 and C_1 in parallel, for thereby we can make $M_1 = \infty$ or $C_1 = 0$ or remove them from consideration without obstructing the transmission of the line. But even with these limitations the number of filters is infinite, for Z_2 can be formed, theoretically, in any manner one chooses.

B. Low Frequency, High Frequency, and Single Band Pass Filters.

In order to determine the construction of low frequency and high frequency pass and single band filters in as simple a manner as possible, an additional provisional limitation will be made, viz., to filters in which Z_2 consists of an equivalent M_2 and C_2 connected either in series or parallel. We shall then have four quantities M_1, M_2, C_1 and C_2. If either M_1 or M_2 is absent, i.e., removed from consideration, its value is infinity. Under similar conditions the value of a capacitance is zero. We have then to ascertain the possible combinations of

$$M_1 = \infty, \quad M_2 = \infty, \quad C_1 = 0, \quad C_2 = 0.$$

Taken singly there are four combinations in pairs, six in triplets, four with one only, and one with none, making a total of fifteen possible combinations. If we remove those combinations which remove the *line* or the branches entirely, we have left nine designated as follows:

1. All present,
2. $M_1 = \infty$,
3. $M_2 = \infty$,
4. $C_1 = 0$,
5. $C_2 = 0$,
6. $M_1 = M_2 = \infty$,
7. $M_1 = \infty$, $C_2 = 0$,
8. $M_2 = \infty$, $C_1 = 0$,
9. $C_1 = 0$, $C_2 = 0$.

By the preceding, our values of acoustical impedances are limited to the following:

In the transmission line,
$$Z_1 = \frac{iM_1\omega}{1 - M_1C_1\omega^2}. \qquad (13)\text{ bis}$$

In the branch,
$$Z_2 = \frac{iM_2\omega}{1 - M_2C_2\omega^2} \qquad (13)\text{ tris}$$

or
$$Z_2 = i\left(M_2\omega - \frac{1}{C_2\omega}\right). \qquad (16)\text{ bis}$$

If the limits of no attenuation are now ascertained by the application of equations (5) and (6) we find the values in Table I. for the nine cases, each having the possibility of Z_2 in parallel or in series.

The explanation of most of the blanks in the fourth and fifth columns is that, assuming M_2 and C_2 to be in series, our original arrangement of Z_2 providing for the same pressure in common at the termini, requires the following:

1. There must always be an M_2 at the junction point, for otherwise there could be no $\dot{X}$ and hence no use of the side branch. Hence M_2 is at the junction point and C_2 is next in the branch.

2. M_2 cannot be infinity, for if infinity, the side branch would not be used at all.

3. C_2 cannot be zero for this would prevent any value of $\dot{X}$ and therefore any use of the side branch. Thus six of the nine cases are eliminated from consideration leaving only cases 1, 2 and 4. Case 1 leads to the possibility of an imaginary frequency and introduces two values of f_2,

TABLE I.

Case.	Parallel.		Series.	
	f_1	f_2	f_1	f_2
1	$\dfrac{1}{2\pi}\sqrt{\dfrac{1}{M_2C_2}}$	$\dfrac{1}{2\pi}\sqrt{\dfrac{M_1+4M_2}{M_1M_2(4C_1+C_2)}}$	0	
2	$\dfrac{1}{2\pi}\sqrt{\dfrac{1}{M_2C_2}}$	$\dfrac{1}{2\pi}\sqrt{\dfrac{1}{M_2(4C_1+C_2)}}$	∞	$\dfrac{1}{2\pi}\sqrt{\dfrac{C_2+4C_1}{4M_2C_1C_2}}$
3	0	$\dfrac{1}{2\pi}\sqrt{\dfrac{4}{M_1(4C_1+C_2)}}$		
4	$\dfrac{1}{2\pi}\sqrt{\dfrac{1}{M_2C_2}}$	$\dfrac{1}{2\pi}\sqrt{\dfrac{M_1+4M_2}{M_1M_2C_2}}$	0	$\dfrac{1}{2\pi}\sqrt{\dfrac{4}{C_2(M_1+4M_2)}}$
5	∞	$\dfrac{1}{2\pi}\sqrt{\dfrac{M_1+4M_2}{4M_1M_2C_2}}$		
6	0	0		
7	∞	$\dfrac{1}{2\pi}\sqrt{\dfrac{1}{4M_2C_1}}$		
8	0	$\dfrac{1}{2\pi}\sqrt{\dfrac{4}{M_1C_2}}$		
9	∞	∞		

which are not found in experience. Knowing that the assumption of C_1 in parallel with M_1 is arbitrary, we are justified in acknowledging the incorrectness of the assumption in this case and omitting the formulæ from consideration.

These considerations lead to the development of the three following types of filters:

1. Low-frequency pass filters; case 3, parallel; case 4, series; and case 8, parallel.

2. High-frequency pass filters; case 2, series; case 5, parallel; case 7, parallel.

3. Single-band pass filters; case 1, parallel; case 2, parallel; case 4, parallel.

Conditions of construction suggest that an additional formula for the single-band filter be determined. The three cases in which we have a single-band filter as above stated, are those in which we have M_2 and C_2 in parallel. But in actual construction it is impractical to have truly a C_2 without an orifice into a volume and in this orifice we have an M_2' in series with C_2. A simple construction is thus suggested of taking *Case 1* and adding an M_2' as an orifice into C_2. The impedance of the orifice and C_2 will be, according to (16)

$$Z_2' = i\left(M_2'\omega - \frac{1}{C_2\omega}\right).$$

This impedance is in parallel with the inertance M_2, the impedance of which is,

$$Z_2'' = i\omega M_2.$$

As readily follows from the definition of impedance and the fact that the sum of the two currents is the resultant branch current, the combined impedance, Z_2, is determined from the following relation,

$$\frac{1}{Z_2} = \frac{1}{iM_2\omega} + \frac{1}{i\left(M_2'\omega - \dfrac{1}{C_2\omega}\right)}$$

or

$$Z_2 = i\frac{M_2\omega\left(M_2'\omega - \dfrac{1}{C_2\omega}\right)}{M_2\omega + M_2'\omega - \dfrac{1}{C_2\omega}} = i\frac{M_2\omega(M_2'C_2\omega^2 - 1)}{M_2C_2\omega^2 + M_2'C_2\omega^2 - 1}. \quad (27)$$

According to equation (13) if we make C_1 arbitrarily zero for the sake of simplicity, we have

$$Z_1 = iM_1\omega. \quad (28)$$

If we now use the values of (27) and (28) for the conditions (5) and (6) we will have, respectively,

$$f_1 = \frac{1}{2\pi}\sqrt{\frac{1}{C_2(M_2 + M_2')}}, \quad (29)$$

$$f_2 = \frac{1}{2\pi}\sqrt{\frac{M_1 + 4M_2}{C_2(M_1M_2 + M_1M_2' + 4M_2M_2')}}. \quad (30)$$

C. *Computation of Inertance and Capacity.*

By comparison of (23) with (7) we see that the inertance for a straight tube, assumed to move as a whole, is,

$$M = \frac{m}{S^2} = \frac{\rho l S}{S^2} = \rho\frac{l}{S}. \quad (31)$$

A Study of the Regular Combination of Acoustic Elements, with Applications to Recurrent Acoustic Filters, Tapered Acoustic Filters, and Horns

By W. P. MASON

Bell Telephone Laboratories, Incorporated

SYNOPSIS: The use of combinations of tubes to produce interference between sound waves and a suppression of certain frequencies originates with Herschel (1833), and was applied by Quincke to stop tones of definite pitch from reaching the ear. Following the development of electrical filters, G. W. Stewart showed that combinations of tubes and resonators could be devised which would give transmission characteristics at low frequencies similar to electrical filters. The assumptions made by Stewart in the development of his theory are that no wave motion need be considered in the elements, and that the lengths of the elements employed are small compared to the wave-length of sound.

The present paper considers primarily regular combinations of acoustic elements, such as straight tubes, and shows that the equations for recurrent filters, tapered filters and horns can be obtained in this manner. The assumption of no wave motion in the elements, made by Stewart, is removed and also account is taken of the viscosity and heat conduction dissipation. The principal difference between acoustic and electric filters is that the former have an infinite number of bands. The effect of using filters between varying terminal impedances is also determined.

Studying next the combination of filters having the same propagation characteristics but in which the conducting tube areas increase in some regular manner, it is shown that a tapered filter results which has a transforming action in addition to its filtering properties. It is shown that if straight tubes are employed and the distance between successive changes in areas is made small we obtain the horn equations first developed by Webster. The general combination of acoustic elements is then considered, and a proof of several theorems has been given.

STEWART, in a series of papers,[1] has studied the recurrent acoustic filter as an analogue of the electric filter with lumped constants. If due account is taken of the wave motion occurring in the individual elements themselves, it appears that the nearest electrical analogue of the acoustic filter is a combination of electric lines.

In the present paper we study primarily regular combinations of acoustic elements, such as straight tubes, and show that the equations for recurrent filters, tapered filters, and horns can be obtained in this manner. The effect of viscosity and heat conduction dissipation has been taken into account, and a consideration of the effect of varying terminal impedances has been included.

I. EQUATIONS OF PROPAGATION OF A PLANE WAVE IN A UNIFORM TUBE

The propagation of plane waves of sound in uniform tubes has been discussed in a number of places,[2] but generally the results obtained are

---

[1] *Phys. Rev.*, 20, 528 (1922); 23, 520 (1924); 25, 90 (1925).

[2] Rayleigh's "Theory of Sound," Vol. II, p. 318. Lamb's "The Dynamical Theory of Sound," p. 193.

1

only a determination of the propagation constant, that is, a determination of the attenuation and phase change per unit length, or as more often stated, the attenuation and velocity characteristics. If we solve the differential equations in the manner first employed by Heaviside in the solution of the equation of the electric line, we obtain one more parameter, namely, the characteristic impedance of the tube.

The differential equation, given by Rayleigh,[2] for the propagation of plane waves of sound in a tube of uniform cross-section is

$$\left(1 + \frac{R}{S}\sqrt{\frac{\mu}{2\omega\rho}}\right)\frac{\partial^2 \xi}{\partial t^2} + \frac{R}{S}\sqrt{\frac{\mu\omega}{2\rho}}\frac{\partial \xi}{\partial t} = c^2 \frac{\partial^2 \xi}{\partial x^2}, \qquad (1)$$

where ξ denotes the displacement of the fluid at a distance x from one end of the tube,

μ = the coefficient of viscosity of the medium,
ρ = the density of the medium,
R = perimeter and S = cross-sectional area of pipe,
ω = frequency of vibration times 2π,
$C = \sqrt{\frac{P_0\gamma}{\rho}}$ = velocity of sound in medium,
γ = ratio of specific heats of medium.

This equation is valid for tube diameters and frequencies such that

$$\sqrt{\frac{\rho\omega}{2\mu}}\frac{S}{R} > 1$$

and hence can be used for all frequencies of interest in connection with acoustic filters.

Kirchoff[3] extended the theory to take account of the losses due to heat conduction in the medium. His results indicate that in order to take account of this effect, the square root of the coefficient of viscosity should be replaced by a quantity γ', given by

$$\gamma' = \sqrt{\mu} + \left(\sqrt{\gamma} - \frac{1}{\sqrt{\gamma}}\right)\sqrt{v},$$

where v is the coefficient of heat conductivity of the medium. By the kinetic theory of gases v has the value $5/2\ \mu$.

The most useful solution for our present purpose is obtained by writing

$$\xi = e^{i\omega t}(A \cosh \alpha x + B \sinh \alpha x), \qquad (2)$$

[3] Rayleigh, "Theory of Sound," Vol. II, p. 325.

where A and B are constants and α by analogy with an electric line is the propagation constant of the tube. Substituting (2) in (1), we see that (2) is a solution provided

$$\alpha^2 = -\frac{\omega^2}{C^2}\left[\left(1 + \frac{R}{S}\sqrt{\frac{\gamma'^2}{2\omega\rho}}\right) - i\frac{R}{S}\sqrt{\frac{\gamma'^2}{2\omega\rho}}\right]. \tag{3}$$

Now α can be written $\alpha = a + ib$, where a is the attenuation constant and b the phase constant. If we solve for a and b, assuming

$$\frac{R}{S}\sqrt{\frac{\gamma'^2}{2\omega\rho}}$$

is a small quantity, we obtain

$$\alpha = a + ib = \frac{1}{2}\frac{R}{CS}\sqrt{\frac{\gamma'^2\omega}{2\rho}} + i\frac{\omega}{C}\left[1 + \frac{1}{2}\frac{R}{S}\sqrt{\frac{\gamma'^2}{2\omega\rho}}\right]. \tag{4}$$

We are generally interested in the volume velocity $S\xi = V$, so we can rewrite equation (2) as

$$V = i\omega S e^{i\omega t}[A \cosh \alpha x + B \sinh \alpha x]. \tag{5}$$

To determine one constant of equation (5), let x equal zero. Then

$$V_{x=0} = V_1 = i\omega e^{i\omega t} S A$$

or

$$A = \frac{V_1}{i\omega S e^{i\omega t}}. \tag{6}$$

We have the additional relation

$$P - P_0 = -P_0\gamma\frac{\partial \xi}{\partial x} = p, \tag{7}$$

where p denotes the excess pressure. Substituting (2) in (7), and differentiating, we have

$$p = -P_0\gamma e^{i\omega t}(A\alpha \sinh \alpha x + B\alpha \cosh \alpha x).$$

Putting $x = 0$, we have

$$p_{x=0} = p_1 = -P_0\gamma e^{i\omega t}(B\alpha)$$

or

$$B = -\frac{p_1}{\alpha P_0\gamma e^{i\omega t}}. \tag{8}$$

Substituting the value of A and B in (5) and (7), we have

$$V = V_1 \cosh \alpha x - \frac{p_1 i\omega S \sinh \alpha x}{P_0 \gamma \alpha},$$

$$p = p_1 \cosh \alpha x - V_1 \frac{(P_0 \gamma \alpha)}{i\omega S} \sinh \alpha x. \qquad (9)$$

$(P_0\gamma\alpha)/(i\omega)$ is, by analogy with the electric line, the characteristic impedance [4] per square centimeter of the tube. It is the ratio of p_1/ξ_1 for an infinitely long tube. For since $\cosh \alpha x = \frac{1}{2}(e^{\alpha x} + e^{-\alpha x})$ while $\sinh \alpha = \frac{1}{2}(e^{\alpha x} - e^{-\alpha x})$, then when x approaches infinity, and dissipation exists in the tube, $\cosh \alpha x$ approaches $\sinh \alpha x$, and both approach infinity. Hence the ratio of P_1/V_1 equals $P_0\gamma\alpha/i\omega S$. The propagation constant α has the physical significance that $e^{-\alpha x}$ equals the ratio of V to V_1 or p to p_1, when we are dealing with an infinitely long tube, as can be seen by substituting $p_1/V_1 = P_0\gamma\alpha/i\omega S$ in (9) and solving for the above ratios. The real part of α, i.e. a, determines the rate at which the linear or volume velocity, or pressure, decreases with distance, while the imaginary part b determines the phase of pressure or velocity with respect to the initial values, and hence is known as the phase constant and gives the phase rotation per unit length of pipe. Now since the velocity of propagation C' is

$$C' = \frac{\omega}{b},$$

we have by equation (4)

$$C' = C\left[1 - \frac{1}{2}\frac{R}{S}\sqrt{\frac{\gamma'^2}{2\omega\rho}}\right].$$

The attenuation constant and the velocity reduce to the familiar Helmholtz formulæ, for circular sections.[5]

We write (9) as

$$\left.\begin{array}{l} V = V_1 \cosh \alpha x - \dfrac{p_1 S}{Z_L} \sinh \alpha x, \\[6pt] p = p_1 \cosh \alpha x - \dfrac{V_1 Z_L}{S} \sinh \alpha x, \end{array}\right\} \qquad (10)$$

where Z_L represents the specific characteristic impedance $P_0\gamma\alpha/i\omega$.

[4] The analogy between pressure and electromotive force, volume velocity and current, and impedance to ratio of pressure and volume velocity was first pointed out by Webster[9]. Another system in which force and e.m.f., and linear velocity and current are related, is very convenient when we are dealing with combinations of mechanical elements such as masses and elasticities and no account has to be taken of the area. In the first system, the total impedance is Z_L (per sq. cm.) divided by S whereas in the second sysetm it is $Z_L S$. We follow the first system expressing, however, the impedance in terms of the impedance per square centimeter, which is the same on either systems of units.

[5] See Lamb, "Dynamical Theory of Sound," p. 193, or Rayleigh, "Theory of Sound," Vol. II, p. 319.

The value of the specific characteristic impedance $P_0\gamma\alpha/i\omega$ becomes on substituting in the value of α

$$Z_L = \sqrt{P_0\gamma\rho}\left[\left(1 + \frac{1}{2}\frac{R}{S}\sqrt{\frac{\gamma'^2}{2\omega\rho}}\right) - i\frac{1}{2}\frac{R}{S}\sqrt{\frac{\gamma'^2}{2\omega\rho}}\right]. \quad (11)$$

If we assume no dissipation, $\gamma' = 0$ and $Z_L = \sqrt{P_0\gamma\rho}$. In any case at fairly high frequencies Z_L approaches $\sqrt{P_0\gamma\rho}$. For example, for air in a circular tube 1 centimeter in diameter, Z_L departs from its final value $\sqrt{P_0\gamma\rho}$ by less than 5 per cent at 100 cycles. The attenuation constant a increases as the square root of the frequency, while the phase constant b is little affected by the dissipation and at high frequencies approaches the value ω/C.

II. Effect of a Junction or of a Change in Area of the Conducting Tube

Suppose that we have a straight conducting tube, with a sidebranch as shown in Fig. 1. Let the excess pressure of the incoming plane wave

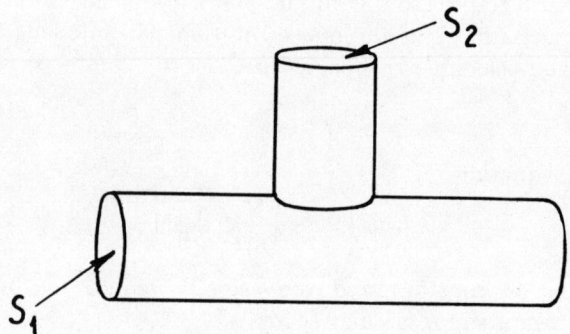

Fig. 1—An acoustic junction

be p_1. The ordinary assumption is that the width of the junction is small compared with a wave-length and hence the pressure is practically constant in the sidebranch, and main branch over the portion in immediate contact with the sidebranch. It states also that the algebraic sum of the volume displacements at a junction of tubes is zero. If S_1 is the area of the main conducting tube, S_2 the area of the branch tube, $\dot{\xi}_1$ the linear velocity of the incoming wave in the conducting tube, $\dot{\xi}_2$ the linear velocity of the outgoing wave from the junction and $\dot{\eta}$ the linear velocity in the branch tube at the junction, we can write the equation

$$\dot{\xi}_1 S_1 = \dot{\xi}_2 S_1 + \dot{\eta} S_2 \quad \text{or} \quad V_1 = V_2 + V'.$$

We have now that $\dot{\eta} = p_1/Z_S$ where Z_S is the impedance per unit area of the sidebranch, or the ratio of the excess pressure to the linear velocity. Substituting this value in the above equation, we have

We have also
$$\left. \begin{array}{c} V_2 = V_1 - \dfrac{p_1 S_2}{Z_S}. \\ p_2 = p_1, \end{array} \right\} \quad (12)$$

where p_2 is the excess pressure in the conducting tube on the outgoing side. The equations are exactly equivalent to Kirchoff's laws, and hence any equation for a combination of acoustic elements will also apply to the combinations of equivalent electric elements.

A slightly better approximation than the above has been obtained by solving completely the case of three pistons placed in the sides of a rectangular box. This corresponds closely to the condition considered here, if we have rectangular tubes, since the waves can be considered plane up to the junction point with little possibility of error. The solution obtained indicates that the main effect of the junction point is to add an end correction to all the tubes entering the junction. For example, we will measure the length of the main conducting tube, between sidebranches, from the center of the sidebranches rather than the edge, as the approximation given first would imply. Also the length of the sidebranch should be measured from the center of the conducting tube, rather than the edge. For other types of junctions, different end corrections will apply to the sidebranch tubes. For example if the width of the junction is large compared to the width of the sidebranch, we should expect Rayleigh's theoretical value of .82 R to apply where R is the radius of the sidebranch tube. Hence the equations for a junction are equivalent to Kirchoff's laws with the additional proviso that end corrections shall be added to tubes entering a junction.

The effect of a change of area of the conducting tube can be obtained with the same assumptions as above. If we have one conducting tube of area S_1, joined to a second of area S_2, we can write

$$\xi_1 S_1 = \xi_2 S_2 \quad \text{or} \quad V_1 = V_2, \quad (13)$$

where $\dot{\xi}_1$ is the linear velocity in the first tube and $\dot{\xi}_2$ in the second tube. We have also that the pressures in the adjoining tubes are equal. Hence

$$p_2 = p_1 \quad \text{and} \quad V_2 = V_1. \quad (14)$$

This equation is of the same order of approximation as the second approximation given above for a junction, since we measure the length from one change of area to the next change.

Equation 14 has been found to hold well as long as the change in area is small while equation 12 holds well as long as the length of a junction is less than half of a wave-length.

III. Recurrent Filters

With the aid of equations (10), (12), and (14), we can obtain the propagation characteristics of any structure employing straight tubes, sidebranches, and changes in area of conducting tubes.

Among the simplest of these are recurrent filters. Fig. 2 shows an

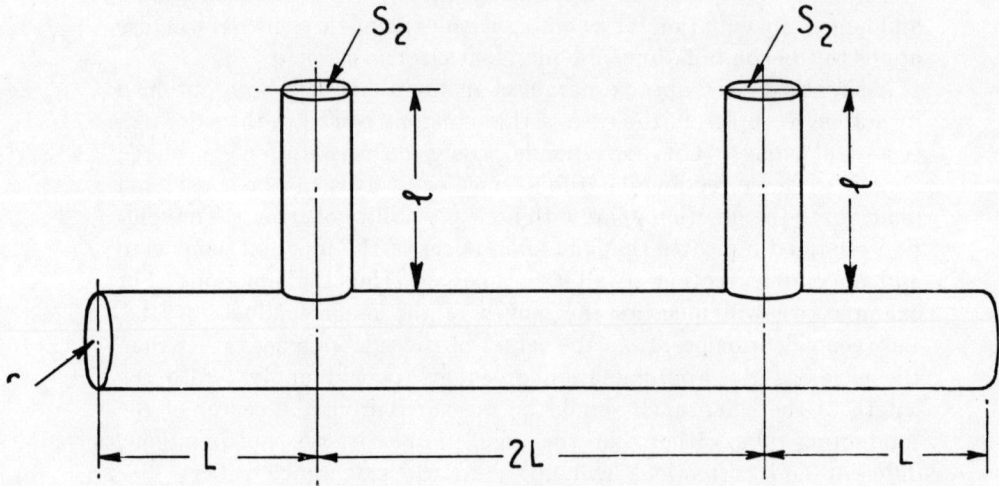

Fig. 2—A typical acoustic filter

example of this type of structure, a main conducting tube, with equally spaced sidebranches. In order to make the structure symmetrical, we let the distance L between one end and the first sidebranch equal one half the distance between two sidebranches. We can then write with regard to the first tube

$$\left. \begin{array}{l} V_2 = V_1 \cosh \alpha_1 L - \dfrac{p_1}{Z_{L_1}} S_1 \sinh \alpha_1 L, \\[2mm] p_2 = p_1 \cosh \alpha_1 L - V_1 \dfrac{Z_{L_1}}{S_1} \sinh \alpha_1 L, \end{array} \right\} \quad (15)$$

where α_1 and Z_{L_1} refer to the conducting tube. For the junction, we have by (12)

$$\left. \begin{array}{l} V_3 = V_2 - \dfrac{p_2}{Z_S} S_2, \\[2mm] p_3 = p_2. \end{array} \right\} \quad (16)$$

Combining with (15), we have

$$\begin{aligned}V_3 &= V_1\left(\cosh \alpha_1 L + \frac{Z_{L_1}S_2}{Z_S S_1}\sinh \alpha_1 L\right) \\ &\quad - p_1 S_1\left[\frac{\sinh \alpha_1 L}{Z_{L_1}} + \frac{S_2 \cosh \alpha_1 L}{Z_S S_1}\right], \\ p_3 &= p_1 \cosh \alpha_1 L - \frac{V_1 Z_{L_1}}{S_1}\sinh \alpha_1 L.\end{aligned} \qquad (17)$$

The pressures and volume velocities p_4 and V_4 at one half the distance between the first and second sidebranches are again

$$\begin{aligned}V_4 &= V_3 \cosh \alpha_1 L - \frac{p_3 S_1}{Z_{L_1}}\sinh \alpha_1 L, \\ p_4 &= p_3 \cosh \alpha_1 L - V_3 \frac{Z_{L_1}}{S_1}\sinh \alpha_1 L.\end{aligned} \qquad (18)$$

Combining with (17), we obtain

$$\begin{aligned}V_4 &= V_1\left(\cosh 2\alpha_1 L + \frac{Z_{L_1}S_2}{2Z_S S_1}\sinh 2\alpha_1 L\right) \\ &\quad - \frac{p_1 S_1}{Z_{L_1}}\left(\sinh 2\alpha_1 L + \frac{Z_{L_1}S_2}{Z_S S_1}\cosh^2 \alpha_1 L\right)\cdot \\ p_4 &= p_1\left(\cosh 2\alpha_1 L + \frac{Z_{L_1}S_2}{2Z_S S_1}\sinh 2\alpha_1 L\right) \\ &\quad - \frac{V_1 Z_{L_1}}{S_1}\left(\sinh 2\alpha_1 L + \frac{Z_{L_1}S_2}{Z_S S_1}\sinh^2 \alpha_1 L\right)\cdot\end{aligned} \qquad (19)$$

These equations apply to the first section of the filter. By comparison with equation (10) we see that we can write equation (19) as

$$\begin{aligned}V_4 &= V_1 \cosh \Gamma - \frac{p_1 S_1}{Z_0}\sinh \Gamma, \\ p_4 &= p_1 \cosh \Gamma - V_1 \frac{Z_0}{S_1}\sinh \Gamma,\end{aligned} \qquad (20)$$

where

$$\cosh \Gamma = \left(\cosh 2\alpha_1 L + \frac{Z_{L_1}S_2}{2Z_S S_1}\sinh 2\alpha_1 L\right),$$

$$Z_0 = Z_{L_1}\sqrt{\frac{1 + \frac{Z_{L_1}S_2}{2Z_S S_1}\tanh \alpha_1 L}{1 + \frac{Z_{L_1}S_2}{2Z_S S_1}\coth \alpha_1 L}} \qquad (21)$$

and

$$\sinh \Gamma = \sinh 2\alpha_1 L \sqrt{\left(1 + \frac{Z_{L_1}S_2}{2Z_S S_1} \tanh \alpha_1 L\right)\left(1 + \frac{Z_{L_1}S_2}{2Z_S S_1} \coth \alpha_1 L\right)}.$$

Z_0 and Γ are sometimes called the equivalent line parameters. If we have n sections of the type discussed above, we can write n equations of the kind given by (20). If we eliminate all the terms except for the first and last sections, it can be shown that

$$\left.\begin{aligned} V_n &= V_1 \cosh n\Gamma - \frac{p_1 S_1}{Z_0} \sinh n\Gamma, \\ p_n &= p_1 \cosh n\Gamma - \frac{V_1 Z_0}{S_1} \sinh n\Gamma. \end{aligned}\right\} \quad (22)$$

We see then that Γ represents the propagation constant of one section and Z_0 its specific characteristic impedance. They have the physical interpretation, that Z_0 represents the specific impedance looking into an infinite sequence of these sections, while Γ represents the ratio of excess pressure or volume velocity between one section and the next, when we are dealing with an infinite number of sections, or with a finite number, terminated in the characteristic impedance of the filter.

It is customary in electric filter design to determine the characteristics of a dissipationless filter, and to regard dissipation as causing a slight change in the filter characteristic, which usually occurs most prominently in the pass bands. If we neglect dissipation, equation (21) becomes

$$\cosh \Gamma = \left[\cos\left(\frac{2\omega L}{C}\right) + \frac{i\sqrt{P_0\gamma\rho}\,S_2}{2Z_S S_1} \sin\left(\frac{2\omega L}{C}\right)\right],$$

$$Z_0 = \sqrt{P_0\gamma\rho}\,\sqrt{\frac{1 + \dfrac{i\sqrt{P_0\gamma\rho}\,S_2}{2Z_S S_1} \tan\left(\dfrac{\omega L}{C}\right)}{1 - \dfrac{i\sqrt{P_0\gamma\rho}\,S_2}{2Z_S S_1} \cot\left(\dfrac{\omega L}{C}\right)}}. \quad (23)$$

The propagation constant Γ is in general a complex number $A + iB$. The real part represents a diminution of the volume velocity or the pressure, while the imaginary part represents a phase change, as can be seen from the fact that the ratio of pressure or volume velocity is

$$\frac{p_2}{p_1} = e^{-\Gamma} = e^{-(A+iB)} = e^{-A}(\cos B - i \sin B).$$

Now $\cosh \Gamma = \cosh (A + iB) = \cosh A \cos B + i \sinh A \sin B$. Hence we see from equation (23), if Z_S is an imaginary quantity, the expression for $\cosh \Gamma$ is always real, and hence either $\sinh A$ or $\sin B$

is always zero. Hence either the attenuation constant A is zero, or the phase shift is zero, π radians or some multiple of π radians. Now since cosh A can never be less than 1 while cos B must lie between $+1$ and -1, then when the expression for cosh Γ is between -1 and $+1$, the attenuation constant A is zero and cos B equals the expression in (23). When the value of cosh Γ is outside the limits ± 1, the phase shift is 0, π, or some multiple and the attenuation constant A is given by the expression in (23).

The specific characteristic impedance Z_0, given in (23), can be shown to be a real quantity within the transmitted band and an imaginary quantity outside the transmitted band.

The type of filter obtained with the structure shown in Fig. 2 depends on the sidebranch impedance Z_S. As long as Z_S is of such a value as to make the expression for cosh Γ greater in magnitude than 1, an attenuation band occurs, while if cosh Γ is less than 1, a pass band occurs. The cut-off frequencies of the band occur when cosh $\Gamma = \pm 1$. From equation (23) the cut-off frequencies occur when

$$Z_S = \frac{i\sqrt{P_0\gamma\rho}S_2}{2S_1}\cot\left(\frac{\omega L}{C}\right) \quad \text{or} \quad Z_S = -\frac{i\sqrt{P_0\gamma\rho}S_2}{2S_1}\tan\left(\frac{\omega L}{C}\right). \quad (24)$$

A. Low Pass Filter

The model shown in Fig. 2 can be used to obtain the different types of recurrent filters possible by acoustic means. One of the simplest types of filters in the electrical case is the low pass filter. No exact analogue of this filter exists in the acoustic case, as every acoustic filter has more than one band, but a filter which passes low frequencies and attenuates high frequencies can be designed.

Suppose that the sidebranch used is a straight tube closed at one end. Then by equation (10), the impedance Z_S, when the tube is terminated in an infinite impedance, is

$$Z_S = Z_{L_2}\coth\alpha_2 l,$$

where Z_{L_2} and α_2 are respectively the specific characteristic impedance and propagation constant of the sidebranch, and l its length measured to the center of the conducting tube. Substituting this in the expression for cosh Γ and Z_0, we have

$$\left. \begin{array}{l} \cosh \Gamma = \left(\cosh 2\alpha_1 L + \dfrac{Z_{L_1}S_2 \sinh 2\alpha_1 L}{2Z_{L_2}S_1 \coth \alpha_2 l}\right), \\[2ex] Z_0 = Z_{L_1}\sqrt{\dfrac{1 + \dfrac{Z_{L_1}S_2 \tanh \alpha_1 L}{2Z_{L_2}S_1 \coth \alpha_2 l}}{1 + \dfrac{Z_{L_1}S_2 \coth \alpha_1 L}{2Z_{L_2}S_1 \coth \alpha_2 l}}}. \end{array} \right\} \quad (25)$$

If we assume no dissipation, and substitute the values of α_1 and Z_L given in section (I), we have

$$\cosh \Gamma = \left[\cos\left(\frac{2\omega}{C}L\right) - \frac{S_2 \sin\left(\frac{2\omega}{C}L\right)}{2S_1 \cot\left(\frac{\omega}{C}l\right)} \right], \tag{26}$$

$$Z_0 = \sqrt{P_0 \gamma \rho} \sqrt{\frac{1 - \frac{S_2}{2S_1}\left(\frac{\tan\frac{\omega}{C}L}{\cot\frac{\omega}{C}l}\right)}{1 + \frac{S_2}{2S_1}\left(\frac{\cot\frac{\omega}{C}L}{\cot\frac{\omega}{C}l}\right)}}. \tag{27}$$

An example of the type of filter obtained by acoustic means, is given when we let $l = 3L$. Fig. 3 gives a plot of the value of Γ for several ratios of S_2/S_1. Fig. 4 shows the corresponding values of the specific characteristic impedance Z_0.

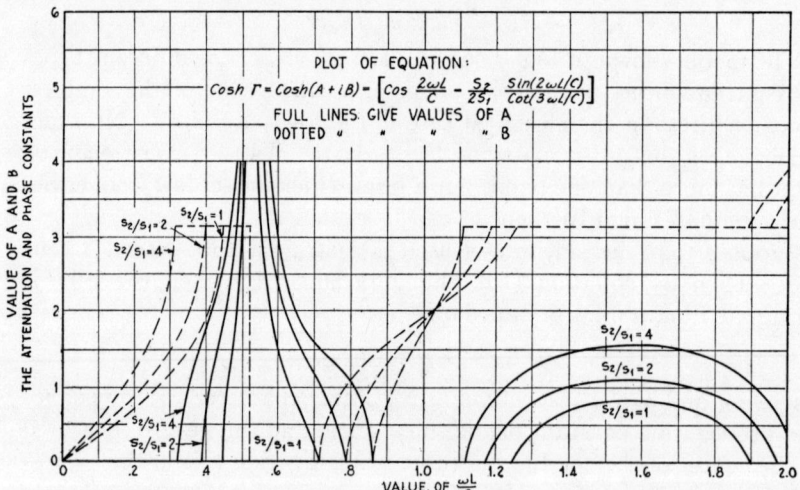

Fig. 3—Propagation constants for a low pass type of filter

A knowledge of Γ will determine the ratio of pressures or volume velocities, if we have an infinite sequence of sections, or if we terminate a finite sequence in the impedance Z_0. If however the terminating impedance is not the characteristic impedance, $e^{-\Gamma}$ no longer represents the ratios of pressures between adjacent sections.

What is generally desired is a knowledge of the effect produced by inserting the filter in a given acoustic system. With the aid of Thévenin's theorem, which is proved for an acoustic system in Appendix I, and equations (20) and (21), this effect can be obtained. Thévenin's

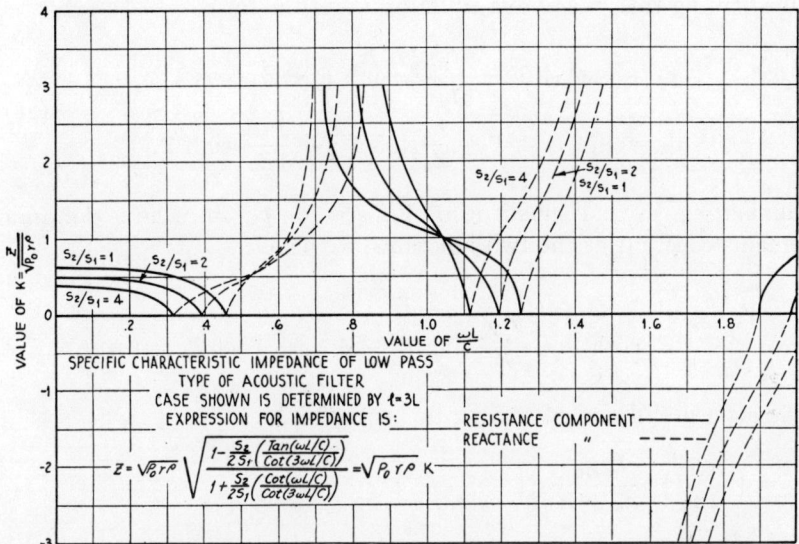

Fig. 4—Specific characteristic impedances for a low pass type of filter

theorem states: If a source of simple harmonic pressure p_0 and of internal impedance Z_T, per square centimeter, is connected to an acoustic system, and if the specific impedance Z_R terminates the system, the volume velocity at the termination of the system will be $p_0'/[(Z_T'/S_1) + (Z_R/S_n)]$, where p_0' is the pressure at the terminating end when this is closed through an infinite impedance, and Z_T' is the impedance per sq. cm. looking back into the acoustic system when this terminated in the impedance Z_T. S_1, and S_n are the areas at the input and output junctions, respectively.

Making use of Thévenin's theorem, the effect of inserting a filter in a given system is the same as the effect obtained by inserting this filter between a source of pressure p_0, with an internal impedance of Z_a/S_1 and a terminating impedance Z_b/S_n, where Z_a/S_1 and Z_b/S_n are respectively the total impedances looking toward the source, and away from the source at the insertion junction of the acoustic system. We have from equation (20)

$$V_2 = V_1 \cosh \Gamma - \frac{p_1 S_1}{Z_0} \sinh \Gamma.$$

$$p_2 = p_1 \cosh \Gamma - \frac{V_1 Z_0}{S_1} \sinh \Gamma,$$

Making use of the above, we can write

$$p_0 = p_1 + \frac{V_1 Z_a}{S_1}.$$

Substituting this, the above equation takes the form

$$\left.\begin{aligned} p_2 &= p_0 \cosh \Gamma - \frac{V_1}{S_1}[Z_0 \sinh \Gamma + Z_a \cosh \Gamma], \\ V_2 &= V_1\left(\cosh \Gamma + \frac{Z_a}{Z_0} \sinh \Gamma\right) - \frac{p_0 S_1}{Z_0} \sinh \Gamma. \end{aligned}\right\} \quad (28)$$

Eliminating V_1 and substituting $V_2 Z_b/S_1$ for p_2, since here the area remains constant at the two junctions, we have

$$V_2 = \frac{p_0 S_1}{\left[Z_b \cosh \Gamma + \frac{Z_a Z_b}{Z_0} \sinh \Gamma + Z_a \cosh \Gamma + Z_0 \sinh \Gamma\right]}.$$

The most useful way of writing this equation is

$$V_2 = \left(\frac{p_0 S_1}{2 Z_b}\right)\left(\frac{2 Z_0}{Z_0 + Z_a}\right)\left(\frac{2 Z_b}{Z_0 + Z_b}\right)(e^{-\Gamma})$$
$$\times \left[\frac{1}{1 - e^{-2\Gamma}\left(\frac{Z_0 - Z_a}{Z_0 + Z_a}\right)\left(\frac{Z_0 - Z_b}{Z_0 + Z_b}\right)}\right]. \quad (29)$$

The volume velocity in the termination of the acoustic system, if the filter were not inserted, is obviously $p_0/[(Z_a/S_1) + (Z_b/S_1)]$. Hence the effect of inserting the filter at any junction is to change the volume velocity of the system by the factor

$$\left(\frac{Z_a + Z_b}{2 Z_b}\right)\left(\frac{2 Z_0}{Z_0 + Z_a}\right)\left(\frac{2 Z_b}{Z_0 + Z_b}\right)(e^{-\Gamma})$$
$$\times \left[\frac{1}{1 - e^{-2\Gamma}\left(\frac{Z_0 - Z_a}{Z_0 + Z_a}\right)\left(\frac{Z_0 - Z_b}{Z_0 + Z_b}\right)}\right]. \quad (30)$$

A physical interpretation of equation (30) can be obtained in terms of the transmission and reflection factors first introduced by Heaviside.[6] Heaviside showed that at a junction, a reflection of a wave takes place if the impedances looking towards the source and away from the source are not equal. He showed that the current reflected on striking a junction, will be the unmodified current in the line multiplied by the

[6] Heaviside "Electromagnetic Theory" Vol. II, page 79.

factor, $(Z_I - Z_T)/(Z_I + Z_T)$, while the current transmitted to the terminating side of the junction will be the unmodified current in the line multiplied by the factor $2Z_T/(Z_I + Z_T)$ where Z_I and Z_T are respectively the impedances looking towards and away from the source at the junction. We see then that the second and third factors are transmission factors, determining respectively the transmission from the input impedance Z_a to the inserted structure, and from the inserted structure to the output impedance Z_b. The first factor is the inverse of the transmission factor determining the transmission from the impedance Z_a to the impedance Z_b. The fourth factor is the transfer factor and gives the reduction in volume velocity due to attenuation. The fifth factor has been called the interaction factor, and it gives the change in volume velocity in the termination due to repeated reflections of the volume velocity within the structure. All of these factors reduce to 1 except the transfer factor when $Z_a = Z_b = Z_0$. It will be noted that all factors except the transfer factor cancel out if $Z_a = Z_0$, or $Z_b = Z_0$.

The effect on the pressure due to inserting a filter can be shown to be given also by equation (30).

If the terminating impedances are resistances about equal to an average of the resistance value of Z_0, the effect of these is generally to introduce some loss in the pass band, when the characteristic impedance differs materially from the terminating impedances due to a reflection of the sound wave at the junction points. Since the characteristic impedance of a non-dissipative filter goes either to zero or infinity at the cut-off frequency, the effect of the reflection loss is generally to narrow the pass bands of the filter.

The effect of dissipation, when we take account of the viscosity effects by equations (20) or (21), is two-fold. It changes slightly the position of the band in the frequency range, due to a small change in the velocity of propagation. This is generally negligible. The other effect is to introduce attenuation in the pass band, due to absorption and dissipation of the sound wave.

B. *High Pass Filter*

An analogous type of high pass filter, which will attenuate the low frequencies and pass the high frequencies, can be made from the structure shown in Fig. 2 by using side tubes which are open on the outer end. The termination at the end of an open tube has been shown by Rayleigh [7] to be a mass with some resistance due to radiation. We could substitute this relation in equation (10) to determine

[7] Rayleigh, "Theory of Sound," Vol. II, p. 106.

the impedance Z_S looking into the sidebranch. Another approximation used with organ pipes is to consider the tube extended by a length .57 times the radius of the tube, and to consider this extended tube terminated in a zero impedance.

The impedance Z_S for this case is from (10)

$$\frac{p_1 S_2}{V_1} = Z_S = Z_{L_2} \tanh \alpha_2 l',$$

where l' is the corrected length of the pipe. Substituting this value in equation (21), we have

$$\cosh \Gamma = \left[\cosh 2\alpha_1 L + \frac{Z_{L_1} S_2 \sinh 2\alpha_1 L}{2 Z_{L_2} S_1 \tanh \alpha_2 l'} \right],$$

$$Z_0 = Z_{L_1} \sqrt{\frac{1 + \dfrac{Z_{L_1} S_2 \tanh \alpha_1 L}{2 Z_{L_2} S_1 \tanh \alpha_2 l'}}{1 + \dfrac{Z_{L_1} S_2 \coth \alpha_1 L}{2 Z_{L_2} S_1 \tanh \alpha_2 l'}}}. \quad (31)$$

For no dissipation these equations become

$$\cosh \Gamma = \left[\cos\left(\frac{2\omega L}{C}\right) + \frac{S_2}{2 S_1} \left[\frac{\sin\left(\dfrac{2\omega L}{C}\right)}{\tan\left(\dfrac{\omega l'}{C}\right)} \right] \right],$$

$$Z_0 = \sqrt{P_0 \gamma \rho} \sqrt{\frac{1 + \dfrac{S_2 \tan\left(\dfrac{\omega L}{C}\right)}{2 S_1 \tan\left(\dfrac{\omega l'}{C}\right)}}{1 - \dfrac{S_2 \cot\left(\dfrac{\omega L}{C}\right)}{2 S_1 \tan\left(\dfrac{\omega l'}{C}\right)}}}.$$

Fig. 5 shows a plot of Γ for several ratios of S_2/S_1, when $l' = 3L$.

C. Band Pass Type of Filter

The high pass type of filter discussed above can also be considered as a band pass type of filter, in that an attenuation occurs at zero frequency, then a pass band, and a second attenuation band. A different arrangement of the pass bands can be obtained from the structure shown in Fig. 2, by inserting two sidebranches at one junction point, one of which is open at the outside end and the other closed.

An example of the type of characteristic obtained, is given by the special case where the lengths of both tubes are the same and equal to

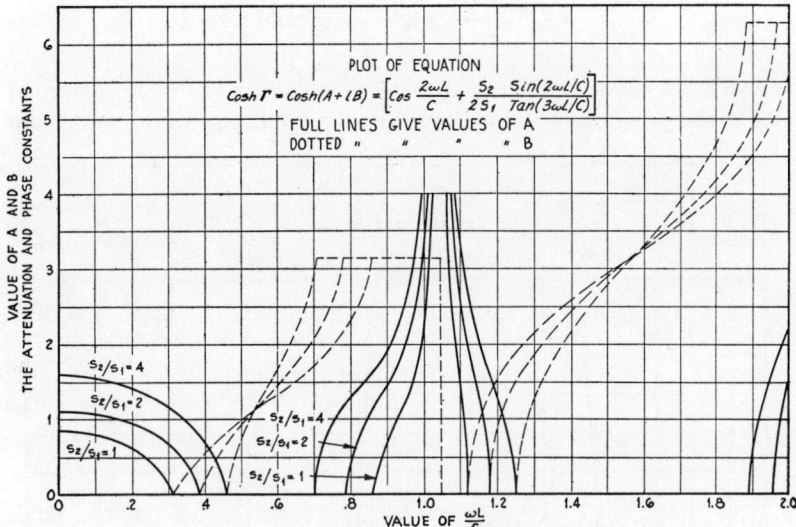

Fig. 5—Propagation constants for a high pass type of filter

$3L$. If S_2 is the area of the open tube and S_3 that of the closed tube, then neglecting dissipation, we find

$$\cosh \Gamma = \cos \frac{2\omega L}{C} + \frac{1}{2S_1}\left[\frac{S_2}{\tan\frac{3\omega L}{C}} - \frac{S_3}{\cot\frac{3\omega L}{C}}\right]\sin\frac{2\omega L}{C}.$$

A plot of A, the attenuation constant, for several values of S_2/S_1 and S_3/S_1 is given in Fig. 6.

D. Other Types of Sidebranches

We have so far considered only the characteristics obtained where we employ straight tubes. A number of cases can be solved in which the elements employed are not straight tubes although we cannot take account of the viscosity dissipation in these cases. As an example, the characteristics of a filter will be worked out, which employs a straight tube for the conducting tube and conical tubes closed on the end for the sidebranches. We can make use of equation (21) to determine Γ and Z_0, if we insert the proper value of Z_S for the conical tube.

It is evident that for a conical tube, the proper type of wave is a spherical wave, in place of the plane wave employed for a straight tube. For this case we can write [8] for a simple harmonic wave

$$\frac{\partial^2(r\varphi)}{\partial t^2} = C^2 \frac{\partial^2(r\varphi)}{\partial r^2} \; ; \; \dot{\eta} = -\frac{\partial \varphi}{\partial r} \text{ and } \frac{p}{\rho_0} = \dot{\varphi},$$

where φ is the velocity potential, $\dot{\eta}$ the linear velocity for the spherical wave, p the pressure, ρ the average density of the medium, and r the

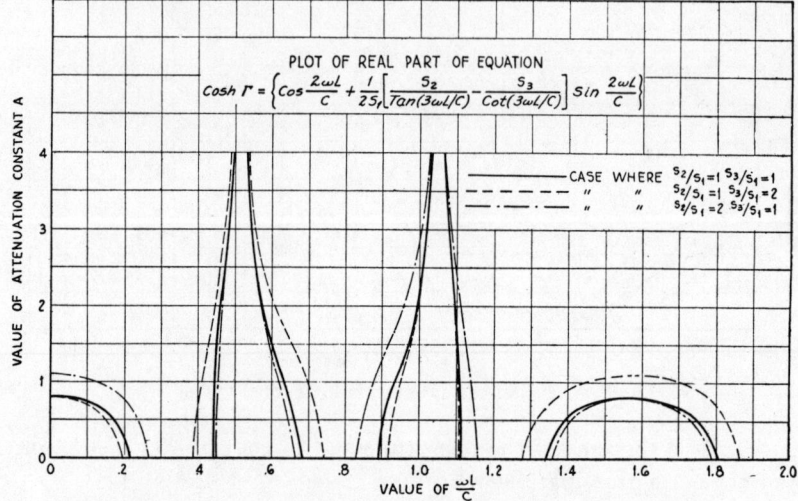

Fig. 6—Attenuation constants for a band pass type of filter

distance from the apex of the cone. The solution for this case is

$$r\varphi = A \sin \frac{\omega}{C} r + B \cos \frac{\omega}{C} r.$$

Hence we can determine $\dot{\eta}$ and p as

$$\dot{\eta} = A \left[\frac{\sin \frac{\omega}{C} r}{r^2} - \frac{\frac{\omega}{C} \cos \frac{\omega}{C} r}{r} \right] + B \left[\frac{\cos \frac{\omega}{C} r}{r^2} + \frac{\frac{\omega}{C} \sin \frac{\omega}{C} r}{r} \right]$$

and

$$p = i\omega\rho \left[\frac{A \sin \frac{\omega}{C} r + B \cos \frac{\omega}{C} r}{r} \right].$$

If now we set $\dot{\eta} = 0$ when $r = x_2$ and determine the ratio of $p/\dot{\eta}$ at

[8] Lamb, "Dynamical Theory of Sound," p. 206. Rayleigh, "Theory of Sound," Vol. II, p. 114.

$r = x_1$, we find

$$Z_S = \frac{p}{\dot{\eta}}$$

$$= -i\sqrt{P_0\gamma\rho} \left[\frac{\cos\frac{\omega}{C}(x_2-x_1) - \frac{\sin\frac{\omega}{C}(x_2-x_1)}{\frac{\omega}{C}x_2}}{\cos\frac{\omega}{C}(x_2-x_1)\left[\frac{1}{\frac{\omega}{C}x_2} - \frac{1}{\frac{\omega}{C}x_1}\right] + \sin\frac{\omega}{C}(x_2-x_1)\left[1 + \frac{1}{\frac{\omega^2}{C^2}x_1x_2}\right]} \right]. \quad (32)$$

If we substitute this value of Z_S in equation (21), we can readily determine the value of Γ and Z_0. Fig. 7 shows a plot of A and B for this case assuming $(x_2 - x_1) = L$.

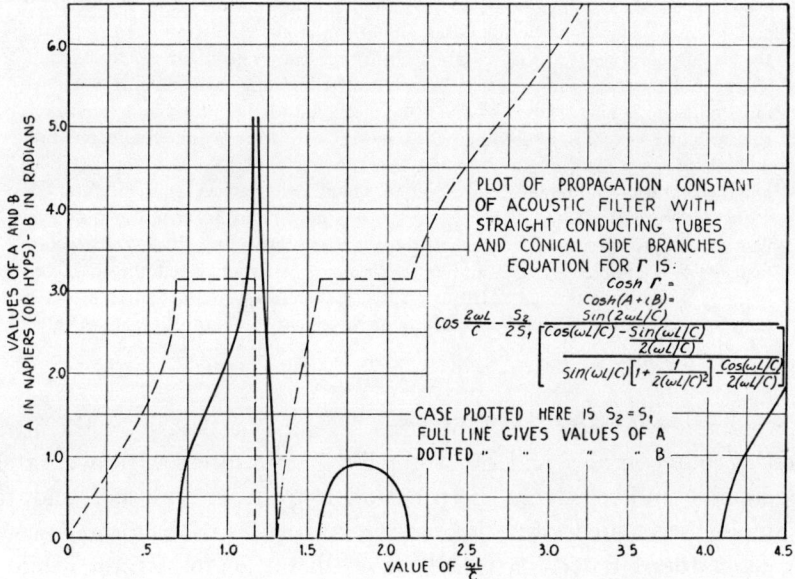

Fig. 7—Propagation constant of a low pass type of filter.

Copyright © 1932 by the Acoustical Society of America
Reprinted from *J. Acoust. Soc. Amer.*, 155–165 (1932)

THE THEORY OF ACOUSTIC FILTRATION IN SOLID RODS

By R. B. LINDSAY AND F. E. WHITE
Brown University

ABSTRACT

The transmission of longitudinal waves through an infinitely long solid rod loaded at equal intervals with equal heavy masses is treated by a method analogous to the so-called "branch transmission" theory of acoustic filtration in air. The masses are assumed to be so concentrated that they move as a whole and longitudinal motions alone are considered. Transmission bands are found for frequencies satisfying the inequality relation
$$-1 \leq \cos 2kl - \omega m/2\rho_0 cS \cdot \sin 2kl \leq +1, \text{ where } k = \omega/c = 2\pi\nu/c$$
ν being the frequency and c the velocity of sound in the rod, m the mass of the concentrated load, S the cross-sectional area of the rod, ρ_0 the density of the rod material and $2l$ the distance between consecutive loads. The loaded rod is thus found to act like a low-pass filter with the cut-off frequency given by the first root of the transcendental equation $\omega m/2\rho_0 cS = \cot kl$. As the frequency increases there are alternate transmission and attenuation bands. Comparison is made between the theory and the experimental results of H. F. Olson (1925) and rather good agreement is found. The case of the transverse waves in a stretched string loaded at equal intervals with mass particles is also treated by the same method and the filter characteristics of this structure very simply obtained in agreement with the previous results of Crandall and others using different methods. Investigation is made of the propagation of longitudinal waves through a solid rod to which are attached at equal intervals strings loaded with masses at their free ends. It is shown that in general such a structure acts like a low-pass filter but that in the limiting case where the loaded ends are fastened to a completely rigid, motionless support the structure has the characteristics of a high-pass filter. Unfortunately the transmission and attenuation bands are of such a nature as to render such an arrangement of doubtful practical value.

INTRODUCTION

Some ten years ago, G. W. Stewart developed a theory for the filtration of sound in air,[1] and constructed acoustic filters with accurately predictable and relatively sharp cut-off frequencies. It was found, for example, that a tube with side branch attachments in the form of acoustic resonators regularly spaced behaves like a low-pass filter, while if the branch attachments are simple orifices, the filter is of the high-pass type. Stewart also envisaged the possibility of constructing solid filter structures. He and his students built and experimentally tested several of these.[2] Thus he found that a metal rod loaded at equal intervals with heavy masses (attached as collars to the rod) acts like a low-

---

[1] G. W. Stewart, Phys. Rev. **20**, 528 (1922); also Stewart and Lindsay, *Acoustics*, p. 159 ff.
[2] See, for example, *Acoustics*, p. 186 with the references there cited. Also V. C. Hall, Phys. Rev. **23**, 116A (1924).

pass filter to the passage of longitudinal waves through the rod. With the help of the lumped impedance theory which he had developed for filtration in air, he was able to account in a semi-empirical fashion for the main experimental results. But a more exact theoretical treatment promised to be rather difficult and was not undertaken. It is the purpose of the present paper to sketch a theoretical discussion of filtration in solids which may be reasonably compared with the available experimental data, and to suggest further possibilities for new structures.

Theory of the Propagation of Longitudinal Waves through a Loaded Rod

We shall consider the propagation of dilatational waves through a solid rod loaded at equal intervals with heavy masses as indicated schematically in Fig. 1. The material of which the rod is made has mean

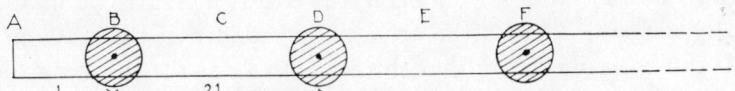

Fig. 1. *Schematic diagram of a loaded rod.*

equilibrium density ρ_0 and each load is of mass m. It will be assumed that the structure extends to infinity on the right. The method to be used envisages the effect of each load as that of a branch line on the propagation through the rod. This is effectively the method first proposed by Mason[3] for the treatment of the problem of filtration in air and presented in a somewhat modified form by Stewart and Lindsay[4] under the name "branch transmission theory." The longitudinal displacement in the rod due to a compressional disturbance ξ, satisfies the wave equation, viz.,

$$\partial^2 \xi / \partial x^2 = \ddot{\xi}/c^2 \qquad (1)$$

where the velocity of propagation, $c = (Y/\rho_0)^{1/2}$, and Y is Young's modulus. The solution of (1) for the special case of a harmonic wave if the x-axis is taken along the rod may be written in the form

$$\xi = \xi_0 e^{i(\omega t - kx)} + \xi_1 e^{i(\omega t + kx)} \qquad (2)$$

where $\omega = 2\pi\nu$, with ν the frequency of the wave, and $k = \omega/c = 2\pi/\lambda$, if λ = wave-length. The two parts of the solution correspond to waves progressing in the positive and negative directions, respectively. ξ_0 and

[3] W. P. Mason, Bell System Tech. Jour. **6**, 258 (1927).
[4] *Acoustics*, p. 334.

ξ_1 are arbitrary amplitude constants to be determined in terms of the boundary conditions. The analogy with acoustic waves in a fluid is more emphatically brought out if we use the volume displacement $X = S\xi$ and the volume current $\dot{X} = S\dot{\xi} = i\omega S\xi$, where S is the area of cross section of the rod. Both X and $\dot{X}$, of course, satisfy the Eq. (1). Corresponding to the *excess pressure* in the case of a fluid and the electromotive force in the case of electric wave transmission we now have the *excess stress* or linear tension per unit area due to the passage of the dilatational wave through the rod. We shall denote this by T. From Hooke's law we have at once

$$\frac{T}{\partial \xi/\partial x} = Y = \rho_0 c^2. \tag{3}$$

It may be remarked that while acoustic impedance for waves in a fluid is defined by the ratio $p/\dot{X}$, p being the excess pressure, we may consistently define $-T/\dot{X}$ as the acoustic impedance of the wave in the rod. It can be easily shown that this leads to $\rho_0 c$ as the specific acoustic resistance for a plane progressive wave.[5]

From Eqs. (2) and (3) we obtain

$$\dot{X} = i\omega S [\xi_0 e^{i(\omega t - kx)} + \xi_1 e^{i(\omega t + kx)}]$$
$$T = i\omega \rho_0 c [-\xi_0 e^{i(\omega t - kx)} + \xi_1 e^{i(\omega t + kx)}]. \tag{4}$$

If now we let the end of the rod denoted by A correspond to $x=0$ and suppose the excess stress and volume current there to be given by $T_1 e^{i\omega t}$ and $\dot{X}_1 e^{i\omega t}$, respectively, we can evaluate ξ_0 and ξ_1 in terms of T_1 and $\dot{X}_1$. Substitution into (4) with subsequent reduction then yields

$$\dot{X} = [\dot{X}_1 \cos kx + iT_1/Z \cdot \sin kx] e^{i\omega t}$$
$$T = [T_1 \cos kx + i\dot{X}_1 Z \cdot \sin kx] e^{i\omega t} \tag{5}$$

wherein we have replaced $\rho_0 c/S$ by Z, the acoustic resistance for a plane progressive wave with wave front S.

The problem now is to compare $\dot{X}_{n+1}$ and $\dot{X}_n$, where $\dot{X}_n$ is the volume current half way between the $(n-1)$st and n-th loads. This will express the filtration characteristics of the structure. It will be assumed in our analysis that each load moves as a whole.

We have therefore to write the boundary conditions at each branch, expressing the continuity of displacement and stress. For this purpose we shall denote the displacement and stress in the rod immediately to

[5] See *Acoustics*, p. 106.

the left of the load at B by the subscript 12, while the subscript characteristic of these quantities in the rod immediately to the right of B will be 21. At the mid-branch points A, C, E, $\cdots$, the corresponding quantities will be characterized by the subscripts 1, 2, 3 $\cdots$. If now we denote the displacement of the load by ξ_L, the boundary conditions at B may be written

$$\xi_{12} = \xi_{21} = \xi_L \qquad (6)$$

$$S(T_{21} - T_{12}) = m\ddot{\xi}_L = i\omega m \dot{X}_L/S \qquad (7)$$

where $\dot{X}_L = S\dot{\xi}_L$, to unify the notation. From Eqs. (5) if we leave off the time factor $e^{i\omega t}$ we have

$$\dot{X}_{12} = \dot{X}_1 \cos kl + iT_1/Z \cdot \sin kl$$
$$T_{12} = T_1 \cos kl + i\dot{X}_1 Z \cdot \sin kl. \qquad (8)$$

Likewise

$$\dot{X}_2 = \dot{X}_{21} \cos kl + iT_{21}/Z \cdot \sin kl$$
$$T_2 = T_{21} \cos kl + i\dot{X}_{21} Z \cdot \sin kl. \qquad (9)$$

Our task now is to apply the boundary conditions to Eqs. (8) and (9) so as to express $\dot{X}_2$ and T_2 in terms of $\dot{X}_1$ and T_1. On performing the necessary substitutions and reductions, we are finally led to

$$\dot{X}_2 = \dot{X}_1[\cos 2kl - \omega m/2ZS^2 \cdot \sin 2kl]$$
$$+ iT_1/Z \cdot [\sin 2kl - \omega m/ZS^2 \cdot \sin^2 kl] \qquad (10)$$

$$T_2 = T_1[\cos 2kl - \omega m/2ZS^2 \cdot \sin 2kl]$$
$$+ i\dot{X}_1 Z[\sin 2kl + \omega m/ZS^2 \cdot \cos^2 kl]. \qquad (11)$$

It has already been assumed that the structure is infinite toward the right. It therefore follows that the impedance is the same to the right of every section. Thus we write

$$- T_1/\dot{X}_1 = - T_2/\dot{X}_2 = \cdots = Z_0 \qquad (12)$$

where the constant impedance has been represented by Z_0, usually referred to as the *characteristic* impedance. Substituting from (10) and (11) into (12) we solve for Z_0 and obtain

$$Z_0 = Z(1 + \omega m/2ZS^2 \cdot \cot kl)^{1/2}(1 - \omega m/2ZS^2 \cdot \tan kl)^{-1/2}. \qquad (13)$$

It will now be convenient to let

$$\cos W = \cos 2kl - \omega m/2ZS^2 \cdot \sin 2kl, \qquad (14)$$

whence it may be readily shown that

$$\sin W = Z/Z_0 \cdot (\sin 2kl + \omega m/ZS^2 \cdot \cos^2 kl)$$
$$= Z_0/Z \cdot (\sin 2kl - \omega m/ZS^2 \cdot \sin^2 kl). \tag{15}$$

We may then write (10) and (11) in the form

$$\dot{X}_2 = \dot{X}_1 \cos W + iT_1/Z_0 \cdot \sin W \tag{16}$$
$$T_2 = T_1 \cos W + i\dot{X}_1 Z_0 \cdot \sin W. \tag{17}$$

These may at once be generalized to apply to any two successive sections such as the n-th and $(n+1)$st. Thus to concentrate on the volume current in particular

$$2\dot{X}_{n+1} = \dot{X}_n(e^{iW} + e^{-iW}) + T_n/Z_0(e^{iW} - e^{-iW}) \tag{18}$$

which since $Z_0 = -T_n/\dot{X}_n$, immediately reduces to the simple form

$$\dot{X}_{n+1} = \dot{X}_n e^{-iW}. \tag{19}$$

There is a similar relation between T_{n+1} and T_n. Inspection of (19) now shows that when W is *real*, the difference between $\dot{X}_{n+1}$ and $\dot{X}_n$ is one of phase only, i.e., there is no *attenuation* of the wave by the structure. On the other hand if W is complex, there will be attenuation whose magnitude will depend on the imaginary part of W. It is clear that for all frequency regions satisfying the inequality

$$+1 \geqq \cos W \geqq -1 \tag{20}$$

the structure will transmit longitudinal waves without attentuation (save that due to the ordinary dissipative processes which are neglected here). The upper and lower frequency limits of the transmission regions are given by setting $\cos W$ equal to the limiting values 1 and -1. We thus obtain, using Eq. (14) and replacing Z by $\rho_0 c/S$ the transcendental equations

$$\omega m/2\rho_0 cS = -\tan kl \tag{21}$$
$$\omega m/2\rho_0 cS = \cot kl. \tag{22}$$

Recalling that $k = \omega/c$, it is seen that Eq. (21) is satisfied by $\omega = 0$. Hence the structure under consideration is a *low-pass* filter, the first transmission region extending from $\omega = 0$ to the lowest value of ω satisfying Eq. (22). The latter gives the so-called *cut-off* frequency. The transmission characteristics of the system may perhaps be most clearly shown by plotting $\cos W$ directly as a function of frequency. Fig. 2 shows the plot of such a curve for the special case of a brass rod 0.32 cm^2 in cross-sectional area with $2l = 25.08$ cm and $m = 875$ grams. For brass $\rho_0 = 8.5$ grams/cm^3 and $c = 3.5 \times 10^5$ cm/sec. The cross-hatched regions are the transmission regions. It is seen that in this particular case we get transmission up to 1225 cycles before an attenuation region begins.

Before passing to a specific comparison of the theory with the available experimental data we may notice some general facts about the cut-off frequency given by (22). Writing the latter in the form

$$m/2\rho_0 Sl = \cot kl/kl \qquad (22')$$

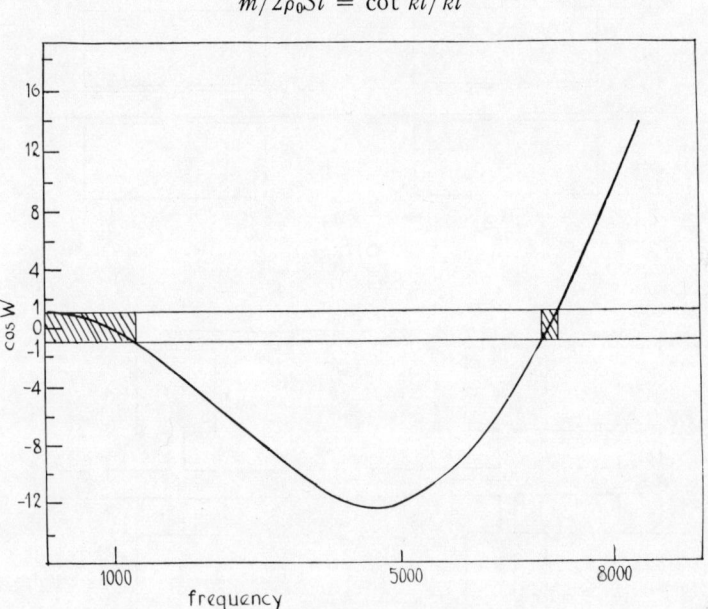

FIG. 2. *Transmission characteristic curve for a solid filter.*

we note that if for a given rod l is kept constant an *increase* in m, the mass of the load, *decreases* the cut-off frequency. Moreover if m is kept constant, we see that an *increase* in l also *decreases* the cut-off frequency.

H. F. Olson, one of G. W. Stewart's students, in 1925 constructed several solid filters and measured their cut-off frequencies.[6] The loads were in the form of cylindrical metal collars and the construction is indicated in the diagrams of Fig. 3 which show the two types used, viz., the O type with short necks and the N type with longer necks. We shall not give details about the methods of experimental measurement which were somewhat difficult to carry out, but shall merely compare some of the results with the corresponding theoretically calculated values. In

---

[6] So far as the authors are aware this work has not been published, the details being available in a master's thesis at the University of Iowa.

the case of each filter the cut-off frequency is indicated by plotting the transmission ratio P_r (= ratio of intensity after two or more sections to incident intensity) as a function of the frequency and noting at what frequency it becomes negligible. All the filters used were brass with $\rho_0 = 8.5$ g/cm³ and $c = 3.5 \times 10^5$ cm/sec. The dimensions given refer to Fig. 3 and all distances are in cm.

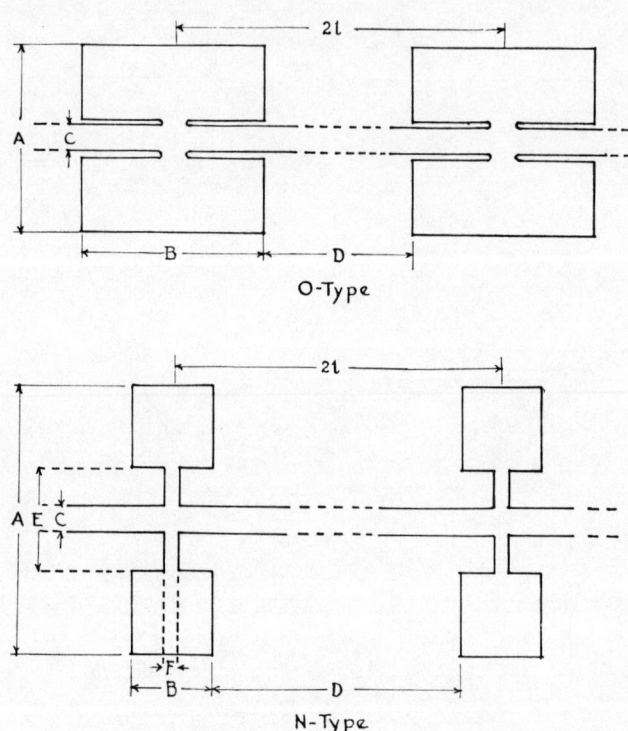

FIG. 3. *Diagrams of experimental solid filters.*

Filter O–3–20

$A = B = 5.08, C = 0.64, D = 20.0$

$2l = 25.08, S = 0.32$ cm², $m = 875$ g

ν	500	600	700	800	1000	1100	1200	1400
P_r	0.20	0.29	0.28	0.35	0.22	0.22	0.045	0.0005

Cut-off observed 1200–1300 cycles
" " calculated 1225 "
(from Eq. 22)

Filter 0–3–40

Same as 0–3–20, save $D = 40$, and $2l = 45.08$

ν	500	600	700	800	900	1000	1100
P_r	0.27	0.42	0.29	0.26	.056	.017	.003

Cut-off observed   900–1000 cycles
" " calculated   910   "

Filter N_1–2–20

$A = 6.95, B = 2.15, C = 0.64, D = 20$

$E = 2.65, F = 0.45, 2l = 22.15$

$S = 0.32 \text{ cm}^2, m = 614 \text{ gm}$

ν	500	600	800	1000	1200	1400	1600
P_r	0.28	0.22	0.35	0.32	0.20	0.0012	0.0008

Cut-off observed   1400 cycles (approx.)
" " calculated   1550   "

Filter N_3–2–20

$A = 8.50, B = 1.80, C = 0.64, D = 20$

$E = 4.90, F = 0.45, 2l = 21.8, S = 0.32 \text{ cm}^2, m = 652 \text{ g}$

ν	500	600	800	900	1000	1100	1200
P_r	0.32	0.35	0.35	0.40	0.14	0.004	0.007

Cut-off observed   1100 cycles (approx.)
" " calculated   1510   "

In interpreting these results it must be borne in mind that the experimental results cannot hope to be known with extreme accuracy. The agreement between theory and experiment, therefore, especially in the case of the O-type structures, must be considered rather satisfactory. It will be noted that the only serious disagreement occurs for the case of an N-type structure. It is just here that we should expect the theory in its present form to be inadequate since it contemplates the motion of each load as a whole, while if the masses are relatively far from the rod, as in the N-type case, different portions of the load will move differently, i.e., there will be transverse as well as longitudinal waves set up in them. From the standpoint of the simple theory the general effect of this would be to make the effective mass of each load larger than the actual mass. From what has been said above this would tend to lower the calculated cut-off frequencies and bring them more into line with the measured values. However, we do not wish to press these considerations for a more elaborate theory will be undertaken later. Nevertheless, it is felt that the simple theory here developed may prove to be of considerable utility.

The Loaded String as a Transverse Wave Filter

Before a further discussion of the filtration of longitudinal waves in solids, we may pause to recall that a stretched string loaded at equal intervals with mass particles acts as a low-pass filter for transverse waves. This has already been treated by Crandall[7] as a case of the forced oscillations of a system of many degrees of freedom. He used the method of Lagrange. It will, however, be of interest to apply the simpler method of the present paper to the same problem.

We shall assume that the string is infinite and stretched with mean tension τ_0. It is loaded with particles of mass m placed at equal intervals of length $2l$. If ξ now denotes the transverse displacement of the string, this will be propagated according to Eq. (1) with velocity $c = (\tau_0/\rho_0)^{1/2}$ where ρ_0 is here the *line density* or mass per unit length. The solution for the case of harmonic waves with the equilibrium position of the string along the x axis is then given by Eq. (2). Corresponding to Eq. (3) we now have the equation relating the instantaneous *normal component* of the tension in the string to the displacement. If we denote the former by τ, we have

$$\tau = \tau_0(\partial \xi/\partial x), \tag{23}$$

the mathematical connection with (3) being sufficiently obvious. The mathematical development now follows closely that for the solid rod with τ taking the place of T and $\dot{\xi}$ used in place of $\dot{X}$, since there is now no particular significance to be attached to a volume current. We thus have in place of Eqs. (4)

$$\begin{aligned} \dot{\xi} &= i\omega[\xi_0 e^{i(\omega t - kx)} + \xi_1 e^{i(\omega t + kx)}] \\ \tau &= i\omega\rho_0 c[-\xi_0 e^{i(\omega t - kx)} + \xi_1 e^{i(\omega t + kx)}] \end{aligned} \tag{24}$$

where ξ_0 and ξ_1 are now *transverse* displacement amplitudes. With $\dot{\xi} = \dot{\xi}_1 e^{i\omega t}$ and $\tau = \tau_1 e^{i\omega t}$ at $x=0$ we have corresponding to (5)

$$\begin{aligned} \dot{\xi} &= [\dot{\xi}_1 \cos kx + i\tau_1/\rho_0 c \cdot \sin kx] e^{i\omega t} \\ \tau &= [\tau_1 \cos kx + i\rho_0 c \cdot \dot{\xi}_1 \sin kx] e^{i\omega t}. \end{aligned} \tag{25}$$

[7] See *Theory of Vibrating Systems and Sound*, p. 64.

Making use of the same notation with regard to subscripts as in the case of the rod, we may write the boundary conditions analogous to (6) and (7) as follows

$$\xi_{12} = \xi_{21} = \xi_L \tag{26}$$

$$\tau_{21} - \tau_{12} = m\ddot{\xi}_L \tag{27}$$

where ξ_L is the vertical displacement of the loading particle. The rest of the analysis is so completely analogous to that for the solid rod that we may omit it, merely noting that corresponding to (12) we may define the characteristic impedance of the infinite loaded string by

$$-\tau_1/\dot{\xi}_1 = -\tau_2/\dot{\xi}_2 = \cdots = Z_0'. \tag{28}$$

Setting

$$\cos W = \cos 2kl - \omega mc/2\tau_0 \cdot \sin 2kl \tag{29}$$

we finally arrive at

$$\dot{\xi}_{n+1} = \dot{\xi}_n e^{-iW} \tag{30}$$

$$\tau_{n+1} = \tau_n e^{-iW}. \tag{31}$$

It thus appears that the infinite loaded string transmits only those frequencies for which

$$-1 \leq \cos 2kl - \omega mc/2\tau_0 \cdot \sin 2kl \leq +1 \tag{32}$$

The limiting frequencies of the transmission region are then given by setting $\cos W$ equal to $+1$ and -1, respectively. We get thus the two transcendental equations

$$\omega mc/2\tau_0 = -\tan kl \tag{33}$$

$$\omega mc/2\tau_0 = \cot kl. \tag{34}$$

Since Eq. (33) is satisfied by $\omega=0$, the loaded string behaves like a low-pass filter. The cut-off frequency for the first transmission band is given by (34). Let us suppose that kl is small; then to a first approximation $\cot kl = kl$, and (34) becomes

$$\nu = \omega/2\pi = 1/2\pi \cdot (2\tau_0/ml)^{1/2}. \tag{35}$$

Now the n characteristic frequencies of a finite string loaded with n particles of mass m separated by equal distances $2l$ are[8]

$$\nu_1 = \frac{1}{2\pi}\left(\frac{2\tau_0}{ml}\right)^{1/2} \sin \frac{\pi}{2(n+1)}, \cdots \nu_n = \frac{1}{2\pi}\left(\frac{2\tau_0}{ml}\right)^{1/2} \sin \frac{n\pi}{2(n+1)}. \tag{36}$$

[8] See, for example, Crandall, *Theory of Vibrating Systems and Sound*, p. 67.

Crandall[8] has shown that under forced oscillation the finite string cuts off all frequencies above $\nu = (1/2\pi)(2\tau_0/ml)^{1/2}$, which is the limit of ν_n as n increases indefinitely. It is seen that this frequency agrees precisely with the cut-off frequency (35) obtained by the transmission theory used in this paper.

Editor's Comments on Paper 9

9 Rayleigh: *On Waves Propagated Along the Plane Surface of an Elastic Solid*

In the article presented here, Rayleigh effectively laid the foundation for the study of surface waves in solids, known since then by his name. Further work in this area has proved to be very important in the field of solid-state physics. The reader should note that in this paper μ is the shear modulus of elasticity and that λ is the Lamé coefficient, which is equal to the bulk modulus minus two-thirds the shear modulus.

John William Strutt, third Baron Rayleigh (1842–1919), was perhaps the most versatile British physical scientist of the second half of the nineteenth and the first quarter of the twentieth centuries. Both an experimentalist and a theoretician, he turned his attention during a long and busy professional career to practically every branch of physics and many phases of chemistry. Although perhaps best known as the codiscoverer (with William Ramsay) of the rare gas argon, he devoted much attention to acoustics. His book *The Theory of Sound* (1877–1878) was a landmark in the development of the subject and has remained a standard treatise since its publication.

Of the 130 articles that he published on acoustics, five were reproduced in the Benchmark Series volume *Acoustics: Historical and Philosophical Development* (R. B. Lindsay, ed., Dowden, Hutchinson & Ross, Stroudsburg, Pa., 1973). Four more are reprinted in this volume, of which this one is the first.

130.

ON WAVES PROPAGATED ALONG THE PLANE SURFACE OF AN ELASTIC SOLID.

[*Proceedings of the London Mathematical Society*, XVII. pp. 4—11, 1885.]

It is proposed to investigate the behaviour of waves upon the plane free surface of an infinite homogeneous isotropic elastic solid, their character being such that the disturbance is confined to a superficial region, of thickness comparable with the wave-length. The case is thus analogous to that of deep-water waves, only that the potential energy here depends upon elastic resilience instead of upon gravity*.

Denoting the displacements by α, β, γ, and the dilatation by θ, we have the usual equations

$$\rho \frac{d^2\alpha}{dt^2} = (\lambda + \mu)\frac{d\theta}{dx} + \mu \nabla^2 \alpha, \quad \&c., \quad \ldots\ldots\ldots\ldots\ldots\ldots(1)$$

in which

$$\theta = \frac{d\alpha}{dx} + \frac{d\beta}{dy} + \frac{d\gamma}{dz}. \quad \ldots\ldots\ldots\ldots\ldots\ldots\ldots\ldots(2)$$

If α, β, γ all vary as e^{ipt}, equations (1) become

$$(\lambda + \mu)\frac{d\theta}{dx} + \mu \nabla^2 \alpha + \rho p^2 \alpha = 0, \quad \&c. \quad \ldots\ldots\ldots\ldots(3)$$

Differentiating equations (3) in order with respect to x, y, z, and adding, we get

$$(\nabla^2 + h^2)\theta = 0, \quad \ldots\ldots\ldots\ldots\ldots\ldots\ldots\ldots(4)$$

in which

$$h^2 = \rho p^2 / (\lambda + 2\mu). \quad \ldots\ldots\ldots\ldots\ldots\ldots\ldots\ldots(5)$$

---

* The statical problem of the deformation of an elastic solid by a harmonic application of pressure to its surface has been treated by Prof. G. Darwin, *Phil. Mag.* Dec. 1882. Jan. 1886.—See also Camb. Math. Trip. Ex. Jan. 20, 1875, Question IV.

Again, if we put
$$k^2 = \rho p^2/\mu, \quad\quad\quad (6)$$
equations (3) take the form
$$(\nabla^2 + k^2)\alpha = \left(1 - \frac{k^2}{h^2}\right)\frac{d\theta}{dx}, \quad \&c. \quad\quad\quad (7)$$
A particular solution of (7) is*
$$\alpha = -\frac{1}{h^2}\frac{d\theta}{dx}, \quad \beta = -\frac{1}{h^2}\frac{d\theta}{dy}, \quad \gamma = -\frac{1}{h^2}\frac{d\theta}{dz}; \quad\quad (8)$$
in order to complete which it is only necessary to add complementary terms u, v, w satisfying the system of equations
$$(\nabla^2 + k^2)u = 0, \quad (\nabla^2 + k^2)v = 0, \quad (\nabla^2 + k^2)w = 0, \quad\quad (9)$$
$$\frac{du}{dx} + \frac{dv}{dy} + \frac{dw}{dz} = 0. \quad\quad\quad (10)$$

For the purposes of the present problem we take the free surface as the plane $z = 0$, and assume that, as functions of x and y, the displacements are proportional to e^{ifx}, e^{igy}. Thus (4) takes the form
$$(d^2/dz^2 + h^2 - f^2 - g^2)\theta = 0;$$
so that
$$\theta = Pe^{-rz} + Qe^{+rz}, \quad\quad\quad (11)$$
where
$$r^2 = f^2 + g^2 - h^2. \quad\quad\quad (12)$$

In (11), r is supposed to be real; otherwise the dilatation would penetrate to an indefinite depth. For the same reason, we must retain only that term (say the first) for which the exponent is negative within the solid†. Thus $Q = 0$, and we will write for brevity $P = 1$, or rather $P = e^{ipt} e^{ifx} e^{igy}$; but the exponential factors may often be omitted without risk of confusion, so that we may take
$$\theta = e^{-rz}. \quad\quad\quad (13)$$

At the same time the particular solution becomes
$$\alpha = -\frac{if}{h^2}e^{-rz}, \quad \beta = -\frac{ig}{h^2}e^{-rz}, \quad \gamma = \frac{r}{h^2}e^{-rz}. \quad\quad (14)$$

For the complementary terms, which must also contain e^{ifx}, e^{igy} as factors, equations (9) become
$$(d^2/dz^2 + k^2 - f^2 - g^2)u = 0, \quad \&c.; \quad\quad (15)$$

* Lamb on the Vibrations of an Elastic Sphere, *Math. Soc. Proc.* May 1882.
† By discarding these restrictions we may deduce the complete solution applicable to a plate, bounded by parallel plane free surfaces; but I have not obtained any results which seem worthy of quotation.

whence, as before, on the assumption that the disturbance is limited to a superficial stratum,

$$u = Ae^{-sz}, \quad v = Be^{-sz}, \quad w = Ce^{-sz}, \quad \ldots\ldots\ldots\ldots(16)$$

where

$$s^2 = f^2 + g^2 - k^2. \quad \ldots\ldots\ldots\ldots\ldots\ldots\ldots(17)$$

In order to satisfy (10), the coefficients in (16) must be subject to the relation

$$ifA + igB - sC = 0. \quad \ldots\ldots\ldots\ldots\ldots\ldots(18)$$

The complete values of α, β, γ may now be written

$$\alpha = -\frac{if}{h^2} e^{-rz} + Ae^{-sz}, \quad \beta = -\frac{ig}{h^2} e^{-rz} + Be^{-sz}, \quad \gamma = \frac{r}{h^2} e^{-rz} + Ce^{-sz},$$
$$\ldots\ldots\ldots\ldots(19)$$

in which A, B, C are subject to (18); and the next step is to express the boundary conditions for the free surface. The two components of tangential stress must vanish, when $z = 0$, and these are proportional to

$$\frac{d\beta}{dz} + \frac{d\gamma}{dy}, \quad \frac{d\gamma}{dx} + \frac{d\alpha}{dz}$$

respectively. Hence

$$sB = \frac{2igr}{h^2} + igC, \quad sA = \frac{2ifr}{h^2} + ifC. \quad \ldots\ldots\ldots\ldots(20)$$

Substituting from (20) in (18), we find

$$C(s^2 + f^2 + g^2) h^2 + 2r(f^2 + g^2) = 0. \quad \ldots\ldots\ldots\ldots(21)$$

We have still to introduce the condition that the normal traction is zero at the surface. We have, in general,

$$N_3 = \lambda \theta + 2\mu \frac{d\gamma}{dz};$$

or, if we express λ in terms of μ, h, k,

$$N_3 = \mu \left\{ \left(\frac{k^2}{h^2} - 2 \right) \theta + 2 \frac{d\gamma}{dz} \right\};$$

so that the condition is

$$k^2 - 2h^2 - 2(+ r^2 + h^2 sC) = 0,$$

or, on substitution for r^2 of its value from (12),

$$k^2 - 2(f^2 + g^2) - 2h^2 sC = 0. \quad \ldots\ldots\ldots\ldots(22)$$

By eliminating C between (21) and (22), we obtain the equation by which the time of vibration is determined as a function of the wave-lengths and of the properties of the solid. It is

$$\{k^2 - 2(f^2 + g^2)\} \{s^2 + f^2 + g^2\} + 4rs(f^2 + g^2) = 0,$$

or, by (17),

$$\{2(f^2 + g^2) - k^2\}^2 = 4rs(f^2 + g^2). \quad \ldots\ldots\ldots\ldots(23)$$

If we square (23), and introduce the values of r^2 and s^2 from (12), (17), we get
$$\{2(f^2+g^2)-k^2\}^4 = 16(f^2+g^2)^2(f^2+g^2-h^2)(f^2+g^2-k^2).$$

As f and g occur here only in the combination (f^2+g^2), a quantity homogeneous with h^2 and k^2, we may conveniently replace (f^2+g^2) by unity. Thus
$$k^8 - 8k^6 + 24k^4 - 16k^2 - 16h^2k^2 + 16h^2 = 0. \quad\ldots\ldots\ldots\ldots(24)$$

Since the ratio $h^2 : k^2$ is known, this equation reduces to a cubic and determines the value of either quantity.

If the solid be incompressible ($\lambda = \infty$), $h^2 = 0$, and the equation becomes
$$k^6 - 8k^4 + 24k^2 - 16 = 0. \quad\ldots\ldots\ldots\ldots\ldots\ldots(25)$$

The real root of (25) is found to be ·91275, and the equation may be written
$$(k^2 - \cdot 91275)(k^4 - 7 \cdot 08725 k^2 + 17 \cdot 5311) = 0.$$

The general theory of vibrations of stable systems forbids us to look for complex values of k^2, as solutions of our problem, though it would at first sight appear possible with them to satisfy the prescribed conditions by taking such roots of (12), (17), as would make the *real* parts of the exponents in e^{-rz}, e^{-sz} negative. But, referring back to (23), which we write in the form
$$(2-k^2)^2 = 4rs,$$
or, in the present case of incompressibility, by putting $r=1$,
$$(2-k^2)^2 = 4s,$$
we see that we are not really free to choose the sign of s. In fact, from the complex values of k^2, viz., $3 \cdot 5436 \pm 2 \cdot 2301 i$, we find
$$4s = -2 \cdot 7431 \pm 6 \cdot 8846 i;$$
so that the real part of s is of the opposite sign to r, and therefore e^{-rz}, e^{-sz} do not both diminish without limit as we penetrate further and further into the solid.

Dismissing then the complex values, we have, in the case of incompressibility, the single solution
$$k^2 = \frac{\rho p^2}{\mu} = \cdot 91275 (f^2 + g^2). \quad\ldots\ldots\ldots\ldots\ldots(26)$$

From (19), (20), (21), we get in general
$$h^2\alpha = if\left\{-e^{-rz} + \frac{2rs}{s^2+f^2+g^2}e^{-sz}\right\}, \quad\ldots\ldots\ldots\ldots(27)$$

$$h^2\beta = ig\left\{-e^{-rz} + \frac{2rs}{s^2+f^2+g^2}e^{-sz}\right\}, \quad\ldots\ldots\ldots\ldots(28)$$

$$h^2\gamma = r\left\{+e^{-rz} - \frac{2(f^2+g^2)}{s^2+f^2+g^2}e^{-sz}\right\}. \quad\ldots\ldots\ldots\ldots(29)$$

In the case of incompressibility, we have k^2 given by (26), and
$$r^2 = f^2 + g^2, \quad s^2 = \cdot 08725 (f^2 + g^2).$$
Hence
$$\begin{aligned} h^2\alpha &= if \{- e^{-rz} + \cdot 5433 e^{-sz}\} e^{ipt} e^{ifx} e^{igy} \\ h^2\beta &= ig \{- e^{-rz} + \cdot 5433 e^{-sz}\} e^{ipt} e^{ifx} e^{igy} \\ h^2\gamma &= \sqrt{(f^2 + g^2)} \{e^{-rz} - 1\cdot 840 e^{-sz}\} e^{ipt} e^{ifx} e^{igy} \end{aligned} \right\} \quad \ldots\ldots\ldots\ldots (30)$$

If we suppose the motion to be in two dimensions only, we may put $g = 0$; so that $\beta = 0$, and
$$\left. \begin{aligned} h^2\alpha/f &= i \{- e^{-fz} + \cdot 5433 e^{-sz}\} e^{ipt} e^{ifx} \\ h^2\gamma/f &= \{ \ e^{-fz} - 1\cdot 840 e^{-sz}\} e^{ipt} e^{ifx} \end{aligned} \right\}, \quad \ldots\ldots\ldots (31)$$
in which
$$k = \cdot 9554 f, \quad s = \cdot 2954 f. \ldots\ldots\ldots\ldots\ldots (32)$$

For a progressive wave we may take simply the real parts of (31). Thus
$$\left. \begin{aligned} h^2\alpha/f &= (e^{-fz} - \cdot 5433 e^{-sz}) \sin (pt + fx) \\ h^2\gamma/f &= (e^{-fz} - 1\cdot 840 e^{-sz}) \cos (pt + fx) \end{aligned} \right\}. \quad \ldots\ldots\ldots (33)$$

The velocity of propagation is p/f, or $\cdot 9554 \sqrt{(\mu/\rho)}$, in which $\sqrt{(\mu/\rho)}$ is the velocity of purely transverse plane waves. The surface waves now under consideration move, therefore, rather more slowly than these.

From (32), (33), we see that α vanishes for all values of x and t when $e^{(s-f)z} = \cdot 5433$, *i.e.*, when $fz = \cdot 8659$. Thus, if λ' be the wave-length ($2\pi/f$), the horizontal motion vanishes at a depth equal to $\cdot 1378 \lambda'$. On the other hand, there is no finite depth at which the vertical motion vanishes.

To find the motion at the surface itself, we have only to put $z = 0$ in (33). We may drop at the same time the constant multiplier (h^2/f) which has no present significance. Accordingly,
$$\alpha = \cdot 4567 \sin (pt + fx), \quad \gamma = - \cdot 840 \cos (pt + fx), \ldots\ldots (34)$$
showing that the motion takes place in elliptic orbits, whose vertical axis is nearly the double of the horizontal axis.

The expressions for stationary vibrations may be obtained from (30) by addition to the similar equations obtained by changing the sign of p, and similar operations with respect to f and g. Dropping an arbitrary multiplier, we may write
$$\left. \begin{aligned} \alpha &= -f \{- e^{-rz} + \cdot 5433 e^{-sz}\} \cos pt \sin fx \cos gy \\ \beta &= -g \{- e^{-rz} + \cdot 5433 e^{-sz}\} \cos pt \cos fx \sin gy \\ \gamma &= \ r \{+ e^{-rz} - 1\cdot 840 e^{-sz}\} \cos pt \cos fx \cos gy \end{aligned} \right\}, \ldots\ldots (35)$$
in which
$$r = \sqrt{(f^2 + g^2)}, \quad s = \cdot 2954 \sqrt{(f^2 + g^2)}. \ldots\ldots\ldots (36)$$

As before, the horizontal motion vanishes at a depth such that

$$\sqrt{(f^2 + g^2)}\, z = \cdot 8659.$$

We will now examine how far the numerical results are affected when we take into account the finite compressibility of all natural bodies. The ratio of the elastic constants is often stated by means of the number expressing the ratio of lateral contraction to longitudinal extension when a bar of the material is strained by forces applied to its ends. According to a theory now generally discarded, this ratio (σ) would be $\frac{1}{4}$; a number which, however, is not far from the truth for a variety of materials, including the principal metals. In the extreme case of incompressibility σ is $\frac{1}{2}$, and there seems to be no theoretical reason why σ should not have any value between this and -1*.

The accompanying table will give an idea of the progress of the values of $k^2/(f^2 + g^2)$ as dependent upon λ/μ, or upon σ. It will be observed that the value diminishes continuously with λ, in accordance with a general principle†.

λ	σ	h^2/k^2	$k^2/(f^2+g^2)$	$k/\sqrt{(f^2+g^2)}$
∞	$\frac{1}{2}$	0	·9127	·9554
μ	$\frac{1}{4}$	$\frac{1}{3}$	·8453	·9194
0	0	$\frac{1}{2}$	·7640	·8741
$-\frac{2}{3}\mu$	-1	$\frac{3}{4}$	·4746	·6896

As an example of finite compressibility, we will consider further the second case of the table. From (12), (17),

$$r^2 = \cdot 7182\,(f^2 + g^2), \qquad r = \cdot 8475\,\sqrt{(f^2 + g^2)},$$
$$s^2 = \cdot 1547\,(f^2 + g^2), \qquad s = \cdot 3933\,\sqrt{(f^2 + g^2)}.$$

Hence, from (27), (28), (29), in correspondence with (30), we have

$$\left.\begin{aligned}
h^2\alpha &= if\{-e^{-rz} + \cdot 5773\,e^{-sz}\}\,e^{ipt}\,e^{ifx}\,e^{igy} \\
h^2\beta &= ig\{-e^{-rz} + \cdot 5773\,e^{-sz}\}\,e^{ipt}\,e^{ifx}\,e^{igy} \\
h^2\gamma &= \cdot 8475\,\sqrt{(f^2+g^2)}\,\{e^{-rz} - 1\cdot 7320\,e^{-sz}\}\,e^{ipt}\,e^{ifx}\,e^{igy}
\end{aligned}\right\} \quad \ldots\ldots(37)$$

---

* Prof. Lamb, in his able paper, seems to regard all negative values of σ as excluded *a priori*. But the necessary and sufficient conditions of stability are merely that the resistance to compression ($\lambda + \frac{2}{3}\mu$) and the resistance to shearing (μ) should be positive. In the second extreme case of a medium which resists shear, but does not resist compression, $\lambda = -\frac{2}{3}\mu$, and $\sigma = -1$. The velocity of a dilatational wave is then $\frac{4}{3}$ of that of a distortional plane wave. (Green, *Camb. Trans.* 1838.) The general value of σ is $\lambda/(2\lambda + 2\mu)$.

† *Math. Soc. Proc.* June 1873, vol. IV. p. 359 [vol. I. p. 171]. *Theory of Sound*, t. I. p. 85. Lamb, *loc. cit.* p. 202.

For a progressive wave in two dimensions, we shall have

$$\left.\begin{array}{l}h^2\alpha/f = (e^{-rz} - \cdot 5773\,e^{-sz}) \sin{(pt + fx)} \\ h^2\gamma/f = (\cdot 8475\,e^{-rz} - 1\cdot 4679\,e^{-sz}) \cos{(pt + fx)}\end{array}\right\} \quad \ldots\ldots\ldots\ldots(38)$$

At the surface,

$$\left.\begin{array}{l}h^2\alpha/f = + \cdot 4227 \sin{(pt + fx)} \\ h^2\gamma/f = - \cdot 6204 \cos{(pt + fx)}\end{array}\right\}, \quad \ldots\ldots\ldots\ldots\ldots\ldots(39)$$

so that the vertical axes of the elliptic orbits are about half as great again as the horizontal axes.

It is proper to remark that the vibrations here considered are covered by the general theory of spherical vibrations given by Lamb in the paper referred to. But it would probably be as difficult, if not more difficult, to deduce the conclusions of the present paper from the analytical expressions of the general theory, as to obtain them independently. It is not improbable that the surface waves here investigated play an important part in earthquakes, and in the collision of elastic solids. Diverging in two dimensions only, they must acquire at a great distance from the source a continually increasing preponderance.

Macrosonics, Nonlinear Acoustics, and Shock Waves

II

Editor's Comments on Paper 10

10 Rayleigh: *Aerial Plane Waves of Finite Amplitude*

Although Lord Rayleigh did not, himself, introduce the study of high-intensity, finite-amplitude sound radiation, macrosonics (see Paper 29 by S. Earnshaw in the Benchmark volume *Acoustics: Historical and Philosophical Development*, R. B. Lindsay, ed., Dowden, Hutchinson & Ross, Stroudsburg, Pa.), in this paper, "Aerial Plane Waves of Finite Amplitude," he provided a remarkably comprehensive review of the subject as it stood at that time. He also included the results of pioneer research on the formation and behavior of shock waves. In this, he went considerably beyond the earlier work in the field by Rankine and Hugoniot by considering the effects of viscosity and heat conduction on waves of discontinuity. This set the stage for all subsequent work in the field of macrosonics, or what has come to be called nonlinear acoustics. This important paper is presented here in full.

346.

AERIAL PLANE WAVES OF FINITE AMPLITUDE.

[*Proceedings of the Royal Society*, A, Vol. LXXXIV. pp. 247—284, 1910.]

Waves of Finite Amplitude without Dissipation.

IN the investigations which follow, we are concerned with the motion of an elastic fluid in one dimension, say, parallel to x. It is implied not only that there are no component velocities perpendicular to x, but that the motion is the same in any perpendicular plane, so that it is a function of x and of the time (t) only. If u be the velocity at any point x, p the pressure, ρ the density, X an impressed force, the dynamical equation for an inviscid fluid is

$$\frac{du}{dt} + u\frac{du}{dx} = X - \frac{1}{\rho}\frac{dp}{dx}. \qquad (1)$$

At the same time the "equation of continuity" takes the form

$$\frac{d\rho}{dt} + \frac{d(\rho u)}{dx} = 0. \qquad (2)$$

The first step, and it was a very important one, in the treatment of waves of finite amplitude is due to Poisson[*]. Under the assumption of Boyle's law, $p = a^2\rho$, he proved that for waves travelling in one direction (positive) the circumstances of the propagation are expressed by

$$u = f\{x - (a+u)t\}, \qquad (3)$$

in which f denotes an arbitrary function. When u can be neglected in comparison with a, this reduces to the familiar law of undisturbed propagation applicable to infinitesimal waves.

Poisson does not discuss the significance of (3) further than to show that the boundaries of a continuous wave, limited to a finite range along x, are propagated with the ordinary velocity a, and that accordingly the length of the wave does not alter as it advances. The meaning of (3) is that in general u advances with a velocity equal, not to a, but to $a + u$, and that this might be expected is very easily seen (Earnshaw). From the ordinary

[*] "Mémoire sur la Théorie du Son," *Journ. de l'École Polytechnique*, 1808, Vol. VII. p. 319.

theory we know that an infinitely small disturbance is propagated with a certain velocity a, which velocity is relative to the parts of the medium undisturbed by the wave. Let us consider now the case of a wave so long that the variations of velocity and density are insensible for a considerable distance along it, and at a place where the velocity (u) is finite let us imagine a small secondary wave to be superposed. The velocity with which the secondary wave is propagated through the surrounding medium is a, but on account of the local motion of the medium itself the whole velocity of advance is $a+u$, and depends upon the part of the long wave at which the small wave is placed. What has been said of the secondary wave applies also to the parts of the long wave itself, and thus we see that after a time t the place where a certain velocity u is to be found is in advance of its original position by a distance equal, not to at, but to $(a+u)t$, or, as we may express it, u is propagated with velocity $(a+u)$.

A closer discussion of the solution represented by Poisson's integral was given by Stokes*, who pointed out the difficulty which ultimately arises from the motion becoming discontinuous. If we draw a curve to represent the distribution of velocity, taking x for abscissa and u for ordinate, we may find the corresponding curve after the lapse of time t by the following construction:—Through any point on the original curve draw a straight line in the positive direction parallel to x, and of length equal to $(a+u)t$, or, as we are concerned with the shape of the curve only, equal to ut. The locus of the ends of these lines is the velocity-curve after a time t.

But this law of derivation cannot hold good indefinitely. The crests of the velocity-curve gain continually on the troughs and must at last overtake them. After this the curve would indicate two values of u for one value of x, ceasing to represent anything that could actually take place. In fact we are not at liberty to push the application of the integral beyond the point at which the velocity becomes discontinuous, or the velocity-curve has a vertical tangent. In order to find when this happens, let us take two neighbouring points on any part of the curve which slopes downwards in the positive direction, and inquire after what time this part of the curve becomes vertical. If the difference of abscissae be dx, the hinder point will overtake the forward point in the time $-dx/du$. Thus the motion, as determined by Poisson's integral, becomes discontinuous after a time equal to the reciprocal, taken positively, of the greatest negative value of du/dx.

For example, let us suppose that

$$u = U \cos \frac{2\pi}{\lambda} \{x - (a+u)t\}, \quad\quad\quad\quad\quad (4)$$

* "On a Difficulty in the Theory of Sound," *Phil. Mag.* November, 1848.

where U is the greatest initial velocity. When $t=0$, the greatest negative value of du/dx is $-2\pi U/\lambda$, so that discontinuity will commence at the time $t = \lambda/2\pi U$.

The only kind of wave travelling in the positive direction which can escape ultimate discontinuity is one which has no forward slope. This is the case of a wave forming the transition between a larger constant value of u when x exceeds a certain value, and a smaller constant value when x falls short of a certain value. As time passes, the slope everywhere becomes easier. We shall see presently that this wave is a wave of rarefaction, in the sense that during its passage the gas passes from a greater to a less density.

It is worthy of remark that, although we may of course conceive a wave of finite disturbance to exist at any moment, there is in general a limit to the duration of its previous independent existence. By drawing lines in the negative instead of in the positive direction we may trace the history of the velocity-curve; and we see that as we push our inquiry further and further into past time the forward slopes become easier and the backward slopes steeper. At a time equal to the greatest positive value of dx/du, antecedent to that at which the curve is first contemplated, the velocity would be discontinuous. The exception is now a wave of condensation, involving a passage always from a less to a greater density.

When discontinuity sets in, a state of things exists to which the usual differential equations are inapplicable; and the subsequent progress of the motion has not been determined. It is probable, as suggested by Stokes, that some sort of reflection would ensue. In regard to this matter we must be careful to keep purely mathematical questions distinct from physical ones. We shall see later how the tendency to discontinuity may be held in check by forces of a dissipative character. But this has nothing directly to do with the mathematical problem of determining what would happen to waves of finite amplitude in a medium, free from viscosity, whose pressure is under all circumstances proportional to the density*. To suppose that the problem has no solution would seem to be tantamount to admitting an inherent contradiction in the assumption, usually made in hydrodynamics, of a continuous fluid subject to Boyle's law. It would be strange if the necessity of a molecular constitution for gases could be established by such an argument.

With Poisson's integral (3), showing how the velocity is propagated, there is associated another law connecting the velocity and the density in a positive progressive wave. In the case of a fluid obeying Boyle's law this relation is

$$u - a \log \rho = \text{const.} \quad\quad\quad\quad\quad\quad\quad\quad (5)$$

It does not occur explicitly in Poisson's memoir, and Earnshaw considers that Poisson did not discover it. Certainly it is remarkable that he omitted

---

* *Theory of Sound*, 1878, § 251.

to formulate the law, but at the same time it is difficult to suppose him ignorant of it, seeing that it follows by simple subtraction from two of his equations*. A formula equivalent to (5) was given explicitly, so far as I know, for the first time by Airy†, who attributes it to De Morgan.

The assumption that in a progressive wave there is a definite relation between u and ρ forms the basis of Earnshaw's investigation‡. That such a relation is to be expected may be shown by a line of argument analogous to that already employed in connection with Poisson's integral.

Whatever may be the law of pressure as a function of density, the velocity of propagation of small disturbances is according to the usual theory equal to $\sqrt{(dp/d\rho)}$, and in a positive progressive wave the relation between velocity and condensation (s) is

$$u : s = \sqrt{(dp/d\rho)}, \quad \ldots\ldots\ldots\ldots\ldots\ldots\ldots\ldots\ldots\ldots(6)$$

where $s = \delta\rho/\rho$. If this relation be violated at any point, a wave will emerge, travelling in the negative direction. Let us now picture to ourselves the case of a positive progressive wave in which the changes of velocity and density are very gradual but become important by accumulation, and let us inquire what condition must be satisfied in order to prevent the formation of a negative wave. It is clear that the answer to the question whether or not a negative wave will be generated at any point will depend upon the state of things in the immediate neighbourhood of the point, and not upon the state of things at a distance from it, and will therefore be determined by the criterion applicable to small disturbances. In applying this criterion we are to consider the velocities and condensations, not absolutely, but relatively to those prevailing in the neighbouring parts of the medium, so that the form of (6) proper for the present purpose is

$$du = \sqrt{\left(\frac{dp}{d\rho}\right)} \cdot \frac{d\rho}{\rho}, \quad \ldots\ldots\ldots\ldots\ldots\ldots\ldots\ldots\ldots\ldots(7)$$

whence
$$u = \int \sqrt{\left(\frac{dp}{d\rho}\right)} \cdot \frac{d\rho}{\rho}, \quad \ldots\ldots\ldots\ldots\ldots\ldots\ldots\ldots\ldots\ldots(8)$$

which is the relation between u and ρ generally necessary for a positive progressive wave, as laid down by Earnshaw§.

Earnshaw worked with the so-called Lagrangian form of the equations, in which the motions of particular particles are followed, and he obtained complete solutions for a wave progressive in one direction. In the case of

---

* Equations (1), p. 364, and (b), p. 367.
† *Phil. Mag.* 1849, Vol. XXXIV. p. 402. The corresponding formula for long tidal waves of finite amplitude was also given.
‡ *Roy. Soc. Proc.* January 6, 1859; *Phil. Trans.* 1860, p. 133.
§ *Theory of Sound*, 1878, § 251.

Boyle's law the relation between velocity and density is that already given (5), and in the case of the adiabatic law, where

$$\frac{p}{p_0} = \left(\frac{\rho}{\rho_0}\right)^\gamma, \quad \ldots\ldots\ldots\ldots\ldots\ldots\ldots\ldots(9)$$

Earnshaw finds
$$\left(\frac{\rho}{\rho_0}\right)^{\frac{1}{2}(\gamma-1)} = 1 + \frac{(\gamma-1)u}{2a}, \quad \ldots\ldots\ldots\ldots\ldots(10)$$

where u is the velocity of infinitesimal disturbances under the condition represented by p_0, ρ_0, viz. $a^2 = \gamma p_0/\rho_0$. In (9) γ denotes as usual the ratio of the two specific heats; and in (10), applicable to a *positive* progressive wave, the constant of integration has been so chosen that $u = 0$ corresponds to $\rho = \rho_0$.

The generalised form of Poisson's integral, appropriate when p is *any* given function of ρ, does not appear quite explicitly in Earnshaw's memoir. The line of argument already used shows that it must be

$$u = f[x - \{u + \sqrt{(dp/d\rho)}\} t]. \quad \ldots\ldots\ldots\ldots\ldots(11)$$

In the case of a gas obeying Boyle's law,

$$\sqrt{(dp/d\rho)} = \text{const.}, \quad \ldots\ldots\ldots\ldots\ldots\ldots\ldots(12)$$

and (11) reduces to Poisson's form.

In the case of the adiabatic law, we have from (9), (10),

$$\sqrt{\left(\frac{dp}{d\rho}\right)} = a\left(\frac{\rho}{\rho_0}\right)^{\frac{1}{2}(\gamma-1)} = a + \frac{\gamma-1}{2}u, \ldots\ldots\ldots\ldots\ldots(13)$$

so that (Earnshaw)

$$u + \sqrt{(dp/d\rho)} = a + \tfrac{1}{2}(\gamma+1)u. \quad \ldots\ldots\ldots\ldots\ldots(14)$$

Thus (11) assumes the form

$$u = f[x - \{a + \tfrac{1}{2}(\gamma+1)u\} t], \quad \ldots\ldots\ldots\ldots\ldots(15)$$

and this with (10) may be considered to constitute the solution of the problem up to the point where discontinuity sets in. We may fall back upon Boyle's law by putting in (15) $\gamma = 1$.

It appears that whether the relation of pressure to density be isothermal or adiabatic, there is a change of type as the wave advances. There can be no escape from such a change unless $u + \sqrt{(dp/d\rho)}$ be constant. Using (8), we may deduce in this case

$$\sqrt{(dp/d\rho)} = B/\rho,$$

B being a constant, whence

$$p = A - B^2/\rho \quad \ldots\ldots\ldots\ldots\ldots\ldots\ldots(16)$$

expresses the only law of pressure under which waves of finite amplitude can be propagated without undergoing a change of type (Earnshaw). A simpler derivation of (16) will be given presently.

Earnshaw further considers the genesis of disturbance in a gas originally at rest by the motion of a piston (supposed to be contained in a tube), but some of his conclusions appear to need revision. All that is required in these problems is virtually contained in (8), (11). If X denote the position of the piston at time T, the velocity of its motion is $U = dX/dT$, and this velocity is shared by the gas in contact with it. On the positive side the velocity of propagation of U (equal to u) is, by (11),

$$U + \sqrt{(dp/d\rho)};$$

so that if at time t (greater than T), U is to be found at x, we must have

$$x = X + \{U + \sqrt{(dp/d\rho)}\}(t - T). \quad\ldots\ldots\ldots\ldots\ldots(17)$$

Among the problems which naturally suggest themselves would be to determine what happens when the piston originally at rest at $X = 0$ begins to move at time $T = 0$ with a constant velocity. But if this velocity be positive, the discontinuity, which immediately ensues, causes the failure of our equations. On the other hand, if the constant velocity be negative, say $-V$, the initial discontinuity disappears forthwith, and the subsequent motion may be traced.

Take, for example, the case of Boyle's law. We have

$$X = UT, \qquad U + \sqrt{(dp/d\rho)} = U + a,$$

so that
$$x = Ut + a(t - T) \qquad (t > T),$$

and we have to consider where the velocities 0 and $-V$ are to be found at time t. Now $U = 0$ corresponds to the range of T from $-\infty$ to 0, so that x ranges from at to ∞. Again, $U = -V$ corresponds to the range of T from 0 to t, so that x ranges from $(a - V)t$ to $-Vt$. The whole range of x on the positive side of the piston is now accounted for, except the interval from $x = (a - V)t$ to $x = at$. This is occupied by the transition of velocity from 0 to $-V$, and we infer, from what was said in the discussion of Poisson's integral, that this transition must take place linearly.

Under Boyle's law the relation between velocity and density (5) is such that, however fast the piston may recede (u negative), a complete vacuum can never be formed behind it. It is otherwise under the adiabatic law (10), where $\rho = 0$ corresponds to

$$u = -2a/(\gamma - 1) \quad \text{(Earnshaw)}.$$

It may be of interest to consider further a few examples of (17). Still assuming Boyle's law, let us suppose that the piston is at rest ($X = 0$) until $T = 0$, and then moves with uniform acceleration (g), so that $(T +)$

$$U = gT, \qquad X = \tfrac{1}{2}gT^2 = U^2/2g.$$

The use of these in (17) gives

$$x = \frac{a^2}{2g} + (U+a)t - \frac{(U+a)^2}{2g}, \quad \ldots\ldots\ldots\ldots\ldots(18)$$

showing that the relation between x and U is parabolic. In (18) we see that dx/dU vanishes, when $t = (U+a)/g$. Thus, if g be positive, *i.e.* if the wave be one of condensation, discontinuity sets in at the front after an interval, reckoned from the beginning of the motion, equal to a/g. But if g be negative, there is no discontinuity, and (18) remains valid for an indefinite time.

In general, as in the last example ($g+$), the discontinuity sets in locally at one point of the velocity-curve, while other parts are temporarily exempt. It is of interest to inquire under what law the piston must advance so as to generate a linear velocity-curve. For then, under the adiabatic law, which includes Boyle's, a velocity-curve, once linear, remains linear, and if discontinuity enters, it must affect the whole curve simultaneously.

It may be worth while to pause here for a moment to inquire what law of pressure is implied in the permanence of the linear character of the velocity-curve. By (7)

$$\rho \frac{du}{d\rho} = \sqrt{\left(\frac{dp}{d\rho}\right)};$$

so that if $u + \sqrt{(dp/d\rho)}$ is a linear function of u, $du/d\log\rho$ must also be a linear function of u. This requires that

$$\rho\, du/d\rho = C\rho^n,$$

where C and n are constants, and the most general relation between p and ρ consistent with the requirements is

$$p = A + B\rho^\gamma, \quad \ldots\ldots\ldots\ldots\ldots\ldots\ldots(19)$$

where A, B, γ are constants. The relation (19) may be regarded as a kind of generalised adiabatic law; it includes the special law (16) under which a velocity-curve is absolutely permanent in type.

Supposing the motion to commence at $T = 0$, we have

$$X = \int_0^T U\, dT = UT - \int_0^U T\, dU,$$

and hence from (17), under the supposition of Boyle's law,

$$x = -\int T\, dU - (U+a)t - aT; \quad \ldots\ldots\ldots\ldots\ldots(20)$$

and the question before us is so to determine T as a function of U that (20) may be linear in U. From (20) when t is constant

$$\frac{dx}{dU} = -T + t - a\frac{dT}{dU} = t_0, \text{ say,}$$

where t_0 is a constant, whence
$$T = t - t_0 + He^{-U/a},$$
H being the constant of integration. But, since $U = 0$ when $T = 0$, this assumes the form
$$T = T'(1 - e^{-U/a}), \quad \text{...........................(21)}$$
T' being written for $t - t_0$. Or, if we express U in terms of T,
$$U = -a \log\left(1 - \frac{T}{T'}\right). \quad \text{........................(22)}$$

In (21), (22), T' is positive, if U is positive; and U becomes infinite when $T = T'$. We must therefore regard the law as limited to values of T less than T'.

From (22) we find
$$X = \int_0^T U\,dT = a\left\{(T' - T)\log\left(1 - \frac{T}{T'}\right) + T\right\}, \quad \text{............(23)}$$
which completely expresses the motion of the piston. The corresponding velocity-curve at time t may be verified by means of (20), (21). It is expressed by
$$U = \frac{at - x}{T' - t}, \quad \text{................................(24)}$$
exhibiting the linear character of the slope of velocity. Evidently the slope becomes vertical throughout when $t = T'$.

In the above example it is not necessary to suppose the law of motion of the piston continued up to $T = T'$. On the contrary, we may imagine U to increase up to some prescribed finite value and then to remain constant. In this case the slope expressed by (24) forms the transition between $U = 0$ for values of x greater than at and the finite value at which the acceleration of the piston stops.

If the wave be one of rarefaction we must take T' negative, say $-T''$. In this case the analogue of (22) shows that U constantly increases in numerical value but does not become infinite in any finite time. The analogue of (24) is
$$U = \frac{x - at}{t + T''}, \quad \text{................................(25)}$$
representing a slope which ever grows easier as time passes.

The problem also admits of solution when the gas follows the adiabatic law (9). As in (15), (20),
$$x = X + \{a + \tfrac{1}{2}(\gamma + 1)U\}(t - T); \quad \text{..................(26)}$$

and (26) is to satisfy the condition of making dx/dU constant. In this

$$dX/dU = U \cdot dT/dU,$$

so that
$$\frac{dT}{dU}\{(\gamma-1)U + 2a\} + (\gamma+1)(T-t) = \text{const}.$$

On integration we obtain, under the condition that U and T vanish together,

$$\left(1 - \frac{T'}{T'}\right)\left[1 + \frac{\gamma-1}{2a}U\right]^{\frac{\gamma+1}{\gamma-1}} = 1, \quad\ldots\ldots\ldots\ldots\ldots(27)$$

T' being a constant which, if positive, corresponds to $U = \infty$, thereby determining U as a function of T.

For the value of $X = \int_0^T U\,dT$, we have

$$\frac{\gamma-1}{2a}\frac{X}{T'} = -\frac{\gamma+1}{2}\left[\left(1 - \frac{T}{T'}\right)^{\frac{2}{\gamma+1}} - 1\right] - \frac{T}{T'}$$

$$= \frac{\gamma-1}{2} - \frac{\gamma+1}{2}\left(1 + \frac{\gamma-1}{2a}U\right)^{-\frac{2}{\gamma-1}} + \left(1 + \frac{\gamma-1}{2a}U\right)^{-\frac{\gamma+1}{\gamma-1}}. \ldots(28)$$

It may be observed that, although U is infinite when $T = T'$, X remains finite. Using this in (26), we get finally, on reduction,

$$x = at - \tfrac{1}{2}(\gamma+1)U(T'-t), \quad\ldots\ldots\ldots\ldots\ldots(29)$$

or
$$U = \frac{2}{\gamma+1}\frac{at-x}{T'-t}. \quad\ldots\ldots\ldots\ldots\ldots\ldots\ldots(30)$$

If $x < at$, U is a linear function of x, and (T' being positive) the slope of the velocity-curve increases until it becomes vertical when $t = T'$.

If T' is negative, the wave is one of rarefaction, and (30) applies however great t may be.

By putting $\gamma = 1$ we fall back from (30) to (24), and less simply from (28) to (23)*.

Riemann's work† is of somewhat later date than Earnshaw's, but in one important respect is more general. It may be convenient briefly to recall the principal result.

Taking p a given function of ρ, say $\phi(\rho)$, and putting $X = 0$, we have from (1) and (2)

$$\frac{du}{dt} + u\frac{du}{dx} = -\phi'(\rho)\frac{d\log\rho}{dx} \qquad \frac{d\log\rho}{dt} + u\frac{d\log\rho}{dx} = -\frac{du}{dx}.$$

---

* I have since found that this problem was successfully treated by Hugoniot.
† *Göttingen Abhandlungen*, 1860, Vol. VIII.

If the second of these equations be multiplied by $\pm \sqrt{\{\phi'(\rho)\}}$ and be added to the first, we find

$$\frac{dr}{dt} = -\{u + \sqrt{\phi'(\rho)}\}\frac{dr}{dx}, \qquad \frac{ds}{dt} = -\{u - \sqrt{\phi'(\rho)}\}\frac{ds}{dx}, \quad \ldots\ldots(31)$$

where
$$2r = f(\rho) + u, \qquad 2s = f(\rho) - u, \quad \ldots\ldots\ldots\ldots\ldots(32)$$

and
$$f(\rho) = \int \sqrt{\phi'(\rho)} \cdot d\log \rho. \quad \ldots\ldots\ldots\ldots\ldots\ldots(33)$$

From these follow

$$dr = \frac{dr}{dx}[dx - \{u + \sqrt{\phi'(\rho)}\}\,dt],$$

$$ds = \frac{ds}{dx}[dx - \{u - \sqrt{\phi'(\rho)}\}\,dt]:$$

so that r remains constant when x and t change in such a manner that $dx = \{u + \sqrt{\phi'(\rho)}\}\,dt$, and s remains constant when x and t change so that $dx = \{u - \sqrt{\phi'(\rho)}\}\,dt$. In the case of a positive progressive wave $s = 0$, whence $f(\rho) = u$ and also $r = u$. The velocity with which u travels in such a wave is accordingly $u + \sqrt{(dp/d\rho)}$, of which fact (11) is merely another form of statement. Riemann's equations are more general than anything previously given, as not limited to a single progressive wave.

Since Riemann's equations do not seem to have been applied in any example of continuous motion, I have thought it worth while to inquire whether they can be satisfied when r and s are both linear functions of x, Boyle's law being assumed, so that in (32), (33)

$$\sqrt{\phi'(\rho)} = a, \qquad f(\rho) = a \log \rho.$$

If we suppose
$$r = Ax + B, \qquad s = Cx + D, \quad \ldots\ldots\ldots\ldots\ldots(34)$$

we obtain, on substitution in (31), equations for the determination of A, B, C, D, as functions of the time. In the first instance we find

$$A - C = 1/t, \quad \ldots\ldots\ldots\ldots\ldots\ldots\ldots(35)$$

in which to t a constant may be added, and further

$$A = \frac{H+1}{2t}, \qquad C = \frac{H-1}{2t}, \quad \ldots\ldots\ldots\ldots\ldots(36)$$

when H is an arbitrary constant. Also

$$B - D = -aH + L/t, \quad \ldots\ldots\ldots\ldots\ldots\ldots(37)$$

L being another arbitrary constant, and thence

$$B = \tfrac{1}{2}a(H^2-1)\log t + \frac{L(H+1)}{2t} + \tfrac{1}{2}M, \quad \ldots\ldots\ldots(38)$$

$$D = \tfrac{1}{2}a(H^2-1)\log t + \frac{L(H-1)}{2t} + \tfrac{1}{2}N, \quad \ldots\ldots\ldots(39)$$

with
$$M - N = -2aH. \quad \ldots\ldots\ldots\ldots\ldots\ldots(40)$$

If we allow the origin of x to be arbitrary, as well as that of t, we may write

$$2r = (H+1)x/t + (H^2-1)a \log t + M, \quad \ldots (41)$$

$$2s = (H-1)x/t + (H^2-1)a \log t + N, \quad \ldots (42)$$

$$u = r - s = x/t - aH, \quad \ldots (43)$$

$$a \log \rho = r + s = Hx/t + (H^2-1)a \log t + \tfrac{1}{2}(M+N). \quad \ldots (44)$$

If $H = \pm 1$, the logarithmic term disappears, and either r or s is constant. In these cases we fall back upon single progressive waves.

If $H = 0$, $u = x/t$ in (43), and (44) gives $1/\rho$ proportional to t. The density is thus uniform with respect to x, but the volume of a given mass grows proportionally with t. The uniform expansion occurs in such a manner that the gas remains unmoved at the origin of co-ordinates. Since in this case $du/dt + u\,du/dx = 0$, we see that every part of the gas moves with unaccelerated velocity.

Waves of Permanent Regime.

When waves are propagated in one dimension without change of type, the circumstances are dynamically the same as in steady motion, as appears at once by impressing on the system a velocity equal and opposite to that of wave-propagation. The problem may conveniently be considered under this form.

From the general equation of continuity (2), by making $d\rho/dt$ equal to zero, or independently, we have

$$\rho u = \rho_0 u_0 = m, \quad \ldots (45)$$

where m is a constant which Rankine called the mass-velocity. The dynamical equation (1) reduces to

$$u \frac{du}{dx} + \frac{1}{\rho}\frac{dp}{dx} = X. \quad \ldots (46)$$

If $X = 0$, (46) may be written

$$\int_{p_0}^{p} \frac{dp}{\rho} = \tfrac{1}{2}u_0^2 - \tfrac{1}{2}u^2, \quad \ldots (47)$$

where u_0 is the velocity corresponding to p_0.

Eliminating u, we get

$$\int_{p_0}^{p} \frac{dp}{\rho} = \tfrac{1}{2}u_0^2 \left(1 - \frac{\rho_0^2}{\rho^2}\right), \quad \ldots (48)$$

determining the law of pressure under which alone it is possible for a stationary wave to maintain itself in fluid moving (outside the wave) with velocity u_0. From (48)

$$\frac{dp}{d\rho} = u_0^2 \frac{\rho_0^2}{\rho^2}, \quad \ldots\ldots\ldots\ldots\ldots\ldots\ldots\ldots\ldots\ldots (49)$$

or
$$p + m^2/\rho = p_0 + m^2/\rho_0, \quad \ldots\ldots\ldots\ldots\ldots\ldots\ldots\ldots\ldots (50)$$

the law found by Earnshaw.

Since, under the adiabatic law, the relation between density and pressure differs from (50), we conclude that a self-maintaining stationary aerial wave is an impossibility, unless it be in virtue of impressed forces, or of viscosity, or other dissipative agencies not now regarded.

When the changes of density concerned are *small*, (50) may be satisfied approximately; and we see from (49) that the velocity of the stream (outside the wave) necessary to keep the wave stationary is given by

$$u_0 = \sqrt{(dp/d\rho)},$$

which is the same as the velocity of the wave reckoned relatively to the fluid at a distance.

This way of regarding the subject shows, perhaps more clearly than any other, the nature of the relation between velocity and density. In a stationary wave-form a loss of velocity accompanies an augmented density, according to the principle of energy, and therefore the fluid composing the condensed parts of a wave moves forward more slowly than the undisturbed portions. Relatively to the fluid at a distance, the motion of the condensed parts is in the same direction as that in which the waves travel.

By means of (46), we can find what impressed force is required in order to ensure a stationary wave-form when (50) is not satisfied. For example, if $p = a^2\rho$, we find from (45), (46),

$$X = u\frac{du}{dx} + a^2 \frac{d\log \rho}{dx} = (u^2 - a^2)\frac{d\log u}{dx}, \quad \ldots\ldots\ldots\ldots (51)$$

showing that an impressed force is necessary at every place where u is variable and unequal to a. In (51) X is the accelerating force so called. The actual force operative upon the element of mass ρdx is $X\rho dx$. Thus, on integration,

$$\int X\rho dx = m\int \left(1 - \frac{a^2}{u^2}\right) du = m(u_2 - u_1)\left\{1 - \frac{a^2}{u_1 u_2}\right\}, \quad \ldots\ldots\ldots (52)$$

if the range of integration extend from the place where the velocity is u_1 to the place where it becomes equal to u_2. The integral applied force vanishes if the terminal velocities are such that their geometric mean is a. We may apply this to the case of a velocity-curve giving a simple gradual transition from one constant velocity u_1 to another constant velocity u_2. Under the above condition the integral force vanishes, but finite forces are required at all points of the slope, except the particular point where $u = a$.

It is of some importance to notice that although, under the condition $u_1 u_2 = a^2$, the applied forces contribute on the whole no *momentum*, yet they do contribute *energy*, positive or negative. To find the work done in unit of time by the forces we have

$$2/m \cdot \int X\rho \cdot u \cdot dx = u_2^2 - u_1^2 - 2u_1 u_2 \log (u_2/u_1). \quad \ldots\ldots\ldots\ldots(53)$$

The better to interpret this let us suppose that u_1 and u_2 are positive, and in the first instance that $u_2 > u_1$, so that the fluid passes from a less to a greater velocity, or by (45) from a greater to a less density. In the case of a wave of rarefaction we have therefore to consider the sign of

$$y^2 - 1 - 2y \log y, \quad \ldots\ldots\ldots\ldots\ldots\ldots\ldots\ldots\ldots(54)$$

when $y > 1$. It is not difficult to prove that this sign is always positive. When $y - 1$ is small, the approximate value of (54) is $\frac{1}{3}(y-1)^3$, and is therefore positive when $y > 1$. Again, if we remove the positive factor y from (54) and then differentiate, we obtain $(1 - 1/y)^2$, which is positive. Hence, when $y > 1$, (54) is necessarily positive. The propagation of the wave of rarefaction without change of type requires that the impressed forces, contributing on the whole no momentum, should nevertheless do work upon, *i.e.* communicate energy to, the gas.

In like manner, if the wave be one of condensation, *i.e.*, if the gas passes from a less to a greater density, the operation of the impressed forces is to remove energy from the gas forming the wave. It follows that although dissipative forces, such as those arising from viscosity, may possibly constitute a machinery capable of maintaining the type of a wave of condensation, in no case can they maintain the type of a wave of rarefaction.

It is desirable to extend this argument to waves propagated under the adiabatic law. In general, from (45), (46),

$$X\rho = m \frac{du}{dx} + \frac{dp}{dx};$$

so that $\quad \int X\rho \, dx = m(u_2 - u_1) + p_2 - p_1 = m^2/\rho_2 - m^2/\rho_1 + p_2 - p_1.$

As in (50), the condition that on the whole no momentum is communicated is

$$m^2/\rho_2 - m^2/\rho_1 + p_2 - p_1 = 0. \quad \ldots\ldots\ldots\ldots\ldots\ldots(55)$$

Again,

$$m^{-1} \cdot \int X\rho \cdot u \cdot dx = \tfrac{1}{2}(u_2^2 - u_1^2) + \int_{p_1}^{p_2} \frac{dp}{\rho} = \int_{p_1}^{p_2} \frac{dp}{\rho} - \frac{p_2 - p_1}{2} \left(\frac{1}{\rho_2} + \frac{1}{\rho_1}\right). \ldots(56)$$

The question in which we are interested is the sign of (56). If we regard v (the volume of unit mass), viz., $1/\rho$, as the ordinate, and p as the abscissa of a curve, the first term on the right of (56) represents the area of the curve bounded by two ordinates and the axis of p, while the second term is what the area would be if the ordinate retained throughout the mean of the terminal values.

So far the argument is general. If the relation between p and v be adiabatic, and $p_2 > p_1$, the expression (56) is negative, since v proportional to $p^{-1/\gamma}$ makes d^2v/dp^2 positive*. The final pressure exceeding the initial pressure denotes a wave of condensation, and we conclude, as before, that maintenance of type in such a wave requires removal of energy from the wave, while in the contrary case of a wave of rarefaction additional energy would need to be supplied.

The problem now under discussion is closely related to one which has given rise to a serious difference of opinion. In his paper of 1848 already referred to, Stokes considered the *sudden* transition from one constant velocity to another, and concluded that the necessary conditions for a permanent regime could be satisfied. Results equivalent to his may be deduced from (45) in connection with the condition $(u_1 u_2 = a^2)$ already found from (52) to express that there is no change of momentum on the whole. Thus,

$$u_1 = a\sqrt{(\rho_2/\rho_1)}, \qquad u_2 = a\sqrt{(\rho_1/\rho_2)}. \quad\dots\dots\dots\dots\dots(57)$$

Similar conclusions were put forward by Riemann in 1860 (*loc. cit.*). Commenting on these results in the *Theory of Sound* (1878), I pointed out that although the conditions of *mass* and *momentum* were satisfied, the condition of *energy* was violated, and that therefore the motion was not possible; and in republishing this paper† Stokes admitted the criticism, which had indeed already been made privately by Kelvin. On the other hand, Burton‡ and H. Weber§ maintain, at least to some extent, the original view.

Inasmuch as they ignored the question of energy, it was natural that Stokes and Riemann made no distinction between the cases where energy is gained or lost. As I understand, Weber abandons Riemann's solution for the discontinuous wave (or *bore*, as it is sometimes called for brevity) of rarefaction, but still maintains it for the case of the bore of condensation. No doubt there is an important distinction between the two cases; nevertheless, I fail to understand how a loss of energy can be admitted in a motion

---

* Compare Lamb's *Hydrodynamics*, § 280.
† *Collected Works*, Vol. II. p. 55.
‡ *Phil. Mag.* 1893, Vol. XXXV. p. 316.
§ *Die Partiellen Differentialgleichungen der Mathematischen Physik*, Braunschweig, 1901, Vol. II. p. 496.

which is supposed to be subject to the isothermal or adiabatic laws, in which no dissipative action is contemplated. In the present paper the discussion proceeds upon the supposition of a *gradual* transition between the two velocities or densities. It does not appear how a solution which violates mechanical principles, however rapid the transition, can become valid when the transition is supposed to become absolutely abrupt. All that I am able to admit is that under these circumstances dissipative forces (such as viscosity) that are infinitely small may be competent to produce a finite effect.

If we suppose that under the influence of small dissipative forces the bore of Stokes and Riemann can be propagated, at least approximately, we naturally inquire whether it can be regarded as the complete outcome of the simple progressive wave with a straight velocity slope which, as we have found, tends after a definite interval of time to assume the character of a bore. It would seem that the answer must be in the negative. Taking Boyle's law, we recognise from (5) that in the progressive wave, just before the formation of the bore, the relation between the velocities and densities is

$$u_2 - u_1 = a \log \rho_2 - a \log \rho_1,$$

while in (57) the relation is

$$u_2 - u_1 = a \sqrt{(\rho_2/\rho_1)} - a \sqrt{(\rho_1/\rho_2)}.$$

The two functions of ρ on the right, which are independent of any common addition to u_1 and u_2, cannot be identified (unless $\rho_2 = \rho_1$), as we have found already in discussing (54). This incompatibility may be regarded as a confirmation of Stokes' opinion that something of the nature of reflection must ensue.

Permanent Regime under the influence of Dissipative Forces.

The first investigation to be considered under this head is a very remarkable one by Rankine "On the Thermodynamic Theory of Waves of Finite Longitudinal Disturbance*," which (except a limited part expounded by Maxwell in his *Theory of Heat*) has been much neglected†. Conduction of heat is here for the first time taken into account and although there are one or two serious deficiencies, not to say errors, presently to be noticed, the memoir marks a very definite advance.

The first step is the establishment of an equation equivalent (when the wave is reduced to rest) to (45), and of Earnshaw's relation (50), in which

* *Phil. Trans.* 1870, Vol. CLX. Part II. p. 277.

† I must take my share of the blame. Rankine is referred to by Lamb (*Hydrodynamics*, 1906, p. 466). The body of Rankine's memoir seems to have been composed without acquaintance with the writings of his predecessors; but in a supplement he notices the work of Poisson, Stokes, Airy, and Earnshaw.

equations we shall usually substitute v, the volume of unit mass, for $1/\rho$. Rankine remarks that "no substance yet known fulfils the condition expressed by (50) between finite limits of disturbance, at a constant temperature, nor in a state of non-conduction of heat (called the *adiabatic* state). In order, then, that permanency of type may be possible in a wave of longitudinal disturbance, there must be both change of temperature and conduction of heat during the disturbance." However, we shall see later that even under Boyle's law *viscosity* is competent to endow a wave with permanency.

The question is, how can Earnshaw's law be satisfied? Obviously not (in the absence of viscosity) if the expansions are adiabatic; but if at every stage the right quantity of heat is added or subtracted, the gas may be made to follow any prescribed law. This is the idea underlying Rankine's investigation. For the unit mass of a *perfect gas* we have, as usual, $pv = R\theta$, θ denoting absolute temperature. The condition of the gas is defined by any two of the three quantities p, v, θ, and the third may be expressed in terms of them. The relation between simultaneous variations is

$$d\theta/\theta = dp/p + dv/v. \tag{58}$$

In order to effect the changes specified by dp and dv, it is in general necessary to communicate heat to the gas. Calling the necessary quantity of heat dQ, we may write

$$dQ = \left(\frac{dQ}{dv}\right)dv + \left(\frac{dQ}{dp}\right)dp. \tag{59}$$

Suppose now (*a*) that $dp = 0$. Equations (58), (59), give

$$\frac{dQ}{d\theta}(p \text{ constant}) = \left(\frac{dQ}{dv}\right)\frac{v}{\theta},$$

where $dQ/d\theta$ (p constant) expresses the specific heat of the gas under constant pressure. Denoting this by C, we have

$$C = \left(\frac{dQ}{dv}\right)\frac{v}{\theta}.$$

Again, suppose (*b*) that $dv = 0$. We find in a similar manner that if c denote the specific heat under constant volume,

$$c = \left(\frac{dQ}{dp}\right)\frac{p}{\theta}.$$

Thus, in general, $\qquad dQ/\theta = Cdv/v + cdp/p. \tag{60}$

If between (58) and (60) we eliminate dp, there results

$$dQ = (C-c)\frac{pdv}{R} + cd\theta. \tag{61}$$

In (61) $dQ = 0$ corresponds to adiabatic expansion, when according to

Mayer's principle the cooling effect $-cd\theta$ is equal to the external work done by the gas pdv. Hence
$$C - c = R, \quad\quad\quad\quad\quad\quad\quad\quad (62)$$
and therefore
$$\gamma = \frac{C}{c} = \frac{C}{C - R}, \quad\quad\quad\quad\quad\quad\quad\quad (63)$$
a relation discovered by Rankine himself in 1850.

Rankine then applies (60) to find what heat must be communicated in order that the gas may follow Earnshaw's law making $dp = -m^2 dv$. With regard to (62), it appears that
$$dQ = \frac{dp}{m^2(\gamma - 1)} \{p_0 + m^2 v_0 - (\gamma + 1) p\}. \quad\quad\quad\quad (64)$$
It will be understood that under the condition now imposed of Earnshaw's relation, as well as of the ordinary gas law, there remains but one independent variable, and that the state of the gas may be expressed in terms of any *one* of the three quantities p, v, θ.

We have next to consider how far the necessary supply of heat defined by (64) can be effected by *conduction*. If the initial state (distinguished by suffix 1) and final state (with suffix 2) be of uniformity with respect to x, the total quantity of heat received by the gas during its passage must be zero, or $\int dQ = 0$. Hence from (64) Rankine finds
$$p_0 + m^2 v_0 = \tfrac{1}{2}(\gamma + 1)(p_1 + p_2). \quad\quad\quad\quad\quad (65)$$
This is a necessary condition; but of course there is nothing so far to show that it is sufficient.

In (65) p_0, v_0 are any corresponding values of p and v within the wave, and we may identify them with p_1, v_1*.

Thus
$$m^2 v_1 = \tfrac{1}{2}(\gamma - 1) p_1 + \tfrac{1}{2}(\gamma + 1) p_2. \quad\quad\quad\quad (66)$$
The velocity u_1, equal to mv_1, is that with which the wave advances relatively to the fluid in state (1). And
$$u_1^2 = m^2 v_1^2 = v_1 \{\tfrac{1}{2}(\gamma - 1) p_1 + \tfrac{1}{2}(\gamma + 1) p_2\} \quad\quad\quad\quad (67)$$
gives the square of the velocity of wave-propagation relatively to fluid (1). The velocity of propagation of infinitely small disturbances (p_2 nearly equal to p_1) is given by $u_1^2 = \gamma p_1 v_1$; and thus a wave of finite condensation is propagated faster than an infinitesimal wave, and according to (67) a wave of finite rarefaction would be propagated slower than an infinitesimal wave. Moreover, there is no limit to the velocity of a wave of condensation.

---

* We may, of course, [alternatively] identify p_0, v_0 with p_2, v_2.

Rankine proceeds to express the absolute temperature (θ) at a point where the pressure is p in a wave of permanent type. By Earnshaw's law (50) in combination with (65)

$$\frac{\theta}{\theta_0} = \frac{pv}{p_0 v_0} = \frac{p}{p_0} \cdot \frac{(\gamma+1)(p_1+p_2) - 2p}{(\gamma+1)(p_1+p_2) - 2p_0}, \quad \ldots\ldots\ldots\ldots(68)$$

and for the ratio of terminal temperatures

$$\frac{\theta_2}{\theta_1} = \frac{p_2}{p_1} \cdot \frac{(\gamma+1)p_1 + (\gamma-1)p_2}{(\gamma+1)p_2 + (\gamma-1)p_1}. \quad \ldots\ldots\ldots\ldots\ldots(69)$$

The second fraction on the right of (69) obviously represents the ratio of volumes v_2/v_1, or of densities ρ_1/ρ_2.

In order to justify (65), it is not necessary that the terminal states be states of absolute uniformity. It will suffice that the temperature be there stationary ($d\theta/dx = 0$), which secures that no conduction of heat takes place there, and a state of stationary temperature usually involves a stationary pressure. To make the most of (65) we must apply it to the smallest ranges, *i.e.* between consecutive places where dp/dx vanishes.

But here a question arises which Rankine does not seem to have considered. In order to secure the necessary transfers of heat by means of conduction it is an indispensable condition that the heat should pass from the hotter to the colder body. If maintenance of type be possible in a particular wave as the result of conduction, a reversal of the motion will give a wave whose type cannot be so maintained. We have seen reason already for the conclusion that a dissipative agency can serve to maintain the type only when the gas passes from a less to a more condensed state. If this be so, the application which Rankine makes to a periodic wave is evidently prohibited.

According to the *second* law of thermodynamics, the criterion whether the transformation is possible as the result of dissipative action is the sign of $\int dQ/\theta$. If this be negative, the transformation is not possible. From (64) with use of (65)

$$dQ = \frac{(\gamma+1)\, dp}{2m^2(\gamma-1)} (p_1 + p_2 - 2p). \quad \ldots\ldots\ldots\ldots\ldots(70)$$

In (68) we may give p_0, θ_0 any corresponding values found in the wave. Thus p_0 lies between p_1 and p_2, and ($\gamma > 1$)

$$(\gamma+1)(p_1+p_2) - 2p_0 \text{ is positive.}$$

Accordingly $\int dQ/\theta$ takes the same sign as

$$\int_{p_1}^{p_2} \frac{dp\,(p_1 + p_2 - p)}{p\,\{(\gamma+1)(p_1+p_2) - 2p\}}. \quad \ldots\ldots\ldots\ldots\ldots(71)$$

The integral (71) is evaluated without difficulty. Dropping the factor $1/(\gamma+1)$, and writing $\varpi = p_2/p_1$, we get

$$\log \varpi + \gamma \log \frac{\gamma+1+(\gamma-1)\varpi}{\gamma-1+(\gamma+1)\varpi} \quad \ldots \ldots \ldots \ldots \ldots \ldots (72)$$

It is evident that (72) changes sign when we substitute $1/\varpi$ for ϖ.

If we expand (72) in powers of $(\varpi - 1)$ we find

$$(72) = \frac{(\gamma^2-1)(\varpi-1)^3}{12\gamma^2} + \ldots,$$

the terms in $(\varpi-1)$, $(\varpi-1)^2$, disappearing. Thus, when $\varpi - 1$ is positive, (72) begins positive. Differentiating (72) with respect to ϖ, we get

$$\frac{1}{\varpi} + \frac{\gamma(\gamma-1)}{\gamma+1+(\gamma-1)\varpi} - \frac{\gamma(\gamma+1)}{\gamma-1+(\gamma+1)\varpi} \quad \ldots \ldots \ldots (73)$$

When (73) is reduced to a single fraction, the denominator is positive, and the numerator is

$$(\gamma^2-1)(\varpi-1)^2.$$

We infer that if $\varpi > 1$, (72) is always positive, and that if $\varpi < 1$, (72) is always negative. Hence if $p_2 > p_1$, *i.e.* if the wave be one of condensation, the communications of heat required are such as may arise from conduction; but if the wave be one of rarefaction, its permanency can in no wise be attained as the result of conduction. A wave of condensation here means a wave such that during its progress the gas passes always from a less dense to a more dense state, and the most important case is when the limits are finite, so that the passage constitutes the transition from one uniform density to a greater uniform density.

Rankine proceeds to examine more particularly under what conditions a wave can be permanent. " In order that a particular type of disturbance may be capable of permanence during its propagation, a relation must exist between the temperatures of the particles and their relative positions, such that the conduction of heat between the particles may effect the transfers of heat required by the thermodynamic conditions of permanence of type."

The equation of conduction is readily found. The heat conducted in unit time across a layer of the gas is represented by $k d\theta/dx$, where k is a coefficient of conductivity which may be a function of the condition of the gas, here dependent on one variable. The equation of conduction is $(u = +)$

$$v \frac{d}{dx}\left(k \frac{d\theta}{dx}\right) = \frac{D}{Dt} Q = u \frac{dQ}{dx} = mv \frac{dQ}{dx},$$

whence, if we reckon Q from the initial condition of constant pressure p_1,

$$k \frac{d\theta}{dx} = mQ. \quad \ldots \ldots \ldots \ldots \ldots \ldots \ldots \ldots \ldots (74)$$

And from (70)
$$Q = \frac{\gamma+1}{2m^2(\gamma-1)} \int_{p_1}^{p} (p_1 + p_2 - 2p)\, dp = \frac{\gamma+1}{2m^2(\gamma-1)} (p - p_1)(p_2 - p). \quad \ldots(75)$$

Also from (50), (65),
$$p + m^2 v = \tfrac{1}{2}(\gamma+1)(p_1 + p_2), \quad \ldots\ldots\ldots\ldots\ldots\ldots(76)$$

whence
$$\theta = \frac{pv}{R} = \frac{p}{2m^2 R}\{(\gamma+1)(p_1 + p_2) - 2p\}, \quad \ldots\ldots\ldots\ldots\ldots\ldots(77)$$

and
$$\frac{d\theta}{dp} = \frac{(\gamma+1)(p_1 + p_2) - 4p}{2m^2 R}. \quad \ldots\ldots\ldots\ldots\ldots\ldots(78)$$

Using these, we find with regard to (62)
$$dx = \frac{k}{mQ} \frac{d\theta}{dp} dp = \frac{k\, dp}{mc(\gamma+1)} \frac{(\gamma+1)(p_1 + p_2) - 4p}{(p - p_1)(p_2 - p)}, \quad \ldots\ldots\ldots(79)$$

by which is determined the distribution of pressure (and thence of density and temperature) along the line of propagation.

On the supposition that k is constant, Rankine integrates (79) in terms of logarithms. Writing
$$p - \tfrac{1}{2}(p_1 + p_2) = q, \qquad \tfrac{1}{2}(p_2 - p_1) = q_1,$$

he obtains
$$\frac{dx}{dq} = \frac{k}{mc(\gamma+1)} \frac{(\gamma-1)(p_1 + p_2) - q}{q_1^2 - q^2},$$

and
$$x = \frac{k}{mc(\gamma+1)} \left\{ \frac{(\gamma-1)(p_1 + p_2)}{2q_1} \log \frac{q_1 + q}{q_1 - q} + 2 \log\left(1 - \frac{q^2}{q_1^2}\right) \right\}, \quad \ldots(80)$$

x being measured from the place where $q = 0$. Mathematically the wave is infinitely long; but practically the transition of pressure is effected in a distance comparable with $k/mc(\gamma+1)$, which may be small in terms of ordinary standards. It is to be observed that the general character of the result does not depend upon the constancy of k.

Reverting to (79), we see that the denominator on the right is positive, and that the numerator is also positive for that part of the wave where p is nearly equal to $\tfrac{1}{2}(p_1 + p_2)$. Thus, for this part of the wave at any rate, p and x increase together; or, since u is positive, the gas passes to a condition of greater density—the wave must be one of condensation. This consideration, as we have seen, Rankine overlooked. And a further limitation presents itself: since there cannot be two pressures in one place, it is evident that dx/dp must not change sign. The numerator in (79) must be positive over its *whole* range from p_1 to p_2, and this will not be the case if p_2/p_1 exceed $(\gamma+1)/(3-\gamma)$, equal for common gases to 1·61. The conclusion is that the only kind of wave, involving a transition from one uniform pressure to another, which can be maintained with the aid of conduction is a wave of condensation, and then only when the ratio of pressures does not exceed a moderate value.

The next contribution to the subject upon which I have to comment is contained in a long and ably written memoir by Hugoniot*. This author, though he covers to a great extent the same ground, makes no reference to Stokes, Earnshaw, Riemann, or Rankine, and but a very slight one to Poisson—a circumstance which increases the difficulty of comparison. Since Hugoniot uses the Lagrangian form of equation, his investigation runs naturally on the same lines as Earnshaw's, whose general solution for a single progressive wave is reproduced. I have already alluded to the solution of special problems relating to the propagation of a wave of variable type.

The most original part of Hugoniot's work has been supposed to be his treatment of discontinuous waves involving a sudden change of pressure, with respect to which he formulated a law often called after his name by French writers. But a little examination reveals that this law is *precisely the same* as that given 15 years earlier by Rankine, a fact which is the more surprising inasmuch as the two authors start from quite different points of view. Rankine's investigation, as we have seen, is expressly based upon conduction of heat in the gas, but Hugoniot supposes his gas to be non-conducting. A question of some delicacy is here involved, which will repay careful examination. It will be convenient to give a paraphrase of Hugoniot's argument†.

This argument depends upon an application of the principle of energy to a region bounded by two fixed planes, including the place of discontinuity. The work done by the fluid as it emerges with volume v_2 against the pressure p_2 is $p_2 v_2$. On the whole, therefore, the external work done by the passage of the unit of mass is $p_2 v_2 - p_1 v_1$. The increase of kinetic energy of the fluid is

$$\tfrac{1}{2}(u_2^2 - u_1^2) = \tfrac{1}{2} m^2 (v_2^2 - v_1^2) = \tfrac{1}{2}(v_2 + v_1)(p_1 - p_2),$$

in virtue of (50), which requires that

$$p_1 - p_2 + m^2 (v_1 - v_2) = 0. \quad\quad\quad\quad\quad\quad (81)$$

The sum of these is

$$p_2 v_2 - p_1 v_1 + \tfrac{1}{2}(p_1 - p_2)(v_2 + v_1) = \tfrac{1}{2}(v_2 - v_1)(p_2 + p_1). \quad\quad (82)$$

We have next to consider the internal energy of unit of mass in the initial and final states. For this purpose we suppose the gas to expand adiabatically from its actual volume v to an infinite volume. In this expansion the work done by the gas is

$$\int_v^\infty p\, dv = \frac{pv}{\gamma - 1}, \quad\quad\quad\quad\quad\quad (83)$$

---

* *Journal de l'École Polytechnique*, 1887, 1889.
† Compare Lamb's *Hydrodynamics*, 1906, § 280.

so that the difference of internal energy in the two states is

$$\frac{p_2 v_2}{\gamma - 1} - \frac{p_1 v_1}{\gamma - 1}. \quad\quad\quad\quad\quad\quad (84)$$

The principle of energy requires that the sum of this and (82) be zero, whence

$$\gamma = \frac{(p_2 - p_1)(v_1 + v_2)}{(p_1 + p_2)(v_1 - v_2)} \quad\quad\quad\quad\quad\quad (85)$$

is the relation between the pressures and volumes in the two states. The result thus found by Hugoniot is the same as Rankine's. From Rankine's equation (65)

$$p_1 + m^2 v_1 = p_2 + m^2 v_2 = \tfrac{1}{2}(\gamma + 1)(p_1 + p_2) = \tfrac{1}{2}(p_1 + p_2) + \tfrac{1}{2} m^2 (v_1 + v_2),$$

it follows that
$$m^2 = \gamma \frac{p_1 + p_2}{v_1 + v_2} = \frac{p_2 - p_1}{v_1 - v_2}, \quad\quad\quad\quad\quad\quad (86)$$

which is identical with (85).

The first remark that I will make is that, although Hugoniot assumes that the transition between the two states is sudden, there is nothing in his argument which requires this, all that is really necessary being that the *régime* is permanent. The next remark is that, however valid (85) may be, its fulfilment does not secure that the wave so defined is possible. As a matter of fact, a whole class of such waves is certainly impossible, and I would maintain, further, that a wave of the kind is never possible under the conditions, laid down by Hugoniot, of no viscosity or heat-conduction.

A closer examination of the process by which (85) was obtained will show that while the first law of thermodynamics has been observed, the second law has been disregarded. The crux of the matter lies in the comparison of the internal energies of the incoming and outgoing gas expressed in (84). If (p_2, v_2) and (p_1, v_1) lie upon the same adiabatic, the work corresponding to the passage from the one state to the other is given without ambiguity by (84). But in the present case the two states do not lie upon the same adiabatic, and the work required is deduced upon the assumption that nothing is involved in the passage at $v = \infty$ from one adiabatic to the other. What is actually there required is the communication (positive or negative) of an infinitesimal quantity of heat. From the point of view of the first law the infinitesimal quantity of heat may be neglected, but not so from the point of view of the second law, since the transfer is supposed to take place at the zero of temperature. When heat and work are distinguished, infinitesimal heat at zero may have a finite value. The imaginary passage to infinity has the advantage of leading rapidly to the required conclusion, but it rather tends to obscure the real nature of the process. While all the other items of the account are mechanical work, the passage from one adiabatic to the other (which may take place at constant finite volume) is a question of *heat* as distinguished from work. If during a complete cycle work would be lost and corresponding heat gained, the operation is dissipative and there need be no contradiction if viscosity or heat-conduction enter, but the opposite contingency of a gain of work at the expense of heat is excluded in all cases. The conclusion is the same as before. While a wave of condensation may, perhaps, maintain a permanent regime as the result of dissipative agencies, a permanent wave of rarefaction is excluded.

It is remarked by Hugoniot that even when the ratio p_1/p_2 is infinite, v_2/v_1 does not exceed $(\gamma+1)/(\gamma-1)$, which for common gases is equal to about 6. A similar remark is made by Duhem*, who discusses the whole question with great generality. With regard to perfect gases "lorsqu'une quasi-onde de choc se propage au sein d'un gaz parfait, le fluide le plus condensé est toujours en amont de l'onde et le fluide le moins condensé en aval." But, so far as I see, neither of these authors proves that the propagation is possible in any case.

It is a question of great interest to inquire what is the influence of viscosity and especially whether alone, or in co-operation with heat-conduction, it allows a wave of condensation to acquire a permanent regime. We proceed to consider this question on the basis of the usual equations, although it must be admitted that their application to conditions which are somewhat extreme raises points of uncertainty.

Reverting to our original equations, we recognise that (45) is unaffected by the inclusion of viscosity, and that the change required in (46) is represented by writing†

$$X = \frac{4}{3\rho} \frac{d}{dx}\left(\mu \frac{du}{dx}\right),$$

so that (46) takes the form

$$m\frac{du}{dx} + \frac{dp}{dx} - \frac{4}{3}\frac{d}{dx}\left(\mu \frac{du}{dx}\right) = 0,$$

whence ($v = 1/\rho$)

$$p + m^2 v - \frac{4}{3} m\mu \frac{dv}{dx} = p_1 + m^2 v_1 = p_2 + m^2 v_2, \quad \ldots\ldots\ldots\ldots(87)$$

the terminal states (p_1, v_1), (p_2, v_2), being of uniformity, so that dv/dx there vanishes. From this it appears that (81), relating to the terminal states, holds good equally when viscosity is regarded.

* *Zeitschrift f. Physikal. Chem.* Vol. LXIX. p. 169 (1909).
† Lamb's *Hydrodynamics*, §§ 314, 316.

A simple example under the head of viscosity is to suppose the temperature maintained uniform, as by a powerful radiation, so that the gas follows Boyle's law, making
$$pv = p_1v_1 = p_2v_2 = a^2.$$

From this and (87) we get
$$m^2v_1v_2 = a^2,$$
and
$$p_1 + m^2v_1 = a^2/v_1 + a^2/v_2.$$

Using these in (87), we find
$$\frac{3m\,dx}{4\mu} = -\frac{v\,dv}{(v_1 - v)(v - v_2)}, \qquad\qquad\qquad(88)$$

as governing the distribution of v along the line of propagation. In a wave of condensation $v_1 > v > v_2$, so that the denominator on the right of (88) is positive. Thus when m is positive, dv/dx is negative, as should be the case. On integration (μ constant)
$$\frac{3mx}{4\mu} = \frac{1}{v_1 - v_2} \{v_1 \log(v_1 - v) - v_2 \log(v - v_2)\}, \qquad(89)$$

the origin of x being chosen suitably.

The transition of volumes from v_1 to v_2 occupies, mathematically speaking, the whole range from $x = -\infty$ to $x = +\infty$, but practically it may be very sudden. Since in (88) dx/dv never changes sign, the condition of permanency for a condensational wave can always be satisfied, whatever may be the value of the ratio v_1/v_2* or p_1/p_2, contrasting in this respect with the limitation found to be necessary on Rankine's conclusion relative to heat-conduction.

As regards the velocity of wave propagation into the rarer medium, we have for its square
$$u_1^2 = m^2v_1^2 = a^2v_1/v_2. \qquad\qquad\qquad(90)$$

Returning to the case where heat development and viscosity are both regarded, we see that in virtue of (81) Hugoniot's reasoning is still applicable without change, and it leads to the same final relation (85) as was found by Rankine when heat-conduction is alone considered.

In endeavouring to apply Rankine's method to the more general case where viscosity is retained, we shall find it more convenient to treat v, or $(1/\rho)$, rather than p, as independent variable. If, as before, dQ denotes the total quantity of heat received by unit mass of the gas, we have from (60), (62),
$$(\gamma - 1)\frac{dQ}{dx} = \gamma p \frac{dv}{dx} + v \frac{dp}{dx};$$

---

* But the limitation pointed out by Hugoniot still obtains; otherwise, one of the pressures would be negative.

or, on elimination of p by means of (87),

$$(\gamma-1)\frac{dQ}{dx} = \gamma\left\{p_1 + m^2v_1 - m^2v + \tfrac{4}{3}m\mu\frac{dv}{dx}\right\}\frac{dv}{dx} + v\left\{-m^2\frac{dv}{dx} + \tfrac{4}{3}m\frac{d}{dx}\left(\mu\frac{dv}{dx}\right)\right\}.$$
...(91)

In (91) dQ consists of two parts, the first (dQ_1), with which alone Rankine dealt, the heat received by conduction, and the second (dQ_2) the heat developed internally under viscosity. As regards the latter, the heat developed in volume v and time dt is $\tfrac{4}{3}v\mu(du/dx)^2 dt$*, in which we are to replace dt by dx/u, and u by mv, so that

$$\frac{dQ_2}{dx} = \tfrac{4}{3}m\mu\left(\frac{dv}{dx}\right)^2. \tag{92}$$

Multiplying this by $(\gamma-1)$ and subtracting it from (91), we get

$$(\gamma-1)\frac{dQ_1}{dx} = \gamma(p_1 + m^2v_1)\frac{dv}{dx} - (\gamma+1)m^2v\frac{dv}{dx} + \tfrac{4}{3}m\frac{d}{dx}\left(\mu v\frac{dv}{dx}\right). \quad\ldots(93)$$

As in Rankine's investigation, the whole heat received by conduction in passing from one uniform state v_1 to another uniform state v_2 must vanish. Hence, on integrating between these limits, and dividing out the factor $(v_2 - v_1)$, we have

$$(\gamma+1)m^2(v_1 + v_2) = 2\gamma(p_1 + m^2v_1) = 2\gamma(p_2 + m^2v_2)$$
$$= \gamma(p_1 + p_2) + \gamma m^2(v_1 + v_2),$$

or, as in (86), $\qquad m^2(v_1 + v_2) = \gamma(p_1 + p_2),$

the same relation as was found by Rankine. Introducing it into (93), we get

$$(\gamma-1)\frac{dQ_1}{dx} = \tfrac{1}{2}(\gamma+1)m^2\left\{(v_1 + v_2)\frac{dv}{dx} - \frac{dv^2}{dx}\right\} + \tfrac{4}{3}m\frac{d}{dx}\left(\mu v\frac{dv}{dx}\right). \quad\ldots(94)$$

A particular case arises when we suppose the conductivity to be zero, so that dQ_1/dx vanishes throughout. We have then

$$\tfrac{4}{3}\mu v\frac{dv}{dx} + \tfrac{1}{2}(\gamma+1)m(v_1 - v)(v - v_2) = 0, \tag{95}$$

differing from (88) only by the factor $\tfrac{1}{2}(\gamma+1)$. On the supposition that μ is constant, the solution is nearly the same as in (89), and, in fact, reduces to it when $\gamma = 1$, which represents Boyle's law. This case of no conduction is thus satisfactorily disposed of. Whatever be the ratio of pressures, a wave of condensation is always possible.

It should be remarked, however, that the supposition of constant μ does not consist with the facts as known for actual gases when γ differs from

* Lamb's *Hydrodynamics*, § 341.

unity. For such gases viscosity, though independent of *density*, varies with *temperature*, so that μ will not be constant in (95). But since μ is always positive, this complication merely affects the particular form of the integral and not the general conclusion as to the possibility of a permanent wave.

In general, from (94), if we reckon Q_1 from the terminal state v_1,

$$(\gamma - 1) Q_1 = \tfrac{1}{2}(\gamma + 1) m^2 (v_1 - v)(v - v_2) + \tfrac{4}{3}m\mu v \frac{dv}{dx} \quad \ldots\ldots\ldots(96)$$

The equation of conduction is the same (74) as before. And for θ, from (87),

$$\theta = \frac{pv}{R} = \frac{v}{R} \left\{ \frac{\gamma + 1}{2\gamma} m^2 (v_1 + v_2) - m^2 v + \tfrac{4}{3}m\mu \frac{dv}{dx} \right\};$$

so that with regard to (62) the equation of conduction becomes

$$\frac{k}{mc}\left[m \frac{dv}{dx}\left\{ \frac{\gamma+1}{2\gamma}(v_1 + v_2) - 2v \right\} + \tfrac{4}{3}\frac{d}{dx}\left(\mu v \frac{dv}{dx}\right)\right]$$
$$= \tfrac{1}{2}(\gamma + 1) m (v_1 - v)(v - v_2) + \tfrac{4}{3}\mu v \frac{dv}{dx}. \quad \ldots(97)$$

By omitting the terms containing μ we may of course fall back on Rankine's problem.

Equation (97), in its general form, is much more complicated than when either viscosity or heat-conduction is alone regarded, in consequence of the occurrence of the differential coefficient of the second order. In general, both k and μ are functions of temperature, and therefore of v; but, according to Maxwell's theory, which assumes a molecular repulsion inversely as the fifth power of the distance, $c\mu/k$ is independent of temperature (as well as of density), and takes the value $\tfrac{2}{5}$. And it would seem that this independence of temperature and density is general, seeing that the ratio is of no dimensions, at least so long as the repulsive force can be represented by an inverse power of the distance*. We shall write h for the above ratio and assume that for a given gas it is an absolute constant. Thus, μ' being written for μ/m, (97) takes the form—

$$\mu' \frac{d}{dx}\left(\mu' \frac{dv^2}{dx}\right) + \mu' \frac{dv^2}{dx}\left\{ \frac{3(\gamma+1)}{8\gamma}\frac{v_1 + v_2}{v} - \tfrac{3}{2} - h \right\} = \tfrac{3}{4}h(\gamma+1)(v_1 - v)(v - v_2);$$
$$\ldots(98)$$

in which v^2 may be regarded as the dependent variable. For v^2 we shall write ξ, and if

$$U = \mu' d\xi/dx, \quad \ldots\ldots\ldots\ldots\ldots\ldots\ldots\ldots\ldots(99)$$

* Compare *Roy. Soc. Proc.* Vol. LXVI. p. 68 (1900); *Scientific Papers*, Vol. IV. p. 453.

our equation, since it contains x only through dx, may be reduced to one of the first order in U and ξ, i.e.,

$$U\frac{dU}{d\xi} + Uf(\xi) = F(\xi), \quad \quad \quad (100)$$

where
$$f(\xi) = \frac{3(\gamma+1)}{8\gamma}\frac{\sqrt{\xi_1}+\sqrt{\xi_2}}{\sqrt{\xi}} - \tfrac{3}{2} - h, \quad \quad \quad (101)$$

and
$$F(\xi) = \tfrac{3}{4}h(\gamma+1)(\sqrt{\xi_1}-\sqrt{\xi})(\sqrt{\xi}-\sqrt{\xi_2}). \quad \quad \quad (102)$$

If U can be found as a function of ξ from (100), x follows by simple integration of (99).

In considering equation (100) we may conveniently regard ξ as the linear co-ordinate of a material particle of unit mass moving in a straight line with velocity U. The first term, $UdU/d\xi$, then represents the acceleration of the particle; the second, $Uf(\xi)$, may be regarded as a *resistance*, proportional to the velocity, and at the same time variable with the position (ξ); and the third term on the right hand represents a force, which is also a function of position. If t be the time in this subsidiary problem, $U = d\xi/dt$, and (100) may be written

$$\frac{d^2\xi}{dt^2} + f(\xi)\frac{d\xi}{dt} = F(\xi), \quad \quad \quad (103)$$

while by (99)
$$dx = \mu' dt. \quad \quad \quad (104)$$

If μ' be constant, the substitution of $\mu' t$ for x in (98) is obvious.

If we take $h = 0.4$, $\gamma = 1.41$, (101), (102), become

$$f(\xi) = 0.641\frac{\sqrt{\xi_1}+\sqrt{\xi_2}}{\sqrt{\xi}} - 1.900, \quad \quad \quad (105)$$

$$F(\xi) = 0.723(\sqrt{\xi_1}-\sqrt{\xi})(\sqrt{\xi}-\sqrt{\xi_2}). \quad \quad \quad (106)$$

It will be observed that, over the range from ξ_1 to ξ_2, $F(\xi)$ is positive, but that the sign of $f(\xi)$ is doubtful. If ξ has the greater terminal value ξ_1, f is negative; but, when it has the smaller terminal value ξ_2, the sign depends upon the ratio ξ_1/ξ_2. If this ratio < 1.21, f is negative; otherwise it is positive.

I suppose that a complete analytical solution of our equation is not to be expected, and it is, indeed, hardly necessary for our purpose. What we most wish to know is whether a solution is possible which satisfies the prescribed conditions. Among these is the requirement that U in (100) vanish at both limits; and even then the manner of evanescence must be such as to secure that x, as determined by (99), shall be infinite at these limits. As the problem originally presents itself, we should have the representative particle travelling in the negative direction from ξ_1 to ξ_2, starting with no velocity and arriving with no velocity. It seems simpler to consider it in a modified form, *i.e.* with the motion reversed, so that it takes place in the direction of the force F. There is, then, no question of the particle stopping between the limits and returning upon its course. We may make this change, if in (105) we reverse the sign of f. We consider, then, the motion of the particle to be in the positive direction, from ξ_2 to ξ_1, with zero velocity at both limits, the motion between ξ_2 and ξ_1 being aided by the force F, which itself vanishes at these limits, and being also subject to a force of the nature of resistance, proportional to velocity. When ξ_1/ξ_2 does not exceed 1·21, the force is a resistance in the ordinary sense, *i.e.* it opposes the motion, and, in any case, it has this character near (and beyond) the arrival end ξ_1. But when ξ_1/ξ_2 exceeds 1·21, the force becomes what we may call a counter-resistance, and aids the motion near the initial end ξ_2. As regards F, in the neighbourhood of each limit it becomes a force proportional to distance therefrom, repulsive near ξ_2, and attractive near ξ_1. Thus, when ξ is nearly equal to ξ_2,

$$F(\xi) = 0{\cdot}723 \frac{\sqrt{\xi_1} - \sqrt{\xi_2}}{2\sqrt{\xi_2}} (\xi - \xi_2); \quad\ldots\ldots\ldots\ldots\ldots(107)$$

and when ξ is nearly equal to ξ_1,

$$F(\xi) = 0{\cdot}723 \frac{\sqrt{\xi_1} - \sqrt{\xi_2}}{2\sqrt{\xi_1}} (\xi_1 - \xi). \quad\ldots\ldots\ldots\ldots\ldots(108)$$

The particle, starting from ξ_2, is bound to go through to ξ_1. If it arrives at ξ_1 with zero velocity, we shall have, presumably, a solution of our problem. It is possible, however, that on first arrival at ξ_1, it may pass through, and only settle down after a number of oscillations. To this there does not appear to be any objection; but if on the return from ξ_1 it overshoots ξ_2, it can never again return to ξ_1, since on the left of ξ_2 the sign of f is negative; and then our problem has no solution. On the other hand, from the nature of F, it is not possible for the particle passing through ξ_1 in the positive direction to escape returning.

The character of the start from ξ_2 can be investigated with the aid of approximate equations. Thus, in (100) we may treat $f(\xi)$ as constant, say 2α, where α may be either positive or negative, and take, as in (107), $F(\xi) = \beta(\xi - \xi_2)$, where β is positive, so that

$$U \frac{dU}{d\xi} + 2\alpha U - \beta(\xi - \xi_2) = 0. \quad\ldots\ldots\ldots\ldots\ldots(109)$$

If in (109) we assume

$$U = \lambda(\xi - \xi_2), \quad\ldots\ldots\ldots\ldots\ldots(110)$$

we find that the equation is satisfied provided that

$$\lambda = -\alpha \pm \sqrt{(\alpha^2 + \beta)}, \quad\ldots\ldots\ldots\ldots\ldots(111)$$

one value (λ_1) being positive and one (λ_2) negative. In the present case, where U must be positive when $\xi > \xi_2$, λ_1 is to be chosen.

The differential equation (109) can be made homogeneous, and its general solution*, when λ is real, can be put into the form

$$\frac{\{U - \lambda_1(\xi - \xi_2)\}^{\lambda_1}}{\{U - \lambda_2(\xi - \xi_2)\}^{\lambda_2}} = C, \quad\quad\quad\quad (112)$$

where C is an arbitrary constant. This solution, of course, covers the cases where the particle starts from ξ_2 with a finite velocity (U_0), and it appears that $C = U_0^{\lambda_1 - \lambda_2}$. We might conclude from this that when $U_0 = 0$, then $C = 0$, but the conclusion is not safe. If, however, U and $\xi - \xi_2$ are of the same order of magnitude, we may write $U = r(\xi - \xi_2)$, where r is not infinite. Substituting this in (112), we get

$$(r - \lambda_1)^{\lambda_1} \cdot (r - \lambda_2)^{-\lambda_2} \cdot (\xi - \xi_2)^{\lambda_1 - \lambda_2} = C. \quad\quad\quad\quad (113)$$

When $\xi - \xi_2$ vanishes, the third factor on the left is zero, and, since the first and second factors are not infinite, the conclusion follows that $C = 0$. This takes us back to (110), (111), the second solution (involving λ_2) relating to the case where U is negative, the motion being one of *approach* from the positive side to ξ_2.

These conclusions may be arrived at more easily from (103), of which the general solution in the present case is

$$\xi - \xi_2 = A e^{\lambda_1 t} + B e^{\lambda_2 t}, \quad\quad\quad\quad (114)$$

giving
$$U = \lambda_1 A e^{\lambda_1 t} + \lambda_2 B e^{\lambda_2 t}. \quad\quad\quad\quad (115)$$

From these we may deduce

$$U - \lambda_1(\xi - \xi_2) = (\lambda_2 - \lambda_1) B e^{\lambda_2 t},$$
$$U - \lambda_2(\xi - \xi_2) = (\lambda_1 - \lambda_2) A e^{\lambda_1 t};$$

whence (112) follows by elimination of t. For our present purpose, $\xi - \xi_2$ is to vanish when $t = -\infty$, so that $B = 0$; and (110) follows with $\lambda = \lambda_1$. It will be remarked that (110) makes x, as determined by (99), infinite when $\xi = \xi_2$. The circumstances of the start from ξ_2 are thus definite and suitable. The question is as to the arrival at ξ_1.

In the neighbourhood of ξ_1 the approximate equation is

$$U \frac{dU}{d\xi} + 2\alpha' U - \beta'(\xi - \xi_1) = 0, \quad\quad\quad\quad (116)$$

where α' is positive and, by (108), β' negative. If now

$$U = \lambda'(\xi - \xi_1), \quad\quad\quad\quad (117)$$

the values of λ' are
$$\lambda' = -\alpha' \pm \sqrt{(\alpha'^2 + \beta')}; \quad\quad\quad\quad (118)$$

---

* See, for example, Boole's *Differential Equations*, p. 33.

so that both values, if real, are negative. On the supposition of reality, (112) retains its form (with ξ_1 for ξ_2). If the velocity of arrival (U_0) be finite, $C = U_0^{\lambda'_1 - \lambda'_2}$, as before; and it might be supposed that, if U_0 vanishes when $\xi - \xi_1 = 0$, C would have to vanish or become infinite. Such a conclusion would be incorrect. If in (113) we suppose λ'_1 to be numerically the smaller of the two values, the third factor indeed vanishes with $(\xi - \xi_1)$ as before; but the conclusion that $C = 0$ is evaded if ultimately $r = \lambda'_1$.

The situation is most easily understood from the solution in terms of t as in (114), (115). Since λ'_1, λ'_2 are *both* negative, the condition that $\xi - \xi_1$ and U shall vanish together when $t = \infty$ is satisfied, whatever may be the values of A and B. There are now an infinite number of possible types of solution, instead of only one as in the former case. And it appears that the two simple types included under (117) are not at all upon an equal footing. Except in the *particular* case where $A = 0$, the solution always tends ultimately to the form $U = \lambda_1 (\xi - \xi_1)$, and of course it may assume this form throughout. All these solutions satisfy the condition as to the infinitude of x when $U = 0$.

Whether the values of λ' be real or not, the particle must ultimately settle down at ξ_1, unless it escape from the region to which the approximate equation applies. For in (118) the *real part* of λ' is always negative.

Returning to (100) in its general form, let us consider the variation of U for a given ξ as dependent upon variations in f and F. We have

$$U \frac{d\delta U}{d\xi} + \delta U \frac{dU}{d\xi} + f \cdot \delta U + U \cdot \delta f - \delta F = 0,$$

or

$$\frac{d\delta U}{d\xi} + P\delta U - Q = 0, \quad \ldots\ldots\ldots\ldots\ldots\ldots(119)$$

where P and Q are supposed to be known functions of ξ, viz.,

$$P = \frac{f + dU/d\xi}{U}, \qquad Q = \frac{\delta F - U \cdot \delta f}{U}. \quad \ldots\ldots\ldots\ldots(120)$$

The solution of the linear equation (119) is

$$\delta U = e^{-\int P d\xi} \left(\int e^{\int P d\xi} Q \, d\xi + c \right). \quad \ldots\ldots\ldots\ldots\ldots(121)$$

If $Q = 0$, δU has the same sign as c, so that an increment of velocity communicated at any point remains throughout of the same sign. Again, if Q be throughout of one sign, δU, as dependent upon it, has the same sign. For example, if U be positive over the range considered, δf positive, and δF negative, then δU is certainly negative. The increments δf, δF may be local, vanishing over any part of the range.

The application to the present problem is obvious. If the particle passing any point between ξ_2 and ξ_1, with velocity U, arrives at ξ_1 for the first time without velocity, it will still arrive at ξ_1 without velocity (it must in any case arrive, since F is positive), if U be diminished, or if f be increased, or if F be diminished, or if all these changes occur together. And in the limit, when ξ_1 is closely approached, the ratio of U to $(\xi_1 - \xi)$ is in general the same.

By use of this principle we may assure ourselves as to the possibility of a solution in certain cases where ξ_1/ξ_2 does not greatly exceed unity. We imagine a simplified problem which admits of analytical solution and is derived from the actual one by alterations which everywhere (over the range from ξ_2 to ξ_1) increase F and diminish f. If this modified problem admits of the solution required, *a fortiori* will the original problem do so.

If we consider the curve which according to (106) represents F as a function of ξ, we see that it is concave downwards, and that F will everywhere be increased if we substitute for the curve the two terminal tangents at ξ_2 and ξ_1, whose equations are given in (107), (108). The abscissa of K, the point of intersection, is $\xi = \sqrt{(\xi_2 \xi_1)}$. (Fig. 1.)

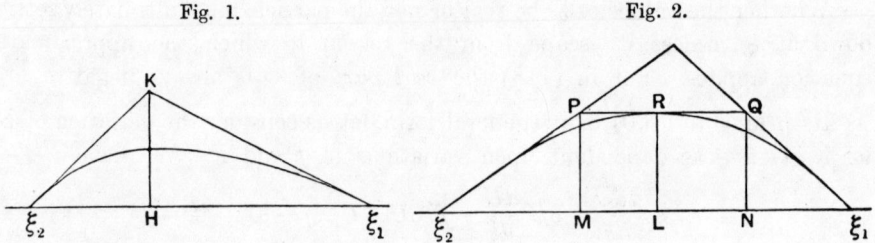

As regards f, its value is given by (105) with sign reversed, and is diminished when ξ is diminished. The changes will therefore be in the required direction if we represent f over $\xi_2 H$ by its value at ξ_2, viz.,

$$f_2 = 2\alpha = 1\cdot259 - 0\cdot641 s, \qquad \qquad (122)$$

and from H to ξ_1 by its value at H, viz.,

$$f_1 = 2\alpha' = 1\cdot900 - 0\cdot641 (s^{\frac{1}{2}} + s^{-\frac{1}{2}}), \qquad (123)$$

if for brevity we write s for $\sqrt{(\xi_1/\xi_2)}$, so that s is the ratio of terminal densities in the original problem.

As regards the first portion of the course, the solution already given (110), (111), determines the value of U on arrival at H. We have

$$U = \lambda_1 \xi_2 (s-1), \quad \text{where} \quad \lambda_1 = \sqrt{(\alpha^2 + \beta)} - \alpha,$$

in which α is given by (122) while by (107) $\beta = 0\cdot361 (s-1)$. Using these, we get

$$U = (s-1) \xi_2 [\sqrt{\{(0\cdot630 - 0\cdot320 s)^2 + 0\cdot361 (s-1)\}} - 0\cdot630 + 0\cdot320 s]. \quad \ldots (124)$$

If $s = 1 + \sigma$, where σ is small, (124) becomes

$$U = \sigma \xi_2 \times \frac{\beta}{2\alpha} = 0{\cdot}584 \xi_2 \sigma^2. \quad\quad\quad\quad\quad\quad (125)$$

As regards the second portion of the course, the appropriate solution is provided by (117), (118), where λ' is restricted to be real. And in accordance with our suppositions α' is given by (123) while from (108) $\beta' = -0{\cdot}361(1 - s^{-1})$. In choosing between the values of λ' we are at liberty to take that which gives the largest value of U at H consistent with $U = 0$ at ξ_1, viz.,

$$\lambda'_2 = -\alpha' - \sqrt{(\alpha'^2 + \beta')}.$$

Thus at H
$$U = (s^2 - s)\xi_2 \{\alpha' + \sqrt{(\alpha'^2 + \beta')}\}, \quad\quad\quad\quad\quad\quad (126)$$

or approximately, in terms of σ (supposed small),

$$U = \sigma \xi_2 (0{\cdot}618 + 0{\cdot}033\sigma). \quad\quad\quad\quad\quad\quad (127)$$

From (125), (127) we may infer that when σ is small, *i.e.* when s does not much exceed unity, the particle, starting from ξ_2, arrives at H with a velocity small enough to admit of its being stopped on arrival at ξ_1, even under the simplifying conditions that have been imposed, and therefore *a fortiori* under the actual conditions of the problem. Hence when ξ_1 does not too much exceed ξ_2, a wave of permanent regime *is* possible as the result of viscosity and heat-conduction.

But the range of ξ_1/ξ_2 thus proved admissible is rather severely limited. The postulated reality of λ' requires that $\alpha'^2 + \beta'$ be positive, and this again requires that $s < 1{\cdot}34$. For values of s greater than this the motion in the simplified problem would become oscillatory. Calculation shows that, for $s = 1{\cdot}34$, (124) gives
$$U = (s - 1)\xi_2 \times 0{\cdot}20,$$
while (126) gives
$$U = (s - 1)\xi_2 \times 0{\cdot}43;$$
so that up to this limit the particle starting from ξ_2 in the simplified problem, and therefore also in the actual problem, would arrive at ξ_1 with zero velocity. Up to a ratio of densities equal to $1{\cdot}34$, the wave of permanent regime is certainly possible.

The next step in this line of procedure will be to replace the curve representing F by the broken line $\xi_2 P Q \xi_1$ formed by *three* of its tangents, of which two are the same terminal tangents as before, while the third, PQ, may be taken to be the horizontal tangent parallel to $\xi_1 \xi_2$. (Fig. 2.) By (106) the point L where F is a maximum is determined by

$$\sqrt{\xi} = \tfrac{1}{2}(\sqrt{\xi_1} + \sqrt{\xi_2}) = \tfrac{1}{2}\sqrt{\xi_2}(s + 1),$$

and the corresponding value of F is

$$0{\cdot}1875 \xi_2 (s - 1)^2.$$

The abscissae of the points of intersection of the horizontal tangent with the terminal tangents are given by (107), (108). For M

$$\xi = \tfrac{1}{2}\xi_2 (s + 1),$$

and for N

$$\xi = \tfrac{1}{2}\xi_2 (s^2 + s).$$

As for f we are to take along $\xi_2 M$ the value at ξ_2; along MN the value at M; and along $N\xi_1$ the value at N, as given by (105), with sign changed.

For the two terminal portions the solutions are of the same form as before, but for the middle portion a new form is required. Making F constant in (100) and writing $2\alpha''$ for f, we find on integration

$$-4\alpha''^2 \xi = F \log (F - 2\alpha'' U) + 2\alpha'' U + C, \ldots\ldots\ldots\ldots(128)$$

where C is an arbitrary constant, to be determined so as to suit the velocity with which the particle arrives at M.

As before, the limiting value of s is determined by the consideration that, for our purpose, the arrival at ξ_1 must not be oscillatory. We get $s = 1.633$ about, and with this value of s it is not difficult to show that the velocity of arrival at N is below the value prescribed by the solution for $N\xi_1$. The necessary conditions are thus fulfilled and we infer that the wave of uniform regime is possible so long as $s < 1.63$. Although this ratio is moderate, it exceeds that found admissible in Rankine's problem, which leaves viscosity out of account. We there found that the greatest admissible ratio of *pressures* was 1.61, which by (68) corresponds to 1.40 for the ratio of *densities*. Of course, in Rankine's problem the solution definitely fails at this point, while in the present problem all that we have so far proved is that the limit exceeds 1.63.

From the low limiting values of s found necessary in these two cases in order to secure the reality of the roots of (118), it might be inferred that the reality would fail for the limiting motion in the neighbourhood of ξ_1, when s had a considerable value, but such is not the case. If we use the value of f from (105) appropriate to the terminal point ξ_1 itself, we have

$$2\alpha' = 1.900 - 0.641 (1 + s^{-1}) = 1.259 - 0.641 s^{-1};$$

and

$$\beta' = -0.3615 (1 - s^{-1}).$$

Thus, even when $s = \infty$,

$$\alpha'^2 + \beta' = 0.3963 - 0.3615 = + 0.0048;$$

and the roots are always real. Hence, subsidence at ξ_1 is not ultimately oscillatory, but there is nothing in this argument to exclude a finite number of oscillations before subsidence into the region governed by the approximate equation.

The method of approximation already followed might be pushed further, but it seems preferable to use the general method of numerical calculation for the solution of differential equations as formulated by Runge*. The equation is that numbered (100), in which f (whose sign is to be reversed) and F are given by (105), (106). If we write $U' = U/\xi_2$, $\xi' = \xi/\xi_2$, our equation takes the form

$$\frac{dU'}{d\xi'} = \frac{0\cdot 723}{U'}(s - \sqrt{\xi'})(\sqrt{\xi'} - 1) - 1\cdot 900 + \frac{0\cdot 641(1+s)}{\sqrt{\xi'}} \ldots\ldots(129)$$

The value of s being given, it is required to trace the connection between U' and ξ', simultaneous values being denoted respectively by a and b. If a receive the increment h, we have to calculate the corresponding increment k for b. If we call the function on the right of (129) $\phi(\xi', U')$, Runge gives as a first approximation to k

$$k_1 = \phi(a + \tfrac{1}{2}h,\ b + \tfrac{1}{2}\phi_0 . h)h, \ldots\ldots\ldots\ldots(130)$$

where $\phi_0 = \phi(a, b)$. The next approximation is

$$k = k_1 + \tfrac{1}{3}(k_2 - k_1), \ldots\ldots\ldots\ldots\ldots\ldots(131)$$

where
$$k_2 = \tfrac{1}{2}(k' + k''')\quad \text{and}\quad k' = \phi_0 . h,$$
$$k'' = \phi(a + h,\ b + k')h,\quad k''' = \phi(a + h,\ b + k'')h. \ldots\ldots(132)$$

Having determined the new simultaneous values $a + h, b + k$, we make a fresh departure therefrom, and so trace out the function step by step.

In the present application the starting point is $a = 1$ (i.e. $\xi = \xi_2$), $b = 0$, and we have to trace the function until $\xi' = s^2$. The initial value of ϕ is to be found from (129) by putting $\xi' = 1$. Writing $U' = \phi_0(\xi' - 1)$, we get

$$\phi_0^2 + \{1\cdot 900 - 0\cdot 641(1+s)\}\phi_0 - \tfrac{1}{2} \times 0\cdot 723(s-1) = 0, \ldots\ldots(133)$$

of which the positive root is to be chosen.

The extreme admissible value of s in our present problem is 6, and the first and rather elaborate calculation that I have made relates to this case. From (133) $\phi_0 = 3\cdot 1591$, so that taking $a = 1$, $b = 0$, $h = 1$, we have $k' = 3\cdot 1591$. Calculating from (130) we find $k_1 = 2\cdot 255$, and from (132)

$$k'' = 1\cdot 7076, \qquad k''' = 2\cdot 0772,$$

making $k_2 = 2\cdot 6182$. Hence the correction to k_1, viz., $\tfrac{1}{3}(k_2 - k_1)$, is equal to $0\cdot 121$, and $k = 2\cdot 376$. Thus, corresponding to $\xi' = 2$, we get $U' = 2\cdot 376$. The following are the values of U' obtained successively in this way:—

ξ'	U'	ξ'	U'	ξ'	U'
1	0·000	14	7·502	33	1·2942
2	2·376	18	6·807	34	0·8532
3	3·903	22	5·674	35	0·4160
4	4·986	26	4·241	$35\tfrac{1}{2}$	0·2043
6	6·380	30	2·610	$35\tfrac{3}{4}$	0·1012
10	7·526	32	1·736	36	0·0004

* See Forsyth's *Differential Equations*, p. 51.

The correction to k_1 is everywhere subordinate. In the last step from $35\frac{3}{4}$ to 36 no correction to k_1 is applied.

There is a little difficulty in tracing by this method and with full accuracy the final progress to zero when $\xi' = 36$. If any doubt be left, it may be removed by applying the former method to the course from 35 to 36, using the value of f appropriate to 35 and the terminal tangent as the representative of the curve for F. From this it appears that even if U' at $\xi' = 35$ were as great as 0.7468, the moving particle could not pass $\xi' = 36$. The conclusion is that even in this extreme case of $s = 6$ the solution exists, and that a wave of permanent regime is possible. Further, from (99) we see that since U and μ' are both positive, $d\xi/dx$ is positive throughout, and the transition from the one density to the other takes place *without alternation*.

After what has been proved little doubt could remain but that a solution is possible when s has any value lower than 6. I have, however, thought it desirable to add a rough calculation (rough on account of the relative magnitude of the steps) for the case of $s = 3$:—

ξ'	1	2	3	5	7	8	$8\frac{1}{2}$	9
U'	0.00	0.90	1.19	1.26	0.82	0.40	0.19	0.01

It is a question of some importance to consider what is the thickness of the transitional layer in the waves of uniform regime which have been proved to be possible. Mathematically speaking, the transition occupies an infinite space; but if we understand the expression to refer to a transition approximately complete, the thickness involved is finite, and indeed extremely small. Reference to (98) shows that x is of the order μ', or μ/m, or $\mu/\rho u$, where u is the velocity of the wave. For the present purpose we may take u as equal to the usual velocity of sound, *i.e.* 3×10^4 cm. per second. For air under ordinary conditions the value of μ/ρ in C.G.S. measure is 0.13; so that x is of the order $\frac{1}{3} \times 10^{-5}$ cm. That the transitional layer is in fact extremely thin is proved by such photographs as those of Boys, of the aerial wave of approximate discontinuity which advances in front of a modern rifle bullet; but that according to calculation this thickness should be well below the microscopic limit may well occasion surprise.

Resistance to Motion through Air at High Velocities.

According to the adiabatic law the pressures and velocities in a compressible fluid free from external force, see (47), are related by

$$\left(\frac{p_2}{p_1}\right)^{\frac{\gamma-1}{\gamma}} = 1 + \frac{\gamma-1}{2}\frac{u_1^2 - u_2^2}{a_1^2},\dots\dots\dots\dots\dots(134)$$

in which p_1, ρ_1, u_1 denote the pressure, density, and velocity at one point of the path; p_2, ρ_2, u_2 the corresponding quantities at another point. Also $a_1^2 = \gamma p_1/\rho_1$, so that a_1 is the velocity of infinitesimal disturbances in the condition (1). In an early paper* I suggested the application of this formula to bodies moving through air at high velocities. Regarding the obstacle as stationary and the fluid in motion with velocity u_1 and pressure p_1, the pressure p_2 corresponding to the loss of this velocity is given by putting $u_2 = 0$ in (134), a_1 being the ordinary velocity of sound. This is the pressure which should obtain at the axial point on the nose of a symmetrical bullet, and although this value in strictness represents the *maximum* pressure, the analogy of an incompressible fluid suggests that the mean pressure on a flat surface would not be greatly inferior. But in a recent discussion†, Mr Mallock has shown that this formula immensely overestimates the resistance actually experienced by a bullet, and (so far as I am aware) the discrepancy remains unexplained.

If indeed the adiabatic law really prevailed throughout, there could be no escape from the conclusion formulated. A consideration of the photographs by Boys‡ will suggest the required explanation. At a short constant distance in front of the bullet there is an aerial bore, or place of approximate discontinuity. Along the axis, the fluid moving up to the bullet changes its density, and therefore pressure and temperature, *suddenly*, so that there is here a special opportunity for viscosity and heat-conduction to take effect. The pressures and velocities on the two sides of the bore are related, not according to the adiabatic law, but according to Rankine's law already discussed. The changes which occur may be separated into two stages. The first is the sudden one in which the fluid passes from the atmospheric condition p_0, ρ_0, with velocity u_0 to the condition denoted by p_1, ρ_1, u_1. After passing the bore the fluid changes gradually according to the adiabatic law already stated until at the nose of the bullet the condition is represented by p_2, ρ_2, with $u_2 = 0$.

* *Phil. Mag.* Vol. II. p. 430 (1876); *Scientific Papers*, Vol. I. p. 289.
† *Roy. Soc. Proc.* A, Vol. LXXIX. p. 266 (1907).
‡ *Nature*, Vol. XLVII. p. 440 (1893). The particular photograph reproduced by Mallock does not exhibit well the feature in question.

We are now in a position to calculate the final pressure p_2. For the first stage we have Rankine's formula (67), making

$$\rho_0 u_0^2 = \tfrac{1}{2}(\gamma-1) p_0 + \tfrac{1}{2}(\gamma+1) p_1,$$

or, if $a^2 = \gamma p_0/\rho_0$,
$$\frac{p_1}{p_0} = \frac{2\gamma}{\gamma+1}\frac{u_0^2}{a^2} - \frac{\gamma-1}{\gamma+1}, \quad\quad\quad\quad\quad(135)$$

determining the pressure just inside the bore in terms of u_0 (the velocity of the bullet through quiescent air) and a, the ordinary velocity of sound. When $u_0 = a$, $p_1 = p_0$. For values of u_0 less than a, the first stage does not exist and we may suppose $p_1 = p_0$, $u_1 = u_0$.

In the second stage we use (134) with $u_2 = 0$. Thus

$$\left(\frac{p_2}{p_1}\right)^{\frac{\gamma-1}{\gamma}} = 1 + \frac{\gamma-1}{2}\frac{\rho_1 u_1^2}{\gamma p_1}, \quad\quad\quad\quad\quad(136)$$

in which, by a formula analogous to (66),

$$\rho_1 u_1^2 = \tfrac{1}{2}(\gamma+1) p_0 + \tfrac{1}{2}(\gamma-1) p_1.$$

Hence
$$\left(\frac{p_2}{p_0}\right)^{\frac{\gamma-1}{\gamma}} = \frac{(\gamma+1)^2}{4\gamma}\left(\frac{p_1}{p_0}\right)^{\frac{\gamma-1}{\gamma}}\left\{1 + \frac{\gamma-1}{\gamma+1}\frac{p_0}{p_1}\right\}. \quad\quad(137)$$

When $u_0 > a$, p_1/p_0 is to be calculated from (135). When the resulting value is substituted in (137), p_2/p_0 is determined.

When $u_0 < a$, we have simply

$$\left(\frac{p_2}{p_0}\right)^{\frac{\gamma-1}{\gamma}} = 1 + \frac{\gamma-1}{2}\frac{u_0^2}{a^2}. \quad\quad\quad\quad\quad(138)$$

If u_0/a be *small*, (138) reduces to

$$\frac{p_2}{p_0} = 1 + \frac{\rho_0 u_0^2}{2 p_0},$$

or
$$p_2 - p_0 = \tfrac{1}{2}\rho_0 u_0^2, \quad\quad\quad\quad\quad(139)$$

as for an incompressible fluid.

When $u_0 = a$, both systems give

$$\left(\frac{p_2}{p_0}\right)^{\frac{\gamma-1}{\gamma}} = \frac{\gamma+1}{2}. \quad\quad\quad\quad\quad(140)$$

When u_0/a is *large*, the second terms on the right of (135) and (137) may be neglected, and we obtain

$$\frac{p_2}{p_0} = \frac{\gamma+1}{2}\frac{u_0^2}{a^2}\left\{\frac{(\gamma+1)^2}{4\gamma}\right\}^{\frac{1}{\gamma-1}}; \quad\quad\quad\quad\quad(141)$$

or, when we put $\gamma = 1\cdot 41$,

$$p_2/p_0 = 1\cdot 30 u_0^2/a^2. \quad\quad\quad\quad\quad(142)$$

The following are some corresponding values of p_2/p_0 and u_0/a, calculated from (135), (137), with $\gamma = 1\cdot 41$:—

u_0/a ...	1	2	3	4
p_2/p_0 ...	1·90	4·49	11·7	20·7

From this point onwards the approximate formula (142) may suffice. The values found are in good agreement with Mr Mallock's curve.

The question as to the linear interval between the bore and the nose of the bullet cannot be answered from the results of the present paper. In strictly one-dimensional motion, the bore and the plane wall constituting the obstacle could not move at the same speed.

Editor's Comments on Paper 11

11 Eckart: *Vortices and Streams Caused by Sound Waves*

Currents of air due to vibrating bodies were probably first observed by Savart, who found that fine powder spread over the surface of a vibrator does not collect at the nodes, as does the sand in Chladni's experiments (see Paper 16, *Acoustics: Historical and Philosophical Development*, R. B. Lindsay, ed., Dowden, Hutchinson & Ross, Stroudsburg, Pa., 1973). Rather, the powder hovers over and finally settles on the loops. Faraday was apparently the first to explain this phenomenon as the result of air currents over the vibrating surface (see Paper 23, in the volume referred to above). Lord Rayleigh considered the problem in connection with the figures observed in Kundt's tubes and emphasized the importance of viscosity in the formation of the currents (see Paper 108 in *Scientific Papers of Lord Rayleigh*, Vol. 2, p. 239, Dover Publications, New York, 1964).

In more recent times direct currents of air and liquids have been observed in the neighborhood of sonic and ultrasonic transducers, and this streaming has proved to be a valuable tool in the biological applications of ultrasonics. A good early review of modern work on acoustic streaming is provided by this paper by Carl Eckart, "Vortices and Streams Caused by Sound Waves," reproduced here in full. In this paper, Eckart develops the important roles played in the production of streaming by the viscosity of the medium and the nonlinearity of the radiation equations connected with high-intensity sound. Although criticism has been directed at certain mathematical aspects of Eckart's paper, its fundamental importance for the future development of the field of nonlinear acoustics cannot be questioned. The reader's attention is called to a still more recent review paper on acoustic streaming by W. L. Nyborg (*J. Acoust. Soc. Amer.*, **25**, 68, 1953).

Carl Eckart (1902–1973) was an American physicist and geophysicist, well known for his research in quantum mechanics, hydrodynamics, marine acoustics, and thermodynamics. For many years he was a professor of marine physics at the University of California in San Diego (Scripps Institution of Oceanography). Eckart died on October 23, 1973. On October 31, 1973, he was awarded posthumously the "Pioneers of Underwater Acoustics Award" of the Acoustical Society of America.

Reprinted from *Phys. Rev.*, **73**, 68–76 (1948)

Vortices and Streams Caused by Sound Waves*

CARL ECKART
University of California, Marine Physical Laboratory, San Diego, California
(Received August 28, 1947)

As shown by Rayleigh, a considerable number of acoustic phenomena are known which involve the viscosity of the medium and require the solution of the hydrodynamic equations to a higher degree of approximation than is customary in elementary treatments of the theory of sound. Among these are the fluid streams that occur near intense sources of sound (e.g.: the "quartz wind").

The general equations of these second-order acoustic phenomena are developed in a systematic manner. When viscous forces are neglected, the effects are of three kinds: (1) those that can be ascribed to the inertia of acoustic energy, (2) those arising from radiation pressure, and (3) those caused by the variable compressibility of the medium. All of them result in the production of overtones of the fundamental vibration. In certain cases, this distortion can become very large, being unlimited except by the viscous forces. However, even when the average value of the gradient of the radiation pressure does not vanish, it does not, on the average, cause an acceleration of the fluid. Such gradients are balanced by the elastic rather than by the viscous forces.

When the latter are introduced into the calculation, a fourth effect appears: the irrotational motion in the sound wave generates vorticity as a second-order effect. This vortex motion will ultimately approach a steady state, being generated and resisted by forces that are independent of the time. Both generating and resisting forces are viscous, and consequently the steady motion is independent of the magnitude of the coefficient of viscosity. However, the resisting forces depend only on the shear viscosity of the medium, while the generating forces depend also on the bulk viscosity. It is suggested that the ratio of the bulk and shear coefficients of viscosity can be determined by studying these phenomena.

Calculations of the velocity of the stream generated by a beam of sound show that it is proportional (1) to $b=(4/3)+(\nu'/\nu)$, where ν' and ν are the bulk and shear viscosities, (2) to the power being radiated in the beam, (3) inversely to the square of the wave-length, and (4) inversely to $\rho^2 c^3$, where ρ is the density and c the sound velocity of the medium. The maximum value of the steady-streaming velocity depends on the resistance offered by the walls of the vessel or room in which the experiment is performed. The time required to set up the steady state is, of course, inversely proportional to this resistance, and the flow is apt to become turbulent when the resistance is low.

INTRODUCTION

THE subject matter of this paper cannot be outlined more clearly than by quoting from the first paragraphs of Lord Rayleigh's paper[1] "On the circulation of air observed in Kundt's tubes, and on some allied acoustical problems":

> Experimenters in acoustics have discovered more than one set of phenomena, apparently depending for their explanation upon the existence of regular currents of air resulting from vibratory motion ... such currents, involving as they do *circulation* of the fluid, could not arise in the absence of friction. ... And even when we are prepared to include the influence of friction, we have no chance of reaching an explanation if, as usual, we limit ourselves to the supposition of infinitely small motion and neglect the squares and higher powers [of the velocity]. ... The more important of the problems relates to the currents generated over a vibrating plate, arranged as in Chladni's experiments. It was discovered by Savart that very fine powder does not collect itself at the nodal lines, as does sand in the production of Chladni's figures, but gathers itself into a cloud which, after hovering for a time, settles itself over the places of maximum vibration. This was traced by Faraday[2] to the action of currents of air, rising from the plate at the place of maximum vibration, and falling back to it at the nodes. In a vacuum the phenomena observed by Savart do not take place, all kinds of powder collecting at the nodes. ... [Another] problem relates to the air currents observed by Dvorak in a Kundt's tube, to which is apparently due the formation of the dust figures.

With the advent of piezoelectric generators of sound, these effects were rediscovered. Strong currents of air ("quartz wind") or liquid appear in front of the vibrating surface of the crystal. In the case of liquids, these currents are frequently great enough to disturb its free surface. Unless great care is exercised, they may vitiate intensity measurements with a Rayleigh disk. It is possible that this effect was actually discovered by Rayleigh, who performed the following experi-

---

* This work represents one of the results of research carried out under contract with the Bureau of Ships and Office of Naval Research, Navy Department.

[1] Lord Rayleigh, *Scientific Papers* (Cambridge University Press, Teddington, England), No. 108, p. 239; Phil. Trans. **175**, 1 (1883).

[2] Michael Faraday, Phil. Trans. 299 (1831).

ment: "... when the corresponding fork, strongly excited, was held to the mouth [of the Helmholtz resonator] a wind of considerable force issued from the nipple at the opposite side. This effect may rise to such intensity as to blow out a candle upon whose wick the stream is directed.... Closer examination revealed the fact that at the sides of the nipple the outward flowing stream was replaced by one in the opposite direction, so that a tongue of flame from a suitable placed candle appeared to enter the nipple at the same time that another candle situated immediately in front was blown away."[3]

Although Rayleigh summarized his calculations in *Theory of Sound*,[4] his results appear to be virtually unknown. At lest two different and incorrect explanations of the quartz wind can be found in recent literature, while the correct explanation appears to be virtually unknown to the experimenter. It is even possible that Rayleigh himself gave an incorrect explanation of the resonator experiment just described. Without giving adequate reasons, he says: "The two effects [flow and counter-flow] are of course in reality alternating, and only appear to be simultaneous in consequence of the inability of the eye to follow such rapid changes." It is at least possible that the streams are steady and that this is an effect similar to the others described above.

In the following pages, a systematic account of the theory of second-order acoustic effects will be developed. In the first part, acoustic radiation pressure and the inertia of acoustic energy will be considered. It will be shown that, as Rayleigh knew, these cannot cause the phenomena described above. However, the structure of this mathematical theory will be useful in the second part, where the second-order viscous forces will be considered, and in the third, where a calculation will be given of the steady flow produced by a sound beam of circular cross section. The remarkable fact will appear that the steady flow "is *independent of the value of the coefficient of viscosity*. We cannot, therefore, avoid considering this motion by supposing the coefficient of viscosity to be very small, the maintenance of the vortices becoming easier in the same proportion as the forces tending to produce the vortical motion diminish."[5]

PART I. ACOUSTIC RADIATION PRESSURE AND THE INERTIA OF ACOUSTIC ENERGY

The question is sometimes asked, why the velocity of sound does not function in acoustic theory in the same way that the velocity of light functions in relativity theory. From the standpoint of the latter, the question is foolish, for the velocity of light is both that, and also the maximum velocity with which any kind of signal or object can be transmitted. The velocity of sound is not maximal in the same sense. Still, the principle of the inertia of energy is not very directly connected with the principle of maximal velocity, but rather with the equations of motion of matter. Consequently, it would be expected that acoustic energy will display an inertia that is much greater than the inertia of electromagnetic radiation, in the inverse ratio of the squares of their velocities of propagation. It will be shown that this expectation is correct, provided the proposition be given a suitable interpretation.

The First- and Second-Order Equations of Acoustics

The general equations of hydrodynamics for a non-viscous fluid are

$$(\partial \rho/\partial t) + \nabla \cdot (\rho u) = 0, \qquad (1)$$

$$[\partial(\rho u)/\partial t] + \rho u \cdot \nabla u + u \nabla \cdot (\rho u) = -\nabla p, \qquad (2)$$

where ρ is the density, p the pressure, and u the velocity of the fluid. For the present purposes, it will be supposed that p is a function of ρ only; then

$$\nabla p = C^2 \nabla \rho, \qquad (3)$$

where C is a function of ρ which has the units of a velocity. This presupposes that the motion is isentropic.

The essential idea of Rayleigh's treatment of the problems discussed in the introduction, is that some of the terms in Eqs. (1) and (2) are sometimes much less important than others. In different circumstances, different terms will be

---

[3] Rayleigh, *Theory of Sound* (MacMillan Company, Ltd., London, 1896), Vol. II, p. 217.
[4] Reference 3, Vol. II, p. 333.

[5] Reference 1, p. 246.

negligible and others will be important. Thus, in acoustics, the terms $\partial \rho/\partial t$ and $\nabla \rho$ are important; in hydraulics, the term $\nabla \cdot (\rho u)$ becomes more important than $\partial \rho/\partial t$, but $\nabla \rho$ retains its importance. In order to bring the relative importance of the terms clearly to the attention, it is useful to depart from the c.g.s. system of units and to introduce one that is specially adapted to the problems under consideration.

In such a system, let the unit of length be X cm; of time, T sec.; of velocity, U cm sec.$^{-1}$. The unit of density is immaterial, since the equations are essentially homogeneous in ρ. In these units the equations become

$$(\partial \rho/\partial t) + N\nabla \cdot (\rho u) = 0, \quad (4)$$

$$[\partial(\rho u)/\partial t] + N[\rho u \cdot \nabla u + u\nabla \cdot (\rho u)]$$
$$= -(NC^2/U^2)\nabla \rho, \quad (5)$$

where the numeric

$$N = UT/X. \quad (6)$$

To insure that the system of units will serve its purpose, the units X and T are to be chosen so that $\partial f/\partial t$ and $\partial f/\partial x$ are of the same order of magnitude, f being any of the functions ρ, u. The unit U could be chosen in any of a number of ways: so that $N=1$, or so that $C/U \sim 1$, or so that $u/U \sim 1$. The third is the choice appropriate for most problems. Having thus defined the units, acoustics may be defined as consisting of those hydrodynamic problems for which $N \ll 1$ and $NC/U \sim 1$. (Hydraulics is apparently those hydrodynamic problems for which $N \gg 1$ and $NU^2/C^2 \sim 1$.)

Introducing the quantity

$$c(\rho) = NC(\rho)/U,$$

Eq. (5) becomes

$$N[\partial(\rho u)/\partial t] + N^2[u\nabla \cdot (\rho u) + \rho u \cdot \nabla u]$$
$$= -c^2 \nabla \rho. \quad (7)$$

The numeric N is now to be treated as a perturbation parameter,[**] and the expansions

$$\rho = \rho_0 + N\rho_1 + N^2 \rho_2 + \cdots, \quad (4.0)$$

$$u = u_0 + Nu_1 + N^2 u_2 + \cdots \quad (7.0)$$

[**] After the special units have served their purpose of providing a perturbation parameter, one may always return to the c.g.s. system, for which $N=1$, $C=c$. This will be done in the following pages.

are introduced. The zero-order equations are then

$$\partial \rho_0/\partial t = 0, \quad (4.0)$$

$$c_0^2 \nabla \rho_0 = 0, \quad (7.0)$$

where $c_0 = c(\rho_0)$. Hence, ρ_0 is a constant, which fact may be used to simplify the first- and second-order equations:

$$(\partial \rho_1/\partial t) + \rho_0 \nabla \cdot u_0 = 0, \quad (4.1)$$

$$\rho_0(\partial u_0/\partial t) = -c_0^2 \nabla \rho_1; \quad (7.1)$$

$$(\partial \rho_2/\partial t) + \rho_0 \nabla \cdot u_1 + \nabla \cdot (\rho_1 u_0) = 0, \quad (4.2)$$

$$\rho_0(\partial u_1/\partial t) + (\partial/\partial t)(\rho_1 u_0) + \rho_0(u_0 \cdot \nabla u_0 + u_0 \nabla \cdot u_0)$$
$$= -c_0^2 \nabla \rho_2 + c_0(dc_0/d\rho_0)\nabla \rho_1^2. \quad (7.2)$$

The Eqs. (4.1) and (7.1) are the equations of elementary acoustic theory, while Eqs. (4.2) and (7.2) are less familiar.

Integrals of the First-Order Equations

The equation expressing the conservation of acoustic energy is derived by multiplying Eq. (4.1) by $c_0^2 \rho_1/\rho_0$, and Eq. (7.1) by u_0, and adding. The result is

$$(\partial W/\partial t) + \nabla \cdot J = 0. \quad (8)$$

The quantity

$$W = \tfrac{1}{2}\rho_0 u_0^2 + \tfrac{1}{2}c_0^2 \rho_1^2/\rho_0 \quad (9)$$

is the acoustic energy density, while

$$J = c_0^2 \rho_1 u_0 \quad (10)$$

is the acoustic energy flow.

The conservation law of acoustic "momentum" is obtained by multiplying Eq. (4.1) by u_0, and Eq. (7.1) by ρ_1/ρ_0 and adding: the result is

$$\frac{1}{c_0^2}\frac{\partial J}{\partial t} + \rho_0(u_0 \cdot \nabla u_0 + u_0 \nabla \cdot u_0)$$
$$+ \nabla(W - \rho_0 u_0^2) + \rho_0 u_0 \times (\nabla \times u_0) = 0. \quad (11)$$

It will be noted that the time derivative of J/c_0^2 appears in Eq. (11). Thus the acoustic momentum is related to acoustic energy flow in the same manner as the corresponding electromagnetic quantities. Also that, instead of the total energy density, W, Eq. (11) contains the Lagrangian difference between the potential and kinetic energy densities.

The last term in Eq. (11) usually vanishes,

since only solutions for which $\nabla \times \boldsymbol{u}_0 = 0$ are of interest. This will be assumed in the following.

Simplification of the Second-Order Equations

The second-order equations are seen to contain several combinations of terms that also appear in the conservation laws. The former can therefore be simplified by introducing a quantity ρ_{II}, defined by

$$\rho_2 = \rho_{II} + W/c_0^2. \qquad (12)$$

The quantity ρ_{II} obeys equations that are much simpler than those for ρ_2, and much more analogous to the first-order equations:

$$(\partial \rho_{II}/\partial t) + \rho_0 \nabla \cdot \boldsymbol{u}_1 = 0, \qquad (4\text{-}II)$$

$$\rho_0(\partial \boldsymbol{u}_1/\partial t) = -c_0^2 \nabla \rho_{II}$$
$$-\nabla[\rho_0 u_0^2 + c_0(dc_0/d\rho_0)\rho_1^2]. \qquad (7\text{-}II)$$

The term in $dc_0/d\rho_0$ would disappear if the medium obeyed Hooke's law. Thus, only the term in $\rho_0 u_0^2$ needs discussion. This functions like an additional (known) pressure, and may therefor be called the acoustic radiation pressure:

$$P = \rho_0 u_0^2, \qquad (13)$$

which is thus given by twice the kinetic energy density. In gases, the term

$$H = c_0(dc_0/d\rho_0)\rho_1^2 \qquad (14)$$

will be of the same order of magnitude as P. In liquids, H will be much less than P.

Because of the similarity between the equations for ρ_{II} and ρ_1, physicists will find it easier to think about ρ_{II} than about ρ_2. In particular, propositions arrived at intuitively (that is, derived from experience with elementary acoustic problems) will usually apply to ρ_{II} and not to ρ_2. Fortunately, the necessary correction is simple: it is only necessary to take the "inertia" of acoustic energy into account and add W/c_0^2 to ρ_{II} in order to obtain ρ_2.

The Solution of the Second-Order Equations

Certain general conclusions can be reached about the second-order effects in simple harmonic sound fields. In these cases, neither P nor H will contain terms of the fundamental frequency, but both will contain constant terms and terms with the double frequency. The latter will result in the generation of the second harmonic in the sound field—a phenomenon that has been observed at even moderate sound intensities. In certain cases, notably that of the plane wave, the amplitude of these harmonics will increase until they are limited by the viscous forces which have thus far been omitted.

Because of interference and the divergence of sound rays, the time-constant terms in P and H will, in general, depend on position. Thus, there will be a constant pressure gradient, and one might expect that this constant gradient will produce a constant acceleration of the fluid, whose velocity would thus increase until the viscous forces balance the pressure gradient. The end result would be fluid streaming at a constant velocity, proportional to the acoustic energy-gradient and inversely proportional to the coefficient of viscosity. These might account for the phenomena described in the Introduction. Unfortunately, however, this reasoning is faulty. The constant part of the gradient of $P+H$ is balanced, not by the viscosity of the fluid, but by its elasticity. Thus neither P nor H cause streaming of the fluid.

This is very simply proven: ρ_{II} may be eliminated between Eqs. (4-II) and (7-II) by taking the gradient of the former and the time derivative of the latter; the result is

$$\rho_0(\partial^2 \boldsymbol{u}_1/\partial t^2) - \rho_0 c_0^2 \nabla \nabla \cdot \boldsymbol{u}_1$$
$$= -(\partial/\partial t)[\nabla(P+H)]. \quad (15)$$

Since only the time derivatives of ∇P and ∇H appear, the first-order velocity $\boldsymbol{u}_1$ is independent of the constant parts of these gradients. This result would not be essentially altered by introducing the viscous terms. On the other hand, if $\boldsymbol{u}_1$ is eliminated from the equations, the result is

$$\rho_0(\partial^2 \rho_{II}/\partial t^2) - \rho_0 c_0^2 \nabla^2 \rho_{II} = \nabla^2(P+H). \quad (16)$$

Thus the constant parts of gradients will produce a constant part of ρ_{II}.

Consequently, neither radiation pressure nor the failure of Hooke's law can be invoked to explain the fluid streams mentioned above, and it becomes necessary to proceed to a study of the viscous forces.

PART II. THE EFFECTS OF VISCOSITY

When the viscous forces are included, the equations become (in the X, T, U system of units):

$$(\partial\rho/\partial t)+N\nabla\cdot(\rho u)=0, \quad (17)$$

$$N[\partial(\rho u)/\partial t]+N^2[u\nabla\cdot(\rho u)+\rho u\cdot\nabla u]$$
$$=-c^2\nabla\rho+(NT/X^2)[((4/3)\nu+\nu')\rho\nabla\nabla\cdot u$$
$$-\nu\rho\nabla\times(\nabla\times u)]. \quad (18)$$

The quantity $\nu\rho$ is the ordinary coefficient of shear viscosity, so that ν is measured in cm²/sec. For an ideal gas, the bulk viscosity, ν', is zero; for liquids it is presumably different from zero, although no measurements or calculations of its magnitude have been made.[6] For convenience the abbreviation

$$b=4/3+\nu'/\nu \quad (19)$$

will be used: it is a numerical characteristic of the fluid, and its value will presumably be somewhere between 2 and 10 for liquids, and near 4/3 for gases.

For simplicity, it will be assumed that both b and the product $\nu\rho=\nu_0\rho_0$ are independent of the density of the liquid; effects due to their variation may be of importance, but will not be treated here. The coefficient of the viscous forces may be written

$$NT\nu_0/X^2=N^2/R,$$

where

$$R=UX/\nu_0$$

is the Reynold's number, calculated for the unit of velocity and the unit of length. The values of N and U have above been fixed with respect to the problem, but the value of X has not yet been specified. This freedom could be utilized to assign any desired order of magnitude to the viscous forces; however, there appear to be good physical reasons for supposing them to be of first order. This determines X by means of the equation $N=R$. The unit of length thus calculated has a simple physical significance in the case of gases: it is the mean-free path of their molecules. Since the viscous terms are introduced into the equations in order to take approximate account of molecular processes, this is a very appropriate unit.

It is then obvious that the zero-order equations will be unaffected by the viscous terms, and

---

[6] H. Lamb, *Hydrodynamics* (Cambridge University Press, Teddington, England), sixth edition, pp. 573, 645; Reference 3, Vol II, pp. 314, 320; G. Kirchhoff, Pogg. Ann. **134**, 177 (1868).

that the first-order equations are

$$(\partial\rho_1/\partial t)+\rho_0\nabla\cdot u_0=0, \quad (17.1)$$

$$\rho_0(\partial u_0/\partial t)=-c_0^2\nabla\rho_1+\rho_0\nu_0 b\nabla\nabla\cdot u_0$$
$$-\rho_0\nu_0\nabla\times(\nabla\times u_0), \quad (18.1)$$

and the second-order equations are

$$(\partial\rho_2/\partial t)+\rho_0\nabla\cdot u_1+\nabla\cdot(\rho_1 u_0)=0, \quad (17.2)$$

$$\rho_0(\partial u_1/\partial t)+(\partial/\partial t)(\rho_1 u_0)$$
$$+\rho_0[u_0\cdot\nabla u_0+u_0\nabla\cdot u_0]=-c_0^2\nabla\rho_2$$
$$+\rho_0\nu_0 b\nabla\nabla\cdot u_1-\rho_0\nu_0\nabla\times(\nabla\times u_1). \quad (18.2)$$

Effects due to the failure of Hooke's law have been omitted, although they may be of importance in problems involving distortion.

The conservation (or better, the dissipation) of acoustic energy is then derivable from the first-order equations as before, and results in the equation

$$(\partial W/\partial t)+\nabla\cdot J=\rho_0\nu_0[bu_0\cdot\nabla u_0$$
$$-u_0\cdot\nabla\times(\nabla\times u_0)]. \quad (20)$$

Similarly, the dissipation of acoustic momentum is expressed by

$$(1/c_0^2)(\partial J/\partial t)+\rho_0[u_0\cdot\nabla u_0+u_0\nabla\cdot u_0]$$
$$+\nabla(W-\rho_0 u_0^2)+\rho_0 u_0\times(\nabla\times u_0)$$
$$=\nu_0 b\rho_1\nabla\nabla\cdot u_0-\nu_0\rho_1\nabla\times(\nabla\times u_0). \quad (21)$$

Setting $\nabla\times u_0=0$, and introducing ρ_{II} as before, Eqs. (17.2) and (18.2) are seen to be equivalent to

$$(\partial\rho_{II}/\partial t)+\rho_0\nabla\cdot u_1=(\nu_0 b/c_0^2)u_0\cdot\nabla(\partial\rho_1/\partial t), \quad (17\text{-II})$$

$$\rho_0(\partial u_1/\partial t)=-c_0^2\nabla\rho_{II}-\nabla(\rho_0 u_0^2)$$
$$+\rho_0\nu_0 b\nabla\nabla\cdot u_1-\rho_0\nu_0\nabla\times(\nabla\times u_1)$$
$$-\rho_1\nu_0 b\nabla\nabla\cdot u_0. \quad (18\text{-II})$$

Equations (17.2) and (17-II) show that neither ρ_2 nor ρ_{II} are conserved. In the previous part, it was noted that ρ_{II} obeyed laws that would be expected by physicists who rely on physical intuition, while ρ_2 did not. The reason that ρ_{II} is no longer conserved is clearly to be found in the dissipation of acoustic energy by the viscous forces. There appears to be no convenient way of introducing a relevant quantity that is conserved.

The Second-Order Motion of the Fluid

Elimination of ρ_{II} between the Eqs. (17-II) and (18-II) results in the equation

$$\rho_0(\partial^2 \boldsymbol{u}_1/\partial t^2) - \rho_0 c_0^2 \nabla\nabla\cdot \boldsymbol{u}_1$$
$$+ \rho_0 \nu_0 b \nabla(\partial/\partial t)(\nabla\cdot \boldsymbol{u}_1)$$
$$+ \rho_0 \nu_0 (\partial/\partial t)\cdot \nabla \times (\nabla \times \boldsymbol{u}_1)$$
$$= -\nabla(\partial/\partial t)(\rho_0 u_0^2) - b\nu_0(\partial/\partial t)[\rho_1 \nabla\nabla\cdot \boldsymbol{u}_0]$$
$$-b\nu_0 \nabla[\boldsymbol{u}_0 \cdot \nabla(\partial \rho_1/\partial t)]. \quad (22)$$

This equation can be simplified by introducing the divergence and rotation of the velocity

$$D = \nabla\cdot \boldsymbol{u}, \quad \boldsymbol{R} = \nabla \times \boldsymbol{u}. \quad (23)$$

Taking the divergence of Eq. (22), the rotation $\boldsymbol{R}_1$ is eliminated:

$$\partial^2 D_1/\partial t^2 - c_0^2 \nabla^2 D_1 - \nu_0 b \nabla^2 (\partial D_1/\partial t)$$
$$= -\nabla^2(\partial u_0^2/\partial t) - (b\nu_0/\rho_0)\nabla\cdot[(\partial/\partial t)(\rho_1 \nabla D_0)]$$
$$- (b\nu_0/\rho_0)\nabla^2[\boldsymbol{u}_0 \cdot \nabla(\partial \rho_1/\partial t)]. \quad (24)$$

Taking the rotation of Eq. (18-II) similarly eliminates D_1:

$$(\partial \boldsymbol{R}_1/\partial t) - \nu_0 \nabla^2 \boldsymbol{R}_1 = (b\nu_0/\rho_0^2)\nabla \rho_1 \times \nabla(\partial \rho_1/\partial t). \quad (25)$$

It is worth noting in more detail than above that when the first-order quantities are simple harmonic functions of the time, the right side of Eq. (25) is independent of the time. For, let

$$p_1 = \rho_1 c_0^2 = P' \cos nt + P'' \sin nt,$$

where P' and P'' are functions of the space coordinates only; then

$$c_0^2 \nabla \rho_1 = \nabla P' \cos nt + \nabla P'' \sin nt,$$
$$(c_0^2/n)\nabla(\partial \rho_1/\partial t) = -\nabla P' \sin nt + \nabla P'' \cos nt,$$

whence

$$\nabla \rho_1 \times \nabla(\partial \rho_1/\partial t) = (n/c_0^4)\nabla P' \times \nabla P''.$$

This is perhaps the first indication that the theory of the fluid streams generated by sound sources is governed by Eq. (25). It also indicates that the vorticity generated by a sound wave will approach a steady value after a sufficiently long time. The length of this time depends on the value of the viscosity coefficient ν_0, but the steady state itself does not: it is determined by the equation

$$-\nabla^2 \boldsymbol{R}_1 = (b/\rho_0^2)\nabla \rho_1 \times \nabla(\partial \rho_1/\partial t). \quad (25a)$$

The general procedure to obtain the velocity $\boldsymbol{u}_1$ will involve two steps, the first being the solution of Eqs. (24) and (25). Then $\boldsymbol{u}_1$ itself can be obtained from D_1 and $\boldsymbol{R}_1$ because of the vector identity

$$\nabla^2 \boldsymbol{u}_1 = \nabla D_1 - \nabla \times \boldsymbol{R}_1.$$

This equation is most conveniently handled by introducing the scalar and vector potentials defined by

$$\nabla^2 \phi_1 = -D_1, \quad (26)$$
$$\nabla^2 \boldsymbol{A}_1 = -\boldsymbol{R}_1, \quad (27)$$

in terms of which

$$\boldsymbol{u}_1 = -\nabla \phi_1 + \nabla \times \boldsymbol{A}_1. \quad (28)$$

Equation (28) indicates that the irrotational and incompressible parts of the motion can be treated independently. Only the steady state of the latter will be considered further, the discussion being based on Eq. (25a).

The Diffusion of Vorticity

This equation has the general form of Fourier's equation for the conduction of heat or the diffusion of matter. Hence vorticity is generated in those regions where its right side is different from zero, and diffuses into other regions. However, the analogy to heat conduction and diffusion is not complete.

Being a partial differential equation, Eq. (25) has many solutions, and that solution appropriate to the given problem is determined by the boundary condition. In the case of heat flow, this is usually one of two: if the boundaries are thermally insulated, the normal component of the temperature gradient will be zero on them; if the boundaries are kept at a fixed temperature, this fact serves to determine the special solution.

Neither of these boundary conditions applies to vorticity. This is perhaps to be expected, since vorticity is a vector rather than a scalar. In fact, the accepted boundary condition is not formulated in terms of $\boldsymbol{R}_1$ but in terms of $\boldsymbol{u}_1$, and requires that all components of the latter vanish on the walls of the container.

For simplicity, let it be supposed that there is no free surface, so that $\boldsymbol{u}_1$ vanishes at all boundaries of the fluid. Then an easy application of Stokes' theorem shows that

$$\boldsymbol{n} \cdot \boldsymbol{R}_1 = 0, \quad (29)$$

$\boldsymbol{n}$ being the unit normal at the boundary, while the divergence theorem results in

$$\iiint \boldsymbol{R}_1 dv = 0. \quad (30)$$

These are the only conditions restricting the solutions of Eq. (25) that can be formulated in terms of R_1 alone. They are very weak conditions, and do not suffice to determine the solution uniquely. This latter step must be postponed until u_1 has been found from Eqs. (27) and (28).

Because of this difference in the boundary conditions, vorticity may be generated at the walls of the containing vessel in a manner that is quite different from the influx (or efflux) of heat in the thermal analogy. Consequently also, the steady states of vortex distribution will be quite different from those familiar in the theory of heat. If these general considerations are not kept in mind, the reader may be surprised at some of the results obtained when the above equations are applied to a special case.

PART III. THE FLOW CAUSED BY A BEAM OF SOUND

In order to simplify the calculations, consider a long tube of radius r_0, whose walls are rigid, and whose ends are closed by some material that permits an axial sound beam to enter and leave the tube without reflection. These ends prevent fluid from entering or leaving the tube; the latter is long enough so that at its center all effects due to the ends may be neglected, except that the total flow through any cross section must be zero.

If the axis of the tube is the z axis and r the perpendicular distance from the axis, the pressure variations in the sound beam will be assumed to be

$$p_1 = \rho_1 c_0^2 = P(r)\sin(kz - nt). \qquad (31)$$

This requires some justification, since it neglects both the divergence and the attenuation of the beam. The former is justified if the wave-length ($=2\pi/k$) is very small compared to the diameter of the beam. The latter would not be justified except that only the ultimate steady state is to be investigated here, and by Eq. (25a), this is independent of the viscosity except as the latter enters into ρ_1. It is thus permissible to consider first the case of negligible attenuation, and to reserve until later the complications resulting from attenuation of the sound beam.

Using Eq. (31), Eq. (25a) reduces to

$$+\nabla^2 R_1 = K(d^2P/dr^2)(-\mathbf{i}\sin\phi + \mathbf{j}\cos\phi), \qquad (32)$$

where

$$K = bk/2\rho_0^2 c_0^3, \qquad (33)$$

and ϕ is the azimuthal coordinate. This equation has the special solution

$$R_1 = f(r)(-\mathbf{i}\sin\phi + \mathbf{j}\cos\phi), \qquad (34)$$

where

$$f(r) = (K/r)\int_0^r rP^2 dr + 2\beta r + \gamma/r, \qquad (35)$$

β and γ being constants of integration. The value of γ must be zero, since infinite values of the vorticity are impossible; the value of β remains indeterminate, since both Eq. (29) and Eq. (30) are satisfied for any value of β.

The calculation of the vector potential A_1 can be avoided by noting that if $u_{1x} = u_{1y} = 0$, $u_{1z} = g(r)$, then

$$R_1 = -(dg/dr)(-\mathbf{i}\sin\phi + \mathbf{j}\cos\phi),$$

so that

$$dg/dr = -f(r). \qquad (36)$$

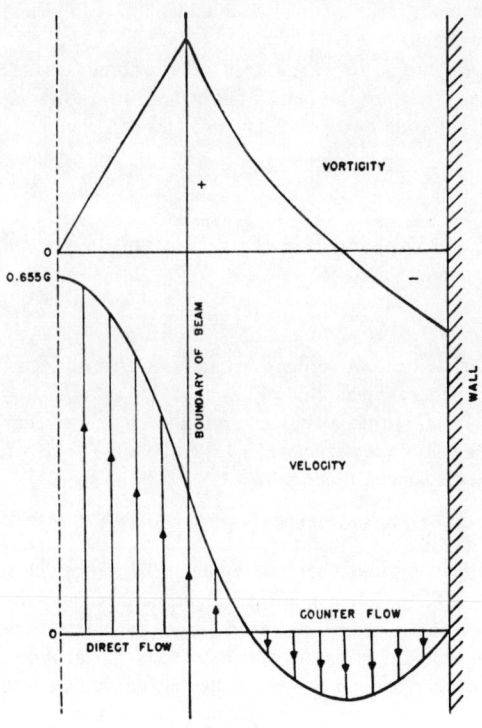

Fig. 1.

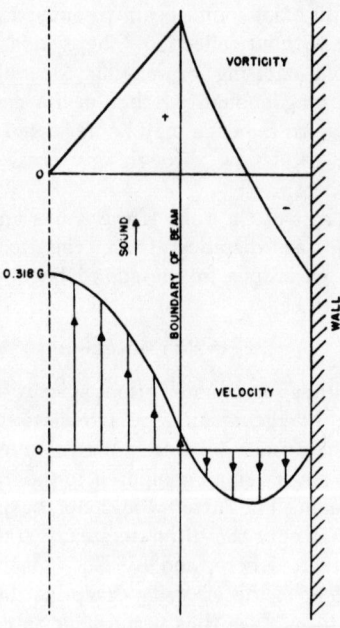

Fig. 2.

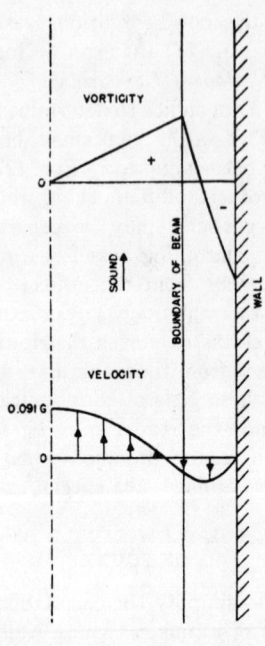

Fig. 3.

That solution of this equation is required which makes $g(r_0) = 0$, since u_1 must vanish on the wall of the tube; this is given by

$$g(r) = \int_r^{r_0} f(r)dr$$
$$= K\int_0^{r_0} \Gamma(s,r)P^2(s)ds + \beta(r_0^2 - r^2), \quad (37)$$

where

$$\Gamma(s,r) = s\log(r_0/r) \quad \text{when} \quad s \leqslant r,$$
$$= s\log(r_0/s) \quad \text{when} \quad s \geqslant r. \quad (38)$$

Until this place, it has not been possible to assign a value to the constant of integration, β. The condition that there be no net flow through the tube can now be imposed:

$$\int_0^{r_0} rg(r)dr = 0,$$

which is equivalent to

$$-\beta = (K/r_0^4)\int_0^{r_0}(rr_0^2 - r^3)P^2(r)dr. \quad (39)$$

This completes the formal solution of the problem; it remains to consider the numerical relations.

In order to carry the calculations further, suppose that the sound beam has the radius r_1, and a constant intensity throughout; then

$$P(r) = P_0, \quad r \leqslant r_1;$$
$$= 0, \quad r > r_1. \quad (40)$$

It will be found that the velocity of the stream is proportional to

$$G = \tfrac{1}{2}KP_0^2 r_1^2, \quad (41)$$

so that it is convenient to calculate this quantity for several special cases.

First, suppose that the medium is water, that the beam has a radius $r_1 = 1.5$ cm, while $P_0 = 10^5 \mu b$ ($= 0.1$ atmos.). Since $c_0 = 1.5 \times 10^5$ cm/sec.

$$G = \tfrac{1}{6}bk^2 \times 10^{-5} \text{ cm/sec. (water)}.$$

If the frequency of the sound is 24 megacycles, $k = 10^3$, and hence $G = 1.57b$ cm/sec. It will be expected, therefore, that these streams will become appreciable only at frequencies above 1 megacycle, and have a negligible velocity at lower frequencies. This is in agreement with observation.

If the medium is air, the value of K is much larger because it is inversely proportional to the square of the density and the cube of the sound velocity. Supposing the sound beam to have the same radius and intensity, it is found that

$$G = 95.0bk^2 \text{ cm/sec. (air)}.$$

For a frequency of 1 kilocycle, $k = 0.19$, and hence $G = 3.4b$ cm/sec. Consequently, these streams should become important in air at frequencies above several hundred cycles. This is perhaps in agreement with Rayleigh's resonator experiment mentioned above, which was performed at 256 cycles per second.

While the velocity of the stream is proportional to G, its value will vary across the section of the tube which confines it. The direction of flow will coincide with that of the acoustic energy on the axis of the beam, and will be compensated by a counter-flow near the walls of the tube. The complete expression for the velocity is most conveniently written in terms of the ratios

$$x = r/r_0, \quad y = r_1/r_0,$$

and is

$$g = G\{\tfrac{1}{2}(1 - x^2/y^2) - (1 - \tfrac{1}{2}y^2)(1 - x^2) - \log y\},$$
$$0 \leqslant x \leqslant y; \quad (42)$$
$$= -G\{(1 - \tfrac{1}{2}y^2)(1 - x^2) + \log x\}, \quad y \leqslant x \leqslant 1.$$

Graphs of g as a function of x, are given, for various values of y, in Figs. 1, 2, and 3.

The maximum value of g occurs on the axis and depends on the resistance offered to the flow by the confining tube. When the radius of the latter is infinite, the maximum value of g becomes logarithmically infinite. On the other hand, when the sound beam fills the whole of the tube, the flow stalls and $g = 0$ everywhere.

Experimental Determination of b

The experimental determination of the numerical constant b has eluded three generations of physicists. Stokes argued that its value must be $4/3$—i.e., that the bulk viscosity must be zero. This hypothesis has been theoretically verified for an ideal gas. Stokes' arguments do not appear to be convincing in the case of liquids, and there is a growing belief (based on discrepancies between the observed absorption of high frequency sound and that calculated on the assumption that $b = 4/3$) that it may have larger values. It is hoped that these calculations may suggest methods for its experimental measurement.

It will be noted, however, that the above discussion shows that the streaming velocity is very sensitive to the geometry of the experiment, and will readily become turbulent. Consequently, the experimental measurement of b may not be easy.

Editor's Comments on Papers 12 and 13

12 Rayleigh: *On the Pressure of Vibrations*

13 Borgnis: *Acoustic Radiation Pressure of Plane Compressional Waves*

Lord Rayleigh, who followed the contemporary literature very closely, was much impressed with the experimental verification by Lebedev of the existence of the pressure of electromagnetic radiation that had been predicted by Maxwell. It seemed natural to inquire whether a similar pressure exists for acoustical radiation. This question is examined in this germinal paper, "On the Pressure of Vibrations."

As was common in Rayleigh's theoretical attacks on problems of some complexity, he begins his analysis with simple cases, here the vibrations of a simple pendulum in which the string is gradually shortened by means of a ring. He then considers the case of sound waves in air. In his pendulum analysis, Rayleigh comes very close to establishing the idea of adiabatic invariance, so important in the adiabatic principle of Ehrenfest in quantum theory.

As noted in the paper by Borgnis, Rayleigh's paper does not present a complete picture of radiation pressure. In fact, his interpretation of such pressure is not the one that is most useful in modern practical applications of radiation pressure, in the "weighing of sound," for example. However, Rayleigh paved the way for further work; hence his influential paper is included here.

Borgnis, in his paper, stresses the complexities associated with the concept of radiation pressure in the case of acoustic radiation and reviews some of the literature produced since the publication of Rayleigh's paper. The need for adequate physical definitions is emphasized.

Fritz E. Borgnis (1906–) is a German-born physicist with a doctorate in physics from the University of Munich. He worked at various institutions in the United States from 1950 to 1957 and is now a professor at the Swiss Federal Institute of Technology in Zurich. His special fields of research have been electrodynamics, gaseous discharges, and ultrasonics.

Reprinted from *Scientific Papers of Lord Rayleigh,* Dover Publications, Inc., New York, Vol. 5, p. 41–48 (1964)

12

276.

ON THE PRESSURE OF VIBRATIONS.

[*Philosophical Magazine,* III. pp. 338—346, 1902.]

THE importance of the consequences deduced by Boltzmann and W. Wien from the doctrine of the pressure of radiation has naturally drawn increased attention to this subject. That æthereal vibrations must exercise a pressure upon a perfectly conducting, and therefore perfectly reflecting, boundary was Maxwell's deduction from his general equations of the electromagnetic field; and the existence of the pressure of light has lately been confirmed experimentally by Lebedew. It seemed to me that it would be of interest to inquire whether other kinds of vibration exercise a pressure, and if possible to frame a general theory of the action.

We are at once confronted with a difference between the conditions to be dealt with in the case of æthereal vibrations and, for example, the vibrations of air. When a plate of polished silver advances against waves of light, the waves indeed are reflected, but the medium itself must be supposed capable of penetrating the plate; whereas in the corresponding case of aerial vibrations the air as well as the vibrations are compressed by the advancing wall. In other cases, however, a closer parallelism may be established. Thus the transverse vibrations of a stretched string, or wire, may be supposed to be limited by a small ring constrained to remain upon the equilibrium line of the string, but capable of sliding freely upon it. In this arrangement the string passes but the vibrations are compressed, when the ring moves inwards.

We will commence with the very simple problem of a pendulum in which a mass C is suspended by a string. B is a ring [which embraces the string] constrained to the vertical line AD and capable of moving along it; $BC = l$, and θ denotes the angle between BC and AD at any time t. If B is held at rest, BC is an ordinary pendulum, and it is supposed to be executing small vibrations; so that $\theta = \Theta \cos nt$, where $n^2 = g/l$. The tension of the string is approximately W, the weight of the bob; and the force tending to push B upwards is at time t $W(1 - \cos\theta)$. Now this expression is closely related to the potential energy of the pendulum, for which

Fig. 1.

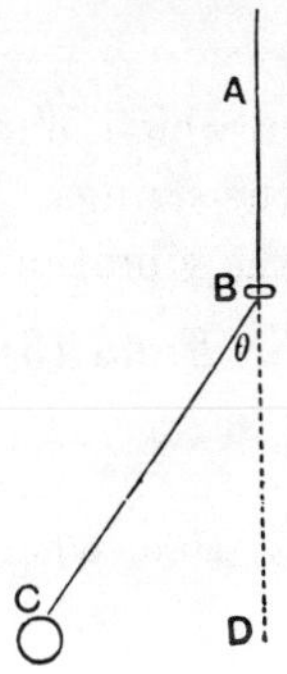

$$V = Wl(1 - \cos\theta).$$

The mean upward force upon B is accordingly equal to the mean value of $V \div l$; or since the mean value of V is half the constant total energy E of the system, we conclude that the mean force (L), driving B upwards, is measured by $\frac{1}{2} E/l$.

From the equation
$$L = \tfrac{1}{2} E/l \quad\quad\quad\quad\quad\quad\quad\quad\quad\quad (1)$$
it is easy to deduce the effect of a *slow* motion upwards of the ring. The work obtained at B must be at the expense of the energy of the system, so that
$$dE = -L\,dl = -\tfrac{1}{2} E\,dl/l.$$

By integration
$$E = E_1 l^{-\frac{1}{2}}, \quad\quad\quad\quad\quad\quad\quad\quad\quad\quad (2)$$
where E_1 denotes the energy corresponding to $l = 1$. From (2) we see that by withdrawing the ring B until l is infinitely great, the whole of the energy of vibration may be abstracted in the form of work done by B, and this by a uniform motion in which no regard is paid to the momentary phase of the vibration.

The argument is nearly the same for the case of a stretched string vibrating transversely in one plane. The string itself may be supposed to be unlimited, while the vibrations are confined by two rings of which one may be fixed and one movable.

If the origin of x be at one end of a string of length l, the transverse displacement [and velocity] may be expressed by
$$y = \phi_1 \frac{\sin \pi x}{l} + \phi_2 \sin \frac{2\pi x}{l} + \ldots, \quad\quad\quad\quad (3)$$
$$\dot{y} = \dot{\phi}_1 \frac{\sin \pi x}{l} + \dot{\phi}_2 \sin \frac{2\pi x}{l} + \ldots, \quad\quad\quad\quad (4)$$
where $\phi_1, \phi_2, \ldots$ are coefficients depending upon the time. For the kinetic and potential energies we have respectively (*Theory of Sound*, § 128)
$$T = \tfrac{1}{4} \rho l \sum_{s=1}^{s=\infty} \dot{\phi}_s^{\,2}, \quad V = \tfrac{1}{4} W l \sum_{s=1}^{s=\infty} \frac{s^2 \pi^2}{l^2} \phi_s^{\,2}, \quad\quad\quad (5)$$
in which W represents the constant tension and ρ the longitudinal density of the string. · For each kind of ϕ the sums of T and V remain constant during the vibration; and the same is of course true of the totals given in (5).

From (3)
$$\frac{dy}{dx} = \frac{\pi}{l} \left(\phi_1 \cos \frac{\pi x}{l} + 2\phi_2 \cos \frac{2\pi x}{l} + \ldots \right),$$
so that when $x = l$
$$\frac{dy}{dx} = \frac{\pi}{l} (-\phi_1 + 2\phi_2 - 3\phi_3 + \ldots).$$

Accordingly the force tending to drive out the ring at $x = l$ is at time t

$$\tfrac{1}{2} W \cdot \frac{\pi^2}{l^2} (-\phi_1 + 2\phi_2 - 3\phi_3 + \ldots)^2,$$

or in the mean taken over a long interval,

$$\tfrac{1}{2} W \cdot \text{Mean } \Sigma \, \frac{s^2 \pi^2}{l^2} \phi_s^2.$$

Comparing with (5), we see that the mean force L has the value $2l \times \text{mean } V$; or since mean $V = \text{mean } T = \tfrac{1}{2} E$, E denoting the constant total energy,

$$L = E/l. \quad\ldots\ldots\ldots\ldots\ldots\ldots\ldots\ldots\ldots(6)$$

The force driving out the ring is thus numerically equal to the *longitudinal density of the energy*.

This result may readily be extended to cases where the vibrations are not limited to one plane; and indeed the case in which the plane of the string uniformly revolves is especially simple in that T and V are then constant with respect to time.

If the ring is allowed to move out slowly, we have

$$dE = -L\,dl = -E\,dl/l,$$

or on integration

$$E = E_1 l^{-1}, \quad\ldots\ldots\ldots\ldots\ldots\ldots\ldots(7)$$

analogous to (2), though different from it in the power of l involved. If l increase without limit, the whole energy of the vibrations may be abstracted in the form of work done on the ring.

We will now pass on to consider the case of air in a cylinder, vibrating in one dimension and supposed to obey Boyle's law according to which $p = a^2 \rho$. By the general hydrodynamical equation (*Theory of Sound*, § 253 a),

$$\varpi = \int \frac{dp}{\rho} = -\frac{d\phi}{dt} - \tfrac{1}{2} U^2, \quad\ldots\ldots\ldots\ldots\ldots\ldots(8)$$

where ϕ denotes the velocity-potential and U the resultant velocity at any point; so that in the present case, if we integrate over a long interval of time,

$$a^2 \int \log p \, dt + \tfrac{1}{2} \int U^2 \, dt \quad\ldots\ldots\ldots\ldots\ldots\ldots(9)$$

retains a constant value over the length of the cylinder. If p_0 denote the pressure that would prevail throughout, had there been no vibrations, $p - p_0$ is small and we may replace (9) by

$$a^2 \int \left\{ \frac{p - p_0}{p_0} - \tfrac{1}{2} \frac{(p - p_0)^2}{p_0^2} \right\} dt + \tfrac{1}{2} \int U^2 \, dt. \quad\ldots\ldots\ldots(10)$$

The expression (10) has accordingly the same value at the piston as

for the mean of the whole column of length l. Now for the mean of the whole column
$$\int (p - p_0) \, dx = 0;$$
and thus if p_1 denote the value of p at the piston where $x = l$,
$$a^2 \int \left\{ \frac{p_1 - p_0}{p_0} - \tfrac{1}{2} \frac{(p_1 - p_0)^2}{p_0^2} \right\} dt = -\frac{a^2}{2l} \iint \frac{(p - p_0)^2}{p_0^2} \, dx \, dt + \frac{1}{2l} \iint U^2 \, dx \, dt. \quad \ldots\ldots\ldots(11)$$

It is not difficult to prove that the right-hand member of (11) vanishes. Thus, expressing the motion in terms of ϕ, suppose that
$$\phi = \cos \frac{s\pi x}{l} \cos \frac{s\pi at}{l}. \quad \ldots\ldots\ldots\ldots\ldots\ldots(12)$$
Then
$$p - p_0 = \rho_0 \, d\phi/dt, \quad U = d\phi/dx;$$
and since $p_0 = a^2 \rho_0$, we get
$$\frac{1}{2l} \iint \left\{ \left(\frac{d\phi}{dx} \right)^2 - \frac{1}{a^2} \left(\frac{d\phi}{dt} \right)^2 \right\} dx \, dt,$$
and this vanishes by (12). Accordingly
$$\int (p_1 - p_0) \, dt = \int \frac{(p_1 - p_0)^2}{2p_0} \, dt. \quad \ldots\ldots\ldots\ldots(13)$$
Again by (12)
$$\int \left(\frac{d\phi}{dt} \right)_l^2 dt = \frac{2}{l} \iint \left(\frac{d\phi}{dt} \right)^2 dx \, dt,$$
so that
$$\int (p_1 - p_0) \, dt = \frac{1}{p_0 l} \iint (p_1 - p_0)^2 \, dx \, dt = \frac{\rho_0}{l} \iint U^2 \, dx \, dt.$$

Now $\rho_0 \iint U^2 \, dx \, dt$ represents twice the mean total kinetic energy of the vibrations or, what is the same, the constant total energy E. Thus if L denote the mean additional force due to the vibrations and tending to push the piston out,
$$L = E l^{-1}. \quad \ldots\ldots\ldots\ldots\ldots\ldots\ldots\ldots\ldots(14)$$

As in the case of the string, the total force is measured by the longitudinal density of the total energy; or, if we prefer so to express it, the additional *pressure* is measured by the volume-density of the energy.

In the last problem, as well as in that of the string, the vibrations are in one dimension. In the case of air there is no difficulty in the extension to two or three dimensions. Thus, if aerial vibrations be distributed equally in all directions, the pressure due to them coincides with *one-third* of the volume-density of the energy. In the case of the string, where the vibrations are transverse, we cannot find an analogue in three dimensions; but a membrane with a flexible and extensive boundary capable of slipping along the surface, provides for two dimensions. If the vibrations be equally distributed in the plane, the force outwards per unit length of contour will be measured by one-half of the superficial density of the total energy.

A more general treatment of the question may be effected by means of Lagrange's theory. If l be one of the coordinates fixing the configuration of a system, the corresponding equation is

$$\frac{d}{dt}\left(\frac{dT}{dl'}\right) - \frac{dT}{dl} + \frac{dV}{dl} = L, \quad \ldots\ldots\ldots\ldots\ldots\ldots(15)$$

where T and V denote as usual the expressions for the kinetic and potential energies. On integration over a time t_1

$$\int \frac{L\,dt}{t_1} = \frac{1}{t_1}\left[\frac{dT}{dl'}\right] + \frac{1}{t_1}\int\left(\frac{dV}{dl} - \frac{dT}{dl}\right)dt.$$

If dT/dl' remain finite throughout, and if the range of integration be sufficiently extended, the integrated term disappears, and we get

$$\int \frac{L\,dt}{t_1} = \frac{1}{t_1}\int\left(\frac{dV}{dl} - \frac{dT}{dl}\right)dt. \quad \ldots\ldots\ldots\ldots\ldots\ldots(16)$$

On the right hand of (16) the differentiations are partial, the coordinates other than l and all the velocities being supposed constant.

We will apply our equation (16) in the first place to the simple pendulum of fig. 1, l denoting the length of the vibrating portion of the string BC. If x, y be the horizontal and vertical coordinates of C,

$$x = l\sin\theta, \quad y = l - l\cos\theta;$$

and accordingly if the mass of C be taken to be unity,

$$T = \tfrac{1}{2}l'^2(2 - 2\cos\theta) + l'\theta' \cdot l\sin\theta + \tfrac{1}{2}\theta'^2 l^2, \quad \ldots\ldots\ldots\ldots(17)$$

l', θ' denoting dl/dt, $d\theta/dt$. Also

$$V = gl(1 - \cos\theta). \quad \ldots\ldots\ldots\ldots\ldots\ldots(18)$$

From (17), (18)

$$\frac{dV}{dl} = g(1 - \cos\theta), \quad \frac{dT}{dl} = l'\theta'\sin\theta + \theta'^2 l. \quad \ldots\ldots\ldots\ldots(19)$$

These expressions are general; but for our present purpose it will suffice if we suppose that l' is zero, that is that the ring is held at rest. Accordingly

$$\frac{dV}{dl} = \frac{V}{l}, \quad \frac{dT}{dl} = \frac{2T}{l},$$

and (16) gives

$$\int \frac{L\,dt}{t_1} = \frac{1}{t_1}\int \frac{V - 2T}{l}dt. \quad \ldots\ldots\ldots\ldots\ldots\ldots(20)$$

On the right hand of (20) we find the mean values of V and of T. But these mean values are equal. In fact

$$\int V\,dt = \int T\,dt = \tfrac{1}{2}Et_1, \quad \ldots\ldots\ldots\ldots\ldots\ldots(21)$$

if E denote the total energy. Hence, if L now denote the mean value,
$$L = -\tfrac{1}{2} E/l, \quad \ldots\ldots\ldots\ldots\ldots\ldots(22)$$
the negative sign denoting that the mean force necessary to hold the ring at rest must be applied in the direction which tends to diminish l, i.e. downwards. In former equations (1), (6), (14), L had the reverse sign.

We will now consider more generally the case of one dimension, using a method that will apply equally whether for example the vibrating body be a stretched string, or a rod vibrating flexurally. All that we postulate is homogeneity of constitution, so that what can be said about any part of the length can be said equally about any other part. In applying Lagrange's method the coordinates are l the length of the vibrating portion, and ϕ_1, ϕ_2, &c. defining, as in (3), the displacement from equilibrium during the vibrations. As functions of l, we suppose that
$$V \propto l^m, \quad T \propto l^n. \quad \ldots\ldots\ldots\ldots\ldots(23)$$
Thus, if L be the force corresponding to l, we get by (16)
$$\int \frac{L\,dt}{t_1} = \frac{1}{t_1} \int \left(\frac{mV}{l} - \frac{nT}{l} \right) dt,$$
in which
$$\int V\,dt = \int T\,dt = \tfrac{1}{2} E \cdot t_1,$$
E representing as before the constant total energy. Accordingly, L now representing the mean value,
$$L = \frac{(m-n)E}{2l}. \quad \ldots\ldots\ldots\ldots\ldots(24)$$

In the case of a medium, like a stretched string, propagating waves of all lengths with the same velocity, $m = -1$, $n = 1$, and $L = -E/l$, as was found before.

In the application to a rod vibrating flexurally, $m = -3$, $n = 1$, so that
$$L = -2E/l. \quad \ldots\ldots\ldots\ldots\ldots(25)$$

If $m = n$, L vanishes. This occurs in the case of the line of disconnected pendulums considered by Reynolds in illustration of the theory of the group velocity*, and the circumstance suggests that L represents the tendency of a group of waves to spread. This conjecture is easily verified. If in conformity with (13) we suppose that
$$V = V_0\, l^m\, \phi_1^2, \quad T = T_0\, l^n\, \dot{\phi}_1^2,$$
and also that
$$\phi_1 = \sin \frac{2\pi t}{\tau}, \quad \dot{\phi}_1 = \frac{2\pi}{\tau} \cos \frac{2\pi t}{\tau},$$

---

* See *Proc. Math. Soc.* ix. p. 21 (1877); this collection, i. p. 322. Also *Theory of Sound*, Vol. i. Appendix.

τ being the period of the vibration represented by the coordinate ϕ_1, we obtain, remembering that the sum of T and V must remain constant,

$$V_0 l^m = T_0 l^n \cdot 4\pi/\tau^2.$$

This gives the relation between τ and l. Now v, the wave-velocity, is proportional to l/τ; so that

$$v \propto l^{1 - \frac{1}{2}n + \frac{1}{2}m}. \quad\quad\quad\quad\quad\quad (26)$$

Thus, if u denote the group-velocity, we have by the general theory

$$u/v = \tfrac{1}{2}n - \tfrac{1}{2}m; \quad\quad\quad\quad\quad\quad (27)$$

and in terms of u and v by (24)

$$L = -\frac{uE}{vl}. \quad\quad\quad\quad\quad\quad (28)$$

Boltzmann's theory is founded upon the application of Carnot's cycle to the radiation inclosed within movable reflecting walls. If the pressure (p) of a body be regarded as a function of the volume v*, and the absolute temperature θ, the general equation deduced from the second law of thermodynamics is

$$\frac{dp}{d\log\theta} = M, \quad\quad\quad\quad\quad\quad (29)$$

where $M\,dv$ represents the heat that must be communicated while the volume alters by dv and $d\theta = 0$. In the application of (29) to radiation we have evidently

$$M = U + p, \quad\quad\quad\quad\quad\quad (30)$$

where U denotes the density of the energy—a function of θ only. Hence†

$$\frac{dp}{d\log\theta} = U + p. \quad\quad\quad\quad\quad\quad (31)$$

If further, as for radiation and for aerial vibrations,

$$p = \tfrac{1}{3}U, \quad\quad\quad\quad\quad\quad (32)$$

it follows at once that

$$d\log U = 4\, d\log \theta,$$

whence

$$U \propto \theta^4, \quad\quad\quad\quad\quad\quad (33)$$

the well-known law of Stefan. It may be observed that the existence of a pressure is demanded by (31), independently of (32).

If we generalize (32) by taking

$$p = \frac{1}{n}U, \quad\quad\quad\quad\quad\quad (34)$$

where n is some numerical quantity, we obtain as the generalization of (33)

$$U \propto \theta^{n+1}. \quad\quad\quad\quad\quad\quad (35)$$

---

* Now with an altered meaning.
† Compare Lorentz, *Amsterdam Proceedings*, Ap. 1901.

It is an interesting question whether any analogue of the second law of thermodynamics can be found in the general theory of the pressure of vibrations, whether for example the energy of the vibrations of a stretched string is partially unavailable in the absence of appliances for distinguishing *phases*. It might appear at first sight that the conclusion already given, as to the possibility of recovering the whole energy by mere retreat of the inclosing ring, was a proof to the contrary. This argument, however, will not appear conclusive, if we remember that a like proposition is true for the energy of a gas confined adiabatically under a piston. The residual energy of the molecules may be made as small as we please, but the completion of the cycle by pushing the piston back will restore the molecular energy unless we can first abolish the infinitesimal residue remaining after expansion, and this can only be done with the aid of a body at the absolute zero of temperature. It would appear that we may find an analogue for temperature, so far as the vibrations of *one* system are concerned; but, so far as I can see, the analogy breaks down when we attempt a general theory.

[1910. See further *Phil. Mag.* Sept. 1905 "On the Momentum and Pressure of Gaseous Vibrations, and on the Connection with the Virial Theorem."]

Acoustic Radiation Pressure of Plane Compressional Waves*

F. E. BORGNIS
California Institute of Technology, Pasadena, California

Both electromagnetic and acoustic waves exert forces of radiation upon an obstacle placed in the path of the wave, the forces being proportional to the mean energy density of the wave motion. In electromagnetics the action of these forces is relatively easily understood through the concept of Maxwell's electromagnetic stress tensor.

The physical processes leading to these forces in a sound wave have been found to be considerably more complex; the difficulties belong to the fact that the acoustic wave equation is not linear and that a beam of finite cross section is subject to effects caused by the surrounding medium.

Though many papers have been devoted to the subject and though various theoretical approaches have been made, some difficulties still seem to stand in the way of a clear understanding of the physics of the problem.

The purpose of the present study is especially to throw light on the physical aspects. The approach adopted, which uses the momentum theorem, is believed to serve this purpose especially well. The expression for the radiation pressure is given in both Eulerian and Lagrangian coordinate systems.

Special consideration is given to liquids of constant compressibility, since in such media the processes involved can be dealt with mathematically in a simple manner. The general case of a plane reflector with arbitrary reflection coefficient is treated; the *modus operandi* of the forces at the interface between liquid and obstacle is explained for some special cases, including the radiation forces on the interface between two nonmiscible liquids.

Finally, a general relation is established between the energy density and the pressure caused by radiation falling normally upon a plane reflector, which, under certain assumptions, is valid in any fluid and at any amplitude.

1. INTRODUCTION

THE concept of radiation pressure originated in electrodynamics. According to Maxwell's equations and his concept of the electrodynamic stress tensor, a plane surface of perfect conductivity emitting normally a plane electromagnetic wave undergoes a reacting force per unit area equal to the total energy density E of the emitted wave. Also, such a plane wave, striking perpendicularly a plane and totally absorbing surface, exerts a radiation pressure equal to E.

Lord Rayleigh was the first to formulate a theory of radiation pressure resulting from compressional acoustic waves in fluids. He established a relation which gives the time-average value of the pressure produced upon a piston by a plane wave of infinite cross section in a fluid, and he showed that this mean pressure is proportional to the mean mechanical energy density $\bar{E}$ of the periodic wave motion. He found that the factor of proportionality was not in general unity, but dependent on the special law relating the pressure p to the density in the fluid under consideration.

The actual radiation pressure, however, as encountered in an acoustic beam under ordinary experimental conditions, is different from Rayleigh's "pressure of vibrations." Rayleigh's result,[1] which is often quoted, applies to a theoretical case rather than to what is usually measured.

The equations describing the motion of acoustic waves are nonlinear (with the one exception of the Lagrangian wave equation in a fluid of constant com-

---

* Work performed under U. S. Office of Naval Research Contract Nonr-220(02).

[1] Lord Rayleigh, Phil. Mag. **3**, 338 (1902) and **10**, 364 (1905). For example, in a gas under adiabatic conditions, Rayleigh's pressure amounts to $(1+\gamma_c)\bar{E}_i$ at a perfect reflector and at small amplitudes. γ_c = ratio of the specific heats and $\bar{E}_i$ = mean energy density of the incident wave = one-half of the total mean energy density in a standing wave.

pressibility). For many purposes, it is sufficient to "linearize" these equations and to retain only first-order terms of the velocities or particle displacements, regarding them as small quantities. Radiation pressure, however, is connected with energy densities, which are quadratic terms containing the squares of velocities or displacements. Any theory dealing with acoustic radiation pressure, therefore, must retain at least all second-order terms, to be valid even at small amplitudes.

The fact that radiation pressure is a second-order quantity and that the vibrational amplitudes are usually very small in comparison with the acoustic wavelength λ, explains its relatively small numerical value in comparison with the values of the periodically alternating first-order pressures encountered in acoustical waves. Whereas the first-order pressure amounts to maximal values up to kilograms per cm^2, acoustic radiation pressure only reaches values of the order of grams or dynes of force per cm^2. Nevertheless, radiation pressure is an important quantity in the experimental determination of acoustic intensity.

Since Rayleigh's investigations, a considerable number of papers have been devoted to the subject, many of

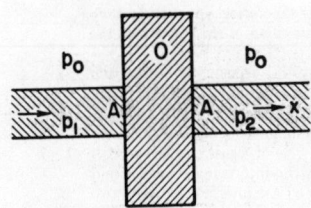

Fig. 1. Acoustic beam of finite cross section A incident upon a plane obstacle 0, and passing through undisturbed regions in which the hydrostatic pressure is p_0.

which, however, do not deal adequately with certain peculiar difficulties inherent in the problem.

L. Brillouin[2] seems to have been the first to give a comprehensive approach to the subject, especially in pointing out the tensorial character of what is usually called radiation "pressure." Indeed, this quantity is not a "pressure" in the sense ordinarily understood in hydrodynamics. Brillouin published his first paper on the subject as early as 1925; various authors, however, in later papers on the same subject, seem not to have paid proper attention to Brillouin's approach.

In what follows, a study is presented of the forces exerted upon a plane obstacle by plane compressional waves in a fluid, meeting the obstacle at normal incidence.[3]

Special attention is devoted to throwing light upon the actual physical processes involved. For this reason, a different approach from that used by Brillouin has been chosen; the study also goes further in the consideration of certain details than has been done hitherto.

[2] L. Brillouin, Ann. phys. (X) **4**, 528 (1925).
[3] The case of oblique incidence can easily be solved, if the forces resulting from radiation pressure at normal incidence are known. See, for example, F. E. Borgnis, J. Acoust. Soc. Am. **24**, 468 (1952).

2. STATEMENT OF THE PROBLEM

In dealing with plane waves, some idealizing assumptions must be made. Plane waves cannot strictly be realized experimentally. Still, if the width of the acoustic beam is large in comparison with the wavelength, the concept of plane waves gives a good approximation, especially in the case of the high frequencies used in ultrasonics. Under most experimental conditions, the acoustic beam in any fluid is surrounded by regions of the same fluid. Reasoning as in geometrical optics, we assume that only the region inside the beam is subject to the acoustical wave motion, whereas the region surrounding the beam is assumed to be undisturbed; that is, we disregard diffraction effects at the edge of the beam and assume uniformity of wave motion over the cross section of the beam. The beam is supposed to fall normally upon a plane obstacle, entirely immersed and at rest in the fluid (Fig. 1). This obstacle causes reflection at its front surface; it may absorb partially or totally the penetrating wave energy. The transmitted part leaves the rear of the obstacle as a purely progressive wave.

The interaction of the obstacle with the plane wave motion can be described completely by an amplitude reflection coefficient γ, an amplitude transmission coefficient δ, and the phase angles θ and θ' of the reflected and transmitted waves with respect to the phase of the incident wave; a plane wave motion in the acoustic beam is thus uniquely determined.

Considerable complications in the measurement of radiation pressure are caused by the fact that the acoustic field sets up a steady streaming in the medium. Two effects have been made responsible for this mass flow: first, the so-called "pumping effect," which may occur at the source of radiation and by which fluid is sucked into the beam and set in motion in the direction of the wave propagation. Second, owing to viscous forces, the wave motion generates what is called the "hydrodynamic flow;" a steady vortical motion is set up in the fluid, causing a steady streaming along the beam in the direction of the incident wave and returning outside the beam. The phenomenon of hydrodynamic flow is inseparably associated with acoustic radiation in a fluid. Like the latter, it is an effect of second order; it cannot be treated without taking into account at least second-order terms in the hydrodynamic equations.[4]

In order to measure the forces due to radiation only, these forces must be separated from those caused by the steady mass flow of the fluid. Usually, some sort of screening is resorted to, consisting of thin films, transparent to radiation but preventing the flow from exerting forces upon the obstacle. Such screens are located close and parallel to the surface of the obstacle,

[4] C. Eckart, Phys. Rev. **73**, 68 (1948); J. J. Markham, Phys. Rev. **86**, 497 (1952); P. J. Westervelt, J. Acoust. Soc. Am. **25**, 60 (1953).

and it is commonly assumed that in this way the forces caused by acoustic radiation can be measured with sufficient accuracy.

A separation of the forces in question may also be achieved by a low-frequency intensity modulation of the beam and measuring the radiation pressure by a device which responds to the modulation but not to the steady forces of the mass flow nor to the unmodulated wave.[5]

Since we are only concerned with forces caused by radiation, viscous forces are not taken into account in this paper.

3. GENERAL FORMULA FOR THE ACOUSTIC RADIATION PRESSURE

In order to obtain the force exerted upon 0 in Fig. 1, we apply Newton's theorem of equivalence of the time rate of change in momentum to the forces acting upon 0. To keep 0 in time average at rest, we have to apply a force F, equal and opposite to that caused by the quantity which we call radiation pressure. Denoting by P this radiation pressure, that is, the force *per unit area* of the beam of total area A, F amounts to PA, since we are assuming uniformity over the cross section A.

The other force acting on 0 is due to the "dynamic" pressure p_d and is given by $\oint p_d d\mathbf{a}$ over the entire surface of 0. In absence of acoustic waves, the pressure p_d is the undisturbed hydrostatic pressure p_0; when variation with depth is ignored, p_0 is constant over the whole surface and $\oint p_d d\mathbf{a} = p_0 \oint d\mathbf{a} = 0$.

If an acoustic wave motion is present, the hydrostatic pressure p_0 becomes "modulated" over the affected parts of 0 by the "excess pressure" p of the acoustic waves; the resultant pressure may be called the "dynamic pressure" $p_d = p_0 + p$. Forces are counted positive in the positive x direction (Fig. 1), which is the direction of propagation of the incident wave; the force upon 0 due to the action of p_d is $\oint p_d d\mathbf{a} = \oint p d\mathbf{a} = (p_1 - p_2)A$.

Next we establish the expression for the time rate of change of momentum of 0. As we have to assume continuity of particle displacement at the interface between the fluid and the surface of 0, mechanical motion is transferred to 0. Let ρ be the density and $\mathbf{u}$ the velocity vector of mass elements in 0; the "momentum-density" then is given by $\rho\mathbf{u}$. The quantities p, ρ, and $\mathbf{u}$ are considered as functions of x, y, z, and t, the coordinates belonging to a system fixed in space. 0 may be regarded as having a large inertia, so that its center of gravity may be assumed to be practically at rest, although parts of 0 undergo small and rapid mechanical movements.

In order to compute the change of momentum and to express it by values of the density and the velocity of the *fluid*, we visualize an imaginary surface S stationary in space and in close neighborhood of the surface of 0,

but at any time entirely within the fluid. Owing to the law of action and reaction, it is permissible to extend the integrations involved over the surface S instead of over the actual surface of 0.

The momentum of 0 can change in two ways: (a) by a "local" change *in time* of $\rho\mathbf{u}$ within volume elements of 0; the rate of change of this momentum is given by $(d/dt)\int_V \rho\mathbf{u} dV$; (b) by a "convectional" change of $\rho\mathbf{u}$ in the course of the displacement *in space* of mass elements of 0.†

The volume integral of the convectional change in

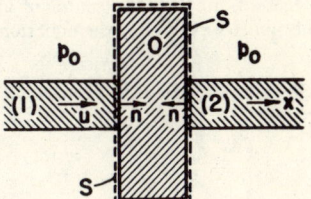

FIG. 2. Imaginary surface S enclosing 0 and stationary in space. n is the inner normal, u the (positive) particle velocity.

momentum of 0 can be transformed into the surface integral over the entire surface of S in Fig. 2 of the "flux of momentum" crossing S per unit time; the integral amounts to $\oint_S (\rho\mathbf{u})(\mathbf{u}\cdot\mathbf{n})da$,[6] Indeed, during a time element dt the "flux of momentum" crossing a surface element da of S is given by $(\rho\mathbf{u})(\mathbf{u}\cdot\mathbf{n})dadt$, $\rho\mathbf{u}$ being the density of momentum and $(\mathbf{u}\cdot\mathbf{n})$ the normal velocity across S. The surface integral results from a gain or loss in momentum of mass elements in the course of their displacements within 0.

In our one-dimensional problem, the total flux of momentum *entering* S per unit time is given by $(\rho_1 u_1^2 - \rho_2 u_2^2)A$; the minus sign of $\rho_2 u_2^2$ expresses the fact that this quantity is the flux *leaving* S, when u is positive in the positive x direction. The parts of S out-

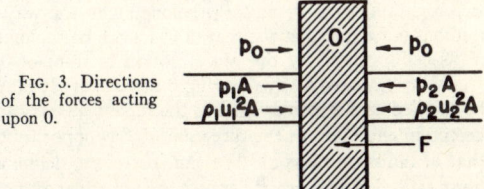

FIG. 3. Directions of the forces acting upon 0.

side the beam give no contribution, as the fluid there is assumed to be at rest.

It may be noted that ρu^2 is always a positive quantity, independent of the sign of u. Owing to the equivalence between force and rate of change of momentum, the

---

† The *total* change of a quantity q is $Dq/Dt = \partial q/\partial t + (\mathbf{u}\cdot\nabla)q$; the first term is called the local change, the latter term the convectional change of q.

[5] A. Barone and M. Nuovo, Ricerca sci. **21**, 516 (1951).

[6] See, for example, L. M. Milne-Thomson, *Theoretical Hydrodynamics* (The Macmillan Company, New York, 1950), p. 72, or *Handbuch der Physik* (Julius Springer, Berlin, 1925), Vol. VII, p. 22.

terms $(\rho_1 u_1^2)$ and $-(\rho_2 u_2^2)$, representing the flux of momentum entering and leaving S per unit time, may be treated as if they represented forces acting upon S and directed always in the sense of the *inner* normal of S (Fig. 2), whatever the direction of motion of the particles crossing S may be. Nothing, however, indicates that such forces *really* act on the front and rear surfaces of S; only their *sum* has a physical meaning, as it integrates the convectional changes of momentum per unit time of all mass elements within S.

Summing up all the forces acting upon 0 (Fig. 3) and assuming that they are balanced by the external force F necessary to keep the center of gravity of 0 in equilibrium,‡ we arrive at the equation

$$\oint_S p_0 \mathbf{da} + (p_1 + \rho_1 u_1^2 - p_2 - \rho_2 u_2^2) A - F - (d/dt)$$

$$\times \int_V \rho u dV = 0, \quad (1)$$

where the last integral is taken over the volume bounded by S.

Since we are interested only in the time average of the radiation pressure, we average Eq. (1) over a full period. Assuming a periodic character of the wave motion and periodicity in time of $\int_V \rho u dV$, the time average of $(d/dt) \int_V \rho u dV$ vanishes; also $\oint_S p_0 \mathbf{da} = 0$, disregarding variation of p_0 with depth. From Eq. (1) we obtain, after replacing F by $P_t A$,

$$\bar{P}_t = \bar{p}_1 + \langle \rho_1 u_1^2 \rangle - \bar{p}_2 - \langle \rho_2 u_2^2 \rangle, \quad (2)$$

where the angular parenthesis denotes the average value.

This is the general expression for the *total* mean radiation pressure $\bar{P}_t$ exerted upon 0 in the present case.

Equation (2) suggests regarding $\bar{P}_t$ as consisting of two separate parts $\bar{P}_1$ and $\bar{P}_2$, belonging to the wave motions in front and at the rear of 0. $\bar{P}_1$ can be thought of as being caused by the wave motion in front of 0, $\bar{P}_2$ by an "emitted" wave leaving 0 at its rear. It may be recalled that F was defined as the force necessary to keep 0 in equilibrium; the direction of F is opposite to that of radiation pressure. We can, therefore, define a mean radiation pressure $\bar{P}$ per unit area of a plane compressional wave by

$$\bar{P} = \bar{p} + \langle \rho u^2 \rangle. \quad (3)$$

As so defined, $\bar{P}$ is always directed in the sense of the *inner* normal of the surface interacting with the acoustic wave, whether the surface is thought of as receiving or as emitting radiation.

As a generalization of the expression (3), the mean radiation pressure upon an obstacle of arbitrary shape

‡ Viscous forces at the interface between 0 and the fluid are excluded. Any viscous forces inside 0 cancel out.

in an acoustic field can obviously be written

$$\langle \mathbf{P} \rangle = \bar{p} \mathbf{n} + \langle \rho \mathbf{u} (\mathbf{u} \cdot \mathbf{n}) \rangle, \quad (4)$$

where $\mathbf{u}$ now stands for the vector of the velocity $\mathbf{u}(x, y, z)$, and $\mathbf{n}$, as before, is the vector of the inner normal on the surface element da.[7] In the general case, the acoustic radiation may also exert shearing forces upon the obstacle, resulting in a torque upon 0.

The procedure of attributing to each surface element of S a radiation force of amount $\bar{P}$, as defined by Eqs. (3) or (4), has its analogy in electrodynamics, where Poynting's radiation vector $\mathbf{S} = \mathbf{E} \times \mathbf{H}$ is assigned to each surface element of an irradiated surface. What is, in fact, derived in both cases is the total force or the total electromagnetic radiation belonging to a closed surface; from the surface integral, however, no rigorous conclusion can be drawn with regard to the distribution over the elements of the surface of derived quantities such as $\langle \mathbf{P} \rangle$ or $\mathbf{S}$. Still the concept of a radiation pressure $\bar{P}$ as defined in Eqs. (3) and (4) is very useful and leads, if properly applied, to correct results.

Incidentally, an expression analogous to Eq. (4) is well known in hydrodynamics in connection with what has been called "Euler's momentum theorem."§

4. BRILLOUIN'S STRESS TENSOR IN FLUIDS

From the foregoing considerations, Brillouin's concept of a tensor describing the dynamical stresses acting upon a fluid under acoustic motion can easily be obtained. The general result of Eq. (4) may equally well be applied to a volume V belonging to the *fluid*. We consider a unit volume in the fluid and describe the force exerted upon it in the x direction by the component T_{xx} of a general stress tensor; the tensor components T_{xx}, T_{yy}, T_{zz} are here positive in the sense of the inner normal, that is, in the sense of the positive pressure. Equation (4), then, gives in the one-dimensional case

$$\oint_S \bar{P} da = (\bar{T}_{xx})_1 - (\bar{T}_{xx})_2 = -\int_1^2 (\partial \bar{T}_{xx}/\partial x) dx$$

$$= \oint_S (\langle p + \rho u^2 \rangle) da,$$

whence by use of a well-known integral transformation

$$-\int_1^2 \frac{\partial \bar{T}_{xx}}{\partial x} dx = -\int_V \frac{\partial}{\partial x} (\langle p + \rho u^2 \rangle) dx dy dz$$

$$= -\int_1^2 \frac{\partial}{\partial x} (\langle p + \rho u^2 \rangle) dx,$$

since $(\langle p + \rho u^2 \rangle)$ is a function of x only and $dydz = 1$. The fluid, therefore, can be thought of as subjected in the

[7] See P. J. Westervelt, J. Acoust. Soc. Am. **23**, 312 (1951), Eq. (8).
§ See L. M. Milne-Thomson, reference 6, p. 74.

x direction to a dynamical stress component

$$\bar{T}_{xx}=p_0+(\langle p+\rho u^2\rangle)=\bar{p}_d+\langle\rho u^2\rangle. \quad (5)$$

Since the particle velocities in the y and z directions are zero, the pressure $\bar{p}_d=p_0+\bar{p}$ is the only force acting in those directions. Thus,

$$\bar{T}_{yy}=\bar{T}_{zz}=\bar{p}_d. \quad (6)$$

These forces acting in the fluid can be described by use of the dynamical stress tensor introduced by Brillouin:[8]

$$\begin{vmatrix} \bar{p}_d+\langle\rho u^2\rangle & 0 & 0 \\ 0 & \bar{p}_d & 0 \\ 0 & 0 & \bar{p}_d \end{vmatrix}. \quad (7)$$

In the absence of acoustic wave motion $\langle\rho u^2\rangle=0$ and $\bar{p}_d=p_0$; the liquid is under the hydrostatic pressure p_0 only, which is the same in every direction. If the acoustic wave is present, the fluid undergoes a non-isotropic state of stress: the stress component $\bar{T}_{xx}$ differs from $\bar{T}_{yy}$ and $\bar{T}_{zz}$ by $\langle\rho u^2\rangle$, that is by the mean flux of momentum density through a stationary area normal to the x direction. Moreover, as is shown later, the pressure p_0 is changed in time average to the mean dynamic pressure $\bar{p}_d=p_0+\bar{p}$, where $\bar{p}$ is found to be proportional to the energy density of the wave motion.

If the fluid is bounded by a plane material surface normal to x, the mean normal stress $\bar{T}_{xx}=\bar{p}_d+\langle\rho u^2\rangle$ is transferred to unit area of this surface; thus, we are led back to the expression for $\bar{P}$, as defined in Eq. (3). At oblique incidence the radiation tensor transforms like the tensor in Eq. (7). In order to attain the final value of $\bar{P}$, the expressions for the tensor components must be completed; since $\bar{p}_d$ will turn out to be different from the pressure p_0 in the undisturbed medium, the interaction between the two regions inside and outside the beam must be considered.

5. A GENERAL RELATION FOR THE MEAN EXCESS PRESSURE, AND FOR THE "RAYLEIGH-PRESSURE"

The mean excess pressure $\bar{p}$ depends upon the properties of the fluid under consideration, that is, upon the relation $p(\rho)$ between pressure and density. $\bar{p}$ finally follows from the solution for the wave motion. Since $\bar{p}$ is related to an area fixed in space, the solution of the so-called *Eulerian* equation of motion is indicated. In this equation the quantities p, ρ, and $\mathbf{u}$ are regarded as functions of x, y, z, and t, the coordinates belonging to a system at rest. A velocity $\mathbf{u}(x, y, z, t)$, for example, means the velocity that would be observed at the point (x, y, z) at the time t; since the fluid is in motion, different particles of the fluid are found at the same point (x, y, z) at a different time t'. In other words,

[8] L. Brillouin, reference 2. Also, "*Les Tenseurs en Mécanique et en Élasticité*" (Dover Publications, New York, 1946), pp. 290, 302. It is proved by Brillouin that the forces can actually be represented by a tensor, that is, that they transform like a tensor.

p, ρ, and $\mathbf{u}$ are *not* associated with *particles*, but with a fixed point in space.

Often, however, a solution for the wave motion can more easily be established by using the *Lagrangian* equation of motion. Here the quantities p, ρ, and $\mathbf{u}$ are related to special particles and the solution describes the change in time of p, ρ, and $\mathbf{u}$, which would be noted by an observer at the instantaneous positions of the particles. In order to mark the difference between Eulerian and Lagrangian quantities, we shall denote the latter by p^*, ρ^*, and $\mathbf{u}^*$. The particles themselves are characterized by their original positions at rest, the coordinates of which may be a, b, c; their instantaneous positions are denoted by the displacements ξ, η, ζ (which are functions of a, b, c, t) from their original positions. The *actual* position of a particle with respect to a fixed system is therefore given by $x=a+\xi$; $y=b+\eta$, $z=c+\zeta$. For example, a solution for $\mathbf{u}(a, b, c, t)$ in Lagrangian coordinates indicates the velocity found at the actual position $a+\xi$, $b+\eta$, $c+\zeta$ of the particle, the rest position of which was a, b, c.

The relation between a Lagrangian quantity q^* and the corresponding Eulerian quantity q is expressed by

$$q^*(a, b, c, t) = q\{a+\xi(t), b+\eta(t), c+\zeta(t), t\}$$
$$= q\{x(t), y(t), z(t), t\} \quad (8)$$

since the positions and velocities of particles are given by

$$\begin{array}{ll} x(t)=a+\xi(a, b, c, t); & dx/dt=u_x=\partial\xi/\partial t \\ y(t)=b+\eta(a, b, c, t); & dy/dt=u_y=\partial\eta/\partial t \\ z(t)=c+\zeta(a, b, c, t); & dz/dt=u_z=\partial\zeta/\partial t. \end{array} \quad (9)$$

If ξ, η, ζ are known functions of (a, b, c, t), the inversion of the system (9) allows one to express a, b, c as functions of (x, y, z, t). By insertion of a, b, c, so found, into q^* of Eq. (8), one obtains the corresponding quantity q in Eulerian coordinates. The inversion of the system (9) cannot ordinarily be accomplished by functions of closed form; one has rather to resort to power developments.∥

Our further considerations will be limited to *small amplitudes* of the acoustic wave motion; that is, terms of third and higher order of amplitudes will be neglected. In compressible liquids, the amplitudes that are so far experimentally obtainable in plane waves are always small; in gases, owing to the nonlinearity of the wave equation, the mathematical difficulties in dealing with finite amplitudes are beyond the scope of the present paper.

Returning to the one-dimensional case under consideration, we will now establish a relation between p and p^*, that is, between the mean pressure p at a fixed position, as needed in Eq. (3) for P, and the mean pressure p^* related to an oscillating particle. Such a relation is very useful, because in acoustics solutions

∥ See F. E. Borgnis, Technical Report No. 1A, March 10, 1953, under U. S. Office of Naval Research Contract Nonr-220(02).

for p^* are often easier to find than solutions for p. According to Eq. (8), $p\{a+\xi(a,t)\}=p^*(a,t)$: by substituting a for $a+\xi$, we have $p(a,t)=p^*\{a-\xi(a-\xi,t)\}$. Regarding ξ, p, and p^* as small first-order quantities and neglecting terms of order higher than second, we may write

$$p(a) = p^*(a-\xi) = p^*(a) - \xi(\partial p^*/\partial a). \quad (10)$$

The Lagrangian equation of motion in one dimension is known to be

$$\rho_0 \frac{\partial^2 \xi(a,t)}{\partial t^2} = -\frac{\partial p^*(a,t)}{\partial a}, \quad (11)$$

where ρ_0 is the *undisturbed* density. This equation is rigorous, with nothing neglected.[9]

From Eqs. (10) and (11), we find, using customary abbreviations denoting partial derivatives,

$$\bar{p} = \bar{p}^* - \langle\langle \xi p_a^* \rangle\rangle = \bar{p}^* + \rho_0 \langle\langle \xi \xi_{tt} \rangle\rangle.$$

Now we can write $\xi \xi_{tt} \equiv (\xi \xi_t)_t - (\xi_t)^2$; taking the time average and assuming ξ and its derivative as periodic in time, we find $\langle(\xi \xi_{tt})\rangle = -\langle(\xi_t)^2\rangle = -\langle u^{*2}\rangle$, since $u^* = \xi_t$. Thus, we obtain

$$\bar{p}(a) = \bar{p}^*(a) - \rho_0 \langle u^{*2}\rangle(a). \quad (12)$$

Denoting first- and second-order terms by the subscripts 1 and 2, we have, excluding a continuous particle velocity u_0,

$$\begin{aligned} u &= u_1 + u_2, & p &= p_1 + p_2, & \rho &= \rho_0 + \rho_1 + \rho_2 \\ u^* &= u_1^* + u_2^*, & p^* &= p_1^* + p_2^*, & \rho^* &= \rho_0 + \rho_1^* + \rho_2^*. \end{aligned} \quad (13)$$

To the second order twice the kinetic energy is $2E_{\text{kin}} = \rho u^2 = \rho_0 u_1^2 = E_{\text{kin}}^* = \rho^* u^{*2} = \rho_0 u^{*2}$ as seen from Eq. (13), because u^2 differs from u^{*2} in terms higher than the second. Also, in this approximation $\bar{E}_{\text{kin}}^* = \bar{E}_{\text{pot}}^*$, and the total energy density $\bar{E}^* = \bar{E}_{\text{kin}}^* + \bar{E}_{\text{pot}}^* = 2\bar{E}_{\text{kin}}^* = \langle \rho u^{*2}\rangle = \langle \rho_0 u^{*2}\rangle$. Consequently, we have at small amplitudes

$$\bar{E} = \bar{E}^* = \langle \rho u^2 \rangle = \langle \rho u^{*2} \rangle = \rho_0 \langle u^2 \rangle = \rho_0 \langle u^{*2} \rangle, \quad (14)$$

and hence, from Eq. (12)

$$\bar{p}(a) = \bar{p}^*(a) - \bar{E}^*(a) = \bar{p}^*(a) - \bar{E}(a). \quad (15)$$

This general relation between the mean pressure $\bar{p}$ at a fixed coordinate a, and the mean pressure $\bar{p}^*$ which would be observed by moving with the particle around a, is independent of the special function $p(\rho)$.[10] Equation (15) enables us to find the mean Eulerian excess pressure $\bar{p}$, when p^* and u^* in Lagrangian coordinates are known.

---

[9] See, for example, H. Lamb, *Theory of Sound* (Edwin Arnold, London, 1910), p. 176; also *Hydrodynamics* (University Press, Cambridge, 1936), p. 479.

[10] See G. Hertz and H. Mende, Z. Physik 114, 354 (1939); also F. Borgnis, Z. Physik 134, 363 (1953), where it is shown that Eq. (15) holds also for finite amplitudes. In the present paper, it is sufficient to know its validity at small amplitudes.

Inserting Eq. (15) in Eq. (3), with consideration of Eq. (14), we obtain

$$\bar{P} = \bar{p} + \langle \rho u^2 \rangle = \bar{p}^*. \quad (16)$$

Thus, the radiation pressure can equally well be expressed by the mean pressure averaged in time over a unit area *moving with the particles* at the interface between fluid and obstacle.

Some authors dealing with the present subject identify radiation pressure with the mean Eulerian excess pressure $\bar{p}$ only. Most problems in acoustics are usually treated only to the first order of approximation; within this approximation the second-order term $\langle \rho u^2 \rangle$ is neglected and $\bar{p}$ becomes identical with $\langle p^* \rangle$. Radiation pressure, however, is a second-order quantity. If Eulerian quantities are used, the term $\langle \rho u^2 \rangle$ (as well as at least second-order terms in $\bar{p}$) are essential; only in the case of a perfectly rigid reflector ($u=0$) does the identification of the radiation pressure with $\bar{p}$ lead to accurate results.[11] If Lagrangian quantities are used, at least second-order terms in $\langle p^* \rangle$ have to be taken into account.

From the physical point of view, it appears natural to reason that the radiation force exerted upon a material surface which (unless it belongs to a perfectly stiff reflector) is in periodic motion, should be obtained by averaging in time the excess pressure on the moving surface, as indicated by Eq. (16). Indeed, as Eq. (16) shows, this is perfectly correct in dealing with a beam of infinite width, or of finite width but not in communication with undisturbed regions of static pressure p_0; in these cases both the Lagrangian quantity $\bar{p}^*$ and the Eulerian quantity $\langle(p+\rho u^2)\rangle$ constitute a correct expression for the radiation force. Nevertheless, the case of infinite width is, of course, purely theoretical; the case of finite width could be conceived as represented by a beam filling completely a closed cylindrical tube with perfectly rigid walls. Experimentally, however, it seems hardly feasible to measure radiation pressure by such a device.

G. Hertz and H. Mende[10] introduced the notation "*Rayleigh pressure*" for the Lagrangian quantity $\bar{p}^*$, that is, the excess pressure averaged in time over a moving surface. Therefore, according to the statements above, the Rayleigh pressure can be identified with the radiation pressure in an *infinitely extended* beam, not in communication with undisturbed regions.

Practically, however, the acoustic beam is of finite width and interacts under almost all conditions with undisturbed regions. This interaction plays a fundamental part in producing the actual radiation forces, because it changes the dynamic pressure in the beam in a way which will be treated below. It is this change in pressure that leads to the actual expression for the radiation pressure. The expressions for $\bar{P}$ in Eq. (16),

---

[11] F. Bopp, Ann. Physik 38, 495 (1940).

therefore, have to be modified, because they fail to take into account this phenomenon of interaction.

6. MEAN EXCESS PRESSURE AND ENERGY DENSITY IN LIQUIDS OF CONSTANT COMPRESSIBILITY

The case of liquids offers the least complicated analytical access to the quantities involved and, therefore, affords the most perspicuous physical insight into the problem. This advantage is due to the fact that in liquids we can assume *constant compressibility*. This concept introduces a simple analytical relation between the hydrodynamic pressure p_d and the relative change in volume $\Delta V/V$ of a volume element having the original volume V, on which a pressure Δp_d is exerted. The compressibility is by definition

$$\beta = -\frac{\Delta V}{V}\frac{1}{\Delta p_d}. \quad (17)$$

For acoustic waves one uses normally the adiabatic value of β, though the processes are certainly not strictly adiabatic. Still, the exact value of β will not be very much affected, even if the process is not strictly adiabatic; this conclusion follows from the fact that the difference between the adiabatic and isothermal compressibility amounts to only a few percent for liquids forming drops and under normal conditions of temperature and pressure. Over *large* ranges of pressure the compressibility is not constant. Still, within the range of excess pressures encountered in acoustic waves, which in plane waves rarely exceed about 30 atmos, the compressibility for both positive and negative pressures can be regarded as practically constant.[12]

In the one-dimensional case, the relative change of a unit volume element is $\Delta V/V = \partial \xi/\partial a$, where ξ denotes the particle displacement in the Lagrangian sense. Hence, we have from Eq. (17) with $\Delta p_d = p^*$,

$$p^* = -(1/\beta) \cdot (\partial \xi/\partial a). \quad (18)$$

In a plane wave, the volume element is stretched and compressed only in the x direction; this causes, however, a dynamic pressure which is a scalar and, therefore, the same in every direction. The relation (18) is exact by definition for media with constant compressibility; no higher terms in ξ are involved.

The exact one-dimensional Lagrangian equation (11) gives, with Eq. (18),

$$\rho_0 \frac{\partial^2 \xi(a,t)}{\partial t^2} = \frac{1}{\beta}\frac{\partial^2 \xi(a,t)}{\partial a^2}. \quad (19)$$

The well-known solution for plane acoustic waves generated by a simple harmonic piston motion of angular frequency ω and located, for example, at any $a = \pm 2\pi n/k$, is

$$\xi(a,t) = \xi_0\{\sin(\omega t - ka) + \gamma \sin(\omega t + ka + \theta)\}, \quad (20)$$

where $\omega/k = c = (1/\beta \rho_0)^{\frac{1}{2}}$ (c = phase velocity). This solution is rigorous, including finite amplitudes, subject only to the limitation that the solution (20) remains unique. This limitation excludes amplitudes so large that one particle can overtake the one in front of it. The amplitude ξ_0 is thus limited to values smaller than $\lambda/4\pi$.

From Eq. (20) we obtain immediately the velocity $u^*(a,t) = \partial \xi/\partial t$ as

$$u^*(a,t) = \omega \xi_0\{\cos(\omega t - ka) + \gamma \cos(\omega t + ka + \theta)\}. \quad (21)$$

The excess pressure, according to Eq. (18), is

$$p^*(a,t) = (k\xi_0/\beta)\{\cos(\omega t - ka) - \gamma \cos(\omega t + ka + \theta)\}. \quad (22)$$

The constants γ and θ are the amplitude reflection coefficient and the phase angle of the reflected wave with respect to the incident wave.

The mean energy density follows from Eqs. (14) and (21):

$$\bar{E}^*(a) = (\rho_0/\tau)\int_0^\tau u^{*2}(a,t)dt, \quad (\tau = 2\pi/\omega),$$

whence

$$\bar{E}^*(a) = \bar{E}_i\{1 + 2\gamma \cos(2ka+\theta) + \gamma^2\}, \quad (23)$$

where the mean energy density of the *incident wave* is denoted by

$$\bar{E}_i = \tfrac{1}{2}\rho_0 \omega^2 \xi_0^2. \quad (24)$$

Since $\bar{p}^*(a) = 0$ according to Eq. (22) in liquids of constant compressibility, Eq. (15) becomes, by use of Eq. (23),

$$\bar{p}(x) = -\bar{E}_i[1 + 2\gamma \cos(2kx+\theta) + \gamma^2], \quad (25)$$

where x is now written for the Eulerian coordinate; it is a matter of notation only whether we call this variable x or a.

With the same change in notation we have, according to Eq. (23), $\langle \rho u^2\rangle = \bar{E} = \bar{E}_i\{1 + 2\gamma \cos(2kx+\theta) + \gamma^2\}$ and with Eq. (25) we obtain $\bar{P} = \bar{p} + \langle \rho u^2\rangle = 0$. This relation, which holds for an infinitely extended beam or a beam not communicating with undisturbed regions, shows that the mean decrease of pressure due to the acoustic field is just compensated by the flux of momentum density at any x; the tensor component $\bar{T}_{xx}$ in Eq. (7) becomes p_0 in this case. Thus, no radiation pressure would be found in liquids of constant compressibility, when the beam does not interact with outside regions. This is obviously in contradiction to experimental results, because, as already pointed out, the radiation pressure actually encountered is essentially due to the interaction of the beam with the surrounding medium; this effect will be treated in the following section.

---

[12] See, for example, the tables in N. E. Dorsey, *Properties of Ordinary Water-Substance* (Reinhold Publishing Corporation, New York, 1940).

7. INTERACTION BETWEEN ACOUSTIC BEAM AND SURROUNDING UNDISTURBED MEDIUM, AND THE RESULTANT RADIATION PRESSURE

Equation (25) indicates that the mean dynamic excess pressure produced by the periodic wave motion varies periodically along x, except when the incident wave falls upon a perfect absorber ($\gamma=0$). Along its circumference the beam is bordered by the parts of the fluid unaffected by the wave motion. The boundary conditions require continuity of stress, that is of $\bar{T}_{yy}$ and $\bar{T}_{zz}$; both amount to $\bar{p}_d = p_0 + \bar{p}$ as seen from the stress tensor (7). The instantaneous stresses T_{yy} and T_{zz} have been replaced by their time averages $\bar{T}_{yy}$ and $\bar{T}_{zz}$, since it is sufficient to postulate that the balance of stresses has to be maintained in time average. Since the pressure outside is assumed to be p_0, the continuity of stress requires that $p_0 = \bar{T}_{yy} = \bar{T}_{zz} = \bar{p}_d = p_0 + \bar{p}$, or: $\bar{p}=0$.

Thus, if no reflected wave is present, $\bar{p}(x) = -\bar{E}_i$ according to Eq. (25), the total dynamic pressure being $\bar{p}_d = p_0 - \bar{E}_i$ inside the beam and p_0 outside. In order to establish the same amount of pressure at the boundary of the beam, the beam will be compressed and its mean density raised by the outside pressure p_0 until the dynamic pressure inside the beam equals p_0.

If a reflected wave is present, the case is more complicated, owing to the second-order periodic change of $\bar{p}$ along x, as expressed by Eq. (25). It is not possible to fulfill exactly the boundary conditions at the interface between beam and undisturbed medium, since the pressure is independent of x in the medium, while periodic in x within the beam; this dilemma results from our idealized assumption of a sharp boundary between the two regions. Actually, the periodic variation of p inside the beam will change gradually into the constant pressure p_0 in the undisturbed medium. However, it is reasonable to assume that this change is essentially performed within a region of transition which is small in comparison with the width of the beam, if the latter is large compared with the acoustic wavelength. Theoretically, this region of transition might extend to infinity. Within this "edge region" of the beam a more complicated (vortical) motion of particles will occur; a closer theoretical investigation of this effect is beyond our scope.

A reasonable way to satisfy the boundary condition at the edge of the beam is the assumption that by reaction of the surrounding medium *the average value in space* of the dynamic pressure $\bar{p}_d = p_0 + \bar{p}$ is brought to p_0; inside the beam $\bar{p}_d$ then varies periodically along the x axis around the value p_0. ¶

If the average value of p in *time and space* is denoted by $\langle\langle p \rangle\rangle$, the boundary condition at the edge of the beam thus leads us to the condition $\langle\langle p \rangle\rangle = 0$ inside the beam. However, if there is no interaction with the surrounding undisturbed medium, it is found from Eq. (25) that $\langle\langle p \rangle\rangle = 1/\lambda \int_0^\lambda \bar{p}(x)dx = -\bar{E}_i(1+\gamma^2)$, where λ denotes the acoustic wavelength. The beam, therefore, will undergo a mean compression, which raises the pressure by the opposite amount $\langle\langle p_0' \rangle\rangle = +\bar{E}_i(1+\gamma^2)$ in order to bring the total *space and time average* of p in the beam to p_0. The mean "effective dynamic pressure" in the beam can then be expressed as

$$(\bar{p}_d)_{\text{eff}} = p_0 + \bar{p} + p_0' = p_0 - 2\gamma \bar{E}_i \cos(2kx + \theta), \quad (26)$$

from which expression we indeed obtain $\langle\langle p_d \rangle\rangle_{\text{eff}} = p_0$. The resultant mean stress tensor in the fluid according to Eq. (7), upon introducing $\langle p_d \rangle_{\text{eff}}$ and $\langle \rho u^2 \rangle = \bar{E}(x)$ from Eq. (23), becomes

$$\begin{vmatrix} p_0 + \bar{E}_i(1+\gamma^2) & 0 & 0 \\ 0 & p_0 - 2\gamma \bar{E}_i \cos(2kx+\theta) & 0 \\ 0 & 0 & p_0 - 2\gamma \bar{E}_i \cos(2kx+\theta) \end{vmatrix}. \quad (27)$$

The stress tensor averaged in time *and space* becomes

$$\begin{vmatrix} p_0 + \bar{E}_i(1+\gamma^2) & 0 & 0 \\ 0 & p_0 & 0 \\ 0 & 0 & p_0 \end{vmatrix}, \quad (27a)$$

thus satisfying the boundary conditions for $\langle\langle T_{yy} \rangle\rangle$ and $\langle\langle T_{zz} \rangle\rangle$, both of which now equal p_0.

Finally, we calculate the value of the radiation pressure, taking into account the additional pressure $\bar{p}_0' = \bar{E}_i(1+\gamma^2)$ resulting from the interaction between beam and undisturbed medium.

From Eqs. (3) and (23) we have

$$\bar{P} = \bar{p} + \langle \rho u^2 \rangle = \bar{p} + \bar{E}_i \{1 + 2\gamma \cos(2kx+\theta) + \gamma^2\}. \quad (28)$$

Replacing $\bar{p}$ by the "effective excess pressure,"

$$\bar{p}_{\text{eff}} = \bar{p} + \bar{p}_0' = \bar{p} + \bar{E}_i(1+\gamma^2) = -2\gamma \bar{E}_i \cos(2kx+\theta), \quad (29)$$

we find for the radiation pressure from Eq. (28):

$$\bar{P} = \bar{E}_i(1+\gamma^2). \quad (30)$$

This result agrees, as it should, with the tensor component $\bar{T}_{xx}$ in Eq. (27).

For a perfect absorber ($\gamma=0$), $\bar{P} = \bar{E}_i$; for a perfect reflector ($\gamma^2=1$), $\bar{P} = 2\bar{E}_i$. The radiation pressure is independent of the phase angle between incident and reflected wave.

One might construe $\bar{P}$ in Eq. (30) as consisting of two parts, $\bar{P}_i = \bar{E}_i$ and $\bar{P}_r = \gamma^2 \bar{E}_i$, $\bar{P}_i$ caused by the incident wave only and $\bar{P}_r$ by the reflected wave, whose mean energy density is $\gamma^2 \bar{E}_i$. In adopting this view,[8] one might say that the incident wave of energy density

¶ F. Bopp (reference 11) uses the same assumption, basing it on the premise that a nonvanishing average value in space of $\bar{p}$ would be neutralized by a lateral flow of fluid into or out from the beam.

$\bar{E}_i$ is completely absorbed by the obstacle, leading to a radiation pressure $\bar{P}_i = \bar{E}_i$, while at the same time, the obstacle re-emits a reflected wave of energy density $\gamma^2 \bar{E}_i$, the obstacle undergoing a *reactional* radiation pressure $\bar{P}_r = \gamma^2 \bar{E}_i$. This concept yields the right numerical value for the radiation pressure; it does not, however, afford an insight into the physical background of the forces really acting at the surface of the obstacle. That mechanism will become apparent from the following section.

8. JOINT ACTION OF THE DYNAMIC PRESSURE AND THE FLUX OF MOMENTUM IN PRODUCING RADIATION PRESSURE IN LIQUIDS

Considering the interface between liquid and obstacle at $x=0$, we find the following Eulerian components acting at the interface by applying Eqs. (23) and (25) to the plane x (or a) $=0$:

$$\bar{p} = -\bar{E}_i(1+\gamma^2+2\gamma\cos\theta), \quad (31)$$

$$\bar{p}_0' = \bar{E}_i(1+\gamma^2), \quad (32)$$

$$\langle \rho u^2 \rangle = \bar{E}_i(1+\gamma^2+2\gamma\cos\theta), \quad (33)$$

the sum of which amounts to the radiation pressure $\bar{P} = \bar{E}_i(1+\gamma^2)$ actually measured. The joint *modus operandi* of these forces may be demonstrated by a discussion of certain examples.

(a) Perfect Absorber ($\gamma = 0$)

An incident wave is absorbed completely by an obstacle, the surface particles of which follow exactly the movement of the fluid particles in the pure progressive wave at the interface. No reflected wave is set up in this case and $\gamma = 0$.**

The mean excess pressure $\bar{p}(x)$ caused by the periodic particle movement is $-\bar{E}_i$ and is constant throughout the beam [Eq. (25)]; by interaction with the surrounding medium $\bar{p}$ is exactly compensated throughout the beam by $\bar{p}_0' = \bar{E}_i$ [Eq. (32)], so that the total mean pressure equals p_0 in the beam. The radiation pressure $\bar{P}$ is solely a consequence of the flux of momentum ρu^2 and therefore equals $\bar{P} = \langle \rho u^2 \rangle = \bar{E}_i$.

(b) Perfect Reflector ($\gamma = 1$)

If the obstacle does not absorb any energy, the entire energy of the incident wave is returned as a reflected wave of the same amplitude. By interference, the two waves cause a standing wave (in a nonviscous liquid),

---

** One practicable approach to a perfect absorber is the acoustic "hohlraum," that is, a cavity with acoustically insulating walls, filled with an absorbing medium, and provided with a small window through which the acoustic beam is admitted. Such a device has been applied for measuring acoustic intensities in water in the form of a cylindrical tube; for frequencies in the megacycle range, the absorption of energy is practically complete in a tube that is not excessively long. The plane of the window, therefore, serves as a totally absorbing surface. Another practicable solution is the use of a 90° wedge (F. Borgnis, J. Acoust. Soc. Am. **24**, 468 (1952).

with periodic variation in excess pressure $\bar{p}(x)$, as well as in $\langle \rho u^2 \rangle(x)$ along the axis of the beam.

Perfect reflection can be achieved by both a perfectly *stiff* and a perfectly *soft* reflector. The particles at the surface of the perfectly stiff reflector are considered as absolutely immovable; the boundary condition at the interface between fluid and reflector is, therefore, expressed as $u(0,t) = u^*(0,t) = 0$ at any t, u being the velocity component normal to the boundary. From Eq. (21) we find that the boundary condition $u^*(0,t) = 0$ requires $\gamma = 1$ and $\theta = \pi$ in the one-dimensional case. On the other hand, a perfectly soft reflecting surface is characterized by $p^*(0,t) = 0$ at any t; a reflection of this kind occurs at the plane free surface of a liquid, where the condition must be satisfied that the pressure shall be continuous as we pass from liquid to air. Since the pressure in air can be assumed to be constant and equal to p_0 (disregarding the negligible wave motion transmitted into the air), the Lagrangian excess pressure p^* must vanish at the free surface of the liquid. The boundary condition $p^*(0,t) = 0$ is satisfied by $\gamma = 1$ and $\theta = 0$, as seen from Eq. (22).

First we consider the *perfectly stiff* reflector: The mean flux of momentum $\langle \rho u^2 \rangle$ vanishes at the interface because here $u = 0$; the radiation pressure is now due entirely to the dynamic pressure in the liquid. According to Eq. (31), $\bar{p} = 0$, since $\gamma = 1$ and $\theta = \pi$. However, by interaction between beam and surrounding medium, all values of $\bar{p}$ distributed along the x axis are raised by an amount $\bar{p}_0' = 2\bar{E}_i$ [Eq. (32)], which leads to the actual radiation pressure $\bar{P} = 2\bar{E}_i$.

At the *perfectly soft* reflector we find from Eq. (31), with $\gamma = 1$ and $\theta = 0$, that $\bar{p} = -4\bar{E}_i$; from Eq. (32), $\bar{p}_0' = 2\bar{E}_i$; and from Eq. (33), $\langle \rho u^2 \rangle = 4\bar{E}_i$. Three effects act jointly in this case: First, the mean pressure $\bar{p}$ amounts to $-4\bar{E}_i$ per unit area of the free surface; second, by interaction between beam and surrounding medium $\bar{p}$ is raised by an amount $\bar{p}_0' = 2\bar{E}_i$, that is, to $-2\bar{E}_i$; third, the mean flux of momentum equals $4\bar{E}_i$, because, owing to the reflection, the resultant velocity amplitude $u(0,t)$ is twice as large as that belonging to the incident wave alone [Eq. (21)]. Thus, in the sum the radiation pressure amounts to $\bar{P} = 4\bar{E}_i - 2\bar{E}_i = 2\bar{E}_i$, just as was found for the perfectly stiff reflector.

Figure 4 illustrates the distribution of pressure and flux of momentum for the two cases of perfect reflection.

(c) General Case ($0 \leq \gamma \leq 1$; $-\pi \leq \theta \leq \pi$)

In general, where the incident energy is partially absorbed and partially reflected, the force upon the obstacle is due to both effects: the effective excess pressure $\bar{p}_{\text{eff}} = \bar{p} + \bar{p}_0'$, and the quantity $\langle \rho u^2 \rangle$. Each of the quantities $\bar{p}$ and $\langle \rho u^2 \rangle$ depends on γ and θ, as indicated by Eqs. (31) and (33), but their *sum* does not; it cancels out in liquids of constant compressibility,

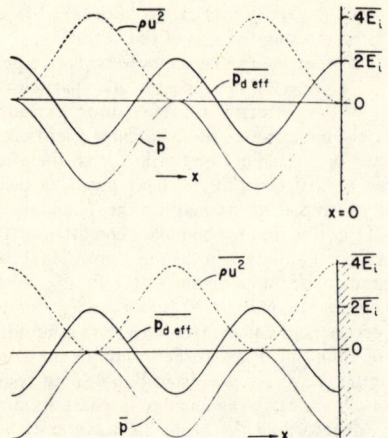

FIG. 4. Distribution of the mean excess pressure $\bar{p}$ resulting from the acoustic wave motion alone; the mean effective dynamic pressure $(\bar{p}_d)_{\text{eff}} = \bar{p} + \bar{p}_0'$ resulting from both the wave motion and the interaction of the beam with the surrounding medium; and the mean flux of momentum $\langle \rho u^2 \rangle$, in the neighborhood of the boundary of a perfectly stiff reflector (above), and a perfectly soft reflector (below), in a liquid of constant compressibility.

reducing the Rayleigh pressure to zero. The radiation pressure encountered in a beam of finite width is a mere consequence of the mean compression of the beam by the outside medium, that is, of the term $\bar{p}_0' = \bar{E}_i(1+\gamma^2)$.

The Rayleigh pressure $\bar{p} + \langle \rho u^2 \rangle$ does not vanish, however, if the compressibility is not constant, as in gases. For example, in a purely progressive wave in a gas under adiabatic conditions and in the neighborhood of the source $\bar{p}^* = \bar{p} + \langle \rho u^2 \rangle = (1+\gamma_e)\bar{E}_i/4$ at small amplitudes (γ_e = ratio of the specific heats).

As an example of a general case, we treat in the following section the radiation pressure produced at the interface between two nonmiscible liquids.

9. RADIATION PRESSURE UPON A PLANE INTERFACE BETWEEN TWO NONMISCIBLE LIQUIDS

Let an acoustic beam of plane waves fall normally upon the plane interface between two nonmiscible liquids 1 and 2. The static pressure p_0 outside the beam may be regarded as constant and the same in both liquids. The "obstacle" is reduced in this case to the interface between the two liquids. In liquid 1 we assume a progressive wave of energy density $\bar{E}_{i1}$; this incident wave causes, in general, a reflected wave in liquid 1 and a transmitted wave in liquid 2. The waves in 1 produce a radiation pressure $\bar{P}_1 = \bar{E}_{i1}(1+\gamma^2)$, while the transmitted wave causes a reactional radiation pressure $\bar{P}_2 = -\bar{E}_{t2}$, where $\bar{E}_{t2}$ is the mean energy density of the progressive wave transmitted into liquid 2. The total radiation pressure, according to the result of Eq. (2), is then given by

$$\bar{P} = \bar{P}_1 + \bar{P}_2 = \bar{E}_{i1}(1+\gamma^2) - \bar{E}_{t2}. \qquad (34)$$

Since the interface does not absorb energy, the balance of the power transmitted requires $c_2 \bar{E}_{t2} = c_1(1-\gamma^2)\bar{E}_{i1}$. Hence, we obtain from Eq. (34):

$$\bar{P} = \bar{E}_{i1}\{1 - c_1/c_2 + \gamma^2(1 + c_1/c_2)\}. \qquad (35)$$

At normal incidence the coefficient of reflection is known to be $\gamma = (1-m)/(1+m)$, where $m = \rho_2 c_2/\rho_1 c_1$, ρ and c being the undisturbed densities and velocities of sound in the respective media.[13]

Equation (35) shows that for special values of c_2/c_1 and ρ_2/ρ_1, the pressure $\bar{P}$ at the interface becomes zero. Inserting the above value of γ in Eq. (35), we obtain the following condition for vanishing radiation pressure:

$$c_2/c_1 = (\rho_1/\rho_2)\{2(\rho_2/\rho_1) - 1\}^{\frac{1}{2}}. \qquad (36)$$

According to this, $\bar{P}$ becomes zero only if the two liquids are arranged so that $\rho_1 < 2\rho_2$; moreover, c_2/c_1 has to obey Eq. (36). If c_2/c_1 is *smaller* than indicated by Eq. (36), $\bar{P}$ is found from Eq. (35) to become *negative*. In this case, the direction of $\bar{P}$ is *opposite* to the direction of propagation of the incident wave. This effect is caused by the fact that c_2/c_1 is now small enough to make the energy density $\bar{E}_{t2}$ exceed $\bar{E}_{i1}(1+\gamma^2)$ in Eq. (35).[14]

These two quantities represent the additional pressures $\bar{p}_0'$ on the two sides of the interface because of the compression of the beam by the surrounding medium. The mean Lagrangian pressure $\bar{p}^*$ on both sides of the interface is zero when the liquids have constant compressibility, and therefore does not contribute to the radiation pressure. The actual physical forces to which the interface is subjected result from the difference in compression of the acoustic beam on both sides of the partition.

10. GENERAL REMARKS ON THE RADIATION PRESSURE IN GASES

The relation between pressure and density in gases leads to a nonlinear Lagrangian wave equation, the rigorous solution of which can be obtained only by series development. As is well known, the wave form in gases becomes distorted in the course of the propagation. This fact is expressed by a variation in space of the amplitudes of higher terms in the series developments; more and more wave energy is transferred from the fundamental mode, which holds in the immediate neighborhood of the source, to higher harmonics in the course of the wave propagation.[15]

It is difficult to obtain a strict solution of the wave

---

[13] H. Lamb, *Theory of Sound* (Edwin Arnold, London, 1910), p. 169.
[14] G. Hertz and H. Mende (reference 10) have demonstrated this effect. If $\rho_1 \geqslant 2\rho_2$, $\bar{P}$ in Eq. (35) is always positive, that is, in the direction of the incident wave.
[15] Reference 13, p. 174.

equation even for small amplitudes, if a reflected wave is present. A rigorous treatment would require consideration of absorption, because the amplitude of each higher harmonic is determined by its particular rate of absorption.[16]

A procedure for computing the radiation pressure for the general case of reflection, such as was applied in the previous sections for compressible liquids, is hardly feasible for gases, owing to the mathematical difficulties involved. Still, the insight into the physical processes that has been gained from the treatment of liquids can be used to establish a very general expression for the radiation pressure in all fluids. It will be shown that the formula $\bar{P} = \bar{E}_i(1+\gamma^2)$ holds in any fluid, that is, also in gases, at small amplitudes, at least under the idealized assumptions introduced in the present treatment.

11. A GENERAL EXPRESSION FOR THE RADIATION PRESSURE IN FLUIDS

By multiplying the one-dimensional Eulerian equation of continuity $\rho_t + (\rho u)_x = 0$ by u and adding the equation so obtained to the Eulerian equation of motion $\rho(u_t + uu_x) + p_x = 0$, a well-known form of the equation of motion in one dimension is obtained:

$$(\rho u)_t + (\rho u^2)_x + p_x = 0. \quad (37)$$

In Eqs. (37) to (39), p may be regarded as representing either the excess pressure or the total dynamic pressure p_d. Considering a purely harmonic motion of the acoustic source and assuming that Eq. (37) has solutions periodic in time, we find by averaging Eq. (37) in time and integrating with respect to x,

$$\langle \rho u^2 \rangle + \bar{p} = C, \quad (38)$$

where C is a constant independent of x and t, but not, in general, of the wave amplitude. Next, averaging Eq. (38) also in *space*, we obtain

$$\langle\langle \rho u^2 \rangle\rangle + \langle\langle p \rangle\rangle = C. \quad (39)$$

Regarding for the moment p as the *total* dynamic pressure, we apply the same conclusions concerning p that were used in Sec. 7, namely, that owing to the interaction between beam and surrounding medium, the total mean Eulerian pressure averaged in *time and space* along the beam may reasonably be assumed equal to the undisturbed outside pressure p_0, or in other words, that $\langle\langle p \rangle\rangle = p_0$. Inserting this condition in Eq. (39) and substituting C so obtained from Eq. (39) in Eq. (38), we find

$$\langle \rho u^2 \rangle + (\bar{p} - p_0) = \langle\langle \rho u^2 \rangle\rangle. \quad (40)$$

Now $(\bar{p} - p_0)$ is what we previously called the mean excess pressure $\bar{p}$, and $\langle\langle \rho u^2 \rangle\rangle = 2\langle\langle E \rangle\rangle_{\text{kin}} = \langle\langle E \rangle\rangle_{\text{total}}$

$+ (\langle\langle E \rangle\rangle_{\text{kin}} - \langle\langle E \rangle\rangle_{\text{pot}})$. Therefore, from Eq. (40) and according to the definition of radiation pressure in Eq. (3), we have[††]

$$\bar{P} = 2\langle\langle E \rangle\rangle_{\text{kin}} = \langle\langle E \rangle\rangle_{\text{total}} + (\langle\langle E \rangle\rangle_{\text{kin}} - \langle\langle E \rangle\rangle_{\text{pot}}). \quad (41)$$

Equation (41), which includes the interaction between beam and surrounding medium, gives the radiation pressure for a beam of finite width in any fluid and on any plane reflecting surface. With the assumption $\langle\langle p \rangle\rangle = p_0$, which led to Eq. (40), Eq. (41) is valid for finite amplitudes, since no restriction was introduced in this respect in the derivation. Whether this assumption is valid or not at *finite* amplitudes is an open question.

At small amplitudes, where we limit ourselves to terms up to the second order, it is sufficient to know the *first*-order solution in u, as already mentioned in Sec. 5. This is correct, at least, within a distance not too far from the origin of the wave motion; or, if absorption is assumed to be exactly zero, within a not too large time interval after the wave motion started. Owing to the transfer of wave energy from the fundamental to higher harmonics, as mentioned in Sec. 10, the amplitudes of these harmonics increase with distance from the origin (Earnshaw's solution), and also with time in absence of absorption. At larger distances, therefore, the terms of higher order may no longer be negligible. On the other hand, absorption is always present, limiting the amplitudes of the harmonics. It is only in a medium with constant compressibility that the fundamental wave is propagated without producing higher harmonics, preserving its original shape everywhere.

To the first order the solution for u of the Lagrangian wave equation is given in any fluid by Eq. (21). Moreover, $\langle\langle E \rangle\rangle_{\text{pot}} = \langle\langle E \rangle\rangle_{\text{kin}}$ at small amplitudes, and therefore $2\langle\langle E \rangle\rangle_{\text{kin}} = \langle\langle E \rangle\rangle_{\text{total}} = \bar{E}_i(1+\gamma^2)$, as seen from Eq. (23). Consequently, we find from Eq. (41) that the expression

$$\bar{P} = 2\bar{E}_{\text{kin}} = \bar{E}_i(1+\gamma^2) = \tfrac{1}{2}\rho_0\omega^2\xi_0^2(1+\gamma^2) \quad (42)$$

is valid both in liquids and gases, when terms of third and higher order in $k\xi_0 = 2\pi\xi_0/\lambda$ are excluded. Eq. (42) agrees with Eq. (30), which was found to hold for liquids of constant compressibility.

Since $\bar{E}_i(1+\gamma^2)$ is the mean total energy density encountered at the surface of the plane reflector, Eq. (42) states that *at small amplitudes the radiation pressure of a finite beam of plane compressional waves equals the mean total energy density at the reflecting surface*. This result is independent of the special law connecting pressure and density in the fluid under consideration. At larger amplitudes, according to Eq. (41), the total energy density must be replaced by $2\langle\langle E \rangle\rangle_{\text{kin}}$, that is twice the average in time and space of the kinetic energy density.

---

[16] See P. J. Westervelt, J. Acoust. Soc. Am. **22**, 319 (1950), Sec. VI, and F. E. Borgnis, Technical Report No. 1A, March 10, 1953, under U. S. Office of Naval Research Contract Nonr-220(02).

[††] An analogous derivation was applied to the special case of a perfectly stiff reflector by F. Bopp (reference 11).

It may be recalled that the expressions for the radiation pressure given in Eqs. (41) and (42) represent only the force that may be attributed to one side of the reflector. A force of the same kind, equal to the energy density behind the reflector, but opposite in sign, has to be attributed to the opposite side.‡‡ Only the *sum* of these two forces has a physical significance; it represents both the change in momentum per unit time to which the wave motion is subjected in passing through a partition, and the effects of interaction between beam and surrounding medium on both sides of the partition.

The *total* radiation pressure $\bar{P}_t$ exerted upon a partition by a finite beam is, therefore, under the assumption leading to Eq. (41), given by

$$\bar{P}_t = (\langle\langle\rho u^2\rangle\rangle)_1 - (\langle\langle\rho u^2\rangle\rangle)_2$$
$$= 2(\langle\langle E\rangle\rangle_{\text{kin}})_1 - 2(\langle\langle E\rangle\rangle_{\text{kin}})_2, \quad (43)$$

where, as before, the indices 1 and 2 denote the two sides of the partition (Fig. 2). At small amplitudes, $2\langle\langle E\rangle\rangle_{\text{kin}}$ can be replaced on both sides by the total energy density $\langle\langle E\rangle\rangle$; the total radiation pressure then equals the difference in energy densities on both sides of the partition. A similar result, namely, that the radiation pressure is equal to the difference between two energy densities, is also well known in electrodynamics.

BIBLIOGRAPHY

A good survey of the literature on the subject, covering the period before 1927, is found in *Handbuch der Physik*, Vol. VIII (Berlin, 1927), pp. 148, 149, 579, and 580. Below is given a selected list of papers of special interest in relation to the present treatment; the list also includes some papers quoted in the text.

---

‡‡ In the special cases of a perfect absorber or of a perfect reflector this force at the rear is zero.

R. T. Beyer, "Radiation pressure in a sound wave," Am. J. Phys. 18, 25 (1950).

P. Biquard, "Ultrasonic waves," Rev. acoust. 1, 93 and 315 (1932).

F. E. Borgnis, (a) "Theory of Acoustic Radiation Pressure," Technical Report No. 1A (March 10, 1953), under U. S. Office of Naval Research Contract Nonr-220(02); (b) "Acoustic radiation pressure of plane compressional waves at oblique incidence," J. Acoust. Soc. Am. 24, 468 (1952); (c) "On the physics of acoustic radiation pressure," Z. Physik 134, 363 (1953).

F. Bopp, "Energetic considerations on acoustic radiation pressure," Ann. Physik 38, 495 (1940).

L. Brillouin, (a) "The tensions of radiation and their interpretation in terms of classical mechanics and relativity," J. phys. et radium 6, 337 (1925); (b) "On the tensions of radiation," Ann. phys. 4, 528 (1925); (c) "The tensions of radiation and their interpretation," Physica 5, 396 (1925); (d) "Radiation pressure and 'tension,'" Rev. acoust. 5, 99 (1936).

G. Hertz, "Acoustic radiation pressure in liquids and gases in connection with the equation of state," Physik. Z. 41, 546 (1940).

G. Hertz and H. Mende, "Acoustic radiation pressure in liquids," Z. Physik 114, 354 (1939).

J. S. Mendousse, "On the theory of acoustic radiation pressure," Proc. Am. Acad. Arts Sci. 78, 148 (1950).

J. H. Poynting, "Radiation pressure," Phil. Mag. 9, 393 (1905).

Lord Rayleigh, (a) "On the pressure of vibrations," Phil. Mag. 3, 338 (1902) or "Collected Papers," Vol. 5, 41; (b) "On the momentum and pressure of gaseous vibrations and on the connection with the virial theorem," Phil. Mag. 10, 364 (1905) or "Collected Papers," Vol. 5, 262.

G. Richter, "On the problem of acoustic radiation pressure," Z. Physik 115, 97 (1940).

C. Schaefer, (a) "On the theory of acoustic radiation pressure," Ann. Physik 35, 473 (1939); (b) "On the discussion concerning acoustic radiation pressure," Z. Physik 115, 109 (1940).

A. Schoch, "On the question of the momentum of a sound wave," Z. Naturforsch. 7a, 273 (1952).

W. Weaver, "The pressure of sound," Phys. Rev. 15, 399 (1920).

P. J. Westervelt, "The theory of steady forces caused by sound waves," J. Acoust. Soc. Am. 23, 312 (1951).

Editor's Comments on Papers 14 and 15

14 Westervelt: *Scattering of Sound by Sound*

15 Bellin and Beyer: *Experimental Investigation of an End-Fire Array*

The nonlinear character of the equations governing the propagation of sound of finite amplitude (macrosonics) leads to many significant physical consequences, as the papers included in this part of the volume amply demonstrate. One of these consequences is the rather curious phenomenon of the scattering of one beam of sound by another. This has been theoretically investigated by Westervelt in a series of papers and some of the results are included in the paper presented here. This has led to further work in the field, including the experimental verification of Westervelt's theory. It has also foreshadowed certain practical consequences of considerable importance in the "end-fire" array for securing sharper directivity of sound beams than can be achieved with conventional piston radiators. Westervelt, (1919–) is an American physicist located at Brown University.

The paper by Bellin and Beyer presented here reports an experimental test of some of Westervelt's more recent theoretical predictions concerning the scattering of sound by sound. The experimental results are in agreement with Westervelt's theory. Comment on the practical significance of the results in connection with acoustical radiation is made.

Beyer (1920–) is an American physicist, currently chairman of the Department of Physics at Brown University. Bellin was one of his graduate students.

Scattering of Sound by Sound*

Peter J. Westervelt
Department of Physics, Brown University, Providence 12, Rhode Island
(Received April 22, 1957)

Earlier studies [J. Acoust. Soc. Am. **29**, 199 (1957)] of the mutual nonlinear interaction of two plane waves of sound with each other are extended to encompass arbitrary directions of travel of one wave with respect to the other; an exact solution to the first-order scattering process is obtained.

1. INTRODUCTION

THE governing wave equation[1] for this study is

$$\Box^2 \rho_s = c_0^{-2}\{\Box^2 E_{12} - \nabla^2 [2T_{12} + \Lambda V_{12}]\}. \quad (1)$$

The symbols have the following meaning: $\Box^2 =$ d'Alembertian operator, $\rho_s =$ excess density of scattered wave, $c_0 =$ ambient velocity of sound, $\nabla^2 =$ Laplacian operator, $(T_{12}, V_{12}, E_{12}) =$ interaction kinetic, potential and total energy densities of primary waves, $\Lambda = \rho_0 c_0^{-2}(d^2p/d\rho_2)_{\rho=\rho_0}$, $\rho_0 =$ equilibrium density, and $p =$ hydrostatic pressure.

Equation (1) was derived by the author[1] from one of Lighthill's equations of motion. Equation (1) has also been derived by Eckart[2] in 1948, so that his work could just as well have formed our starting point. Equation (1) is a direct consequence of the fact that the first-order fields, represented by their excess densities ρ_1 and ρ_2, or their particle velocities $\mathbf{u}_1$ and $\mathbf{u}_2$, satisfy the first-order equation of motion.

Another relation, vital to this study, is

$$\nabla^2 V_{12} = \tfrac{1}{2}\omega_1^{-1}\omega_2^{-1}(\omega_1^2 + \omega_2^2)\Box^2 W_{12} + c_0^{-2}(\partial^2 T_{12}/\partial t^2), \quad (2)$$

in which the quantity W_{12} has the dimensions of energy density and is defined by†

$$W_{12} = \rho_0^{-1} c_0^2 \omega_1^{-1}\omega_2^{-1} \dot{\rho}_1 \dot{\rho}_2 = \tfrac{1}{2}\omega_1^{-1}\omega_2^{-1}$$
$$\times [(\partial^2 V_{12}/\partial t^2) + (\omega_1^2 + \omega_2^2)V_{12}], \quad (3)$$

and ω_1 and ω_2 are the circular frequencies of the two primary waves and the dot stands for time differentiation. A somewhat more convenient form for Eq. (2) will now be derived. If the two relations

$$\nabla^2 V_{12} = 2\rho_0 c_0^{-2} \dot{\mathbf{u}}_1 \cdot \dot{\mathbf{u}}_2 - c_0^{-2}(\omega_1^2 + \omega_2^2) V_{12}$$

and

$$\partial^2 T_{12}/\partial t^2 = 2\rho_0 \dot{\mathbf{u}}_1 \cdot \dot{\mathbf{u}}_2 - (\omega_1^2 + \omega_2^2)T_{12},$$

(which are easily found by straightforward differentiation and use of the first-order equations of motion) are introduced into Eq. (2), there results

$$\tfrac{1}{2}\omega_1^{-1}\omega_2^{-1}c_0^2 \Box^2 W_{12} = T_{12} - V_{12} = L_{12}. \quad (4)$$

* This research was supported by the Office of Naval Research.
[1] P. J. Westervelt, J. Acoust. Soc. Am. **29**, 199 (1957).
[2] Carl Eckart, Phys. Rev. **73**, 68 (1948). See Eq. (16) of this reference.
† This is Eq. (16) of reference 1 which, owing to a typographical error, appears there with the bracket raised to the one half power.

Equation (4) states that the interaction Lagrangian density, L_{12}, is derivable from a d'Alembertian.

2. TWO INTERSECTING PLANE WAVES

The interaction kinetic and potential energy density for two plane waves whose directions of travel form the angle θ with one another, satisfy this relation:

$$T_{12} = V_{12} \cos\theta. \quad (5)$$

Operating on Eqs. (4) and (5) with $c_0^{-2}\nabla^2$ yields the following two equations:

$$\tfrac{1}{2}\omega_1^{-1}\omega_2^{-1}\Box^2[\nabla^2 W_{12}] = c_0^{-2}\nabla^2 T_{12} - c_0^{-2}\nabla^2 V_{12} \quad (6)$$

and

$$c_0^{-2}\nabla^2 T_{12} = c_0^{-2}\cos\theta \nabla^2 V_{12}. \quad (7)$$

The above equations may be solved for the Laplacian's in terms of the d'Alembertian as follows:

$$c_0^{-2}\nabla^2 V_{12} = \tfrac{1}{2}\omega_1^{-1}\omega_2^{-1}(\cos\theta)^{-1}\Box^2[\nabla^2 W_{12}] \quad (8)$$

$$c_0^{-2}\nabla^2 T_{12} = \tfrac{1}{2}\omega_1^{-1}\omega_2^{-1}\cos\theta(\cos\theta - 1)^{-1}\Box^2[\nabla^2 W_{12}]. \quad (9)$$

The foregoing solutions may now be entered into the basic wave equation, Eq. (1):

$$\Box^2 \rho_s = \Box^2 \{c_0^{-2} E_{12} - \tfrac{1}{2}\omega_1^{-1}\omega_2^{-1}(\cos\theta - 1)^{-1}$$
$$\times (2\cos\theta + \Lambda)\nabla^2 W_{12}\}. \quad (10)$$

The solution to Eq. (10) is obtained by operating on both sides with $\Box^{-2}$, which has the effect of canceling the $\Box^2$ operators, thus the solution may be written down by inspection.

Except when $\Lambda = -2$, which represents a physically unrealizable state of matter, the source term of Eq. (10) has a singularity at $\theta = 0$ which corresponds to having the two waves traveling in the same direction. Obviously, Eq. (10) loses significance in this instance; however, this case is well known and the solution to this special problem has been obtained theoretically,[3,4] and experiments[4] agree quite well with theory.

3. SCATTERING FROM TWO COLLIMATED SOUND BEAMS

A perfectly collimated sound beam has a discontinuity in its field quantities, pressure, and velocity at its edge. These discontinuities signify that the first-order

[3] H. Lamb, *Dynamical Theory of Sound* (Arnold, London, 1931).
[4] Thuras, Jenkins, and O'Neill, J. Acoust. Soc. Am. **6**, 173 (1935).

field does not satisfy the homogeneous wave equation at the edge. The edges contain surface source distributions in the form of a force normal to the edges of the wave and having proper phase and magnitude to "confine" the wave. Thus it is not surprising to find the scattered waves also contain discontinuities, which arise from the artificiality of the concept of perfect collimation here employed.

From Eq. (10) one can conclude that no combination waves will be scattered outside of a region traversed by two perfectly collimated plane waves forming a non-zero angle between them. These conclusions extend the earlier findings[1] and contradict none of them. Here, only the lowest order scattering process has been investigated. It is apparent that, as θ becomes quite small (but unequal to zero), some rather large stresses will occur which will bring into effect higher order terms that have here been neglected. The less intense the primary waves are, the smaller θ can be before this order of the perturbation breaks down and ceases to correspond to the physical facts (whatever they may be!).

It was previously stated[1] that the waves scattered away from the region of interaction would vanish when $\theta = -\cos^{-1}(\frac{1}{2}\Lambda)$; this is still the case and the only thing that appears to differentiate this angle from others is the simplicity of the solution for the scattered wave, i.e., $\rho_s = c_0^{-2} E_{12}$ at this particular angle.

4. SOME REMAINING DIFFICULTIES

In addition to the interfering effects of pseudo-sound, mentioned in an earlier paper,[1] other difficulties lie in the path of a clean-cut experiment designed to check the theory presented here. Any solid body in the region of interaction will give rise to scattered waves having the same combination frequencies. This results, in part‡ from the action of time dependent radiation pressure forces which the primary waves exert on the surface of the object. These surface forces, reacting back on the medium will radiate sound. If two loudspeakers are used to obtain the primary waves, one of the speakers might well act as the solid object in question, unless the speakers are well separated.

On the other hand, if the speakers are well separated, care must be taken to account for the fact that the interaction zone may be in the Fraunhofer zone of one or both of the speakers. In this case the waves are only locally plane, and the intersection angle θ must then be thought of as varying with position within the interaction region; this region, itself, thus becomes hazily defined. When θ varies with position, there is every reason to believe that interaction waves will be scattered outside the region common to the two primary waves, unless the opposite can be proved to be the case.

A detailed theoretical analysis of the nonlinear interaction of two waves, when the interaction region lies within the Fresnel zone of both speakers, seems at the moment to be beyond attainment, inasmuch as the first-order fields would need to be known and this latter aspect of the problem is not in a satisfactory state, as anyone who is familiar with diffraction theory knows.

We conclude with some questions. Are some applications of this study to other physical problems worth considering? The problem discussed here is the classical analog of the quantum theoretical phonon-phonon scattering process. What does the correspondence principle have to offer in this connection? What about interpreting certain irreversible phenomena in terms of the scattering of a coherent wave by the random thermal waves? Are there other possible applications in the study of quantum field theory, or nonlinear meson theory?

Professor F. E. Borgnis has called my attention to the fact that the interaction of two plane waves in classical nonlinear electrodynamics has been treated by Smirnov.[5]

‡ In addition to this, the primary waves will scatter directly from the object and in turn interact nonlinearly with themselves and the incident waves.

[5] A. A. Smirnov, J. Phys. U.S.S.R. 3, 447 (1940).

Experimental Investigation of an End-Fire Array

J. L. S. BELLIN* AND R. T. BEYER†
Department of Physics, Brown University, Providence 12, Rhode Island
(Received April 9, 1962)

Experiments are reported on the measurement of the scattered pressure from two finite-amplitude, collinear sound beams. The measurements were carried out at a carrier frequency of 13.5 Mc in water and of 350 kc in air. The slope of the half-pressure angle vs difference frequency curve agrees well with that predicted by Westervelt, although the radiation pattern measured experimentally was in each case more directive than predicted by the theory.

INTRODUCTION

IN recent years there have been several investigations of the mutual nonlinear interaction of two plane waves of sound. In particular, Westervelt[1,2] has demonstrated theoretically that, for the case of two well-collimated beams intersecting at any nonzero angle, no combination wave (of sum or difference frequency) will be scattered outside of the region of intersection of the beams. This prediction has been supported experimentally by the results of an earlier study[3] by the present authors.

More recently, Westervelt[4] has examined the special case of the nonlinear interaction of two collinear sound waves traveling in the same direction. For this particular situation, a first-order scattering occurs, and an exact solution for the difference frequency has been derived.

The primary purpose of our experiment was to examine the validity of Westervelt's expression for the scattering of two collinear sound beams. The positions of these beams is indicated schematically in Fig. 1. The magnitude of scattered pressure obtained by Westervelt for the far-field approximation is given by

$$P_s = \frac{\omega_s^2 P_0^2 S_0}{8\pi R_0 \rho_0 c_0^4}\left(1+\tfrac{1}{2}\rho_0 c_0^{-2}\frac{d^2P}{(d\rho^2)_{\rho=\rho_0}}\right) \times \frac{1}{[\alpha^2+K_s^2\sin^4(\tfrac{1}{2}\theta)]^{\frac{1}{2}}}, \quad (1)$$

where ω_s is the angular modulation frequency, equal to the difference of the carrier-beam frequencies ω_1, ω_2; P_0 is the peak acoustic pressure of each of the carrier beams. The same amplitude was set for each beam in order to give 100% modulation; S_0 is the cross-sectional area of the carrier beams; ρ_0 and c_0 are, respectively, the ambient density and the sound velocity in the transmission medium; R_0 is the distance between the modulation pickup and the carrier-beam transducer; θ is the angular departure of the observation point from the axis of the beams; α is the primary beam-pressure-absorption coefficient for the case of a wave of infinitesimal amplitude; and K_s is the wave number of the modulation signal $[=(\omega_1-\omega_2)/c_0]$. One notices from (1) that the directivity exhibited by the scattered signal is identical with that of Rutherford scattering, and also with the radiation pattern of an end-fire array antenna (and hence the justification for the title of this paper). Experiments were carried out both in water and in air to test the validity of Eq. (1).

MEASUREMENTS IN WATER

The measurements in water were performed in a 22×35×12 in. tank filled with water to within an inch of the top. The transducer for the carrier beams, a 13.5-Mc 1-in. diameter, circular X-cut quartz crystal, was firmly attached to one side of the tank in a watertight housing. As a modulation signal pickup, we used a miniature cylindrical barium titanate probe ($\tfrac{1}{16}$ in. in outer diameter, $\tfrac{1}{16}$ in. long). It could be rotated freely in the horizontal plane of the tank as well as be moved along its length. Care was taken to shield both these transducers electrically. All internal surfaces of the tank were covered with layers of rubberized horsehair to minimize standing waves.

A pair of 60 w cw transmitters, operating simultaneously, drove the 13.5-Mc quartz transducer, thus generating the two collinear carrier beams. In the course of these experiments the two transmitters were tuned in the following manner: One of them was first tuned to a frequency slightly below 13.5 Mc, and then the other to a frequency an equal amount above 13.5 Mc. Advantage was thereby taken of the approximately symmetrical response of the carrier transducer about its

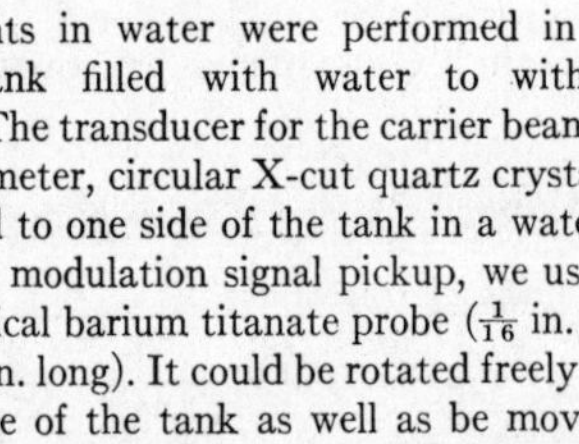

---

* Present address: Minnesota Mining and Manufacturing Company, Electrical Solid State Physics Division, St. Paul 19, Minn.
† Address for the year 1961–1962, I. Physikalisches Institut der Technischen Hochschule, Stuttgart, Germany.
[1] P. J. Westervelt, J. Acoust. Soc. Am. 29, 199 (1957).
[2] P. J. Westervelt, J. Acoust. Soc. Am. 29, 934 (1957).
[3] J. L. S. Bellin and R. T. Beyer, J. Acoust. Soc. Am. 32, 339 (1960).
[4] P. J. Westervelt, J. Acoust. Soc. Am. 32, 934A (1960).

FIG. 1. Orientation of the primary beams and the pickup probe.

resonant frequency to produce sound beams of equal intensity. The barium titanate probe was directly coupled to a sensitive communications receiver (Nc-2-40D), which was tuned to the difference frequency of the carrier signals. An oscilloscope connected to the output of the last i.f. stage in the communications receiver was used to display a sinusoidal signal whose amplitude (since the A.V.C. had been disengaged) was proportional to the modulation signal picked up by the probe.

PROCEDURE AND RESULTS IN WATER

In our first trial the lower-frequency transmitter was tuned to 13.0 and the other to 14.0 Mc. The tuning and coupling controls of both these transmitters had been adjusted to present clean, high intensity rf signals to the carrier transducer. The power of each of these signals, approximately 40 w, produced an acoustic pressure of 3 atm in the water. The pickup probe, tuned to the difference frequency, was then slowly rotated in the horizontal plane to determine the directivity pattern of the scattered signal.

The averaged result of several runs at these frequencies is shown in Fig. 2, together with a theoretical comparison obtained from Eq. (1). The ordinate scale is in arbitrary units and the pressures are set equal to one another at $\theta=0$. The first conclusion from this experiment is that the difference frequency was indeed detectable. However, the experimental directivity pattern is considerably narrower than the theoretical one. In fact, the half-pressure angle of the latter is nearly twice that of the former. The theoretical value of the angle at which the pressure is equal to one half the axial value, $\theta_{\frac{1}{2}}$, can be obtained by Eq. (1), and is given by

$$\theta_{\frac{1}{2}} = 2(3^{\frac{1}{2}})(\alpha/K_s)^{\frac{1}{2}}. \qquad (2)$$

The modulation (i.e., difference) frequency can be decreased or increased by shifting the carrier transmitter frequencies closer to or further from the transducer resonance. By employing this technique, information concerning the directivity of the scattering was obtained for several modulation frequencies within the range 1.00–2.05 Mc. A plot of these data in terms of the half-pressure angles vs modulation frequency is presented in Fig. 3. Here it may be seen that the slope of the experimental plot is extremely close to the value

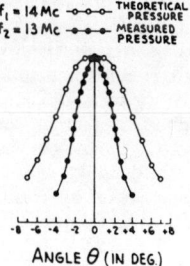

FIG. 2. Experimental and theoretical directivity patterns for end-fire radiation in water.

(−0.5) given by the Westervelt theory. Further discussion of the significance of this fact will be given below. At difference frequencies appreciably lower than 1 Mc, standing-wave interference in the tank, due to the decreased absorption of the modulation signal, made the accurate measurement of directivity impractical. Similarly, for difference frequencies, above 2 Mc, the decreasing radiation efficiency of the 13.5 Mc transducer, operating 1 Mc or more off resonance, was a limiting factor. In addition, the increasing sharpness of the directivity pattern (less than 2°) made accurate measurements increasingly difficult.

An attempt was also made to determine the absolute magnitude of the modulation signal. For this purpose, it was necessary to calibrate the receiving pickup system. A standard substitution radiation-pressure method described in an earlier paper[3] was employed. Thus, for carrier beams of 13.00 and 14.00 Mc with the pickup probe on axis ($\theta=0$) a modulation signal of the order of 20 000 dyn/cm² was observed. The theoretical value determined by inserting the proper parameters

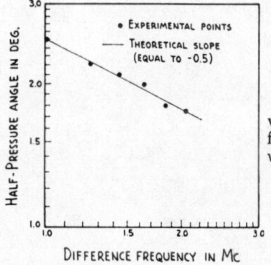

FIG. 3. Half-pressure angle versus modulation frequency for the case of transmission in water.

into Eq. (1) was closer to 35 000 dyn/cm². However, in terms of the accuracy of the calibration procedure and of the parameters in Eq. (1), this rather rough agreement is actually quite satisfactory.

In any finite amplitude investigation it is quite important to ensure that experimental data are not the result of undesired nonlinear interactions. In the above work, in which the pickup probe was subjected to two acoustic beams of fairly large amplitudes, there is the possibility of pseudo-sound,[5] due to the interaction radiation pressure of these waves on the surface of the pickup probe. Making use of the treatment given in reference 1, we have that this pressure is equal to the negative interaction Lagrangian $L_{12}=T_{12}-V_{12}$, where 1, 2 refer to the two carrier waves, and T_{12} and V_{12} are the interaction kinetic and potential energies, respectively. That is, $P_{\text{rad}} = -L_{12} = \rho_0 u_1 \cdot u_2 + \rho_0^{-1} c_0^{-2} P_1 P_2$, where u_i and P_i are the particle velocity and the acoustic pressure due to the carrier beams at the surface of the probe

---

[5] For a discussion of pseudo-sound see D. Blokhintzev, "Acoustics of an Inhomogeneous, Moving Medium" (translated from the Russian by R. T. Beyer and D. Mintzer, Brown University).

($i=1,2$). In this investigation $P_i=10^5$ dyn/cm² and $u_i \cong 1$ cm/sec. It is immediately seen that the contribution of this effect (~ 1 dyn/cm²) can be disregarded in the light of the signal actually observed ($=2\times 10^4$ dyn/cm²).

In addition, the possibility that the pickup signal was actually generated from a nonlinear interaction in the carrier-beam transducer must be considered. If this had been the case, the acoustic pickup signal would have exhibited the radiation properties of a circular piston source. For an ideal plane piston radiator it can be shown[6,7] that the half-pressure angle, for the case of narrow beams (i.e., $\sin\theta_{\frac{1}{2}} \cong \theta_{\frac{1}{2}}$), is inversely proportional to the frequency of radiation. Thus, in a log-log plot, such as Fig. 3, a slope of minus one would be expected. However, the slope of the experimental plot in this figure is, on the contrary, very close to the value of -0.5 predicted by the Westervelt expression [Eq. (2)]. Furthermore, care had been taken with the electronics associated with this investigation to ensure operation within the limits of linear performance. It, therefore,

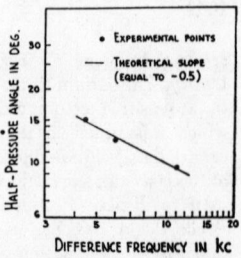

FIG. 5. Half-pressure angle versus modulation frequency for the case of transmission in air.

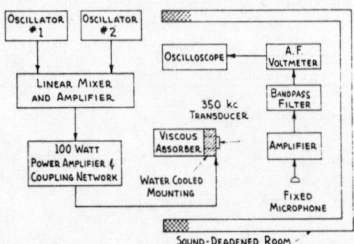

FIG. 4. Block diagram of the experimental arrangement for end-fire radiation in air.

seems clear that the proposition of spurious experimental data due to uncompensated nonlinearities can be dismissed.

The disparity in the widths of the theoretical and experimental values of the directivity pattern is possibly due to the fact that the Westervelt formula holds strictly for the Fraunhofer diffraction region. Because of physical limitations of tank size, it was not possible for us to achieve this region of measurement. Further experimentation in a much larger environment would therefore be useful.

EXPERIMENTAL ARRANGEMENT IN AIR

A second set of measurements was undertaken to test the Westervelt formula for the nonlinear interaction of collinear acoustic beams in air. However, numerous technical difficulties were encountered here, including the general one of producing high-amplitude

[6] T. Hueter and R. Bolt, *Sonics* (John Wiley & Sons, Inc., New York, 1955), Chap. 3.
[7] A. O. Williams, Jr., and L. W. Labaw, J. Acoust. Soc. Am. **16**, 231 (1945).

ultrasound in air. The block diagram of an experimental system which yielded some results is shown in Fig. 4. This setup was essentially the same as the one employed in water. The carrier beam transducer, a barium titanate circular disk (diameter, 1 in.) having a fundamental resonance of 350 kc, was driven by a 100-w, class-a power amplifier. Because of the great acoustic impedance mismatch between this radiator and the air, most of the amplifier's output was dissipated within the barium titanate element. Water-cooling was employed to maintain the transducer at a safe operating temperature. Another difficulty was presented by the production of the difference frequency in the liquid medium backing the source. This signal was much more intense than that in the air, because of matching conditions, and tended to fill the whole room with its sound. This unwanted signal could be virtually eliminated by the use of a highly viscous liquid for the backing medium. Karo syrup was found to be quite successful for this purpose. In addition, the viscous fluid backing sufficiently broadened the resonant response of the carrier transducer to permit the use of modulation frequencies in the range above 1 kc. The carrier transducer was mounted on a spectrometer table so that it could be conveniently rotated in directivity measurements. The pickup system consisted of a microphone, voltage amplifiers, a variable bandpass filter, and an oscilloscope. Both the carrier transducer and pickup microphone were placed in a sound-deadened room to minimize standing-wave interference.

PROCEDURE AND RESULTS IN AIR

In making a directivity run the oscillators were tuned to frequencies appropriate to the particular modulation frequency desired, just as in the water case. Then the variable bandpass filter was adjusted to maximize the signal-to-noise ratio. After several hours of warming-up and adjustment of voltage regulation for greatest stability, the measurements could be recorded. By revolving the platform and carrier transducer and noting the pickup signal's strength the directive properties of the interaction were determined. A plot of half-pressure vs modulation frequency thereby obtained is presented in Fig. 5. Once again, the slope of the $\theta_{\frac{1}{2}}$ vs f_s curve was very close to the value predicted by Westervelt. Because the operating conditions for this particular setup were only barely acceptable, no more than three valid experimental points were obtained. As in the case of the measurements in water considerable caution against spurious nonlinear effects had to be exercised during the investigation in air.

As in the water case, the disagreement between experimental and theoretical half-pressure angles may well be due to our inability to work entirely in the Fraunhofer region, the limitation here being the small size of our sound-deadened room.

Some of the practical significance of this effect can be seen by comparing the directivity obtained from the "end-fire array" with that available from conventional circular piston sources. For instance, in the case of air, it is noticed from Fig. 5 that the half-pressure angle for a frequency of 4.5 kc was 15°. The actual diameter of the carrier transducer employed was 1 in. An ordinary plane piston source possessing a comparable directivity would have to be at least 10 in. in diameter. This phenomenon would seem to possess interesting possibilities for both airborne and underwater signaling. Furthermore, the efficiency of the end-fire radiation is proportional to the primary beam power.[4] Thus, although the efficiencies obtained in the present investigation were small (about 1%), it should be possible to achieve substantially higher values through a sufficient increase of the primary beam strength.

CONCLUSION

Westervelt's theoretical analysis of the interaction of two collinear, finite-amplitude sound beams is supported by the results of this study. In particular, the slopes of the half-pressure angle vs difference frequency plots for both the cases of transmission in water and in air concurred well with the figure derived from Westervelt's treatment. Furthermore, the measured pressure of the difference wave was in substantial agreement with the value calculated from theory.

ACKNOWLEDGMENTS

The authors wish to express their gratitude to Professor Westervelt for his many stimulating and useful contributions. They are also indebted to Jan Toots, Jack Butler, and Lawrence Foley for their capable assistance in the laboratory.

This work was supported in part by the U. S. Office of Naval Research.

Editor's Comments on Paper 16

16 Rayleigh: *The Explanation of Certain Acoustical Phenomena*

   The study of the direct production of sound by means of heated gases probably goes back to C. Sondhauss (*Ann. Physik*, **79**, 1, 1850) and P. L. Rijke (*Ann. Physik*, **107**, 339, 1859), both of whom observed sounds from heated air under somewhat different circumstances. Lord Rayleigh became interested in these phenomena and discussed them with demonstrations at a lecture at the Royal Institution of Great Britain in 1878 and in a brief paper published in *Nature* the same year (*Nature*, **18**, 319–321, 1878). The full paper is reprinted as Paper 55 in *Scientific Papers of Lord Rayleigh* (Dover Publications, New York, 1964, Vol. 1, pp. 348–354) with the title "The Explanation of Certain Acoustical Phenomena." Although the paper contains other things besides aerothermoacoustics, most of it is devoted to this subject. Rayleigh approached the problem from the intuitive physical point of view.

   In recent times much attention, both experimental and theoretical, has been paid to thermoacoustics. A good modern review will be found in a paper by Osman K. Mawardi in *Reports on Progress in Physics*, **19**, 156–187, 1956. In this connection the earlier work by John Tyndall on singing flames should not be overlooked. (See, for example, Paper 31 in *Acoustics: Historical and Philosophical Development*, R. B. Lindsay, ed., Dowden, Hutchinson & Ross, Stroudsburg, Pa., 1973.)

55.

THE EXPLANATION OF CERTAIN ACOUSTICAL PHENOMENA.

[*Roy. Inst. Proc.* VIII. pp. 536—542, 1878; *Nature*, XVIII. pp. 319—321, 1878.]

MUSICAL sounds have their origin in the vibrations of material systems. In many cases, *e.g.* the pianoforte, the vibrations are free, and are then necessarily of short duration. In other cases, *e.g.* organ pipes and instruments of the violin class, the vibrations are maintained, which can only happen when the vibrating body is in connexion with a source of energy, capable of compensating the loss caused by friction and generation of aerial waves. The theory of free vibrations is tolerably complete, but the explanations hitherto given of maintained vibrations are generally inadequate and in most cases altogether illusory.

In consequence of its connexion with a source of energy, a vibrating body is subject to certain forces, whose nature and effects are to be estimated. These forces are divisible into two groups. The first group operate upon the periodic time of the vibration, *i.e.* upon the pitch of the resulting note, and their effect may be in either direction. The second group of forces do not alter the pitch, but either encourage or discourage the vibration. In the first case only can the vibration be maintained; so that for the explanation of any maintained vibration, it is necessary to examine the character of the second group of forces sufficiently to discover whether their effect is favourable or unfavourable. In illustration of these remarks, the simple case of a common pendulum was considered. The effect of a small periodic horizontal impulse is in general both to alter the periodic time and the amplitude of vibration. If the impulse (supposed to be always in the same direction) acts when the pendulum passes through its lowest position, the force belongs to the second group. It leaves the periodic time unaltered, and encourages or discourages the vibration according as the direction of the pendulum's motion is the same or the opposite of that of the impulse. If, on the other hand, the impulse acts when the pendulum is at one or other of the limits of its swing, the effect is solely on the periodic time, and the vibration is neither encouraged nor discouraged. In order to encourage, *i.e.* practically in order to maintain, a vibration, it is necessary that the forces should not depend solely upon the position of the vibrating body. Thus, in the case of the pendulum, if a small impulse in a given direction acts upon it every time that it passes through its lowest position, the vibration is not maintained, the advantage gained as the pendulum makes a passage in the same direction as that in which the impulse acts being exactly neutralized on the return passage when the motion is in the opposite direction.

As an example of the application of these principles the maintenance of an electric tuning-fork was discussed. If the magnetic forces depended only upon the position of the fork, the vibration could not be maintained. It appears therefore that the explanations usually given do not touch the real point at all. The fact that the vibrations are maintained is a proof that the forces do not depend solely upon the position of the fork. The causes of deviation are two—the self-induction of the electric currents, and the adhesion of the mercury to the wire whose motion makes and breaks the contact. On both accounts the magnetic forces are more powerful in the latter than in the earlier part of the contact, although the position of the fork is the same; and it is on this *difference* that the possibility of maintenance depends. Of course the arrangement must be such that the retardation of force *encourages* the vibration, and the arrangement which in fact encourages the vibration would have had the opposite effect, if the nature of electric currents had been such that they were more powerful during the earlier than during the later stages of a contact.

In order to bring the subject within the limits of a lecture, one class of maintained vibrations was selected for discussion, that, namely, of which *heat* is the motive power. The best understood example of this kind of maintenance is that afforded by Trevelyan's bars, or rockers. A heated brass or copper bar, so shaped as to rock readily from one point of support to another, is laid upon a cold block of lead. The communication of heat through the point of support expands the lead lying immediately below in such a manner that the rocker receives a small impulse. During the interruption of the contact the communicated heat has time to disperse itself in some degree into the mass of lead, and it is not difficult to see that the impulse is of a kind to encourage the motion. But the most interesting vibrations of this class are those in which the vibrating body consists of a mass of air more or less completely confined.

If heat be periodically communicated to, and abstracted from, a mass of air vibrating (for example) in a cylinder bounded by a piston, the effect produced will depend upon the phase of the vibration at which the transfer of heat takes place. If heat be given to the air at the moment of greatest condensation, or taken from it at the moment of greatest rarefaction, the vibration is encouraged. On the other hand, if heat be given at the moment of greatest rarefaction, or abstracted at the moment of greatest condensation, the vibration is discouraged. The latter effect takes place of itself, when the rapidity of alternation is neither very great nor very small, in consequence of radiation; for when air is condensed it becomes hotter, and communicates heat to surrounding bodies. The two extreme cases are exceptional, though for different reasons. In the first, which corresponds to the suppositions of Laplace's theory of the propagation of sound, there is not sufficient time for a sensible transfer to be effected. In the second, the temperature remains nearly constant, and the loss of heat occurs during the *process* of condensation, and not when the condensation is effected. This case corresponds to Newton's theory of the velocity of sound. When the transfer of heat takes place at the moments of greatest condensation or of greatest rarefaction, the pitch is not affected.

If the air be at its normal density at the moment when the transfer of heat takes place, the vibration is neither encouraged nor discouraged, but the pitch is altered. Thus the pitch is *raised*, if heat be communicated to the air a quarter period *before* the phase of greatest condensation, and the pitch is *lowered* if the heat be communicated a quarter period *after* the phase of greatest condensation.

In general both kinds of effects are produced by a periodic transfer of heat. The pitch is altered, and the vibrations are either encouraged or discouraged. But there is no effect of the second kind if the air concerned be at a loop, *i.e.* a place where the density does not vary, nor if the communication of heat be the same at any stage of rarefaction as in the corresponding stage of condensation.

The first example of aerial vibrations maintained by heat was found in a phenomenon which has often been observed by glass-blowers, and was made the subject of a systematic investigation by Dr Sondhauss. When a bulb about three-quarters of an inch in diameter is blown at the end of a somewhat narrow tube, 5 or 6 inches in length, a sound is sometimes heard proceeding from the heated glass. It was proved by Sondhauss that a vibration of the glass itself is no essential part of the phenomenon, and the same observer was very successful in discovering the connexion between the *pitch* of the note and the dimensions of the apparatus. But no explanation (worthy of the name) of the production of sound has been given.

For the sake of simplicity, a simple tube, hot at the closed end and getting gradually cooler towards the open end, was first considered. At a quarter of a period *before* the phase of greatest condensation (which occurs almost simultaneously at all parts of the column) the air is moving inwards, *i.e.* towards the closed end, and therefore is passing from colder to hotter parts of the tube; but the heat received at this moment (of normal density) has no effect either in encouraging or discouraging the vibration. The same would be true of the entire operation of the heat, if the adjustment of temperature were instantaneous, so that there was never any sensible difference between the temperatures of the air and of the neighbouring parts of the tube. But in fact the adjustment of temperature takes *time*, and thus the temperature of the air deviates from that of the neighbouring parts of the tube, inclining towards the temperature of that part of the tube *from* which the air has just come. From this it follows that at the phase of greatest condensation heat is received by the air, and at the phase of greatest rarefaction is given up from it, and thus there is a tendency to maintain the vibrations. It must not be forgotten, however, that apart from transfer of heat altogether, the condensed air is hotter than the rarefied air, and that in order that the whole effect of heat may be on the side of encouragement, it is necessary that previous to condensation the air should pass not merely towards a hotter part of the tube, but towards a part of the tube which is hotter than the air will be when it arrives there. On this account a great range of temperature is necessary for the maintenance of vibration, and even with a great range the influence of the transfer of heat is necessarily unfavourable at the closed end where the motion is very small. This is probably the reason of the advantage of a bulb. It is obvious that if the *open* end of the tube were heated, the effect of the transfer of heat would be even more unfavourable than in the case of a temperature uniform throughout.

The sounds emitted by a jet of hydrogen, burning in an open tube, were noticed soon after the discovery of the gas and have been the subject of several elaborate inquiries. The fact that the notes are substantially the same as those which may be elicited from the tube in other ways, *e.g.* by blowing, was announced by Chladni. Faraday proved that other gases were competent to take the place of hydrogen, though not without disadvantage. But it is to Sondhauss that we owe the most detailed examination of the circumstances under which the sound is produced. His experiments prove the importance of the part taken by the column of gas in the tube which supplies the jet. For example, sound cannot be obtained with a supply tube which is plugged with cotton in the neighbourhood of the jet, although no difference can be detected by the eye between the flame thus obtained and others which are competent to excite sound. When the supply tube is unobstructed, the sounds obtainable are limited as to pitch, often dividing themselves into detached groups. In the intervals between the groups no coaxing will induce a maintained sound, and it may be added that, for a part of the interval at any rate, the influence of the flame is inimical, so that a vibration started by a blow is damped more rapidly than if the jet were not ignited.

Partly in consequence of the peculiar behaviour of flames and partly for other reasons, the thorough explanation of these phenomena is a matter of some difficulty; but there can be no doubt that they fall under the head of vibrations maintained by heat, the heat being communicated periodically to the mass of air confined in the sounding tube at a place where, in the course of a vibration, the pressure varies. Although some authors have shown an inclination to lay stress upon the effects of the current of air passing through the tube, the sounds can readily be produced, not only when there is no through draught, but even when the flame is so situated that there is no sensible periodic motion of the air in its neighbourhood. In the course of the lecture a globe intended for burning phosphorus in oxygen gas was used as a resonator, and, when excited by a hydrogen flame well removed from the neck, gave a pure tone of about 95 vibrations per second.

In consequence of the variable pressure within the resonator, the issue of gas, and therefore the development of heat, varies during the vibration. The question is under what circumstances the variation is of the kind necessary for the maintenance of the vibration. If we were to suppose, as we might at first be inclined to do, that the issue of gas is greatest when the pressure in the resonator is least, and that the phase of greatest development of heat coincides with that of the greatest issue of gas, we should have the condition of things the most unfavourable of all to the persistence of the vibration. It is not difficult, however, to see that both suppositions are incorrect. In the supply tube (supposed to be unplugged, and of not too small bore) stationary, or approximately stationary, vibrations are excited, whose phase is either the same or the opposite of that of the vibration in the resonator. If the length of the supply tube from the burner to the open end in the gas-generating flask be less than a quarter of the wave-length in hydrogen of the actual vibration, the greatest issue of gas *precedes* by a quarter period the phase of greatest condensation; so that, if the development of heat is *retarded* somewhat in comparison with the issue of gas, a state of things exists *favourable* to the maintenance of the sound. Some such retardation is inevitable, because a jet of inflammable gas can burn only at the outside; but in many cases a still more potent cause may be found, in the fact that during the retreat of the gas in the supply tube small quantities of air may enter from the interior of the resonator, whose expulsion must be effected before the inflammable gas can again begin to escape.

If the length of the supply tube amounts to exactly one quarter of the wave-length, the stationary vibration within it will be of such a character that a node is formed at the burner, the variable part of the pressure just inside the burner being the same as in the interior of the resonator. Under these circumstances there is nothing to make the flow of gas, or the development of heat, variable, and therefore the vibration cannot be maintained. This particular case is free from some of the difficulties which attach themselves to the general problem, and the conclusion is in accordance with Sondhauss' observations.

When the supply tube is somewhat longer than a quarter of the wave, the motion of the gas is materially different from that first described. Instead of preceding, the greatest outward flow of gas *follows* at a quarter period interval the phase of greatest condensation, and therefore if the development of heat be somewhat retarded, the whole effect is unfavourable. This state of things continues to prevail, as the supply tube is lengthened, until the length of half a wave is reached, after which the motion again changes sign, so as to restore the possibility of maintenance. Although the size of the flame and its position in the tube (or neck of resonator) are not without influence, this sketch of the theory is sufficient to explain the fact, formulated by Dr Sondhauss, that the principal element in the question is the length of the supply tube.

The next example of the production of sound by heat, shown in the lecture, was a very interesting phenomenon discovered by Rijke. When a piece of fine metallic gauze, stretching across the lower part of a tube open at both ends and held vertically, is heated by a gas flame placed under it, a sound of considerable power, and lasting for several seconds, is observed almost immediately *after* the removal of the flame. Differing in this respect from the case of sonorous flames, the generation of sound was found by Rijke to be closely connected with the formation of a through draught, which impinges upon the heated gauze. In this form of the experiment the heat is soon abstracted, and then the sound ceases; but by keeping the gauze hot by the current from a powerful galvanic battery, Rijke was able to obtain the prolongation of the sound for an indefinite period. In any case from the point of view of the lecture the sound is to be regarded as a *maintained* sound.

In accordance with the general views already explained, we have to examine the character of the variable communication of heat from the gauze to the air. So far as the communication is affected directly by variations of pressure or density, the influence is unfavourable, inasmuch as the air will receive less heat from the gauze when its own temperature is raised by condensation. The maintenance depends upon the variable transfer of heat due to the varying *motions* of the air through the gauze, this motion being compounded of a uniform motion upwards with a motion, alternately upwards and downwards, due to the vibration. In the lower half of the tube these motions conspire a quarter period *before* the phase of greatest condensation, and oppose one another a quarter period after that phase. The rate of transfer of heat will depend mainly upon the temperature of the air in contact with the gauze, being greatest when that temperature is lowest. Perhaps the easiest way to trace the mode of action is to begin with the case of a simple vibration without a steady current. Under these circumstances the whole of the air which comes in contact with the metal, in the course of a complete period, becomes heated; and after this state of things is established, there is comparatively little further transfer of heat. The effect of superposing a small steady upwards current is now easily recognized. At the limit of the inwards motion, *i.e.* at the phase of greatest condensation, a small quantity of air comes into contact with the metal, which has not done so before, and is accordingly cool; and the heat communicated to this quantity of air acts in the most favourable manner for the maintenance of the vibration.

A quite different result ensues if the gauze be placed in the *upper* half of the tube. In this case the fresh air will come into the field at the moment of greatest rarefaction, when the communication of heat has an unfavourable instead of a favourable effect. The principal note of the tube therefore cannot be sounded.

A complementary phenomenon discovered by Bosscha and Riess may be explained upon the same principles. If a current of *hot* air impinge upon *cold* gauze, sound is produced; but in order to obtain the principal note of the tube the gauze must be in the upper, and not as before in the lower, half of the tube. An experiment due to Riess was shown in which the sound is maintained indefinitely. The upper part of a brass tube is kept cool by water contained in a tin vessel, through the bottom of which the tube passes. In this way the gauze remains comparatively cool, although exposed to the heat of a gas flame situated an inch or two below it. The experiment sometimes succeeds better when the draught is checked by a plate of wood placed somewhat closely over the top of the tube.

Both in Rijke's and Riess' experiments the variable transfer of heat depends upon the motion of vibration, while the effect of the transfer depends upon the variation of pressure. The gauze must therefore be placed where both effects are sensible, *i.e.* neither near a node nor near a loop. About a quarter of the length of the tube, from the lower or upper end, as the case may be, appears to be the most favourable position.

Ultrasonics. Interaction of Sound with the Medium

III

Editor's Comments on Paper 17

17 Cady: *The Piezo-electric Resonator*

   This paper, some extracts from which are reprinted here, was one of the first to stress the importance of piezoelectric crystal vibrations at high frequencies in electroacoustical applications. Emphasis is laid on the use of these vibrations in establishing standards of frequency and for the frequency stabilization of electrical oscillations. This, and associated work by the author, had considerable influence on the employment of piezoelectrically produced ultrasonic radiation in all kinds of acoustical problems. In this connection, attention is called to the discovery of piezoelectricity by the Curie brothers in France in 1880 as described in their paper, reprinted in English translation, in *Acoustics: Historical and Philosophical Development* (R. B. Lindsay, ed., Dowden, Hutchinson & Ross, Stroudsburg, Pa., 1973, Paper 35).

   We reprint from the Cady paper the Introduction and Sections I, II, III, IV, V (through subsection 11), and VI. The remaining sections deal with applications, knowledge of which is not necessary for an understanding of the author's principal ideas.

   W. G. Cady (1874–   ) is an American physicist who taught for many years at Wesleyan University in Middletown, Connecticut, and became Professor Emeritus there in 1946. Since his retirement, he has remained active in ultrasonic transducer research.

17

THE PIEZO-ELECTRIC RESONATOR*

By
W. G. Cady

(Wesleyan University, Middletown, Connecticut)

In the course of experiments with piezo-electric crystals, extending over a number of years, certain radio frequency phenomena were brought to light, the practical application of which appeared worthy of development. The two applications that seem most promising at present are (1) as a frequency-standard, and (2) as a frequency-stabilizer, or means of generating electric oscillations of very constant frequency. It is with these that this paper is chiefly concerned. The fundamental phenomena will first be described, followed by the mathematical theory, and finally an account of the applications will be given.[1]

I. Fundamental Phenomena

1. A plate or rod suitably prepared from a piezo-electric crystal, and provided with metallic coatings, can be brought into a state of vigorous longitudinal vibration when the coatings are connected to a source of alternating emf. of the right frequency. Under these conditions the plate reacts upon the electric circuit in a remarkable manner. Owing to the piezo-electric polarization produced by the vibrations, and to the absorption of energy in the plate, the apparent electrostatic capacity and resistance of the plate are not constant, but depend upon the frequency somewhat as does the motional impedance of a telephone receiver.[2] Over a certain very narrow range in frequency the capacity becomes negative. An analogy may also be drawn between the vibrating plate and a synchronous motor. The man-

---

*Received by the Editor, October 11, 1921. Presented before The Institute of Radio Engineers, New York, November 2, 1921.

[1] Preliminary reports on this work have appeared in "The Physical Review," 17, page 531, 1921, and 18, page 142, 1921. The writer wishes to acknowledge the aid that he has received thru a grant from the American Association for the Advancement of Science.

[2] For an explanation of the motional impedance of a telephone receiver, see Proceedings of The Institute of Radio Engineers, volume 6, 1918, page 40.—Editor.

ner in which the reactions upon the circuit are utilized will be described below. It is necessary, however, to consider the theory of the phenomenon first.

II. PIEZO-ELECTRIC THEORY

2. Four decades have elapsed since the discovery of piezo-electricity by the Curie brothers, and the prediction of the converse effect by Lippmann, which the Curies promptly verified. During this time much has been accomplished, both theoretically and experimentally, in systematizing and extending our knowledge of the behavior of crystals under static mechanical or electric stress. Only in very recent years, however, has consideration been given to rapidly varying stresses in piezo-electric crystals.

Nicolson[3] has had marked success in the use of suitably treated Rochelle salt crystals at telephonic frequencies, both as transmitters (direct piezo-electric effect) and as receivers (converse effect). The writer has also experimented with crystals at audio frequencies, but has devoted his attention chiefly to radio frequency vibrations in the neighborhood of the natural frequency of the crystal plates or rods.

We now summarize briefly those features of Voigt's theory of which we shall make use hereafter.[4]

When a piezo-electric crystal is mechanically strained, there results a dielectric polarization, the magnitude of which is proportional to the strain, and the direction and magnitude of which depend upon the direction of the strain and upon the class to which the crystal belongs. Except in the case of the class of crystals of lowest symmetry (triclinic), not all of the six components of strain are effective in producing a polarization. The higher the degree of symmetry, the smaller does this number become. Of the 32 classes, ten are devoid of piezo-electric properties.

The only two crystals the piezo-electric applications of which have hitherto been considered important are quartz and Rochelle salt; the latter, because it is far more strongly piezo-electric than any other crystal thus far examined; and quartz, because of its excellent mechanical qualities, which make it for most purposes decidedly preferable to Rochelle salt, in spite of its

[3] Nicolson, "Proceedings of the American Institute of Electrical Engineers," 38, page 1315, 1919; "Electrical World," June 12, page 1358, 1920.
[4] For a more complete statement, see Voigt, "Lehrbuch der Kristallphysik," Leipzig, 1910; Graetz, "Handbuch der Elektricität und des Magnetismus," Leipzig, 1914, volume 1, page 342: or Winkelmann, "Handbuch der Physik," 1905, volume 4, part 1, page 774.

being only moderately piezo-electric. The present paper has to do only with quartz, tho obviously the theory applies to any piezo-electric crystal.

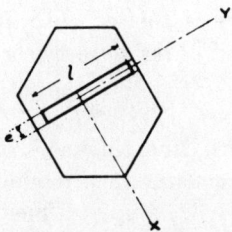

FIGURE 1—Section of a Quartz Crystal perpendicular to the Optical Axis

3. PIEZO-ELECTRIC PROPERTIES OF QUARTZ—Quartz belongs to the trigonal trapezohedral class of crystals. Figure 1 shows a cross-section of a quartz crystal, of which the Z-axis (optical axis) is perpendicular to the paper. The Y-axis is normal to two opposite prismatic faces. Owing to the threefold symmetry of quartz, the Y-axis may be drawn in any one of three directions 120° apart. The three X-axes (electric axes) are perpendicular to the Z- and Y-axes. For piezo-electric experiments, a plate is usually cut from the crystal with its length l, breadth b, and thickness e parallel respectively to the Y-, Z-, and X-axes. The two faces perpendicular to the X-axis are provided with conductive coatings, which may or may not be in actual contact with the quartz.

DIRECT EFFECT—If the plate is compressed in a direction parallel to the X-axis (*longitudinal effect*), the resulting polarization induces equal and opposite charges on the coatings, and the charges change sign with the pressure. Similarly, in the *transverse effect*, an endwise compression of the plate, parallel to the Y-axis, causes the coatings to become charged. A *compresions* of the plate parallel with the X-axis causes a polarization in the same direction as an *extension* parallel with the Y-axis.

CONVERSE EFFECT—In terms of the *converse effect*, if the plate is polarized by an external electric field in the same direction in which it would become polarized by compression along the X-axis, it tends to contract along the X-axis and to expand along the Y-axis.

From what has been said, two important conclusions should

85

be borne in mind: first, that, in quartz, just as the *direct* effect may be produced by compression along either one of two directions (longitudinal and transverse effects), so both of these effects manifest themselves in connection with the *converse* effect; and second, that in both the direct and converse effects, a given strain is always associated with an electric polarization *in the same direction* and *of the same algebraic sign*.

Symbols

$l, b, e,$ length, breadth, and thickness of quartz plate or rod.

$\varepsilon, \delta,$ piezo-electric constant and modulus respectively. From section 11 on, a special meaning is attached to δ.

$M, N, g,$ equivalent mass, resistance, and stiffness of resonator.

$x,$ displacement of end of resonator.

$F,$ equivalent mechanical force on resonator.

$E,$ voltage impressed on circuit.

$V,$ potential difference across resonator.

$D,$ piezo-electric polarization in resonator.

$I, i,$ currents in coil and resonator branches, Figure 3.

$C_1,$ normal capacity of resonator, vibrations damped.

$C_2,$ capacity of tuning condenser.

$C_1', C_1'',$ equivalent series and parallel capacity of resonator.

$R_1', R_1'',$ equivalent series and parallel resistance of resonator.

$C_a,$ "apparent" capacity of resonator.

$C_t, R_t,$ equivalent series capacity and resistance of entire circuit, Figure 3.

$R_{12},$ equivalent series resistance of resonator and C_2, together.

When printed without subscripts, x, F, E, V, D, I, and i denote instantaneous values. x_o and so on, denote maximum values.

$f,$ frequency.

$\omega,$ angular velocity $= 2\pi f$. ω_o and f_o denote resonance values.

4. In the case of quartz, the general polarization-strain equations reduce to the following form:

$$P_1 = \varepsilon_{11} x_x + \varepsilon_{12} y_y + \varepsilon_{14} y_z \tag{1}$$

$$P_2 = \varepsilon_{25} z_x + \varepsilon_{26} x_y. \tag{2}$$

P_1 and P_2 are X and Y components, respectively, of polarization (electric moment per unit volume), and the ε's are the *piezo-electric constants*. x_x and x_y are, in Voigt's notation, the components of extension (elongation or contraction per unit length), and y_z and so on, the components of shearing strain.

If, instead of the components of strain, we have given the components of *stress*, (1) and (2) become

$$-P_1 = \partial_{11} X_x + \partial_{12} Y_y + \partial_{14} Y_z \qquad (3)$$
$$-P_2 = \partial_{25} Z_x + \partial_{26} X_y. \qquad (4)$$

The ∂'s are the *piezo-electric moduli*, which are related to the piezo-electric constants ε by equations involving also the elastic constants.

As is evident from equations (2) and (4), the polarization P_2 is produced only by shears, which may be neglected in the present paper, as may also the third term in (1) and (3). Of the two remaining terms on the right-hand side of (1) and (3) the first expresses the longitudinal effect, the second the transverse effect.

We shall need also the following expressions for the *converse effect*, in which the stresses along the X- and Y-axis are given in terms of the X-component E_1 of impressed electric intensity:

$$-X_x = \varepsilon_{11} E_1 \qquad (5)$$
$$-Y_y = \varepsilon_{12} E_1. \qquad (5a)$$

The other stress-components are of no concern here. The equation (5) expresses the longitudinal effect, and (5a) the transverse. In the applications described in the present paper, only the *transverse effect* is utilized.

One more fundamental equation must be added, namely the strain-equation for the transverse converse effect, which is analogous to (5a):

$$y_y = \partial_{12} E_1 \qquad (6)$$

According to Voigt's theory, in the case of the class of crystals to which quartz belongs, $\varepsilon_{26} = \varepsilon_{12} = -\varepsilon_{11}$, $\varepsilon_{25} = -\varepsilon_{14}$, $\partial_{12} = -\partial_{11}$, $\partial_{25} = -\partial_{14}$, and $\partial_{26} = -2\partial_{11}$. Hence in all only two different numerical values of ∂ and ε have to be known, and of these only one occurs in the present investigation. The following values of ε_{11} and ∂_{11} were determined by Riecke and Voigt:[5]

$$\varepsilon_{11} = -4.77 \times 10^4, \ \partial_{11} = -6.45 \times 10^{-8}.$$

The ε's and ∂'s as indeed all electric and magnetic quantities in this paper, unless otherwise stated, are in c. g. s. electrostatic units. As is evident from (1) and (2), ε has the dimensions of an electrostatic polarization, while from (3) and (4) it may be seen that ∂ has the dimensions of the reciprocal of an electric intensity. Hence

$$\varepsilon = [k^{\frac{1}{2}} M^{\frac{1}{2}} L^{-\frac{1}{2}} T^{-1}], \quad \partial = [k^{\frac{1}{2}} M^{-\frac{1}{2}} L^{\frac{1}{2}} T].$$

Other observers have obtained slightly different values for

[5] Voigt, previous citation, pages 869-870.

ϵ_{11} and δ_{11}. Fortunately, in the practical applications under consideration, the absolute values need not be accurately known.

III. Theory of Longitudinal Vibrations in Rods

5. The theory of electric reactions of vibrating piezo-electric plates is a structure built upon two main piers. First, there is the fundamental piezo-electric theory which has just been set forth; and second, the theory of longitudinal mechanical vibrations in rods, which will now be briefly summarized. The "plates" which the writer uses are, as far as mechanical considerations permit, in the form of thin rods. The advantage of this procedure, in addition to economy of material, is that the fundamental vibration together with harmonics of considerable purity may be secured, free from the disturbing effects of other modes of vibration. The theroretical treatment is also greatly simplified.

In a paper which is to appear in "The Physical Review," the general theory of forced longitudinal vibrations in rods is developed. The characterizing feature is the insertion in the equations of a symbol representing the *viscosity* of the material composing the rod; for that property of the rod whereby it absorbs energy and damps its own vibrations is as important here as is the resistance in an oscillating electric circuit. It is possible to measure the actual value of the viscosity by a purely electrical method, at any desired frequency; this, as well as the effect upon the resultant viscosity of air friction and of restraints imposed by the method of mounting, need not concern us here. It is only necessary to remark that a successful piezo-electric resonator must be prepared and mounted as to reduce the damping to a minimum.

6. When an alternating emf. is applied to the metallic coatings of a rod of this sort, an alternating mechanical stress is set up in the rod in accordance with equation (5a), which is uniform throut the mass of the rod. In this statement we neglect the "edge effect" of the condenser formed by the quartz and its coatings. Considering the thinness of the quartz and its high dielectric constant—about 4.5—this procedure is justifiable as a first approximation. In the paper referred to above, it is shown that the vibrations are the same as if the rod had impressed upon its ends two alternating forces, numerically equal to the actual internal stress, of like amplitude but opposite phases, and it is on this basis that the theory of forced vibrations is developed.

The general equation of motion is

$$\frac{\partial^2 \xi}{\partial t^2} = P \frac{\partial^2 \xi}{\partial u^2} + Q \frac{\partial^3 \xi}{\partial u^2 \partial t}. \qquad (7)$$

ξ is the displacement, at the time t, of that cross-section of the rod whose undisturbed co-ordinate is u. P is defined by the equation $P = G/\rho$, where G is Young's modulus and ρ the density; P is therefore the square of the wave-velocity in absence of damping.[6] For brevity, we call Q the "viscosity," and treat it as a constant of the material, implying thereby that it is independent of the frequency. Its possible dependence upon frequency can be tested experimentally. The dimensions of Q are $[L^2 T^{-1}]$.

7. In the paper referred to, equation (7) is solved, but its application to actual cases of forced vibration is somewhat cumbersome. It is, however, shown that, for the fundamental vibration in the neighborhood of resonance, the rod may be replaced by a *fictitious "equivalent mass"* M possessing one degree of freedom. The equation of motion then has the familiar form

$$M \frac{d^2 x}{dt^2} + N \frac{dx}{dt} + g x = F = F_o \cos \omega t. \qquad (8)$$

Here M is half the actual mass of the rod, or $M = \frac{1}{2} \rho b l e$. In place of Young's modulus G in (7) we use the "equivalent stiffness," $g = M \omega_o^2$, which is related to G by the equation $g = \pi^2 b e G / 2l$. This follows from the equation $\omega_o = 2\pi f_o$, and $\sqrt{G/\rho} = 2l f_o$, $2l$ being the fundamental wave-length.[7]

x is the mechanical displacement at time t of the end of the rod, so that the actual elongation (or contraction) of the entire rod at any instant is $2x$.

M and g correspond to L and $1/C$ in an electric circuit hav-

---

[6]In crystalline media, the elastic constants depend, of course, upon the direction with respect to the axis of the crystal. Slight differences are found between individual crystals. Moreover, in the case of our rods, the elastic modulus is modified by lateral effects, unless the rod is extremely narrow, and by any discrepancy between the axis of the rod and the true Y-axis of the crystal. The effective value of G with the rods employed by the writer ranges from 8×10^{11} to 10×10^{11}. The value for quartz as given by Voigt is 8.51×10^{11}.

[7]Strictly, ω_o is the angular velocity when the amplitude of the *velocity* of the equivalent mass M is a maximum under forced vibrations; it is also the free angular velocity in absence of damping. The maximum amplitude of equivalent *displacement* x (equation (10)) comes (under forced vibrations) at the angular velocity $\sqrt{\dfrac{g}{M} - \dfrac{N^2}{2M^2}}$, while the angular velocity of free damped vibrations is $\sqrt{\dfrac{g}{M} - \dfrac{N^2}{4M^2}}$. The distinction between these three values may under ordinary circumstances be ignored.

89

ing concentrated, as contrasted with distributed, constants. N is the equivalent resistance, and bears to the viscosity Q the relation $N = \pi^2 \rho \, b \, e \, Q/2 \, l$. For the proof of this the paper on longitudinal vibrations must be consulted. F is the equivalent impressed force. If the actual stress acts thruout the entire length of the rod, it may be proven that F is twice the actual force at any cross-section, or $F = 2beX$, where X is the instantaneous stress. The expression for X in terms of the piezo-electric constant is given below, section 11.

We now write the steady-state solution of equation (8), which is of prime importance for the graphical method described in section 12:

$$x = x_o \sin(\omega t - \theta), \tag{9}$$

in which the maximum displacement is

$$x_o = \frac{F_o}{\omega \sqrt{N^2 + \left(\omega M - \frac{g}{\omega}\right)^2}}, \tag{10}$$

and

$$\tan \theta = \frac{\omega M - \frac{g}{\omega}}{N} = \frac{\pi(\omega - \omega_o)}{\omega_o \Delta} \tag{11}$$

approximately, since $g/\omega = \omega_o M$ very nearly, and the logarithmic decrement per period, Δ, is, as in the electrical analogy, $N/2fM$.

The *power expended in maintaining vibrations*, as in the case of the electrical analogy, is easily proved to be

$$p = \frac{1}{2} \cdot \frac{F_o^2}{N} \quad (ergs \ per \ sec.) \tag{12}$$

The *maximum stress when in resonance* may easily become so great as to break the quartz rod. On the assumption that the distribution of stress is sinusoidal, being zero at the ends, and' for the fundamental, a maximum at the center, we find that the maximum stress at the center is $\pi x_o G/l$, where x_o is half the maximum elongation of the rod of length l, and G is Young's modulus.

IV. The Resonance Circle

8. In applying the foregoing theory to investigations with piezo-electric resonators, it is advantageous to employ a graphical method, based on the properties of what may be called, for brevity, the resonance circle. In principle, this curve is similar to the "motional impedance" circle which has been used by

Kennelly and his collaborators in their studies of the telephone receiver.[8]

The equation of the curve in question is obtained by eliminating $\omega M - g/\omega$ between equations (10) and (11):

$$x_o = \frac{F_o}{\omega N} \cos \theta. \qquad (13)$$

If ω were constant, this would be the polar equation of a circle passing thru the origin. In reality, as θ varies from $-90°$ thru zero to $+90°$, ω varies from zero to infinity. Nevertheless, when N is very small, as is the case with quartz, not only is the "diameter" of the "circle" in Figure 2 large, but that portion of the curve corresponding to the neighborhood of resonance comprises nearly the entire curve. For all other values of ω, θ is nearly equal either to $-90°$ or to $+90°$, so that with quartz, to the precision attainable by ordinary graphical methods, the curve cannot be distinguished from a perfect circle. The distortion of the curve owing to varying ω in Figure 2 is very greatly exaggerated in order to illustrate the principle.

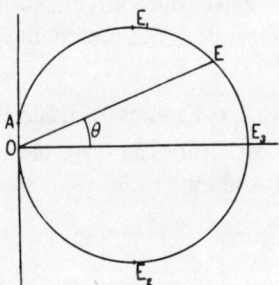

FIGURE 2—The Resonance Circle

OE represents one value of the modulus x_o, with the corresponding argument θ. It has been found most convenient to draw the maximum modulus OE_3 horizontally to the right from the origin, and to lay off positive values of θ *below* the horizontal axis, so that *increasing frequency* is represented by a *clockwise movement* of the point E around the curve. Strictly speaking, the maximum modulus, $x_o = F_o/\omega N$, should be inclined slightly upwards, corresponding to a small negative value of θ. Here

---

[8]"Proc. Am. Acad. Arts and Sci.," 48, page 113, 1912; 51, page 421, 1915; "Proc. Am. Phil. Soc.," 54, page 96, 1915; 55, page 415, 1916. The circle diagram is also used by Hahnemann and Hecht, "Phys. Zeitschr.," 20, page 104, 1919, and 21, page 264, 1920, and by Wegel, "Journal of the American Institute of Electrical Engineers," 40, page 791, 1921.

again the damping in the case of quartz is so slight that the maximum is practically the line $O E_3$.

$O A$ represents the "amplitude" when $\omega = 0$, that is, $O A$ is the equilibrium elongation under static stress at zero frequency, and must therefore have the value $X_o l/2 G$. In the case of a typical quartz plate (Quartz Resonator N 2, to which further reference will be made), $3.07 \times 0.41 \times 0.14$ cm. ($1.21 \times 0.16 \times 0.055$ inch), the fundamental frequency of which is 89,870, the equilibrium elongation at either end under a potential difference of one electrostatic unit (300 volts) is 6.5×10^{-7} cm., while the maximum amplitude of vibration at either end is (by calculation) 0.0025 cm. (0.001 inch). Thus we see that $O E$ is about 4,000 times as large as $O A$: in other words, the growth of amplitude at resonance is 4,000-fold.

For our purposes, the advantage of the resonance circle as outlined above is two-fold: the moduli $O E$, being proportional to elongations of the plate, are thereby also proportional to the piezo-electric polarization; and since the argument θ is a phase angle, the resonance circle can be incorporated into an ordinary alternating current vector diagram in studying the reaction of the plate upon the circuit. We now come to a consideration of the latter.

V. Reaction of Piezo-Electric Resonator Upon Circuit

9. Description of Circuit—The circuit that I have most frequently used in these experiments is that shown in Figure 3.

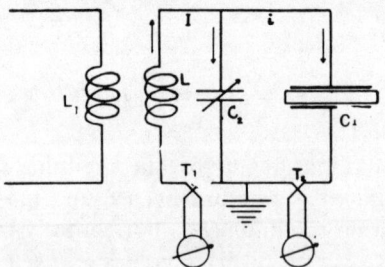

Figure 3—Resonator connected in a Secondary Circuit, with Thermo-elements for measuring Currents. The arrows indicate positive directions in the three branches

The main conclusions derived from the following paragraphs are, however, applicable to any circuit to which the resonator is likely to be connected. L_1 is a coil of about 5 millihenrys in the anode circuit of a vacuum-tube generating set kindly loaned by the Western Electric Company, so designed as to minimize the effect of load upon frequency. The latter is an extremely important requirement, and was found to be admirably fulfilled. Loosely coupled to L_1 is the litzendraht coil L of 3 or 4 millihenrys, in parallel with which are the precision variable condenser C_2 (maximum 1,500 $\mu\mu$f.) and the piezo-electric resonator C_1. Most of the quantitative work has been done with resonators about 3 cm. (1.2 inch) long, having frequencies around 90,000, corresponding to radio wave lengths of the order of 3,000 meters. References to other wave lengths will be made later, and also to the method of mounting C_1. T_1 and T_2 are thermo-elements of resistances 4.5 and 2,810 ohms, respectively, T_2 being a Western Electric vacuum thermo-element. The thermo-elements are in Figure 3 represented as connected to separate galvanometers, but in practice a single Leeds and Northrup high-sensitivity galvanometer of 10 ohms resistance was employed, provided with a change-over switch and suitable shunts. It was thus possible, at each frequency used, to measure the current I in the coil, and the current i flowing to the resonator. C_2 was provided with a worm gear for fine regulation, and readings were taken by means of a lamp and calibrated scale, a mirror being mounted on the condenser handle. By means of a specially constructed parallel-plate air condenser with micrometer control in the generating circuit an extremely fine regulation of frequency was possible. The voltage induced in L was practically constant over the small range in frequency involved.

10. The most instructive data are obtained by observing C_2, I, and i at a number of frequencies, and plotting the results as in Figure 6. These curves will be further discussed in section 15. At each frequency, C_2 is varied until either I or i is a maximum.

At a certain critical frequency, the absorption of energy by the resonator is so great that the coil-current I falls almost to zero, even when C_2 is at its most favorable setting, as represented in Figure 6. Any change in C_2 causes I to decrease still more: but a slight change in frequency causes I to increase enormously, showing that the mechanical tuning of the resonator is very much sharper than the electrical tuning of the circuit. At this point, then, *I is a maximum with respect to C_2, but a minimum with respect to f.*

Further consideration of Figure 6 will be deferred until the theory of the resonator's reactions has been given.

11. THEORY OF THE REACTIONS—We assume that an emf.

$E_o \cos \omega t$ is impressed upon coil L, Figure 3, E_o being constant at all frequencies. Let V represent the instantaneous potential difference across C_2 and C_1, which, of course, varies with frequency, and R the combined resistance of L and T_1. The capacity of the coil should be included in C_2. The impedance of C_1 is so great that the resistance of T_2 may, as a first approximation, be neglected. We then have

$$E = E_o \cos \omega t = V + L \dot{I} + R I. \qquad (14)$$

The link between this equation and that for the vibration of the resonator is the quantity V. For the electric field in the quartz has the value V/e, where e is the thickness of the quartz; and by equation (5b), the mechanical stress is $X = \varepsilon V/e$. From section 7, we see that the value of the equivalent force F in equation (8) must be $F = 2 b e X = 2 \varepsilon b V$. Equation (8) may, therefore, be written thus:

$$2 \varepsilon b V = M \ddot{x} + N \dot{x} + G x. \qquad (15)$$

Next, we consider the components of polarization within the resonator. Let D_1 represent the component due to the potential difference between the coatings. If this p. d. is V, and if the air-gaps between coatings and crystal be neglected, then D_1 is the ordinary displacement in the dielectric, as given by the equation $D_1 = k V/4 \pi e$. k, the dielectric constant of quartz, has the value 4.5, approximately.

A second component of electric polarization, that which is responsible for all the effects described in this paper, is caused by the deformation of the resonator. Using the same notation as heretofore, we assume that at any given instant the p. d. across the resonator is V, and that the total elongation (difference, $+$ or $-$, between instantaneous length and normal length) is $2x$. Of this elongation, a portion, say $2 x_1$ is that which would be produced under static conditions with a constant potential-difference V. This portion is the "equilibrium elongation," and exerts no reaction upon the circuit. The value of $2 x_1$ may be derived from (6) thus:

$$y_y = \delta_{12} E_1 = \delta_{12} \frac{V}{e}.$$

Further, we have the fundamental relation $2 x_1 = l y_y$, hence $2 x_1 = \delta_{12} V l/e$. x_1 is readily seen to be the same as the $OA = X_o l/2 G$ in section 8, since by Hooke's law $y_y = X_o/G$. That portion of the total elongation $2x$ which is due to the vibrations, and therefore, effective in producing an electric reaction, is $2x - 2x_1$. Applying equation (1) (only the second term on the right side of which remains, since x_z and y_z do not appear) to this elongation, we find $P_1 = 2\, \varepsilon_{12}\, (x-x_1)/l$, or, writing D in place of P_1, dropping the subscript from ε_{12}, replacing $2\,x_1$ by its value above, and letting ∂ represent the quantity $\partial_{12}\, l/e$,

$$D = \frac{\varepsilon}{l}(2x - \partial V). \qquad (16)$$

At frequencies sufficiently removed from resonance, $2\,x$ may be so small as to be even less than ∂V; but with quartz, in the neighborhood of resonance, the static elongation ∂V is negligible, as stated already in section 8.

Equations (14), (15), and (16) contain four unknown quantities, namely: V, I, x, and D. The fourth equation necessary to a solution of the problem is found either in the expression for the resonator current:

$$i = b\,l\,(\dot{D}_1 + \dot{D}) = C_1 \dot{V} + b\,l\,\dot{D}, \qquad (17)$$

or preferably that for the coil current:

$$I = C_2 \dot{V} + i = C_2 \dot{V} + b\,l\,(\dot{D}_1 + \dot{D}) = (C_1 + C_2)\dot{V} + b\,l\,\dot{D}. \qquad (18)$$

Equation (18) follows from Kirchhoff's first law.

These equations are simply expressions of the principles that current is time-rate of change of dielectric flux, and that the polarization in C_1 has the two components D_1 and D discussed above. Dielectric flux is of course polarization times area $b\,l$.

From equations (14), (15), (16), and (18) a single differential equation of the fourth order may be formed, giving x as a function of time in terms of impressed voltage E and the circuit constants.

* * * * * * *

VI. Construction of Resonators

20. It was found in the early stages that strongest and most constant results were obtained when the crystal plate was placed between and not quite touching the metallic coatings. For the latter small flat plates of brass are now used, of the same size as the crystal or somewhat shorter.[12] Light contact between quartz and brass at one or two points is, for most purposes, of no consequence. The quartz rod may be supported by a thread

---

[12] When the metallic coatings are shorter than the crystal, the behavior of the resonator is qualitatively unchanged. The consequent small modifications to the theory are easily made, if needed. Strictly, a correction should also be made in the theory on account of the air-space between crystal and the coatings. This is not so simple a matter. Fortunately the essential performance of the crystal is not altered, except in intensity, by the air space, and for quantitative tests of the theory, it is possible so to mount the rod as to make the air-space negligible.

tied about its center, or balanced on edge upon a small block. Successful tests have also been made with a quartz rod or plate silvered on both sides. It is easy to deposit chemically a coating of silver sufficiently thick for the electrical effects, without decreasing the frequency more than one or two tenths of a per cent. The advantages of this method are not sufficient, however, to offset the difficulty of making an electrical contact with the silver that is both permanent and delicate.

For a portable unit, the best method of mounting is to let the crystal plate lie in a small pocket in which it is just free to vibrate. The sides of the pocket are formed by the brass "coatings," the bottom is of glass, and the ends of bakelite or hard rubber. By its own vibrations the plate keeps itself sufficiently free from contact with sides and ends, while any particles of dust on which it may rest serve as roller bearings to reduce friction with the bottom of the pocket.

Figure 7 shows a partially completed unit containing four plates, of which the longest is about 3 cm. (1.2 inch) in length. To the left is seen a small resonator containing one 3 cm. plate.

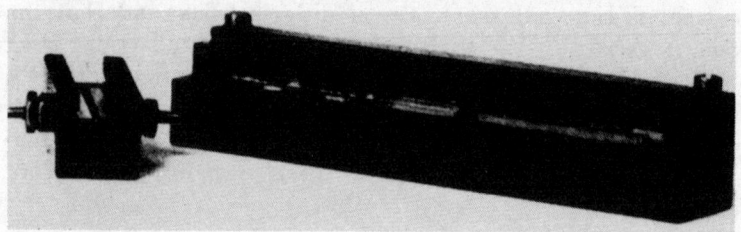

FIGURE 7—Quartz Resonators. Left: a single unit, mounted, for 3,040 m. Right: partially completed mounting for four quartz plates, which are seen lying on the front edge of the bakelite base. The wave-lengths are approximately 3,000, 1,200, 500, and 400 meters. These plates will stand on edge on the glass strip immediately behind them, having as one common "coating" the long brass bar which rests on the glass. The four individual "coatings," and the spacers to keep the quartz plates in place, are not shown.

21. THE CONSTANTS OF QUARTZ are such that the radio wave length in meters is roughly 100 times the length of the resonator in millimeters (2,500 times the length in inches). Concerning the relative dimensions of the crystal plates, the following are determining factors, which follow directly from the theory. Since all the essential phenomena are proportional to the diameter of the resonance circle as drawn in Figure 5, and since this diameter (compare section 13) contains the width b of the plate as a factor, it follows that b should be large. The limit to the width of the plate is set by the fact that disturbing modes of vibration

enter in when the plate is not relatively narrow. In practice, the best value of the ratio l/b has been found to vary from 2 for the smallest plates (length about 2 mm. or 0.08 inch) to 10 or 20 for the largest. Thin plates are always an advantage, since for same voltage the electric field in the quartz is then more intense.

If the air-space is increased, or if the ends of the quartz extend far out beyond the brass coatings, the frequency is slightly raised. In order to understand this, one should compare the polarized plate with a short, wide bar magnet. When the coatings are close to the quartz, they and the associated circuit are analogous to a massive yoke of highly permeable material, allowing the induction to attain the greatest possible value. As the air gap increases, the plate partially depolarizes itself, owing to the turning back of some of the lines of electrostatic induction that are produced by its state of strain. The piezo-electric action of this depolarizing component is always to tend to make the plate regain its normal form; that is, the effective value of Young's modulus is larger, the greater the air gaps. In the most extreme case hitherto observed, the increase in frequency when the coatings, originally touching the quartz, were entirely removed, amounted to about 0.6 per cent. Hence no perceptible inaccuracy need be feared from this direction. Data at present available indicate that the frequency of the larger quartz resonators, when permanently mounted, is constant at least to within one part in 10,000. The quartz-steel resonators are subject to a small correction for temperature, amounting to about 0.01 percent decrease per degree centigrade.

The decrement of the resonator is easily proven to be independent of its cross-section, but directly proportional to the frequency. Resonators for large wave lengths are therefore more efficient and more sharply tuned than those for short waves. This statement assumes that the viscosity Q is constant. The larger resonators are for this reason much more in danger of fracture from excessive voltage. It is doubtful whether a good resonator for 3,000 meters will stand safely as much as 50 volts at the resonant frequency.

The *electrostatic capacity* of the resonators varies from a few micro-micro-farads down to a small fraction of a $\mu\mu f$.

22. QUARTZ-STEEL RESONATORS—For the longer wave lengths used in radio it will hardly be possible to secure sufficiently long quartz rods. The writer has used quartz up to about 4,000 m. (length of rod about 4 cm. or 1.6 inch), and beyond this has had good results with flat rods of tool steel or invar, excited to longitudinal vibration by means of small quartz plates cemented to

the sides with solid shellac. The wave velocity in steel is not very different from that in quartz. For a 10,000 m. resonator, a steel rod about 95×9×3 mm. (3.71×0.35×0.12 inch) is used, quartz plates about 9×10×1 mm. (0.35×0.39×0.04 inch) being cemented to each side at the center, as shown in Figure 8. The steel itself forms one "coating," the quartz plates being so placed that the same polarity of each faces the steel. The other coatings are of tinfoil, to which fine wires are soldered and connected in parallel. The decrement of this combination is not very different from that of the larger quartz resonators. A small hook is screwed into the steel at the exact center, between the quartz plates, to serve as a suspension and as one terminal of the resonator, leaving the rod free to vibrate at its fundamental frequency. Thru the action of the transverse effect the quartz rods expand and contract when connection is made to an alternating current supply, causing an alternating condensation and rarefaction at the center of the steel rod, whereby the longitudinal vibrations are excited. The frequency is essentially that of the steel rod, and the electric reactions take place exactly as with the quartz resonators, with an intensity sufficient to produce a strong response.

FIGURE 8—Steel resonator, the exciting quartz plates having tinfoil coatings electrically connected. The hook by which the steel rod is suspended serves as the other terminal

In the earlier experiments, the steel rods had plates of quartz or Rochelle salt at their ends, but this construction proved less reliable than that with the quartz side-plates.

* * * * * * *

Editor's Comments on Paper 18

18 Wood and Loomis: *The Physical and Biological Effects of High-Frequency Sound-waves of Great Intensity*

Although P. Langevin had used the piezoelectric effect in France to produce high-frequency sound for submarine-detection purposes as early as 1918 and W. G. Cady had employed high-frequency vibrations of quartz rods to stabilize the frequency of oscillating electric circuits, the first extensive study of the physical and biological properties of high-frequency sound at high intensity was carried out by R. W. Wood of Johns Hopkins University and A. L. Loomis at the latter's Tuxedo Park Laboratory. Their work was published in a paper entitled "The Physical and Biological Effects of High-Frequency Sound-waves of Great Intensity." This was the forerunner of a large body of research in the field now known as ultrasonics, which continues to attract considerable attention among acoustical scientists and engineers.

Robert Williams Wood (1868–1955) was an American experimental physicist whose professional career was spent largely at Johns Hopkins University. Although his principal research interests were optics and spectroscopy, he made distinguished contributions to other fields, including acoustics, shock waves, photography, and astronomy. He was a lecturer of unusual brilliance on scientific topics and a writer of fiction and humorous verse.

Alfred Lee Loomis (1887–   ) is an American physicist who began his professional career as a lawyer but switched to physical research as an independent investigator when he was 40 years old. He founded the Tuxedo Park Laboratory in New York State and collaborated with well-known physicists in research projects of great interest. In addition to his work on ultrasonics, he worked on the precision measurement of time and on the development of navigational systems.

Reprinted from *Phil. Mag.*, Seventh Series, **4**, 417–436 (1927)

18

THE

LONDON, EDINBURGH, AND DUBLIN

PHILOSOPHICAL MAGAZINE

AND

JOURNAL OF SCIENCE.

[SEVENTH SERIES.]

SEPTEMBER, 1927.

XXXVIII. *The Physical and Biological Effects of High-frequency Sound-waves of Great Intensity.* *By* Prof. R. W. Wood, *For.Mem.R.S.*, *and* Alfred L. Loomis*. (Communication No. 1 from the Alfred Lee Loomis Laboratory, Tuxedo, N.Y.)

[Plates VII.—XIII.]

Introduction.

In the present paper we shall give an account of a preliminary survey of what appears to be a wide field for investigation, opened up by the study of the very surprising and remarkable effects obtained with sound-waves of high frequency and great intensity generated in an oil-bath by a piezo-electric oscillator of quartz operated at 50,000 volts and vibrating 300,000 times per second.

The radiation pressure exerted against a glass disk 8 cm. in diameter amounts, under certain circumstances, to 150 grams, and when operating against the free surface of the oil (from which the radiation is totally reflected) raises it in a mound 7 cm. in height, surmounted by a fountain of oil drops, some of which are projected to an elevation of 30 or 40 cm.

The waves can be transmitted along a glass thread 0·2 mm. in diameter and a metre or more in length, and the end of the thread, if squeezed between the thumb and finger, burns a groove in the skin. A tapering glass rod, 0·5 mm. in diameter at the tip, can be thrown into vibration of such

* Communicated by the Authors.

intensity that a pine chip smokes and emits sparks when pressed against the tip, the rod burning its way rapidly through the wood, leaving a hole with blackened edges. If a glass plate is substituted for the chip, the rod drills its way through the plate, throwing out the displaced material in the form of a fine powder or minute fused globules of glass.

If the waves are passed across the boundary separating two liquids such as oil and water or mercury and water, more or less stable emulsions are formed. Chemical reactions are accelerated, crystallizations started, and other remarkable effects produced by these very intense super-sonic vibrations.

Preliminary experiments with interference fringes formed between a vibrating plate and one at rest indicate that the amplitude of the vibration is of the order of magnitude of a wave-length of light, yet an enormous amount of energy is delivered. The mean energy and acceleration are both proportional to the square of the frequency, and we are here dealing with what Prof. C. V. Boys very tersely describes as "*All* acceleration and *no* motion."

The method employed in the generation of the waves is essentially the one developed by Professor Langevin in 1917 for the purpose of locating submarines by the echo of a narrow beam of high-frequency sound-waves. Shortly thereafter experiments along similar lines were inaugurated by the British and American navies.

In Langevin's original apparatus the vibrations of the piezo-electric quartz plate were excited by a Poulsen arc in connexion with suitable condensers and coils. Voltages as high as thirty or forty thousand were applied to the plates, and the amplitude of the waves raised to such a degree that small fish were killed by the radiation and pain of considerable severity was experienced when the hand was thrust into the water in the tank. The Poulsen arc proved troublesome, however, owing to its instability, which made it impossible to keep the electrical vibration in tune with the natural frequency of the quartz plate, and it was speedily supplanted by the vacuum tube, which is used exclusively at the present time.

Since the coefficient of viscosity increases with the square of the frequency, comparatively low frequencies only can be employed when beams of sound are to be projected under water over long distances for signalling or other purposes. Voltages of one or two thousand at frequencies of from 30 to 40 thousand are used for the most part in work of this nature.

In our experiments this limitation is not imposed, since

absorption of the radiation by the medium does not interfere to any great degree with the study of effects close to the source, and we operate usually with voltages in the vicinity of 50,000 at frequencies ranging from 200,000 to 500,000.

Description of the Apparatus.

(See Plate VII. fig. 1.)

The apparatus employed in the present work was built in the Research Laboratory of the General Electric Co. at

Fig. 1.

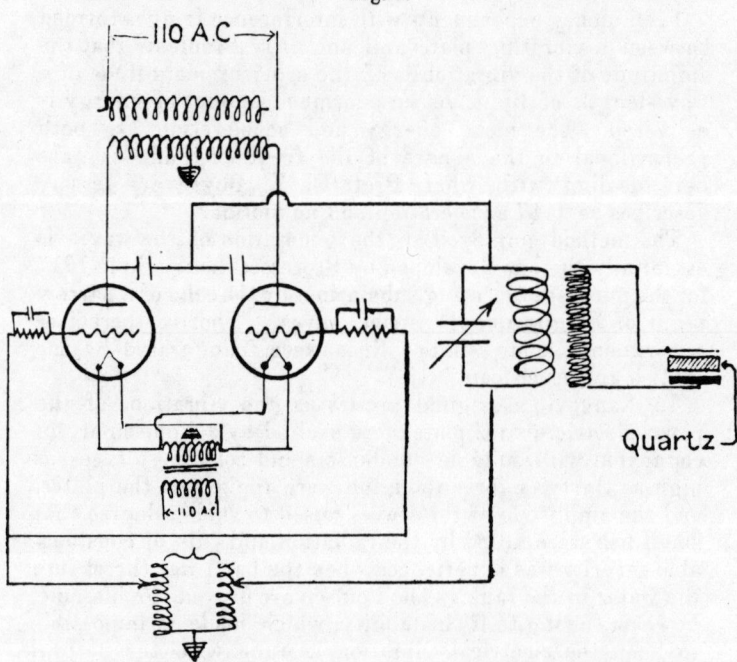

Schenectady. It consists of a two kilowatt oscillator, designed originally for an induction furnace, a bank of oil condensers, giving capacities up to 0·1 microfarad, a large variable air condenser, and several pairs of coaxial coils for raising the voltage. The primary or outer coil consisted of from 7 to 20 turns of Litzendraht cable, the coils varying from 16 to 24 cm. in diameter. The secondary coils were wound on glass cylinders (100 to 250 turns) and mounted within the primaries. Fig. 1 shows in conventional manner the wiring of the various parts. The use of several coils was found to be necessary as we employed quartz plates varying in thickness from 7 to 14 mm., with which we obtained waves with frequencies ranging from 100,000 to 700,000 cycles per second. The quartz plates were circular disks, and when in operation, one of them rested on a disk of sheet lead at the bottom of a dish of transformer oil. The other electrode consisted of a disk of very thin sheet brass resting on the upper surface of the quartz. The coils for raising the potential and the glass dish with oil, in which the quartz oscillator is immersed, are shown on Plate VIII.

Pressure due to the Radiation.

We have already mentioned the pressure developed against the free surface of the oil above the quartz vibrator as a result

Fig. 2.

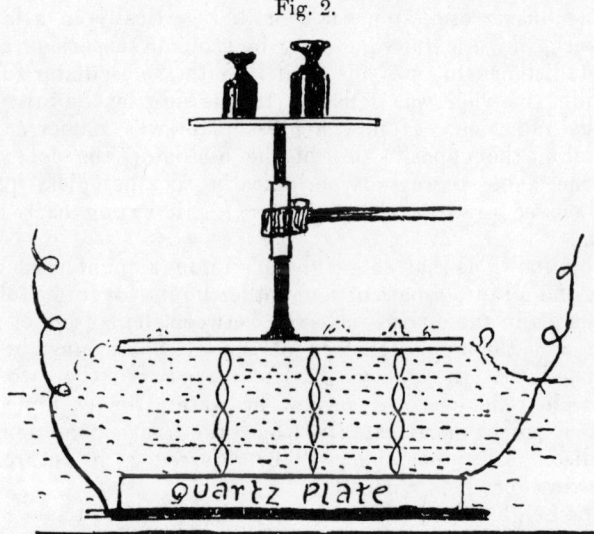

of the reflexion of the radiation. In the case of reflexion from plates of glass or metal the magnitude of the effect can be measured. We found that a glass disk 8 cm. in diameter attached to a glass rod and supported as shown in fig. 2 would support a weight of 150 grams. The pressure is a maximum when the distance between the under surface of the plate and the upper surface of the quartz oscillator is a whole number of half wave-lengths. Under this condition the reflected wave strikes the oscillator when its phase is such as to reflect the wave back to the plate. The energy is thus imprisoned by multiple reflexions between the vibrator and the plate, and the amplitude rises to a very high value, for the same reason that the amplitude of vibration of the stationary waves on a thread attached to a vibrating tuning-fork may be twenty or thirty times the amplitude of the fork when the length of the thread is properly adjusted.

If the rod which supports the glass plate is held in the fingers and the plate pushed down gradually into the oil a strong resistance is encountered periodically, as if the plate were breaking its way through a series of resisting films. In the positions of maximum pressure the energy is reflected back and forth between the oscillator and plate. In the positions of minimum pressure the wave reflected down from the plate meets the oscillator when it is in such a phase as to transmit the reflected wave. That such is the case was shown by the following experiment.

The quartz oscillator was mounted vertically in a large oil bath and a metal vane hung by a bifilar suspension at a short distance to one side of it. With the oscillator functioning, this vane was deflected to one side by the pressure of the radiation. If now a glass plate was immersed in the oil on the opposite side of the oscillator, the deflexion of the vane increased periodically as the glass plate was moved towards the oscillator, *i. e.* it swung back and forth.

This shows us that, to get the maximum amount of energy from the oil into a bath of some other liquid (or into a solid) immersed in the oil, the distance between the bottom of the bath and the upper surface of the oscillator must be so adjusted that the energy builds up between the two by multiple reflexions. A beaker of water, for example, is heated much more rapidly when the above condition is fulfilled. This operation will be referred to in future as adjusting for energy density.

The height of the oil mound raised above the surface over the plate depends in the same way upon the depth of the oil. We have never obtained a smooth uniform mound, as might be expected with low amplitude if the plate were simply expanding and contracting as a whole. With the oscillator operated at low voltage, a number of humps appear on the surface which shift their position with every alteration in the capacity of the condenser. When the frequency of the electrical oscillation is tuned exactly to the natural frequency of the crystal plate the oil mound rises to a height of 7 cm., its summit erupting oil drops like a miniature volcano. Further increase of the power gave a mound 10 cm. in height, but the quartz plate broke into fragments. A photograph of the oil fountain taken against a very bright background with an exposure of 1/500 second is reproduced on Plate VII. fig. 2.

Stationary waves on tubes

If a glass tube a metre long and 2 or 3 cm. in diameter, closed at the bottom like a test-tube, is coated on the inside with a layer of a heavy oil, the oil gathers itself together in rings about 3 mm. apart, which line the tube from top to bottom, as soon as the lower end is dipped into the vibrating oil over the quartz plate. The rings appeared to be interrupted in places by another system of waves, and a permanent record of the pattern was secured by substituting paraffin, coloured with aniline red, for the oil and using it in a warm tube. A photograph of a portion of the tube with the rings cut across into a pattern of regularly spaced dots is reproduced on Plate IX. fig. 3.

The wave-length of the oblique system is about 1·5 times that of the horizontal system.

We at first attributed the rings to a stationary system of compressional waves formed by interference between disturbances reflected down from the top of the tube with those coming up from below, but the velocity deduced from the wave-length and frequency was much less than the velocity of sound in glass. The waves turned out to be transverse vibrations, the wave-length being but a small fraction of the diameter of the tube. If a similar glass tube was used without the oil, and the outer surface heated to the softening point in the flame of a blast-lamp while the vibrations were running up and down the tube, the stationary system was permanently recorded in the glass, and could be made visible by casting a shadow of the tube with sunlight (or light from any concentrated source) on a sheet of paper held at a distance of 10 or 15 cm. from the tube. A shadow photograph, made in this way, is represented on Plate IX. fig. 4.

We have made no careful study of the modes of vibration of a tube for high frequencies, and have no explanation for the oblique system of greater apparent wave-length. It seems evident, however, that the velocity of propagation is greater for the waves forming this system than for the others. It was observed also, as the tube cooled down, that the paraffin remained fluid longer in those portions of the tube where the double system registered than at other places, indicating that the internal heating of glass was greater here than elsewhere.

Transverse waves on glass plates, rods and threads.

If a glass rod is cemented with sealing-wax to the centre of a circular glass disk, dusted with lycopodium, a beautiful system of concentric circular rings forms on the plate as soon as the lower end of the rod is brought into contact with the vibrating oil-bath.

These rings are formed at the nodal lines of a system of stationary waves in the plate, formed by the interference of the waves reflected from the rim with those radiating from the centre. If the disk is thicker at the centre than at the rim (we used the base and stem of a broken wine-glass) the distance between the rings is less at the rim than at the centre, from which the inference can be drawn that the waves are transverse vibrations, which is to be expected considering the arrangement of the rod and disk. The velocity is higher at the thick than at the thin portions, consequently the rings are further apart.

If the rod is cemented to the disk at a point situated at a small distance from the centre, we obtain the complicated pattern reproduced as a negative (*i.e.* the lycopodium lines black) on Plate X. fig. 5. Here we evidently have the waves reflected from the rim coming to a focus, which becomes a second source of radiation on the side of the centre opposite to the rod, and a system of radiating interference fringes is formed, as with two similar sources of light. The pattern in the immediate vicinity of the sources is of especial interest.

A beautiful system of circular rings of variable spacing is produced by dusting the inner surface of a champagne glass with lycopodium, and touching the base to the surface of the vibrating oil. At the rim the rings are closer together than near the centre, where the glass is thicker.

Applying the lycopodium method to a glass rod which has been drawn down in a flame to a long tapering point, the diameter varying from 7 to 0·5 mm., gives us a system of rings, the separation of which decreases rapidly as we pass from the thick to the thin portion. This shows that the velocity of propagation is a function of the diameter of the rod, which will of course be true for transverse, but not for longitudinal disturbances. If the rod terminates in a fine point no rings are formed, since in this case the reflexion from the end is negligible and the stationary wave system is not formed.

More permanent rings better suited for wave-length measurements were made by the following method. A small ball of soft red wax was stuck on the point of the rod held vertically over the oil. This melted and slid down the rod as soon as the lower end was dipped in the oil, owing to heat developed by friction between the vibrating glass and the wax. The wax solidifies in rings above the ball as it descends leaving a permanent record of the wave-length, as shown on Plate X. fig. 6.

In this particular case the rod is drawn down from a glass tube closed at the bottom. The energy abstracted from the oil and thrown into the rod is greater than when a large solid rod forms the collector. Remarkable calorific effects obtained with this type of collector will be described presently.

Heat developed at Contact-point between vibrating rods and matter.

This type of heating was first accidentally observed when taking the temperature of the oil in the erupting mound over the vibrator. Though the mercury registered only 25°, the thermometer tube became so hot at the point where it was held between the thumb and finger that it had to be released. The heat of course is developed by friction between the vibrating glass stem and the skin of the fingers, or rather by the rapid *pounding* of the transverse vibrations, and becomes unbearable only when the glass is squeezed tightly between the thumb and finger. This same heating is observed when any object such as a rod, tube, beaker or flask is held by the fingers and dipped into the vibrating oil-bath. If a glass rod is drawn out into a long thread of the diameter of a horsehair, terminating in a pear-shaped bead, the heat developed, when the end of the thread is squeezed between the fingers and the bead dipped into the oil fountain, is so great that a groove with seared edges is left in the skin. A week later bright red spots similar in appearance to blood-blisters developed, which did not disappear for several weeks. These were perhaps due to an effusion of blood from capillaries deep down in the skin, which were ruptured by the vibration. Still more powerful effects were obtained with rods 0·5 mm. in diameter drawn out from the top of an Erlenmeyer flask the neck of which had been closed by fusion in a blast-lamp. Shown at extreme right of Plate VIII., which also shows rack-and-pinion stand for adjusting a flask of water containing a frog for maximum energy density. A side tube was fused to the flask, by which it was supported in a clamp-stand, furnished with a rack-and-pinion movement, by which the distance between the flat bottom of the flask and the vibrator could be accurately adjusted for the position necessary for securing the maximum density of the radiation imprisoned between the surfaces. With this condition fulfilled a dry pine-chip, pressed against the top of the glass rod, smoked and emitted an occasional spark, while the rod rapidly burned its way through the wood, leaving a hole with charred edges. The heating, of course occurs only at the point of contact, the remainder of the rod being quite cold.

If a plate of glass is pressed lightly against the top of the rod, the surface of the plate is etched at the point of contact, the microscope showing a curious scalloped pattern. If the pressure is increased the rod drills its way rapidly through the plate, and the microscope shows small globules of molten glass, and finely powdered material.

This method of conducting the vibration along threads of glass yields a valuable technique for the investigation of the biological effects of the high-frequency vibration, which can thus be applied at a small point on a living organism, egg or embryo under the microscope. We have felt the heat at the end of a thread a metre long and 0·2 mm. in diameter. With the flask form of collector, with proper adjustment for the system of stationary waves between the top of the quartz vibrator and bottom of the flask, the energy thrown into the glass thread is often so great that the thread breaks into pieces.

Another form of collector by which energy can be abstracted from the oil and conducted into a rod or thread is shown in fig. 6, Plate X.—a tube of glass closed with a round bottom and drawn down to rod or thread at the opposite end. The sloping wall of the bottom appears to facilitate the production of transverse waves and it was found also, in experimenting with a flat collector made of a thin plate of glass, drawn off into a rod at one corner, that the most vigorous vibrations occurred in the rod when the plate was immersed in the oil in a slightly oblique position and not when it was parallel to the surface of the quartz plate.

Velocity and Dispersion of Transverse Waves in Solids.

By the methods just indicated it is possible to measure the velocity of propagation of the transverse waves if the frequency is known. As has been said, the velocity is a function of the diameter of the rod or the thickness of the plate. Micro-photographs of the rings on rods of 0·15, 0·5 and 1 mm. in diameter are reproduced on Plate XI. fig. 7. Observations made with waves generated by quartz vibrators of different thicknesses showed that the velocity was also a function of the frequency, as is the case with light traversing a dispersing medium. We investigated the phenomenon of dispersion employing disturbances of four different frequencies—441, 405, 350, and 285 thousand vibrations per second—forming the rings of red wax or lycopodium on rods and threads of glass varying in diameter from 6 mm. to 0·1 mm. The results are shown graphically in the curves reproduced in fig. 3. The velocity, which at

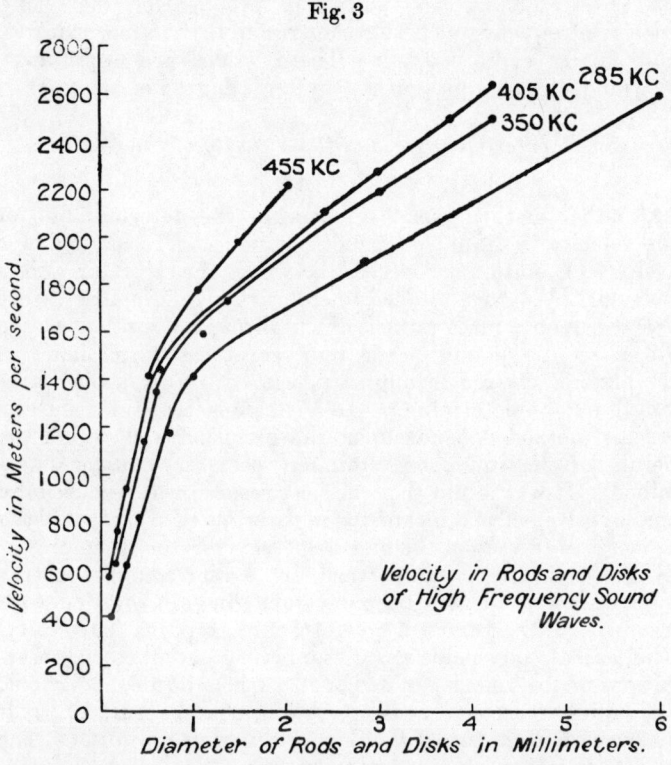

Fig. 3

Velocity in Rods and Disks of High Frequency Sound Waves.

285 kilocycles is 2600 metres per second in a glass rod 6 mm. in diameter, falls off to 400 metres in the case of a glass thread 0·1 mm. in diameter.

As the diameter of the rod is increased it becomes increasingly difficult to form the rings, and for values of the order of magnitude of the half wave-length it becomes impossible to obtain any record of the transverse waves, even by the lycopodium method, which is the more sensitive of the two. With very large energy input we several times obtained under these conditions indications of waves of considerably greater wave-length than the transverse ones which we at first believed represented the longitudinal disturbance. Measurements, however, were not in agreement with the known velocity of sound in glass, which is in the neighbourhood of 5000 metres per second.

The dispersion, for a rod of any given diameter, is given by taking the ordinates from the four curves corresponding to the same abscissa, the velocity of propagation increasing with the frequency at first rather rapidly, then more slowly, and finally rapidly again. These values are preliminary only and do not represent a very high degree of accuracy.

Sonic Interferometer and the Velocity of the Waves in Liquids.

Another method was developed for the determination of the velocity in liquids, depending upon the formation of a system of stationary waves between the vibrator and a reflector. We have alluded to the variable periodic pressure exerted upon a plate pushed down through the oil over the vibrator. By counting the number of resisting planes as the plate is lowered through a measured distance, the wave-length can be determined. It is obvious that the accuracy of this method depends upon the precision with which the points of maximum or minimum pressure can be determined. It was found that the best results were secured by employing electrical means for registering the reaction of the reflected waves upon the piezo-electric vibrator, as employed by Professor Pierce of Harvard, and a very compact instrument of low power has been developed in collaboration with Professor J. C. Hubbard, now of Johns Hopkins University, who has already made a series of very satisfactory observations of the velocity of sound in various liquids, solutions, and liquid mixtures. It is possible to obtain results of great accuracy with only a few cubic centimetres of liquid. The results of this work will be reported in a subsequent paper.

As is apparent this method is analogous to the employment of the Fabry and Perot interferometer for determining the wave-length of light, and the apparatus may be termed a Sonic Interferometer. With the plate in the position for maximum pressure, and multiple to-and-fro reflexion of the waves, the condition is similar to that obtaining with a Fabry and Perot instrument illuminated by *parallel rays* at normal incidence, with its plates at such a distance as to secure the maximum transmission of light. If the wavelength of light or the distance between the plates is slightly altered, the transmission falls to a very low value. Transmission in this case corresponds to the entrance of the energy into a beaker of water tuned for maximum energy density, as previously described. In the usual treatment of the Fabry and Perot interferometer one is apt to overlook the circumstance that under certain conditions transmission of the light is refused, as it is customary to illuminate the instrument by an extended luminous source such as a flame—in which case the amount of light transmitted is independent of wave-length, but transmission is permitted in specified directions only, this limitation giving rise to the rings.

Heating of Liquids and Solids.

The kinematic coefficient of viscosity increases as the square of the frequency. At frequencies from 300 to 400 K.C. the heating of liquids is very pronounced. Thus a test-tube filled with water and immersed in a beaker containing water and cracked ice heats rapidly when the beaker is lowered over the vibrating quartz disk, showing that the energy of the sound-waves (which pass into the water in the test-tube after traversing the intervening glass walls and the ice-water) is converted into heat by absorption, the water in the test-tube rising rapidly in temperature notwithstanding the circumstance that it is surrounded by a layer of water at $0°$. The rise of temperature may be as great as one degree every three seconds, the rate depending upon whether the depth of the water and the distance between the vibrator and the bottom of the beaker are adjusted for multiple reflexions of the radiation, as previously described. With 250 c.c. of water the heat developed amounted to about 900 calories per minute—with 150 c.c., 750 cals.; with 100 c.c.; 700 cals.; and with 50 c.c. (in a test-tube) 430 cals.

These results show that though the temperature rise is higher in the case of small volumes of liquid, the total energy abstracted from the radiation increases with the volume of liquid employed.

We have not as yet made any precise determinations of the heating of various liquids by the radiations. Three determinations with 45 c.c. of ethyl alcohol gave $4°·1$, $3°·6$, and $4°·5$ as the temperature elevation for an exposure of 20 seconds. These observations were, however, made before the necessity of accurate adjustment of the containing vessel had been realized. In work of this nature it will be necessary to devise some method by which it will be possible to throw the same amount (or a measured amount) of energy into the fluids under investigation.

A few observations have also been made of the internal heating of solids. A block of newly formed ice (distilled water frozen in a beaker by ice and salt) was subjected to the action of the sound-waves for two minutes in a beaker of ice water containing numerous small fragments of ice which kept the temperature of the water at 0°. Adjustment for maximum energy density was secured by watching the block of ice which was elevated above its normal position in the water by the radiation pressure. At the end of the exposure the block of ice, on being squeezed between the thumb and fingers, broke up into small fragments showing that liquefaction had taken place throughout the mass, as in the case of so-called "rotten ice" after exposure to the sun's rays.

We found that this experiment could not be duplicated with natural ice (*i. e.* pond ice), and believe that this may be due to the circumstance that in this case we are dealing with a single crystal, whereas in the case of the artificial ice we have a mass of interlocking crystals, the heating taking place at the crystal interfaces.

To eliminate effects due to air bubbles in the ice, distilled water, thoroughly boiled to remove air, and coloured with fluorescein, was covered with a thin layer of paraffin and frozen. The outer portions of the ice block were perfectly transparent, while the central portion was yellowish in colour, somewhat cloudy and devoid of fluorescence, showing that the fluorescein was not in solution. On exposing this block to the radiation for thirty seconds and examining it in sunlight against a black background, it was found to be traversed by innumerable interlacing planes of green fluorescence, caused by the internal melting and consequent solution of the dye. A somewhat analogous effect was noted in one of our earlier experiments, in which a bit of candle (probably stearic acid) was melted on the surface of water in a test-tube and allowed to solidify. Water was then introduced above the solid plug, and the lower end of tube subjected to the vibration. No trace of the radiation appeared in the water above the plug the *under surface* of which melted rapidly and was thrown down into the water as a white emulsion, showing the powerful absorption of the radiation by the solid stearic acid.

It was during an attempt later on to duplicate this experiment with paraffin that we obtained the apparent crystallization of the substance referred to later on in this paper.

Formation of Emulsions and Fogs.

If two non-miscible liquids such as oil and water are simultaneously subjected to the radiation in the same beaker, an emulsion or colloidal solution is formed as a result of the forces acting at the interface between the liquids. This phenomenon was first observed in an earlier arrangement of our apparatus, in which the quartz oscillator operated in a thick layer of oil floating on water. White clouds of finely divided oil were thrown down into the water and occasionally, probably when exact tuning happened to be secured, large masses of oil were projected, with almost explosive violence, down into the water, as a shower of large drops. Stable emulsions resembling milk can be made of stearic acid or paraffin and water. A beaker of water with a layer of mercury at the bottom, under the influence of the vibrations becomes first milky, then brown, and finally black. At the end of twenty-four hours most of the mercury has settled, but a sufficient amount remains in suspension to make the water slightly turbid.

Colloidal solutions of the low melting-point alloys have also been made.

At a liquid-air interface, in the case of less viscous liquids the forces brought into play drive the liquid into the air in the form of a spray of minute droplets, forming a fog. This atomization (to use the popular term) of a liquid by the sound-waves is best shown by pouring a little benzol into a beaker and lowering the beaker into the oil, tuning it for energy density. The beaker fills rapidly with a cloud of white smoke, a benzol fog, the surface of which is in tumultuous motion. A photograph of this phenomenon is reproduced on Plate XII. fig. 10. The beaker was illuminated by sunlight (reflected from a mirror) against a black background, the camera pointing as nearly as possible towards the direction from which the light was coming, to secure the maximum illumination. The filaments rising from the surface are larger droplets moving too rapidly for the camera shutter. Fogs can also be formed over water, but in this case the droplets are larger and settle rapidly.

Prof. C. V. Boys has drawn my attention to the analogy between this experiment and a phenomenon produced by the explosion of a "depth-charge." The first indication of the explosion seen at the surface is the sudden development of a great cloud of fine spray which is projected to a height of ten or fifteen feet. This is followed immediately by the rising mound of water and the great fountain, lifted by the expanding gases. The spray is due to the shock of a "pulse"-wave of almost instantaneous pressure. The spray is never seen following the explosion of a mine, which always contains a large volume of air. This acts as a cushion, and prevents the development of the instantaneous pressure.

A fog of extremely small droplets of heavy transformer oil can be formed by a collector of special construction. This is perhaps the most spectacular experiment of all. A glass tube of about 2·5 cm. diameter is closed at one end, and drawn down to a diameter of about 7 mm. at the other end. The tube is clamped to the rack and pinion stand, and the rounded bottom lowered into the oil, adjustment being made for energy density. This form of collector for compressing the radiation into small volume (if we may use this expression) is the most efficient thus far found. The constricted portion of the tube heats rapidly by internal friction, and if touched with the finger becomes unbearably hot. To distinguish between heating by internal friction and the heat developed when another body is pressed against the vibrating glass it is only necessary to operate the tube for a few seconds, shut off the power and touch the thin constricted portion with the finger. If now, with the tube adjusted for maximum heating and the oscillator working at full power, we apply a little oil with a medicine dropper to the outside of the tube above the narrowed portion, a very surprising thing happens. The oil spreads over the surface and is thrown out in jets of spray resembling smoke and a dense cloud gathers about the tube. If a match flame is brought gradually up to this cloud brilliant flashes of countless small scintillations occur resembling the sparks of the Japanese fireworks, and if the flame is brought closer, the whole cloud goes up in a grand burst of flame and the top of the tube continues to burn fiercely like a torch until the oil supply is exhausted. For some reason not quite clear the oil is unable to run down the thin portion of the tube, gathering in a ring of greater or less width at the top. If the power on the oscillator is reduced or the tube thrown out of adjustment by an up or down movement of a fraction of a millimetre the ring crawls down the tube, rising again as soon as the intensity of the vibration is increased. This driving of the oil layer up the tube is doubtless due to the fact that the energy of the radiation travelling up the tube is greater than the energy reflected down. It is obvious that the walls of the tube are vibrating as stationary waves, since close inspection shows that the oil in the ring has gathered in more or less regularly arranged dots, and that the jets of spray shoot out from these dots (fig. 4).

Fig. 4.

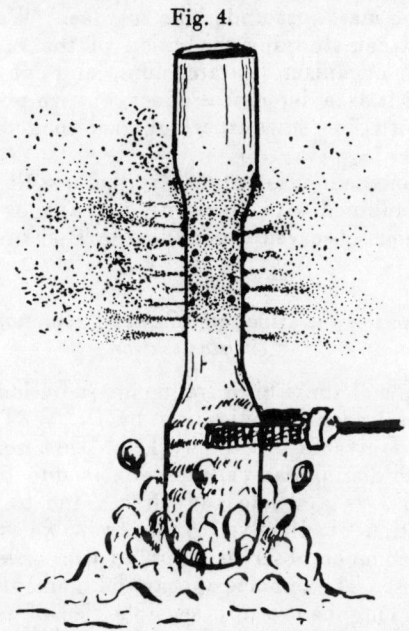

A photograph of the tube in action is reproduced on Plate XIII., the clouds of spray and the jets being illuminated by sunlight and photographed under the same conditions as with the benzol fog in the beaker. Twenty or more of the small jets of spray are visible in the original photograph shooting out from the tube just below the cloud, and three vertical rows of the oil dots referred to above can be seen without difficulty between the jets.

A fog of metallic mercury was also formed with this type of collector.

With a collector of this type the amplitude of the vibration at the constricted portion frequently becomes so great that the tube is fractured in a curious manner, small irregular pieces of glass breaking away from the tube. A photograph of a tube fractured in this way is reproduced on Plate XI. fig. 8.

Flocculation of Suspended Particles in a Liquid.

In the case of particles exceeding a certain size and of a specific gravity not much greater than that of the liquid in which they are suspended, flocculation occurs the moment the liquid is traversed by the waves, the particles rushing together to form clusters which presently gather into a single dense mass just under the surface. We first noticed this effect when studying the action of the radiation on the unicellular organism paramecium, and at the moment interpreted it as a biological effect, but we presently duplicated it with fine sawdust which had soaked until waterlogged.

The phenomenon is probably the result of radiation pressure combined with shielding perhaps, or analogous to the attractions observed and studied by Bjerknes in pulsating liquids.

Effects of the waves on Chemical Reactions and Crystallization.

The effects of these high frequency radiations on Chemical reactions is under investigation by Dr. W. T. Richards of Princeton University. In work of this sort it is very necessary to distinguish between effects due to the heating of the liquid as a whole and effects due to the vibration. The liberation of dissolved gases from water has been a matter of common observation by all who have worked with super-sonics. The bubbles appear the moment the vibration is started, long before any sensible rise of temperature is observed, and instead of rising with a uniform velocity they remain suspended in the nodal planes moving up in an irregular manner, with frequent pauses. Distinct evidence has been found that certain chemical reactions are accelerated by the vibrations, the most striking case being the so-called "clock-reaction" in which the termination of the reaction is marked by the sudden change from a clear transparent solution to a deep blue one. In attempting to repeat the experiment, previously made with stearic acid, with paraffin wax, we obtained what appeared to be a crystallization of the paraffin induced by the vibration. The melted paraffin was allowed to solidify on the surface of hot water in a beaker. When the whole was quite cold, the bottom of the beaker was dipped into the vibrating oil. Small opaque white spots immediately appeared in the layer of translucent paraffin, which increased in size, forming irregular clusters. A photograph of the sheet of paraffin (natural size) by transmitted light is reproduced on Plate XI. fig. 9. Under a Zeiss binocular stereoscopic microscope the white spots appeared to be nodules covered with protruding points which suggested a crystalline structure. They have not yet been carefully examined however. Sir W. Bragg, on seeing the photograph, remarked that paraffin crystallized in two modifications, one at the solidifying point, and another at a temperature a few degrees lower.

Dr. Richards failed to get conclusive results on the crystallization of super-saturated solutions of sodium hyposulphate by subjecting the entire amount to the vibrations, but we made one interesting experiment in which the super-sonic waves were carried to the surface of the solution by a bent glass thread. Crystallization immediately started around the tip of the thread, and also around a minute crystal which we dropped on the surface at a distance from the thread, the *type of crystallization* and *rate of growth* being different in the two cases. At the request of the Bureau of Soils, Dept. of Agriculture, we made some experiments on the dispersion of colloids from soils. In soil analysis it is often a long and tedious process to disperse the colloids adsorbed on the soil grains. It was found that the colloids of a "very difficult" soil sample sent for examination were completely dispersed into the water by a few minutes treatment to the super-sonic vibration, while by the usual methods the process of shaking violently and centrifuging has to be repeated twenty or thirty times before the colloid is completely dispersed.

Biological Effects.

Though the effects of these waves upon living matter might more properly be discussed elsewhere, it may not be out of place to mention briefly a few of the observations which we have made as they have some bearing on the physical processes involved.

In marked contrast to the flocculation, or driving together of small particles of suspended matter, which has been mentioned, we have fragmentation, or the tearing to pieces of small and fragile bodies. Filaments of living spirogyra were torn to pieces and the cells ruptured. Small unicellular organisms such as paramecium were rendered immobile by a short treatment to vibration of moderate intensity, subsequently recovering, but were killed by a longer exposure, many of them being torn open. The circumstance that all are not treated alike is doubtless due to the fact that those which manage to keep out of the nodes of the stationary wave system are less roughly handled by the vibrations. Bacteria apparently are able to survive owing to their small size, for the fragmentation of larger bodies is due to the fact that the forces applied to their surfaces vary in magnitude and direction at different points of the body, while in the case of a bacterium the whole body is subjected to the same treatment.

Red blood corpuscles in physiological salt solution are rapidly destroyed, the turbid liquid becoming as clear as a solution of a red aniline dye.

With vibrations of less intensity the destruction is less complete, a blood count made at the end of each 15 seconds of exposure showing that the percentage destroyed decreases, a point being reached at which no further destruction occurs unless the intensity of the radiation is augmented. This means of course that some of the corpuscles, the recently formed ones perhaps, are more hardy than those of greater age. Small fish and frogs are killed by an exposure of one or two minutes, an observation also made by Langevin at Toulon with his Poulsen arc oscillator (see Plate VIII.) Mice are less sensitive, a twenty-minute exposure not resulting in death, and though at the end of the treatment the animal was barely able to move, the recovery was fairly rapid. Blood counts made with a mouse during exposure showed a diminishing number of corpuscles, until a stationary state (about 60 per cent. normal) was reached. The biologists inform us, however, that the blood count of a mouse is affected by fear, the corpuscles hiding in the liver until the danger is over! We made the count with drops taken from the tip of the tail.

We have not yet determined the cause of death in the case of the fishes and frogs. They were protected against rise of temperature as much as possible by ice fragments, dropped into the water from time to time, but this does not shield them from internal heating, which may be the cause of death, as in the case of small animals introduced into a high-frequency electric field. In the case of a mouse killed by an exposure of two minutes between the plates of an air-condenser operated at about 1000 volts with a frequency of 100 million, we found that the temperature of the body-cavity was over 113° F.

With distilled water or a fairly strong solution of salt in a test-tube between the plates of the condenser, little or no heating occurred; but for small concentrations the heating was very marked, the maximum being for ·8 per cent., which is very nearly the amount found in mammalian blood. At lower frequencies the heating appears to be greater for distilled water, at least with high voltages. We found that one terminal of our 60,000 volt coil could be held in the fingers without the production of any sensation, but if dipped into the open end of a glass tube a metre long and filled with distilled water, caused the water to boil in less than 10 seconds. The introduction of a small amount of salt into the water, prevented the heat entirely in this case, which explains why no thermal discomfort was felt when the wire was held in the hand. The wire must be seized, however, before the current is turned on, otherwise a very vicious arc jumps to the finger producing a burn which is very slow in healing.

Fig. 1.

Two 1000 Watt tubes.   Oil condensers.   Variable air condenser and coils.

Fig. 2.

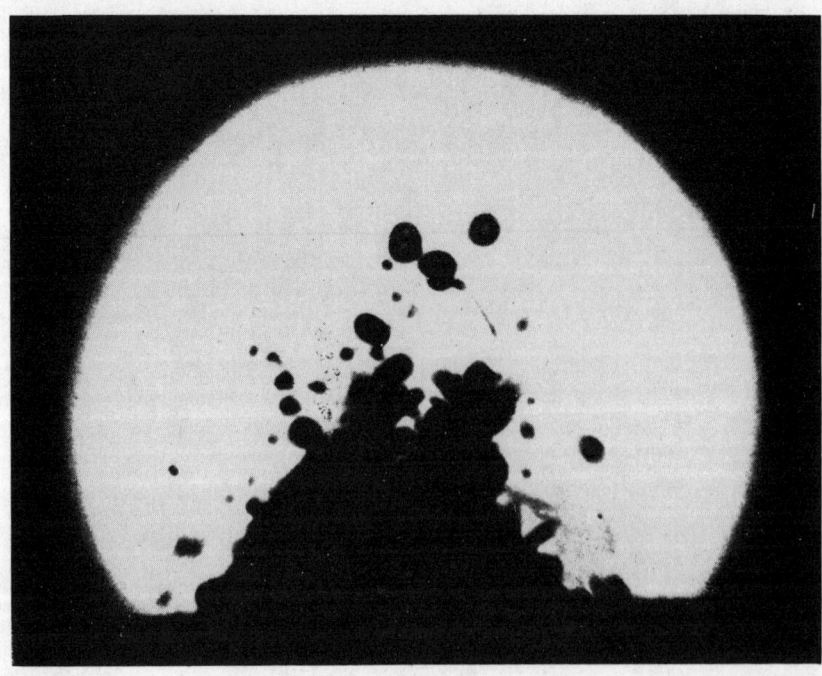

Fig. 3.

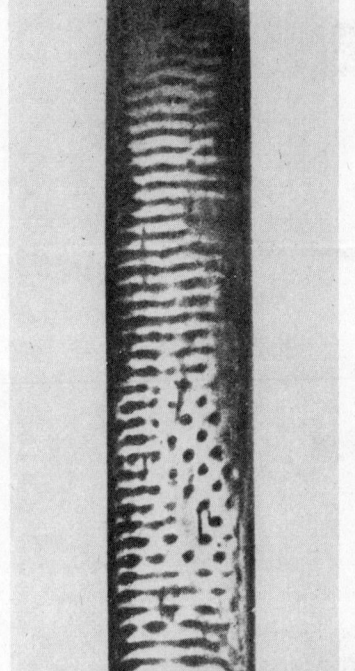

Fig. 4.

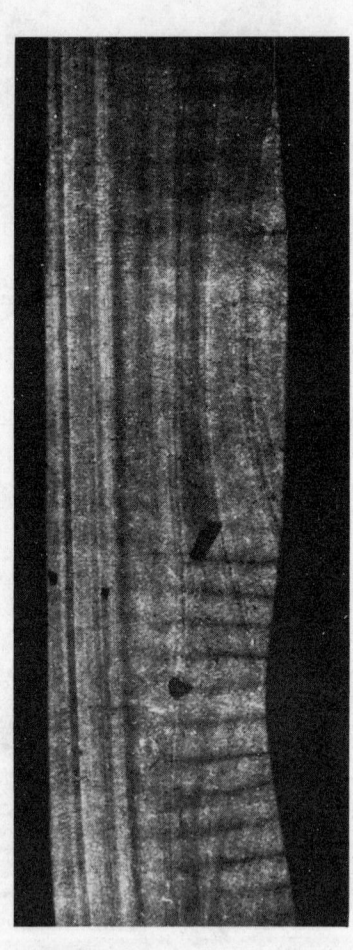

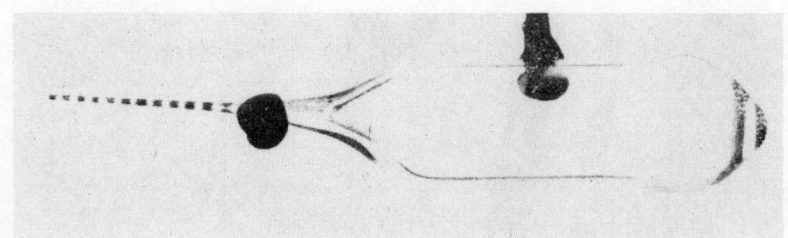

Fig. 6.

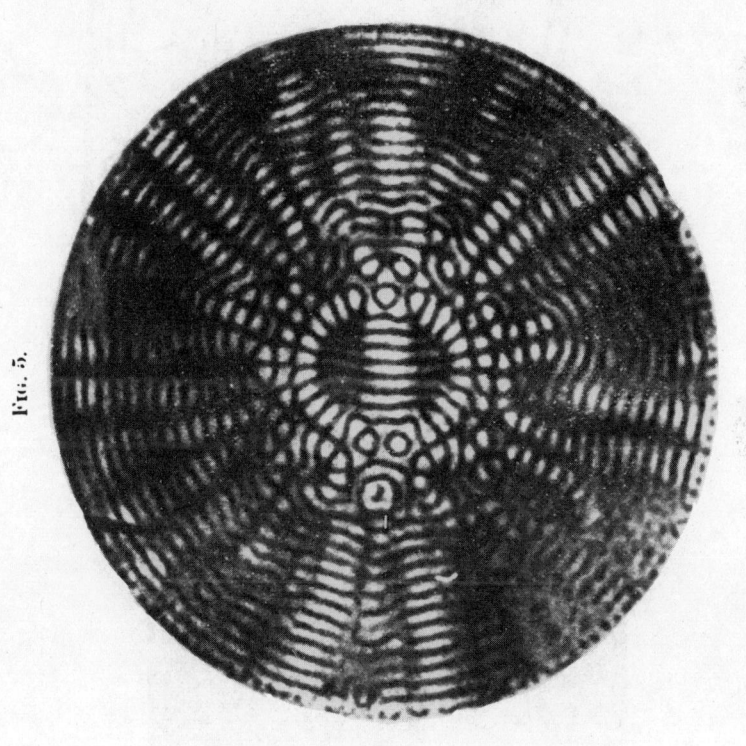

Fig. 5.

Fig. 7.

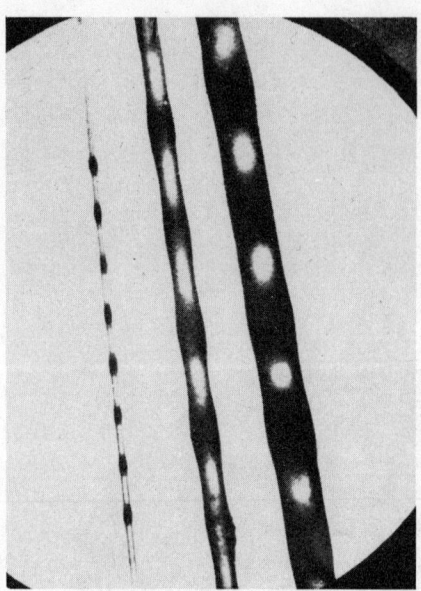

Fig. 8.

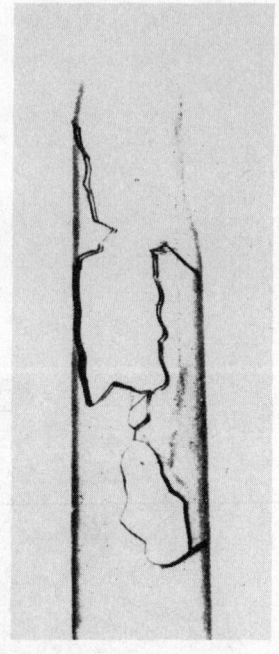

Fig. 9.

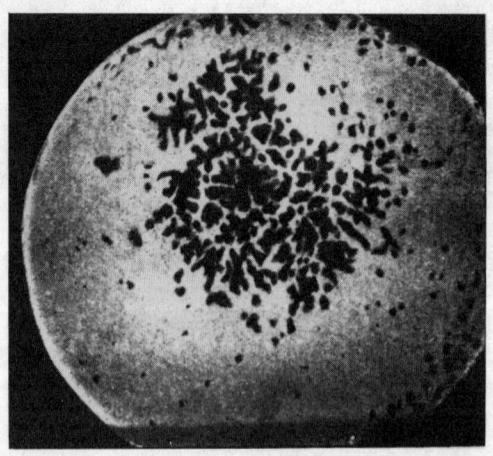

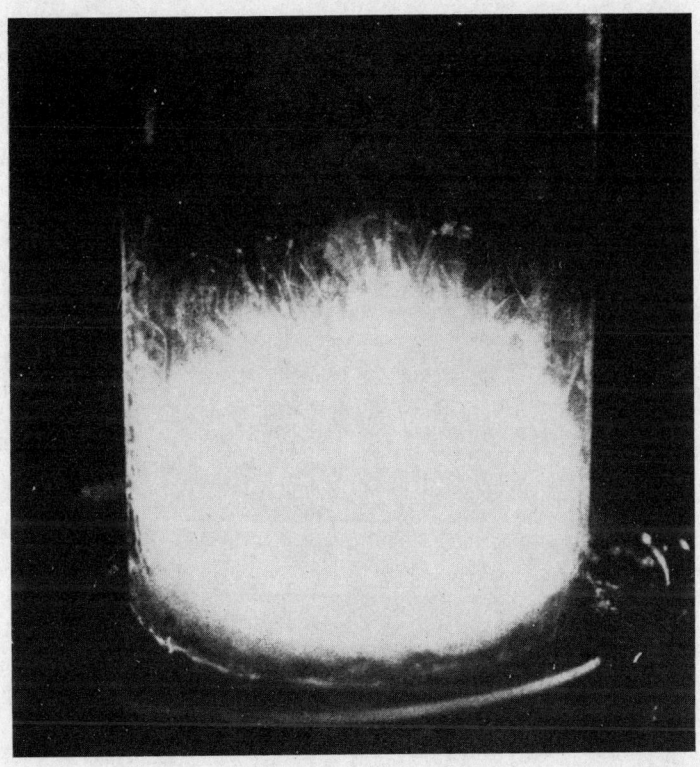

Fig. 10.

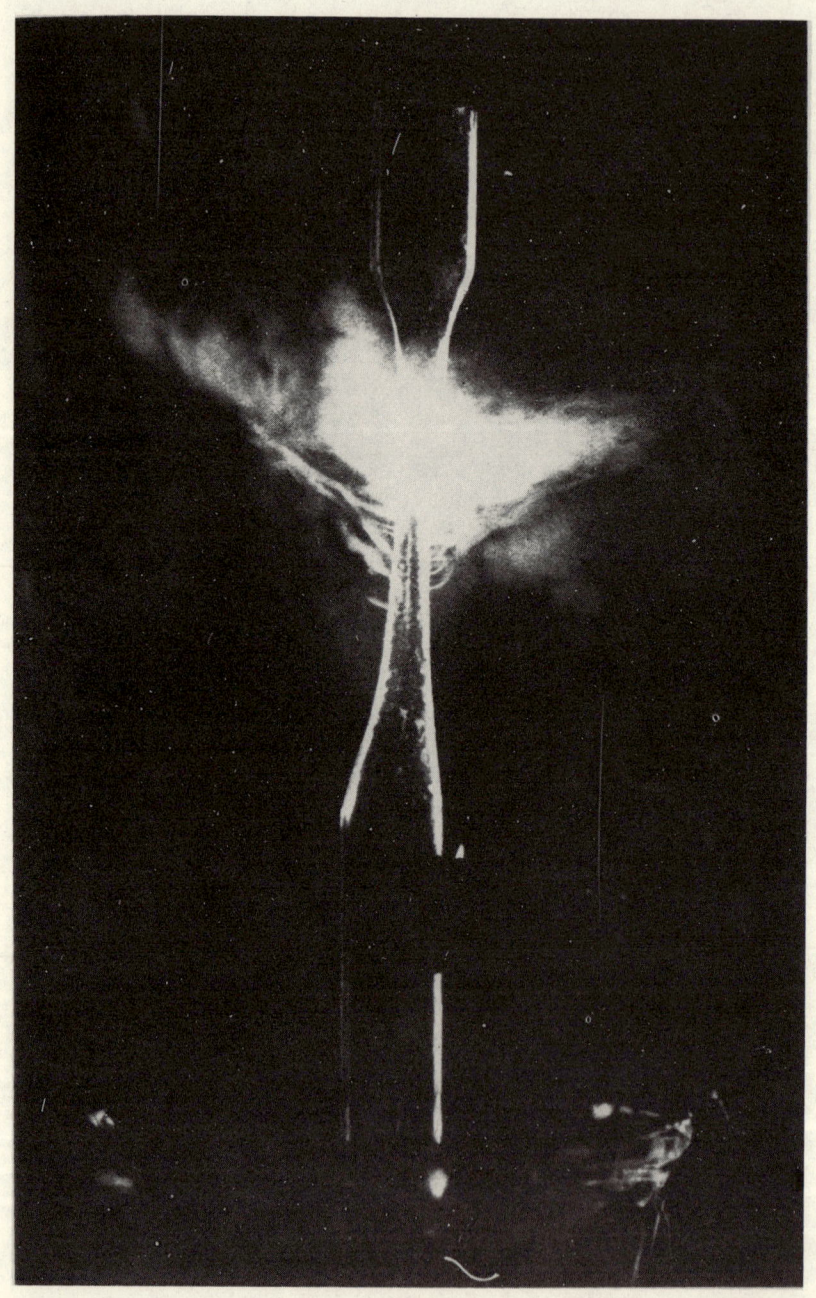

Editor's Comments on Paper 19

19 Einstein: *Sound Propagation in Partially Dissociated Gases*

In the Benchmark Series in Acoustics volume *Acoustics: Historical and Philosophical Development* (R. B. Lindsay, ed., Dowden, Hutchinson & Ross, Stroudsburg, Pa., 1973), we presented Lord Rayleigh's 1899 article "On the Cooling of Air by Radiation and Conduction and on the Propagation of Sound" (Paper 38). In it, he suggested that a more detailed study of the exchange of energy between translational motions of molecules in a gas and rotational and vibrational motions might lead to a better understanding of the phenomenon of attenuation of sound transmission in the gas. Rayleigh apparently never followed up this suggestion. J. H. Jeans, however, did so in his *The Dynamical Theory of Gases* (Cambridge University Press, Cambridge, England, 1904), applying the idea to a gas composed of loaded spheres. His results indicated a contribution to the sound attenuation in the gas much smaller than that due to viscosity. Although he employed the idea of a lag in the adjustment between the translational energy of the molecules and the "internal" energy states, he did not explicitly introduce a lag or relaxation time. He merely related the attenuation due to lag to that due to viscosity and concluded that since it was so small for loaded spheres as molecules, it would also be small in the general case of molecules composed of many atoms. Hence he missed the opportunity to found a genuine theory of molecular acoustics. Incidentally, Jeans made no reference to Rayleigh's suggestion in his book, so it is impossible to know to what extent his work was stimulated by Rayleigh's idea.

It appears that Einstein was the next to consider the problem of sound propagation as affected by molecular relaxation. As a matter of fact, his main interest in the problem was his desire to find a satisfactory way of estimating chemical reaction velocities in gases; he conceived that he could do this by relating these velocities to the dispersion and attenuation of sound in a gas. In the paper, presented here in English translation, he discusses the equilibrium between dissociated and nondissociated molecules of a diatomic gas and derives the expression for the velocity of sound in such a mixture. His resulting formula for velocity as a function of frequency shows how one can work backward from the observed dispersion to calculate the rate of transfer between the two kinds of molecules. Einstein made no attempt to carry out numerical calculations based on this theoretical result; moreover, he did not express sound attenuation as a function of frequency. Nevertheless, this paper set the stage for the important developments of the later 1920s and 1930s.

Extensive biographical notes about Einstein would be superfluous. One of the greatest of twentieth-century physicists, Einstein turned his attention to a host of physical problems, in addition to the creation of the theory of relativity for which he is most famous. This paper on sound propagation is a good indication of the breadth of his interest in physical problems, which also encompassed heat, electricity, electromagnetic radiation, and atomic properties.

19

Sound Propagation in Partially Dissociated Gases

A. EINSTEIN

Translated expressly for this Benchmark volume by R. Bruce Lindsay from "Schallausbreitung in teilweise dissozierten Gasen," Sitzber. Preussischem Akad. Wiss., Berlin, 24, 380–385 (1920)

While our knowledge of the chemical equilibrium of gases is of very long standing, we possess insufficient knowledge concerning the reaction velocity of gas reactions. There exists a particularly great difficulty in the experimental determination of reaction velocities. This is due to the fact that these velocities are influenced catalytically through fixed walls. Moreover, the high temperature with which most gas reactions are associated produces difficulties, particularly in view of the large reaction velocities to be expected therefrom. It now seems to me that these difficulties can be obviated by estimating the reaction velocities indirectly through investigations of sound propagation in partially dissociated gases.

We can understand from the following considerations how it is that such investigations can serve for the determination of the reaction velocity. If we change the volume of a partially dissociated gas adiabatically so fast that during the time of volume change practically no chemical transformation takes place, the gas behaves under these circumstances like an ordinary mixture. If, on the other hand, the volume is changed so slowly that the process consists practically entirely of chemical equilibrium states, the dependence of the pressure on the density will be a different one in such a way that the compressibility of the mixture is less than in the first case. The velocity of sound must therefore increase from an initial value to a limiting value as the frequency increases. For frequencies which lie between the two extremes, the reaction will lag behind the increase in density in such a fashion that there takes place a kind of temporal lag of the pressure curve with respect to the density curve with the concomitant transformation of mechanical energy into heat. In what follows we shall provide a theoretical investigation only of sound propagation in a partially dissociated gas in which a reaction of the most simple conceivable type is in question,[1]

$$J_2 \rightleftharpoons J + J.$$

---

[1] Experimental research of the kind being considered here was carried out in 1910 in Nernst's laboratory on N_2O_4 (see the Berlin dissertation of F. Kenkel, 1910). This already emphasizes the dependence of sound velocity on reaction velocity.

Let us first consider the purely mechanical part of the problem. The Eulerian differential equation of motion for a plane wave with the usual approximations takes the form

$$-\frac{\partial \pi}{\partial x} = \rho \frac{\partial^2 u}{\partial t^2} \tag{1}$$

Here π is the very small deviation of the pressure from its equilibrium value p and ρ is the equilibrium density. Moreover, u is the elongation of a particle of air in the direction of the x-axis or what we have called the wave normal. The excess pressure π has a definite relation to the *condensation* Δ. The latter is connected with the elongation u by the relation

$$\Delta = -\rho \frac{\partial u}{\partial x} \tag{2}$$

We seek to find the propagation law for a damped harmonic plane sine wave, for which we assume

$$\pi = \pi_0 \cos\left[\omega\left(t - \frac{x}{V}\right) + \Phi\right] e^{-\beta x}$$
$$\Delta = \Delta_0 \cos\left[\omega\left(t - \frac{x}{V}\right)\right] e^{-\beta x} \tag{3}$$

in which π_0, Δ_0, ω, V, Φ, and β are real constants. The phase difference Φ corresponds to energy dissipation.

In place of the real expressions (3) we shall use the complex expressions

$$\pi = \pi_0 e^{i(\omega t - ax + \Phi)}$$
$$\Delta = \Delta_0 e^{i(\omega t - ax)} \tag{4}$$

where we employ the abbreviation

$$a = \frac{\omega}{V} - i\beta \tag{5}$$

There will be a corresponding expression for u. Since Eqs. (1) and (2) are linear equations with real coefficients, these equations are also satisfied by the real parts of π, Δ, and u. The simplification of the analysis involved in this complex notation taken out of optics consists not only in the fact that (4) is easier to differentiate than (3) but more particularly in the fact that following (4),

$$\frac{\pi}{\Delta} = \frac{\pi_0}{\Delta_0} e^{i\Phi} = \text{constant} \tag{6}$$

From (1) and (2) and (6) it follows that

$$\frac{\pi}{\Delta} \frac{\partial^2 u}{\partial x^2} = \frac{\partial^2 u}{\partial t^2} \tag{7}$$

This differs from the usual wave equation for linear sound waves only in the fact that on the left side, in place of the real constant

$$S^2 = \left(\frac{dp}{d\rho}\right)_{\text{adiab}}$$

269

we have the complex constant π/Δ.

The quantity π/Δ may be evaluated by an investigation of the cyclic adiabatic process. From π/Δ we can then get the phase velocity V and the dissipation constant β. With due regard to (4) and (5), Eq. (7) yields

$$a = \frac{\omega}{V} - i\beta = \omega\left(\frac{\pi}{\Delta}\right)^{-1/2} \tag{8}$$

If β^2 is small compared with ω^2/V^2 we get from this the more convenient approximation

$$V + \frac{i\beta V^2}{\omega} = \left(\frac{\pi}{\Delta}\right)^{1/2} \tag{8a}$$

We now embark on the calculation of π/Δ by considering cyclic adiabatic volume changes of a partially dissociated gas. Let v be the molar volume and ρ the density of the partially dissociated gas, which we subject to small, time-varying changes (Δv, $\Delta \rho$, Δp, etc.). Then we have

$$v\rho = mn = \text{constant} \tag{9}$$

Here m is the atomic weight of the gas $\mathcal{J}$, while n is the total number of associated and non-associated atoms per mole of the gas. Then we can deduce from (9) the relation

$$\frac{\pi}{\Delta} = \frac{\Delta p}{\Delta \rho} = \frac{1}{\rho}\left[p - \frac{\Delta(pv)}{\Delta v}\right] \tag{10}$$

We can write the equation of state of the gas in the form

$$pv = RT(n_1 + n_2), \tag{11}$$

in which n_1 denotes the number of moles of $\mathcal{J}_2$, and n_2 denotes the number of moles of the dissociated $\mathcal{J}$ gas, so that we have

$$n = 2n_1 + n_2 \tag{12}$$

From (11) and (12) there follows

$$\Delta(pv) = R(n_1 + n_2)\,\Delta T + RT(\Delta n_1 + \Delta n_2)$$

or, taking cognizance of the constancy of n and with due regard to (12),

$$\Delta(pv) = R(n_1 + n_2)\,\Delta T - RT\,\Delta n \tag{13}$$

We still have to look for two relations which allow ΔT and Δn to be expressed in terms of Δv. Then because of (10), our calculation of π/Δ will be complete. Since the process is to be adiabatic, for every time element we must have

$$C\,dT - D\,dn_1 = -p\,dv$$

where C is the sum of the heat capacities of the dissociated and undissociated parts of the gas, and D denotes the heat of dissociation per mole at constant volume. We also have the equation

$$0 = C\,\Delta T - D\,\Delta n + p\,\Delta v \tag{14}$$

which holds to the degree of approximation appropriate to our problem. We have further to take into consideration the chemical transformation taking place in the time interval dt. For this we must make a hypothesis about the dynamics of the dissociation reaction, which then will be tested through the sound observations. The simplest formal assumption, although from the kinetic standpoint by no means the most plausible, is that the dissociation reaction is one of the first order, that is, that $\kappa_1 (n_1/v)$ molecules of J_2 per unit volume dissociate per unit time. This hypothesis assumes that the collisions of the molecules do not lead at once to dissociation. It would be possible, indeed, that molecules of definite internal energy would possess a definite dissociation probability, something like that of radioactive atoms. Or it might be possible that radiation would bring about molecular dissociation, a view that has recently been advanced with great force by J. Perrin. If dissociation is brought about by the collision of two molecules of J_2 or of a molecule J_2 with an atom J, in place of the above expression we should have to write

$$\kappa_1 \left(\frac{n_1}{v}\right)^2 \quad \text{or} \quad \kappa_1 \frac{n_1}{v}\frac{n_2}{v}$$

respectively, in which κ_1 is to be considered independent of the concentrations. We can take into account all these possibilities by sticking to the expression $\kappa_1 (n_1/v)$, but keep in mind the possibility that κ_1 *may* be dependent on the concentrations of both types of molecules. For the velocity of reassociation we have to choose $\kappa_2 (n_2/v)^2$. From this we get the following relation for the time element dt:

$$v\left[\kappa_1 \frac{n_1}{v} - \kappa_2 \left(\frac{n_2}{v}\right)^2\right] dt = -dn_1$$

or

$$\frac{\kappa_1}{\kappa_2} n_1 - \frac{n_2^2}{v} = -\frac{1}{\kappa_2}\frac{dn_1}{dt} \tag{15}$$

Here $\kappa_1/\kappa_2 = \kappa$ is the mass-action-law constant, for which the well-known relation

$$\frac{1}{\kappa}\frac{d\kappa}{dT} = \frac{D}{RT} \tag{16}$$

holds. In order to get some use out of Eq. (15), we apply it to a state which differs only infinitesimally from the equilibrium state. Taking account of (16) and (12), and treating the quantities κ_1, κ_2, κ, n_1, n_2, and v as all relating to the equilibrium state, we arrive at

$$0 = \frac{\kappa D n_1}{RT^2} \Delta T + \left(\kappa + \frac{4n_2}{v}\right) \Delta n_1 + \frac{1}{\kappa_2}\frac{d\Delta n_1}{dt} + \left(\frac{n_2}{v}\right)^2 \Delta v = 0$$

By assuming that the variables ΔT, Δn, and Δv vary cyclically and all contain the complex factor $e^{i\omega t}$, we can carry through the differentiation and write for the previous equation

$$0 = \frac{\kappa D n_1}{RT^2} \Delta T + \left(\kappa + \frac{4n_2}{v} + \frac{i\omega}{\kappa_2}\right) \Delta n_1 + \left(\frac{n_2}{v}\right)^2 \Delta v \tag{17}$$

Through the solution of the Eqs. (13), (14), and (17), ΔT, Δn, and Δv are given as functions of $\Delta(pv)$. We then obtain for the ratio $\Delta(pv)/\Delta v$ the result

$$-\frac{\Delta(pv)}{\Delta v} = \frac{p\left[\frac{\kappa D n_1}{T} + R(n_1 + n_2)\left(\kappa - \frac{4n_2}{v} + \frac{i\omega}{\kappa_2}\right)\right] + \left(\frac{n_2}{v}\right)^2 \left[RD(n_1 + n_2) - CRT\right]}{C\left(\kappa - \frac{4n_2}{v} + \frac{i\omega}{\kappa_2}\right) + \frac{\kappa D^2 n_1}{RT^2}} \quad (18)$$

Taking into account the equilibrium condition

$$\frac{\kappa_1 n_1}{v} = \kappa_2 \left(\frac{n_2}{v}\right)^2$$

we get from (18) and (10),

$$\frac{\pi}{\Delta} = \frac{p}{\rho}\left(1 + \frac{\kappa_1 A + iR\omega}{\kappa_1 B + ic\omega}\right) \quad (19)$$

where

$$c = \frac{C}{n_1 + n_2} = \frac{c_1 n_1 + c_2 n_2}{n_1 + n_2} \quad (20)$$

$$A = \left(\frac{2D}{T} - c\right)\frac{n_1}{n_1 + n_2} + R\left(1 - 4\frac{n_1}{n_2}\right) \quad (21)$$

$$B = \frac{D^2}{RT^2}\frac{n_1}{n_1 + n_2} + c\left(1 - 4\frac{n_1}{n_2}\right) \quad (22)$$

Our problem is completely solved by the use of (19) and (8). For in the first place we have

$$V_{\omega = \infty} = \sqrt{\frac{p}{\rho}\left(1 + \frac{R}{\bar{c}}\right)} \quad (23)$$

By the use of this formula, $\bar{c}$ can be determined experimentally. Moreover, through well-known dissociation formulas, A and B, can be evaluated. Further, from (19) it follows that

$$V_{\omega = 0} = \sqrt{\frac{p}{\rho}\left(1 + \frac{A}{B}\right)} \quad (24)$$

In the frequency range for which sound absorption is sufficiently small we get the approximation formula

$$V = \sqrt{\frac{p}{\rho}\left(1 + \frac{\kappa_1^2 AB + Rc\omega^2}{\kappa_1^2 B^2 + c^2\omega^2}\right)} \quad (25)$$

which includes (23) and (24) as special cases. It can be used for the determination of κ_1. Through investigations at different gas densities we can finally determine whether κ_1 depends on the density.

Editor's Comments on Papers 20 Through 26

20 Pierce: *Piezoelectric Crystal Oscillators Applied to the Precision Measurement of the Velocity of Sound in Air and CO_2 at High Frequencies*

21 Herzfeld and Rice: *Dispersion and Absorption of High Frequency Sound Waves*

22 Kneser: *The Dispersion Theory of Sound*

23 Henry: *The Energy Exchanges Between Molecules*

24 Knudsen: *The Effect of Humidity upon the Absorption of Sound in a Room, and a Determination of the Coefficients of Absorption of Sound in Air*

25 Knudsen: *The Absorption of Sound in Air, in Oxygen, and in Nitrogen—Effects of Humidity and Temperature*

26 Markham, Beyer, and Lindsay: *Absorption of Sound in Fluids*

Nineteenth- and early twentieth-century measurements of the velocity of sound in air and other gases failed to show any sign of dispersion, that is, any change of velocity with frequency. This was scarcely surprising, since the theoretical work of Stokes and Kirchhoff predicted that the effect would be a second-order one, probably observable only at very high frequencies. These were unavailable to the early experimentalists prior to the development of electroacoustic ultrasonic generators.

G. W. Pierce of Harvard University was stimulated to use the new piezoelectric ultrasonic generator in the measurement of sound velocity in air and carbon dioxide at frequencies of the order of 1 MHz. He detected dispersion well in excess of that predicted by Stokes and Kirchhoff on the basis of viscosity and heat conduction. His results are given in the paper "Piezoelectric Crystal Oscillators Applied to the Precision Measurement of the Velocity of Sound in Air and CO_2 at High Frequencies," the first 20 pages of which are reproduced here. It is interesting to note that Pierce made no attempt to explain in any detail what was, at that time, considered to be an anomalous effect. It does not appear that he was familiar with the relaxation idea suggested by Rayleigh and applied by Einstein to the dissociation of gases. Moreover, although Pierce noted the unusually large attenuation of the sound at the high frequencies he was using, he presents no figures for absorption as a function of frequency. However, his pioneer work was the forerunner of a great deal of research on sound dispersion.

George Washington Pierce (1872–1956), an American physicist and electrical engineer, was a professor at Harvard for his entire professional career. He became an authority on electrical communications and set the standard for precision measurements in ultrasonic propagation.

Measurements of sound attenuation in air made in the late nineteenth and early twentieth centuries consistently gave higher values than those predicted on the basis of viscosity and heat conduction. It will be recalled that Rayleigh, in 1899, had suggested as a possible additional mechanism the slowness of energy exchange between the translational mode of the molecules and their internal energy states of vibration and rotation [see Paper 38 in *Acoustics: Historical and Philosophical Development* (Dowden, Hutchinson & Ross, Stroudsburg, Pa., 1973)]. This idea was applied to the problem of sound attenuation in a gas for the first time by Herzfeld and Rice in the paper "Dispersion and Absorption of High Frequency Sound Waves." Relaxation time for the relevant energy exchange is introduced as a basic factor in the theoretical deduction, although it was not given this name. This paper proved to be the forerunner of a host of papers in the field now known as molecular acoustics. Unfortunately, sound attenuation measurements at high frequency had not at that time reached a stage of precision adequate to provide a definitive test of the theory. Thus the paper proved a great stimulus to experimental work in this field.

It is interesting that the authors make no mention of the earlier work of Rayleigh, Jeans, and Einstein in connection with the relaxation concept, although one of the authors (Rice) was evidently familiar with the basic concept of Einstein's work on the velocity of chemical reactions.

Karl Ferdinand Herzfeld (1892–) is an Austrian-born physicist who has spent most of his professional career since 1926 in the United States with professorial posts at Johns Hopkins University and the Catholic University of America. He is an authority on kinetic theory and thermodynamics.

Francis Owen Rice (1890–) is a British-born chemist who has followed a professional career in the United States since 1919 with professorships at Johns Hopkins University and the Catholic University of America. He has recently served as a research scientist at Notre Dame University.

Although by no means the first to apply the molecular theory of sound propagation to the problem of the variation of sound velocity with the frequency, the paper by Kneser proved very influential in encouraging further work in the field that became known as molecular acoustics. It is striking that while the author refers to earlier work by Lorentz, Herzfeld and Rice, Bourgin, and Einstein, he mentions neither the fundamental suggestion by Lord Rayleigh in his 1899 paper nor J. H. Jeans' abortive attempt at a simplified version of the problem. (See Editor's Comments on Paper 19 in this volume.)

The author employs a method that assigns external and internal specific heats to the external and internal energy states in question and then writes a reaction equation for the change in the number of excited molecules per unit time in terms of the total number of molecules per mole. This involves the quantity, later known as *relaxation time*, that measures the average time taken for the restoration of the original equilibrium distribution of energy between external and internal states. Kneser does not use this term in his paper.

The dispersion formula derived in this paper was tested experimentally by Kneser and the results reported in a subsequent paper (*Ann. Physik*, **11**, 777–801, 1931). Generally good agreement was found for sound dispersion in carbon dioxide.

It is also interesting to note that although Kneser realized that his dispersion analysis could also lead to a theoretical expression for sound attenuation in a gas as a function of frequency, he made no effort in this paper to express this explicitly nor did he endeavor in his next paper to measure attenuation and compare it with any theoretical result such as that obtained by Herzfeld and Rice in their earlier paper (Paper 21, this volume). Later work by Kneser and others, notably V. O. Knudsen (Paper 24, this volume), repaired this omission. The history of acoustics, like that of the history of science in general, rarely, if ever, runs as smoothly as one is tempted to reconstruct it after the fact!

Hans Otto Kneser (1901–) is a German physicist. Educated at Breslau and Munich, he spent most of his professional career as a professor of theoretical physics in the Technische Hochschule (now University) of Stuttgart, in which city he lives in retirement. He is particularly noted for his research on relaxation processes in gases.

The high-frequency dispersion and absorption of sound in gases was investigated theoretically by Herzfeld and Rice and Kneser and their results presented in the two preceding papers. They based their calculations on the relaxation involved in the exchange between energy states of the molecules in a gas as effected by the sound propagation, although they did not use the term "relaxation." P. S. H. Henry, then working at Cambridge University, was led to examine sound absorption in gases as a means of explaining anomalies in the specific heats of gases at high temperatures. He was familiar with the work of the investigators named, but he proceeded to calculate for himself the sound absorption by the introduction of a complex specific heat, which is a function of the relaxation time, a term that he introduces and specifically names. His method has been widely used in recent calculations in molecular acoustics. Unfortunately, he was not aware of Knudsen's 1931 paper on the absorption of sound by the air in a room. Henry's paper "The Energy Exchanges Between Molecules" is reproduced in full.

P. S. H. Henry (1906–) is a British physicist, now retired and living in Wales, most of whose professional career was devoted to research in textile physics.

It is interesting that the first precise measurements of the attenuation of sound in air and other gases came as a byproduct of a very practical problem in architectural acoustics, the measurement of reverberation time in a room. Earlier research (e.g., that of Sabine and his followers) had assumed that the sound absorption in a room (as related to reverberation time) was provided entirely by the material on the surfaces of the room. In experimental research conducted about 1930, V. O. Knudsen showed that under certain conditions the absorption of sound by the air in the room could also play an important role. In a series of papers, of which we reproduce the first two, "The Effect of Humidity upon the Absorption of Sound in a Room, and a Determination of the Coefficients of Absorption of Sound in Air" and "The Absorption of Sound in Air, in Oxygen, and in Nitrogen—Effects of Humidity and Temperature," Knudsen made precision measurements of sound absorption coefficients and demonstrated that the values are, in general, considerably in excess of those predicted by viscosity and heat conduction. He indicated his belief that his results could be theoretically interpreted only in terms of molecular absorption, as suggested earlier by Herzfeld and Rice, Kneser, and others. Knudsen's later cooperation with Kneser served to establish molecular acoustics both theoretically and experimentally.

Vern O. Knudsen (1893–) is an American physicist and former university administrator. His entire professional career has been spent at the University of California, Los Angeles, where he has served as professor of physics, Chairman of the Department of Physics, Dean of the Graduate School, Vice-Chancellor, and Chancellor prior to his retirement in 1960. He is an eminent authority on room and building acoustics and noise control.

Paper 26 constitutes a general review of the various modern theories advanced for the explanation of sound attenuation in liquids and gases. The emphasis is on relaxation methods and the different types of relaxation processes leading to sound absorption are reviewed with careful attention to the hypotheses introduced. Experimental data obtained up to 1951 are discussed in some detail. The article constitutes a rather complete picture of the status, experimental and theoretical, of sound attenuation in fluids at the time of its publi-

Editor's Comments on Papers 20 Through 26

cation. Not an original research paper, it is presented here because it has been frequently quoted in the subsequent literature.

J. J. Markham (1916–) is a professor of physics at the Illinois Institute of Technology. His principal field of research is solid-state physics with special reference to color centers in alkali halides.

R. T. Beyer (1920–) is a professor of physics and Chairman of the Department of Physics at Brown University. He is well known for his research investigations in acoustics, especially sound absorption in liquids.

20

PIEZOELECTRIC CRYSTAL OSCILLATORS APPLIED TO THE PRECISION MEASUREMENT OF THE VELOCITY OF SOUND IN AIR AND CO_2 AT HIGH FREQUENCIES.

By George W. Pierce.

Received July 18, 1924.   Presented October 14, 1925.

1. Reference to Previous Paper.— In a previous paper[1] I have described a method of producing sustained high-frequency electric and mechanical vibrations by a novel combination of a plate of piezoelectric crystal with a thermionic vacuum tube, and have shown how to employ the apparatus in the calibration of wavemeters. These vibrations are of extraordinary constancy as to frequency so that it has seemed desirable to apply the apparatus to other measurements.

The present account describes the precision measurement of the velocity of sound at high frequency.

2. Preparation of the Piezoelectric Crystal Plate.— Quartz was used as the piezoelectric substance. A plate was cut from the

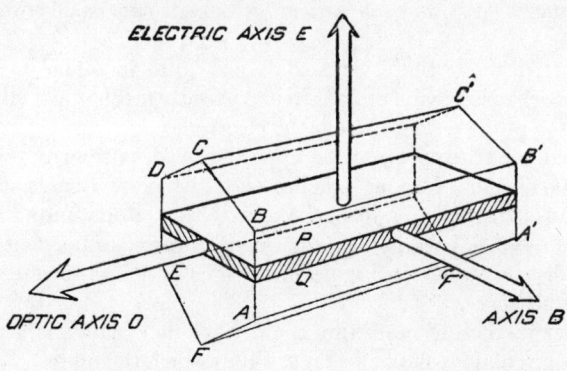

Figure 1. Orientation of axes in natural quartz crystal. Method of sectioning.

natural crystal of quartz with the orientation suggested by P. and J. Curie,[2] as is here shown in Figure 1. After the natural crystal had

---

[1] G. W. Pierce, Piezoelectric Crystal Resonators and Crystal Oscillators Applied to the Precision Calibration of Wavemeters, Proc. Am. Acad. of Arts and Sciences, **59**, No. 4 (1923). (Reprints may be purchased from the Librarian of the Academy, 28 Newbury Street, Boston, Massachusetts.)

[2] Pierre and Jacques Curie, Comptes Rendus, **91**, 383 (1880); also, Oeuvres de Pierre Curie, Paris, 1908.

been trued up by crosswise cuts so as to form a prism with the hexagonal ends ABCDEF and A'B'C'..F', which are respectively perpendicular to the natural edges AA', BB' of the crystal, a rectangular slab was obtained from the prism by two parallel cuts P and Q, which are perpendicular to a natural face such as ABB'A'.

3. Mounting of Plate Between Electrodes.— The rectangular slab so obtained has three axes represented in the diagrams by arrows: the *optic axis* O (parallel to the lengthwise natural edges of the crystal), the *electric axis* E (parallel to two opposite natural faces of the crystal), and the *third axis* B (perpendicular to the optic axis and the electric axis). This slab is placed between two metal electrodes M' and M'', as shown in Figure 2. The crystal may rest on M'', while M' may best

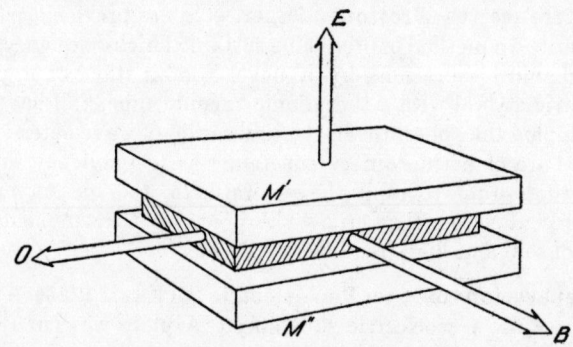

FIGURE 2. Electrodes M' and M'' with plate of crystal between.

be supported above the crystal so as not quite to touch it. This is done in order to leave the crystal free to execute mechanical vibrations, without too much restraint from the electrodes.

The manner of supporting the upper plate M' is shown in Figure 3. M' is attached to an upper bakelite plate, which in turn is supported on columns attached to a lower bakelite base. Bolts, nuts and spiral springs, as shown, permit the adjustment of the clearance between M' and the crystal slab. In Figure 3, the axis B points toward or away from the observer.

Another method of mounting the crystal slab (which is here represented as circular) is shown in Figure 4, in which the plates M' and M'' are in vertical planes, and the plate M'' is perforated with a hole about 1 cm. in diameter, so as to permit the radiation of sound through this hole. In this figure the plate M'' is in the form of a spring clamp and may rest upon the crystal with sufficient pressure to hold the crystal in place, or if desired, the pressure may be removed by the screw S and the crystal supported independently by a small shelf below.

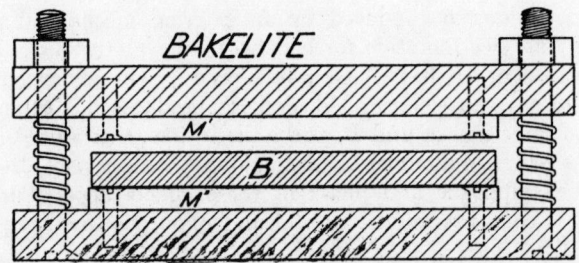

FIGURE 3. Mounting of plate of crystal in clamp with adjustable electrode distances. This arrangement for emitting and receiving sound of frequency determined by dimension in direction of axis B.

In Figure 4, the electric axis E points through the perforate electrode M'', while the axes B and O are in the plane of the crystal slab with orientation that is immaterial for the present purposes.

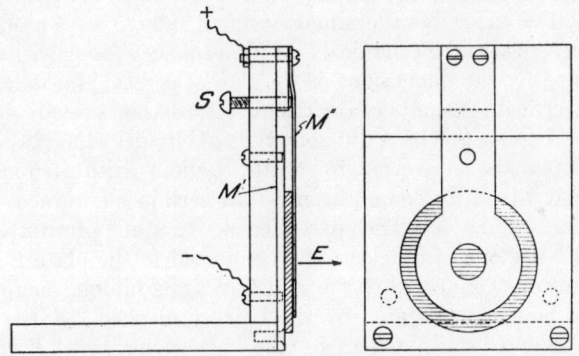

FIGURE 4. Side view and front view of crystal plate mounted in a vertical plane having a perforate electrode for emitting and receiving sound of frequency determined by thickness of plate in direction of electric axis E.

4. The Piezoelectric Action.— As to the action of the piezoelectric crystal, attention is called to the discovery of Curie that a difference of electrical potential of proper sign established between electrodes such as the plate M' and M'' (that is, an electric force of proper sign established in the direction of the electric axis E of Fig. 2) causes the piezoelectric quartz crystal to expand along the axis E and contract along the axis B. No change occurs along the optic axis O. A reversal of the electric force causes a reversal of these expansions and contractions. Curie also found that an expansion along E or contraction along B produced by an external mechanical pressure developed an electromotive force between the plates (i.e., along E) opposite in sign to that which would produce the given expansion. That is to say, an electric field along E produces distortion of the crystal in directions E and B, and a distortion so produced in these directions reacts to diminish the field along E. In case of alternating forces the amplitude and phase of these effects depend upon the mechanical constants of the crystal and upon the frequency of the applied e.m.f.

Langevin[3] with great ingenuity showed how to use such a piezoelectric crystal body as a source of sound particularly in water, and Cady[4] made a thorough and beautiful investigation of the crystal oscillators and crystal resonators, and adapted them to use as constants of electrical frequency.

5. The Electric Circuits for Producing Sustained Vibrations. — In my work above cited I showed a simple form of connections of the crystal vibrator to a thermionic vacuum tube so as to produce sustained electrical and mechanical vibrations of the system with a period determined by the dimensions of the slab of crystal and independent of the electrical constants of the circuit. Cady had already described a means of doing this by a different type of circuit, which is described in his paper above referred to in the Radio Institute Proceedings. My circuit which for some purposes has certain advantages is illustrated in Figure 5. The piezoelectric crystal vibrator marked "crystal" has one of its electrodes connected to the plate P and the other electrode connected to the grid G of a thermionic vacuum tube, having a filament F heated by the battery marked "A Bat." The plate is supplied with current by the battery marked "B Bat." A microammeter, A, and a telephone "Tel" shunted by a bypass condenser C are included in the plate circuit. The element marked "*load*" in the figure was described in my original paper as a resistance of about 30,000 ohms or a large inductance, say 20 milhenries. This

---

[3] Langevin, Brit. Pat. Specifications, N.S., **457**, No. 145,691 (1920).
[4] W. G. Cady, The Piezoelectric Resonator, Proc. Inst. Radio Engineers, **10**, 83 (1922).

system produces sustained electric oscillations in the circuit and mechanical oscillations of the crystal vibrator with a period of one mode of natural mechanical vibration of the crystal body. This period in my original investigation was the mechanical period of compression and recovery of the crystal in its shortest dimension (which was along the electric axis). Such a system radiates sound in the direction of the electric axis E.

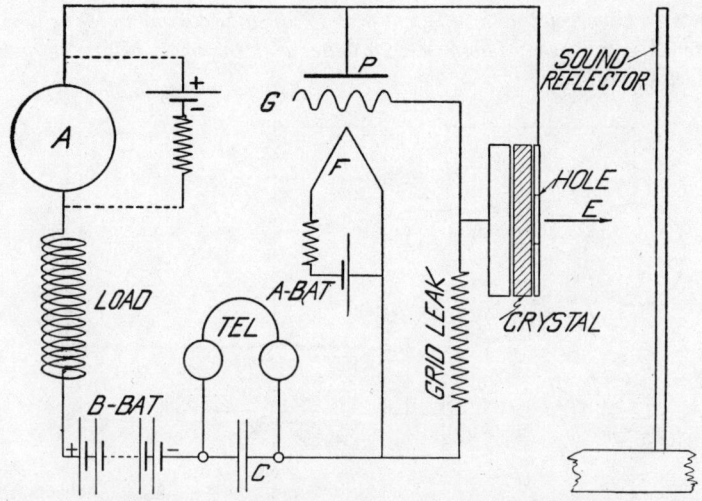

FIGURE 5. The author's type of electric circuit best suited for crystal frequency determined by dimension E.

It is also possible to cause the system to oscillate with the period determined by a larger dimension of the vibrator (that is in the direction of the axis B).

To attain this vibration of longer period I find that the best circuit is that shown in Figure 6 which is similar to Figure 5 except that the crystal vibrator is connected between the grid and some point below the inductance, as for example, the positive end of the B battery. The telephones, or their equivalent, acting as a choke, and the bypass condenser shunting them, are, in this case, usually necessary to give the proper reaction to make the system oscillate. This arrangement causes the crystal to vibrate with a period determined by the mechanical frequency of the crystal slab, expanding and contracting along the direction of the B axis, so that sound is radiated in this direction.

Other types of piezoelectric oscillating circuits will be described elsewhere.

6. Exploration of Sound Waves.— In Figures 5 and 6 is also shown a sound reflector which may be moved toward or away from the crystal so as to explore the standing sound waves. *No additional apparatus for detecting the sound is necessary, for the reflected sound wave falling on the emitting face of the crystal vibrator, even when the reflector in some cases is at a distance of 300 half waves of sound from the vibrator, reacts on the crystal with sufficient force to cause the current in the milliammeter A to fluctuate visibly in accordance with the phase of arrival of the reflected wave.*

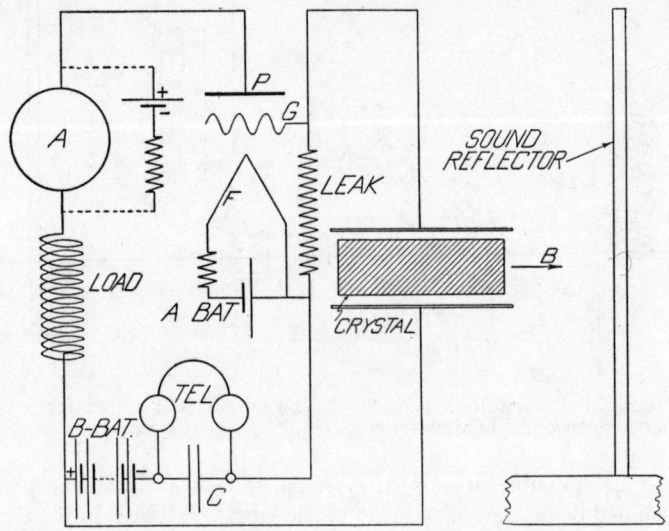

FIGURE 6. Type of electric circuit best suited for crystal frequency determined by dimension B.

When a small dry-battery vacuum tube (UV 199) is used the normal current in A when the crystal is oscillating is about 0.5 milamperes and the fluctuations range from 1.0 milampere (or circuit stops oscillating) when the mirror is close to the crystal down to zero when the mirror is at infinite distance from the crystal. To render the fluctuations more evident for precision measurements, a Weston microammeter, with one division equal to 4 microamperes, is used at A and is shunted by a potentiometer and battery combination to make the normal current through A small, say 10 divisions (40 microamperes). The fluctuations in the current in A, as the reflector is moved, may then amount to the whole additional scale, 90 divisions, when the reflector is 20 half wavelengths of sound away, and to about 25 divisions when the reflector is 100 half wavelengths away. These magnitudes depend on the frequency and area of radiating face of the crystal. The microammeter gives a sharp maximum at each half wavelength of displacement of the reflector.

It is seen that the apparatus oscillates of itself with a highly constant fixed frequency determined by a mode of mechanical vibration of the crystal plate. It thus produces sound waves and at the same time, by the strength of the plate current, determines the relative phase of the direct and reflected sound waves in air, so that the distance be-

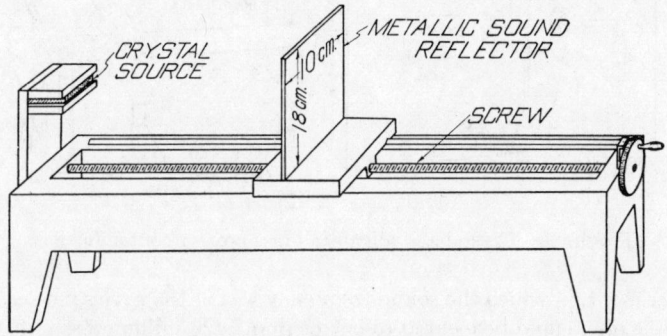

FIGURE 7. Mounting of crystal and mirror with screw drive.

tween successive loops of the standing sound wave for a range of 100 or more half wavelengths may be measured with great precision, by merely changing the distance between the reflector and the radiating face of the crystal, and making proper readings. These distances were changed by a calibrated precision screw, illustrated in Figure 7, by which the mirror could be moved about 50 centimeters and its position read with an accuracy of one one-thousandth of a millimeter, but such accuracy was not ordinarily required or attainable in the location of maxima. A modification of the mounting is shown in Figure 8, in which the mirror and crystal are contained in a gas-tight box.

7. Example of Microammeter Readings Plotted Against Crystal-to-Mirror Distance.— To show the nature of the observations of the standing waves reference is made to Figure 9, which is a plot of divisions of the microammeter in the plate circuit of the oscillator as ordinates, against the scale reading in millimeters of crystal-to-mirror distance. Three half waves are plotted so that an idea may be had of the accuracy of location of the maxima. In this particular

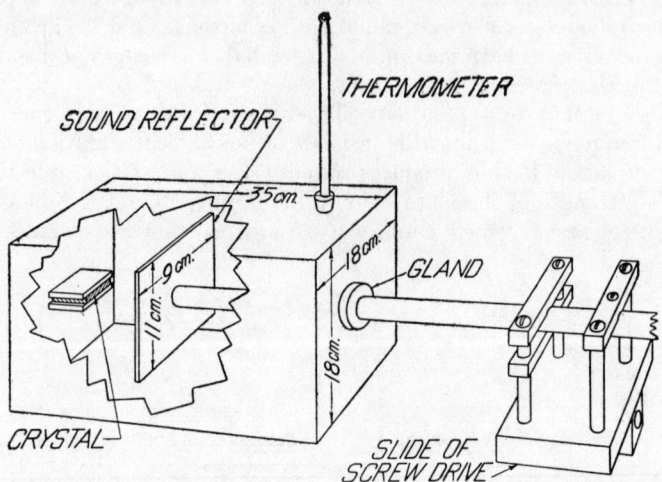

FIGURE 8. Crystal and mirror in brass box for containing gas.

experiment, in which the sound frequency was 98183 cycles per second, the maxima could be located to better than 1/20 millimeter, and since the train of standing waves in this case was explored for 140 millimeters, this degree of precision of locating the maxima gives the

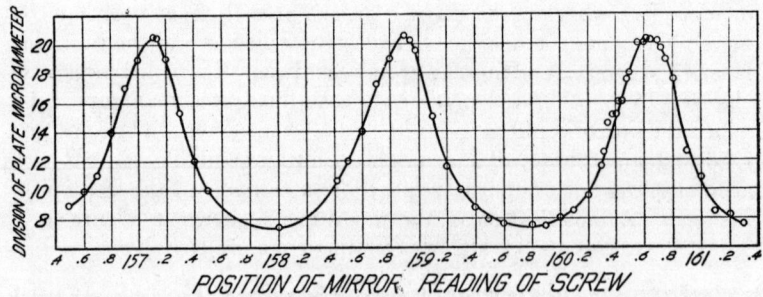

FIGURE 9. Plot of readings of microammeter in plate circuit of piezo-electric oscillator against readings of position of reflecting mirror.

VELOCITY OF SOUND.

wavelength to about 1/30 of one per cent. In the case of some of the other frequencies a still greater degree of precision was attainable.

For the present purpose it was not necessary to make readings in sufficient number to plot the complete standing wave system. It was only necessary to determine the positions of the mirror for maxima of the microammeter, and to locate accurately such of these positions as were to be used in the calculations. Usually maxima at intervals of five, ten, or twenty half wavelengths were employed.

8. Method of Averaging Observations.
— Suppose that we have set the reflecting mirror at a series of positions of maxima schematically represented as *positions* 0, 1, 2, ... n, in Figure 10, and let the dis-

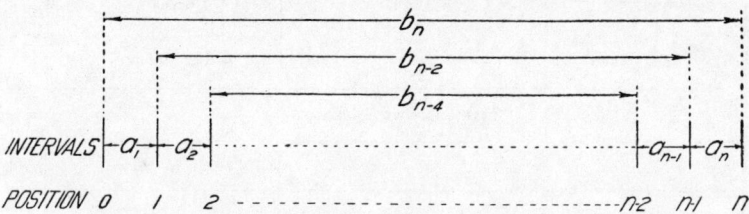

FIGURE 10. Diagram used in discussion of method of averaging.

tances, which we shall call *intervals*, between successive *adjacent positions* be $a_1, a_2, \ldots a_n$. These intervals are attempts to measure the same quantity x (say), and question arises as to the best method of averaging the intervals to get this quantity x. If we merely take

$$x = \frac{a_1 + a_2 + \ldots + a_n}{n}$$ we see that $x = \frac{\text{distance from 0 to } n}{n}$; that is

to say, this method of averaging, if employed, makes the result depend on the settings 0 and n alone and ignores all the intervening settings, which might as well not have been made. This method of averaging is evidently not correct.

Assuming that the positions 0, 1, ... n are all located with equal accuracy we may obtain the proper method of averaging by regarding the distance between any pair of positions, for example, the position 2 and position 8, as data for determining the fundamental interval, giving to this distance a weight proportional to its length ($8-2=6$) and averaging for all such pairs of positions, with due precaution to count every pair only once. Starting with position 0 we have distances 0 to 1 = a_1, 2(0 to 2) = $2(a_1+a_2)$, etc. Next starting with position 1, we have (1 to 2) = a_2, 2(1 to 3) = $2(a_2+a_3)$, etc. These are collected in Table I.

TABLE I.

Weighted Distances for Determining Fundamental Interval.

Starting with Position			
0	1	2	...
a_1	a_2	a_3	...
$2(a_1+a_2)$	$2(a_2+a_3)$	$2(a_3+a_4)$	...
$3(a_1+a_2+a_3)$	$3(a_2+a_3+a_4)$	$3(a_3+a_4+a_5)$	...
....	....	....	...
$n[a_1+a_2+\ldots+a_n]$	$(n-1)[a_2+a_3+\ldots+a_n]$	$(n-2)[a_3+a_4\ldots+a_n]$	...
Starting with Position			
....	$n-3$	$n-2$	$n-1$
....	a_{n-2}	a_{n-1}	a_n
....	$2(a_{n-2}+a_{n-1})$	$2(a_{n-1}+a_n)$	
....	$3(a_{n-2}+a_{n-1}+a_n)$		

The sum of all these quantities Σ, say, is

$$\Sigma = [1+2+3+\ldots+n][a_1+a_2+\ldots+a_n]$$
$$+ [2+3+\ldots+n-1][a_2+a_3+\ldots+a_{n-1}]$$
$$+ [3+4+\ldots+n-2][a_3+a_4+\ldots+a_{n-2}] + \text{etc.}$$

Now, as in Figure 10, putting

$$b_n = a_1+a_2+\ldots+a_n,$$
$$b_{n-2} = a_2+a_3+\ldots+a_{n-1}, \text{ etc.,}$$

and noting that

$$(1+2+3+\ldots+n) = \frac{(1+n)n}{2},$$
$$(2+3+\ldots+n-1) = \frac{(1+n)(n-2)}{2},$$

VELOCITY OF SOUND. 281

we obtain, after dividing out the common factor $\frac{1+n}{2}$,

$$\Sigma = n\, b_n + (n-2)\, b_{n-2} + (n-4)\, b_{n-4} + \ldots \ldots \quad (1)$$

This sum may be otherwise written in the form

$$\Sigma = n\, a_1 + 2\,(n-1)\, a_2 + 3\,(n-2)\, a_3 + \ldots . + n\, a_n \quad (2)$$

If now the fundamental interval is x, which $a_1, a_2, \ldots \ldots a_n$ are attempts to measure, the value of x may be obtained from (1) or (2) by dividing Σ by the sum of the coefficients of the a's, giving

$$x = \frac{n\, b_n + (n-2)\, b_{n-2} + (n-4)\, b_{n-4} + \ldots}{n + 2(n-1) + 3(n-2) + \ldots + n}, \quad (3)$$

or

$$x = \frac{n\, a_1 + 2\,(n-1)\, a_2 + 3\,(n-2)\, a_3 + \ldots \ldots + n\, a_n}{n + 2\,(n-1) + 3\,(n-2) + \ldots \ldots + n}. \quad (4)$$

Equation (3), *or the alternative equation* (4), *gives the weighted mean value of the quantity* x, *of which* $a_1, a_2, \ldots$ *are the observed values.* Equation (4) is the more convenient and is used below.

9. Temperature Reduction.— For convenience in reducing wavelengths and velocities measured at temperature $t°$ C. to the corresponding values at $0°$ C., Table II has been compiled.

TABLE II.
Temperature Reduction Table.

$v_0 = v_t \times \theta$, where $\theta = 1/\sqrt{1 + 0.00367t}$.

$t°$ C.	θ Temperature factor	0.1 × diff. Interpolation
15	0.97365	
16	190	−17
17	025	
18	.96857	
19	692	−16.5
20	527	
21	.96365	
22	198	−16.5
23	036	
24	.95875	
25	713	−16.1
26	553	

10. Sample Set of Observations on Wavelength of Sound in Air.

Table III contains a sample set of measurements of the wavelength of sound in air at 205620 cycles per second. With the mirror 3.5 cm. from the crystal a position of maximum deflection of the microammeter was noted. The mirror was then moved twenty half

TABLE III.

Run No. 144. Wavelength of Sound in Air.
Frequency 205620 Cycles per Second.
Steps 20 $\lambda/2$. Humidity 86%.

Temp. degrees C.	Step $20\lambda/2$ cm.	Weight	Weight × step
22.8	1.6812	17	28.580
	1.6805	2 × 16	53.776
	1.6832	3 × 15	75.744
	1.6830	4 × 14	94.248
	1.6767	5 × 13	108.986
22.92	1.6736	6 × 12	120.499
	1.6830	7 × 11	129.591
	1.6830	8 × 10	134.640
	1.6761	9 × 9	135.764
	1.6962	10 × 8	135.696
23.00	1.6672	11 × 7	128.374
	1.6861	12 × 6	121.399
	1.6814	13 × 5	109.298
	1.6730	14 × 4	93.688
22.98	1.6806	15 × 3	75.627
	1.6754	16 × 2	53.613
22.90	1.6829	17	28.609
22.92		969	Σ = 1628.23

$$\lambda = \frac{1628.23}{969 \times 10} = 0.16803 \text{ cm.}$$

Temperature factor = 0.96046

$\lambda_0 = 0.16138$ cm.

wavelengths farther away. This was found to be a displacement of 1.6812 cm., which is recorded as the first value in the table. The mirror was then displaced a second step of twenty half wavelengths, giving 1.6805, and so on for seventeen steps, encompassing a wavetrain of 340 half wavelengths. This table thus epitomizes the exploration of a stationary wave system of 340 maxima. Averaging results as in § 8 we obtain the wavelength as 0.16803 cm. at 22.92° C., which reduced to 0° C. gives $\lambda = 0.16138$ cm.

The data of Table III are for a single run. Five other similar runs were made at this frequency, and the complete set of values of wavelength at 0° C. are recorded as λ_0 in Table IV.

TABLE IV.

Collection of Results of Wavelength and Velocity of Sound in Air at 0° C.

Frequency 205620 cycles per second.

Humidity per cent	Run no.	λ_0 cm.	Residual
87	141	0.16146	0.00015
80	142	.16118	13
79	143A	.16115	14
86	143B	.16137	6
86	144	.16138	7
84	153	.16132	1
Mean 84		0.16131	0.00009

$$\lambda_0 = 0.16131 \pm 0.00003$$
$$v = 331.69 \pm 0.06 \text{ at } 0° \text{ C.}$$

In the third column is the value of λ_0 for each of the six runs, having an average of $\lambda_0 = 0.16131$ cm. In the fourth column are the residuals, which are the amounts by which the individual values differ from their mean. The probable error of the mean E_a computed by the approximate formula

$$E_a = 0.6745 \frac{\text{mean residual}}{\sqrt{n-1}},$$

where $n =$ number of separate values, is 0.00003 cm. Multiplication of λ_0 (reduced to meters) by the frequency gives for the velocity of sound at 0° C. at this frequency, 205620 cycles per second, the value

$$v = 331.69 \pm 0.06 \text{ meters per second.}$$

This result was obtained in free air without any enclosure about the gas. The mean humidity during the several runs was 84 per cent.

In a similar way other high frequencies were employed. For comparison a determination of the velocity of sound in air at the audio frequencies 995.88 and 2987.6 cycles per second was also made. For this purpose a novel method was employed, a description of which follows.

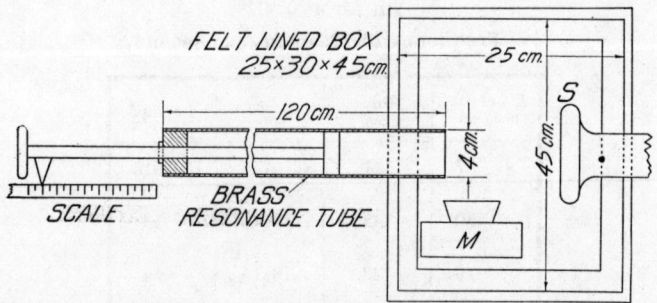

FIGURE 11. Apparatus for measuring velocity of low-frequency sound. $S =$ magnetophone source, $M =$ microphone detector.

11. Low-Frequency Sound Required a Special Method.— The crystal oscillator method with crystals at hand could not be employed at frequencies below about 40000 cycles per second, so that for low frequencies resort was had to a resonance-tube method. This method as ordinarily employed is inaccurate. A modification was introduced that resulted in a considerable improvement. This consisted in the use of a superimposed interference effect that gave a position of practical silence, or null point, in the middle of what would ordinarily be the sound maximum.

12. Null Method.— A sketch of the apparatus for this purpose is given in Figure 11. A brass resonance tube 120 cm. long and 4 cm. internal diameter, and provided with a piston and scale, was employed. The source of sound was a magneto telephone at S, which was driven by a tuning fork of known frequency. The source S, the mouth of the resonance tube, and a microphone receiver M were enclosed in a felt-

VELOCITY OF SOUND. 285

lined wooden box, and the distances between them were adjusted so that the sound direct from S to M just neutralized the sound from the tube to M when the sound emitted by the tube was a maximum. This gave a fiducial point of silence in the middle of what would otherwise be a maximum and permitted very accurate settings with sound of frequency 995.88 cycles per second. The microphone M communicated through a transformer with a head telephone receiver in which the observer listened. By means of an electric filter in this circuit, the fundamental frequency of the sound (995.88 cycles per second) could be eliminated and settings could then be made on the harmonic of $3 \times 995.88 = 2987.6$ cycles per second, with which the settings were still more accurate.

13. Sample Sets of Low-Frequency Sound Measurements.— Table V contains a sample set of observations at 995.88 cycles per second analyzed by the method of § 8.

TABLE V.

Run No. 67 in Air at 995.88 Cycles per Second.

Temp. degree C.	Setting at minima in cm.	Mean setting	$\lambda/2$ cm.	Wt.	$\lambda/2 \times$ wt.
21.7	20.32				
	.32				
	.33	20.323	17.274	4	69.096
	37.58				
	.60				
	.61	37.597	17.339	6	104.034
	54.86				
	.96				
	.96				
	.96	54.936	17.307	6	103.842
	72.26				
	.21				
	.26	72.243	17.190	4	68.760
	84.44				
	.47				
	.39	89.433			
				20	345.732

$$\lambda/2 = 345.732 \div 20 = 17.287 \text{ cm.}$$
$$\lambda = 34.573 \text{ cm.}$$
$$\text{Temperature factor} = 0.96250$$
$$\lambda_0 = 33.276 \text{ cm.}$$

Other values of this wavelength λ_0 are collected in Table VI, in which they are given weights proportional to the number of actual settings at each position of the piston.

TABLE VI.

Collection of Results for Frequency 995.88 Cycles per Second.

Humidity per cent	Run no.	λ_0 cm.	Weight	Residuals
36	66A	33.350	1	0.019
36	66B	33.266	1	.065
36	67	33.276	3	.044
35	69	33.320	3	.011
52	72	33.412	3	.081
52	73	33.328	6	.003
55	74	33.340	6	.009
Weighted mean		33.332		0.027

Whence $v = 331.94 \pm 0.07$ m/sec. at $0°$ C. Not corrected for effect of tube.

TABLE VII.

Collection of Results for Frequency 2987.6 Cycles per Second.

Humidity per cent	Run no.	λ_0 cm.	Weight	Residuals
36	66C	11.105	2	0.007
36	67B	11.110	1	.002
35	68A	11.115	6	.003
35	68B	11.111	3	.001
Weighted mean		11.112		0.003

$v = 331.98 \pm 0.03$ m/sec. at $0°$ C.

VELOCITY OF SOUND. 287

The weighted mean value of velocity of sound in air at 0° C. (Table VI) obtained at this frequency is 331.94 meters per second with a probable error of 0.07 meters per second, uncorrected for effects of the tube.

This apparatus proved to be especially accurate when applied to the measurement of the velocity of the harmonic frequency 2987.6, as is shown by the results in Table VII.

14. Dimensions of Crystal Vibrators.— Figure 12 gives a dimensional sketch of the different crystal vibrators employed. The

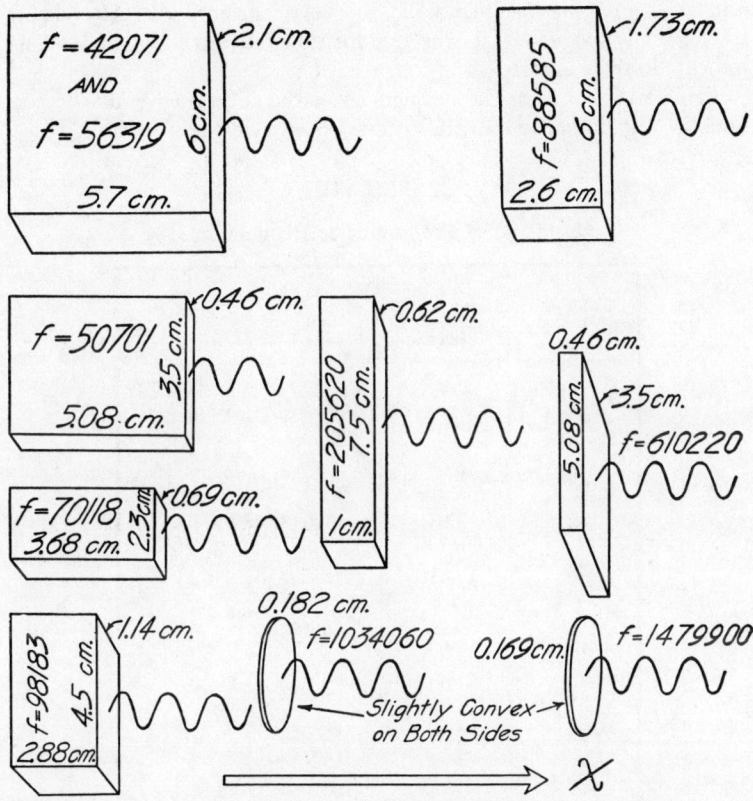

FIGURE 12. Dimensions of crystals for different frequencies.

wavy line emerging from the right-hand face of each crystal vibrator indicates the emitting surface and the direction of radiation of sound. In each specimen the optical axis is vertical in the sketch and the electric axis is perpendicular to the paper for all except those of frequencies 610220, 1034060, and 1479900, in which the electric axis is in the direction of sound radiation (to the right in the sketch). The frequency, designated by f, is marked on each vibrator. The upper left-hand specimen was used for two different frequencies, which could be differentiated one from the other, and each employed at will, by throwing in or out a suitable condenser in shunt to the inductance marked "*load*" of Figure 6. This same vibrator was originally somewhat longer in the dimension running right and left in the sketch and then gave the frequency 41009 cycles per second. After being used at this frequency it was ground down to give the frequencies indicated in the sketch.

Table VIII contains the frequencies of the several crystal vibrators and also the dimension x (in direction right to left in sketch). This x

TABLE VIII.

Relation of Frequency to Dimension x.

Axis of radiation	x Determining dimension in cm.	f cycles per sec.	fx
B	5.7	42071	23.98×10^4
		56319	32.10
B	5.08	50701	25.76
B	3.68	70118	25.80
B	2.61	88585	23.12
B	2.88	98183	28.29
B	1.00	205620	20.56
B	0.46	610220	28.07
E	0.182	1034060	18.82
E	0.169	1479900	25.01
		Average	25.14×10^4

is the dimension in the direction of propagation of the sound. The product of f in cycles per second and x in centimeters is contained in the last column. The average of this product is about 25×10^4, and this may be employed as a very rough guide in cutting the crystal vibrators for a required frequency.

After the crystals are cut the vibrators are ground with coarse carborundum powder and finished with fine carborundum, fine emery, and in the case of thin specimens with rouge, to such fineness of surface as may be required for efficient oscillation. Repeated measurements of frequency during grinding enable a skillful operator to attain a desired frequency to better than 1/100 of 1 per cent, if standards of that precision are available.

15. Standards of Frequencies.— This subject is treated in the previous paper [5] to which reference is made for details. A standard tuning fork, here designated fork "S," made by the Western Electric Company, was used to operate electrically a high-speed siphon recorder made by the General Radio Company, and was thus chronographed four times for a period of 300 seconds each and once for a period of 250 seconds, and was found to have a frequency of 49.916 at 18.2° C., with a probable error of 1/400 of 1 per cent. Immediately preceding and following this operation, the 20th harmonic of the fork was observed to beat 2.44 times per second with a second fork, driven by a vacuum tube circuit, as shown in the previous paper. This second fork, here designated fork "P," was thus found to have a frequency

$$f_{\text{fork } P} = 20 \times 49.916 - 2.44 = 995.88 \pm 0.02$$

at 18° C.

Next a standard piezoelectric crystal oscillator permanently mounted was calibrated in terms of fork P, by the use of an electric oscillator and resonant-circuit wavemeter as intermediary. The wavemeter was calibrated by setting it to resonance with the electric oscillator when the latter was at a beat zero with the 6th to the 18th harmonic of fork P, and immediately thereafter the readings of the wavemeter were again taken at resonance with the electric oscillator when each of the harmonics of the electric oscillator between the 25th and the 58th was at beat zero with the standard piezoelectric crystal oscillator. By careful interpolation this gave

$$f_{713} = 420710 \pm \frac{1}{200} \text{ per cent.}$$

---

[5] Pierce, Proc. Am. Acad. of Arts and Sciences, **59**, No. 4 (1923).

This crystal oscillator will be known as No. 713, which is approximately its electric wavelength.

In terms of this crystal and its harmonics and multiples, the entire wavemeter was now carefully calibrated for the range of frequencies between 6000 and 6,000,000.

16. Measurement of Frequencies of Crystal Vibrators.—Immediately following or preceding the exploration of the sound wave system the frequency of each crystal vibrator was measured by one or more of the following methods:

I. *Harmonic Bracketing with Standard Crystal.* This consists of obtaining a wavemeter setting on an electric oscillator at beat zero with some harmonic of the unknown crystal X and then obtaining, on both sides of this wavemeter setting and as near to it as possible, a wavemeter setting on the electric oscillator at beat zero with some two harmonics of the standard crystal No. 713. Interpolation gives the wavemeter correction and the value of the frequency of the unknown.

II. *Audio-Frequency Beats.* Intercomparison of crystal vibrators was made in some cases by measuring on an audio-frequency meter the beat frequency of the fundamental of one crystal with some harmonic of another crystal.

III. *Superheterodyne Method.* When the beat frequency of II was above audibility, a third oscillator frequency could be made to beat with the rectified beat frequency of the two crystals. Measurement of the third frequency (which was a difference) could then be made with sufficient accuracy by the wavemeter.

IV. *Subheterodyne.* When the beat frequency of II was too low for audibility it could be made audible by an oscillator beating with both crystals (say 1000 per second with one and 1002 per second with the other), then the beating of the two audio frequencies with each other could be counted (as 2 beats per second).

V. *Direct Check of Crystals against Tuning Fork P.* This is a method similar to that used in § 13.

These various methods will be designated in the tables of results. As an example of Method I an electric oscillator beating with the fundamental of a crystal X and two harmonics of the oscillator beating with crystal No. 713 gave readings as in Table IX. The fundamental wavelength of No. 713 is 713.08 meters, based on 3×10^8 meters per second as the velocity of the waves, so that by regarding λ of column three as approximately correct it is found that n of the fourth column is 5 and 14/3, as entered. These last two

VELOCITY OF SOUND.

numbers are then multiplied by 713.08 and entered as $\lambda_{correct}$ for No. 713. A comparison of columns three and five gives the correction increments 0.3 and 0.9 entered as δ for No. 713. Interpolation between these gives $\delta = +0.8$ for X, which added to the observed λ gives $\lambda_{correct} = 3386.6$, for which the frequency is 88585. Note that the wavemeter calibration was in error less than 1/30 of 1 per cent in the range of Table VIII.

TABLE IX.

Determination of Frequency of Crystal X.

Crystal	Condenser divisions of W.M. No. 14 Coil E	λ Meters by calibration table	n	λ correct	δ	f
X	10.409 10.410 10.410	3385.8		3386.6	0.8	88585
713	11.530 11.532 11.530	3565.1	5	3365.4	.3	
713	10.048 10.047 10.047	3326.8	14/3	3327.7	.9	

DISPERSION AND ABSORPTION OF HIGH FREQUENCY SOUND WAVES

By K. F. Herzfeld and F. O. Rice

Abstract

The absorption of sound waves in gases had been explained heretofore by friction and heat condution. A third factor is here introduced, namely, the slow rate of exchange of energy between the translational movement and the internal degrees of freedom of the molecules. The formulas determining the absorption and dispersion of sound waves, due to these three effects, are developed. Comparison with the experimental data available shows that the new effect is either of considerable influence or even predominant. The rate of exchange can be calculated.

I. Introduction

THE propagation of high frequency sound waves in gases has recently received great attention because of the improvement of experimental methods. The first absorption measurements by Neklepajew[1] are unfortunately perhaps affected by a carbon dioxide content of the air, and the same may be true for newer experiments by D. L. Rich and W. H. Pielemeier.[2] Abello[3] has measured the absorptions in mixtures of hydrogen and carbon dioxide with air. While the theory of absorption has been developed at least partially long ago by Stokes[4] and Kirchhoff[5,6] no discussion of the change of velocity with frequency exists. This latter effect has been detected by Pierce[7] for air and carbon dioxide in contradiction to earlier workers who, on account of less accurate methods, had looked in vain for it.

There are three things which can affect the phenomena considered here, namely internal friction, heat conduction, and slowness of energy exchange between the translational and the internal degrees of freedom of a gas molecule. While the first two are taken into account by Stokes and Kirchhoff, the third is considered here for the first time. It had occurred to one of us, (F.O.R.) in connection with some considerations on the velocity of chemical reactions in gases that this velocity may be influenced by the rate at which the energy stored up in the translation of gas molecules could be transferred to the internal degrees of freedom where it is available for activation. In the attempt to find other physical phenomena in which this supposed process would have influence and which could, therefore, be used for the calculation

---

[1] N. Neklepajew, Ann. d. Physik **35**, 175 (1911).
[2] D. L. Rich and W. H. Pielemeier, Phys. Rev. **25**, 117 (1925).
[3] T. P. Abello, Proc. Nat. Acad. **13**, 699 (1927).
[4] C. G. Stokes, Trans. Cambridge Phil. Soc. **8**, 287 (1845).
[5] G. Kirchhoff, Pogg. Ann. **134**, 177 (1868).
[6] See also Lord Rayleigh, Theory of Sound 2nd Ed. Lon. 1896, 2nd Vol. par. 345 and P. Lebedew, Ann. d. Physik **35**, 171 (1911).
[7] G. W. Pierce, Am. Acad. Sci. **60**, 271 (1925).

of the rate of exchange, two such cases appeared: heat conduction[8] and propagation of sound waves.

All three processes mentioned above will increase the absorption, but their effect on the velocity of propagation V is not quite so clear, and will therefore, be discussed first qualitatively.

a. *Internal friction.* For a given geometrical wave-form the effect of internal friction is to retard the movement. Therefore, for a given frequency the wave-lengths will be larger with friction than without. Internal friction, therefore, increases the velocity of propagation.

b. *Heat conduction.*[9] As is well known, Newton's assumption of constant temperature in the gas leads to a velocity, the square of which is given by RT/M where M is the molecular weight, but the experimental results are represented by Laplace's expression, the square of which is $\gamma RT/M$ where γ is the ratio of the specific heats at constant pressure and constant volume, and where an adiabatic state has been assumed. An erroneous opinion on the influence of the frequency upon the completeness of the adiabatic state seems to be often held. It is believed that while ordinary sound frequencies are sufficiently high to guarantee the adiabatic condition, we should expect Newton's velocity for very low frequencies. But a close scrutiny shows that the adiabatic state is best guaranteed for low frequencies, while for higher frequencies the influence of heat conduction is larger; therefore, the velocity should decrease with increasing frequency. This result comes about in the following way:

While it is true that with decreasing frequency the time allowed for heat conduction increases, the amount of heat conducted at a given moment decreases much more rapidly. This latter is proportional to $\partial^2 T/\partial x^2$ and for a given amplitude this expression is inversely proportional to the square of the wave-lengths. Therefore, the heat conducted during one period increases proportionally with the frequency because of the increasing steepness of the temperature gradient. The numerical values, however, are such that for (a) and (b) together the velocity increases with the frequency.

c. A slow rate of exchange between external and internal degrees of freedom keeps the internal degrees from taking up the whole amount of heat, and therefore acts as if the effective specific heat were decreased, and the velocity of sound increased with increasing frequency.

II. Calculation of Dispersion and Absorption

a. *Symbols.* We shall use the following notation: u, velocity of gas; ρ, density of gas, $=pM/RT=\rho_0(1+s)$ where s is the relative compression; T, temperature of the translational kinetic energy of the gas molecules $=T_0(1+\theta)$; αR, specific heat of translational kinetic energy ($\alpha=3/2$); T', temperature of the internal degrees of freedom $=T_0(1+\theta')$ (for slow movements $T=T'$); βR, specific heat of internal degrees of freedom; $cR=(\alpha+\beta)R$, total specific heat at constant volume; μ, coefficient of internal

---

[8] A. Eucken, Phys. Zeits. **14**, 324 (1913).
[9] See also C. G. Stokes, Phil. Mag. (4) **1**, 305 (1851). Lord Rayleigh, l.c. p. 25.

friction; L, coefficient of heat conduction; $\epsilon = LM/cR\mu$, a dimensionless constant (according to the kinetic theory of gases it is 2.5 for monoatomic molecules and smaller for others.)[8]

It may be that the heat conduction under nonstationary conditions is different for the heat belonging to the external and to the internal degrees of freedom.[8] We therefore put $L = L' + L''$ and $\sigma = L''/L'$ or $\sigma/(1+\sigma) = L''/L$.

We shall denote by n the frequency of the sound wave, and set $\xi = 8\pi\mu/3V_0^2 = 8\pi c\mu/3(c+1)p_0$, $p_0 = 1.014 \cdot 10^6$ dynes/cm.^{2} The product ξn is dimensionless and will be considered as small.* τ is a time measuring the rate of exchange between external and internal degrees of freedom. We represent the amount of energy which passes in one mol from the external to the internal degrees of freedom during unit time by $\alpha R dT/dt = \alpha R(T-T')/\tau$. We assume s, u, θ, θ' to be proportional to $e^{i2\pi(nt-kz)}$. We set $k = k_1 - ik_2$ where $k_1 = n/V$ measures the velocity of the propagation of sound. $2\pi k_2$ is the absorption coefficient for the amplitude; therefore $4\pi k_2$ the absorption coefficient for the intensity.

b. *The equations of motion and their integration.* Assuming the amplitude to be small, we have the following equations for a plane wave propagated along the x axis: the equation of motion;

$$\frac{\partial u}{\partial t} - \frac{4}{3}\frac{\mu}{\rho}\frac{\partial^2 u}{\partial x^2} = -\frac{1}{\rho}\frac{\partial p}{\partial x} = -\frac{RT_0}{M}\left(\frac{\partial \ln T}{\partial x} + \frac{\partial \ln \rho}{\partial x}\right)$$

the equation of continuity; $\partial \rho/\partial t = -\rho\, \partial u/\partial x$.

The equation of energy for the degrees of freedom of translation expresses the fact that this energy can be changed by external work, heat conduction and exchange with the internal degrees of freedom.

$$\alpha R\frac{\partial T}{\partial t} = p\frac{M}{\rho^2}\frac{\partial \rho}{\partial t} + \frac{L'M}{\rho}\frac{\partial^2 T}{\partial x^2} - \frac{\alpha}{\tau}R(T-T')$$

The equation of energy for the internal degrees of freedom determines the corresponding change in energy by heat conduction and exchange with the external degrees of freedom.

$$\beta R\frac{\partial T'}{\partial t} = \frac{\alpha R}{\tau}(T-T') + \frac{L''M}{\rho}\frac{\partial^2 T'}{\partial x^2}$$

If we now consider the way in which u, s, θ, θ' depend on t and x we get the following equations:

$$u\left(in + \frac{8\pi\mu}{3\rho_0}k^2\right) = ik\frac{RT}{M}(\theta+s) \qquad ns = ku$$

$$in\alpha\theta = ins - \frac{L'M}{\rho R}2\pi k^2\theta - \frac{\alpha}{2\pi\tau}(\theta-\theta') \qquad in\beta\theta' = -\frac{L''M}{\rho R}2\pi k^2\theta' + \frac{\alpha}{2\pi\tau}(\theta-\theta')$$

* Probably ξ and τ will depend in the same way on temperature and pressure, both being proportional to the time between two collisions, or to (mean free path)/(Velocity).

Eliminating u, s, θ, θ' we get the following "equation of dispersion" for k:

$$-c(c+1)\frac{k^2V_0^2/n^2-1+ik^2V_0^2/n^2}{1-c(k^2V_0^2/n^2-1)-i(c+1)k^2V_0^2/n^2}=\frac{3ic\epsilon\xi nk^2V_0^2}{4(1+\sigma)n^2}$$

$$+\frac{i\beta^2 2\pi\tau n+3\sigma c\epsilon\xi nk^2V_0^2(\alpha i+\beta 2\pi\tau n)/4(1+\sigma)n^2}{\alpha+i\beta 2\pi\tau n+3\sigma c\epsilon\xi n2\pi\tau nk^2V_0^2/4(1+\sigma)n^2}$$

To solve this equation we treat ξn and $2\pi\tau n$ as small quantities. Then it is seen that the solution of zero order will be $k_1=n/V_0$; $k_2=0$. We write then for the next order which will be sufficient

$$k_1-ik_2=(n/V_0)[1+(K\xi^2+P\xi\tau+Q\tau^2)n^2]-in(D\tau+E\xi)$$

Comparison of equal powers of ξ and τ leads to the final solution.

$$4\pi k_2=\frac{n^2}{V_0}\left[2\pi\left(1+\frac{3}{4}\frac{\epsilon}{c+1}\right)\xi+4\pi^2\frac{\beta^2}{\alpha c(c+1)}\tau\right]$$

$$V=V_0\left\{1+n^2\left[\frac{3}{8}\left(1+\frac{5}{2}\frac{\epsilon}{c+1}-\frac{3}{4}\epsilon^2\frac{c-3/4}{(c+1)^2}\right)\xi^2\right.\right.$$

$$+\frac{3\pi}{2}\frac{\beta^2}{\alpha c(c+1)}\left(1-\epsilon\frac{c-1/4}{c+1}+\epsilon\frac{c}{\beta}\frac{\sigma}{1+\sigma}\right)\xi\tau$$

$$\left.\left.+2\pi^2\frac{\beta^3}{\alpha^2 c^2(c+1)}\left(\alpha+\frac{3}{4}\frac{\beta}{c+1}\right)\tau^2\right]\right\}$$

III. NUMERICAL RESULTS AND COMPARISON WITH EXPERIMENTS

For the following gases we use numerical data given in the following table.

	μ	ϵ	γ	β	c
Air	19×10^{-5}	1.91	1.405	0.96	2.46
Hydrogen	88.5×10^{-5}	1.97	1.408	0.93	2.43
Carbon Dioxide	16×10^{-5}	1.628	1.305	1.78	3.28

The values of μ and ϵ have been taken from Eucken,[8] γ from Landolt-Bornstein tables. β and c have been calculated from these values using $\alpha=3/2$. We find then

Air:
$$4\pi k_2=n^2(2.99\times 10^{-13}+0.428\times 10^{-4}\tau)$$
$$V=V_0\{1+n^2[9.32\times 10^{-19}-(0.084-1.86\sigma/(1+\sigma))10^{-9}\tau+0.635\tau^2]\}$$

Hydrogen:
$$4\pi k_2=n^2(3.65\times 10^{-13}+0.106\times 10^{-4}\tau)$$
$$V=V_0\{1+n^2[2.05\times 10^{-19}-(0.42-8.84\sigma/(1+\sigma))10^{-9}\tau+0.595\tau^2]\}$$

Carbon dioxide:
$$4\pi k_2=n^2(3.16\times 10^{-13}+1.16\times 10^{-4}\tau)$$
$$V=V_0\{1+n^2[6.48\times 10^{-19}-(0.108-2.17\sigma/(1+\sigma))10^{-9}\tau+1.96\tau^2]\}$$

For air we have the absorption measurements of Neklepajew.[1] As mentioned in the introduction, Rich and Pielemeier suspect that these are affected by a content of carbon dioxide as the values are higher than the theoretical ones if only friction and heat conduction are taken into account; but, according to Abello's[3] new measurements one would not expect an influence larger than 5 percent as it seems improbable that more than 1 percent of carbon dioxide would be present. We will, therefore, provisionally use Neklepajew's value which gives for the absorption coefficient $6.6 \times 10^{-13} n^2$. Comparing this with our formula we see, as pointed out by Lebedew,[6] that about half this value is due to heat conduction and friction. The other half, therefore, must be due to slow energy exchange between different degrees of freedom. We get from this a value of τ approximately 0.8×10^{-8} sec. For $n = 10^6$ this would modify the velocity of sound by 3 cm per second, a value far below Pierce's[7] results. Neither the shape nor the values of Pierce's curve for the variation of velocity of sound with frequency can, therefore, be explained.

For carbon dioxide, Pierce's curve for the velocity of sound is of the form we should expect according to our formula. Using the value at $n = 2 \times 10^5$ cycles per second, we find that the change of V is primarily due to the exchange rate, and gives a value for $\tau = 2.6 \times 10^{-8}$ seconds. With this we get an absorption coefficient $3 \times 10^{-12} n^2$. This agrees reasonably with his statement that carbon dioxide is opaque for $n = 10^6$, as it gives at this frequency an absorption coefficient 3.3. It agrees, furthermore, with his result that at $n = 10^5$ a column of CO_2 of 17 mm length is equivalent to one of air of 68 mm, as for the former $4\pi k_2 x$ would be 0.056 while for the latter it would be 0.045, using Neklepajew's data. It would not agree with his other statement that for $n = 2 \times 10^5$ a column of 4 mm CO_2 is equivalent to 204 mm of air, as this would give values of 0.056, 0.55 respectively; but the two results of Pierce do not agree with each other if we assume that the absorption coefficient of carbon dioxide and air depend in the same way on the frequency, because then the ratio of equivalent columns of these two gases should be independent of the frequency. Finally, we may try to exterpolate the absorption coefficient of pure carbon dioxide from Abello's results with the help of his statement that the connection between transmission and percentage of carbon dioxide is a logarithmic one (which statement means that the absorption coefficient is a linear function of the composition.) We then get for the absorption coefficient $-5,6 \times 10^{-12} n^2$ which is in sufficient agreement with our value. If we extrapolate, in the same way, Abello's hydrogen measurements we find $3.8 \times 10^{-12} n^2$ giving $\tau \sim 3 \times 10^{-7}$ seconds.

A real test of the theory can be made only if for the same gas absorption and dispersion are measured carefully.

THE JOHNS HOPKINS UNIVERSITY,
  PHYSICAL LABORATORY AND DEPARTMENT OF CHEMISTRY,
    January 9, 1928.

22

The Dispersion Theory of Sound

H. O. KNESER

Translated expressly for this Benchmark volume by R. Bruce Lindsay from "Zur Dispersionstheorie des Schalles," Ann. Physik (Folge 5), 761–776 (1931)

The kinetic theory of gases in combination with the quantum theory teaches that the portions of the total energy of a homogeneous gas, which in the equilibrium state are distributed over the degrees of freedom of the molecules, stand in a very definite, calculatable relation to each other, and that this relation depends only on the temperature. If, as is customary, we distinguish between the external (i.e., the translational) degrees of freedom and the internal degrees of freedom (i.e., all the others) and denote the associated energy portions by E_a and E_i, respectively, we may formulate the above statement in general as $E_a : E_i = f(T)$. In any alteration of the total energy both components of the energy can alter in the same direction only insofar as equilibrium is preserved. But an energy change in which both external and internal degrees of freedom are equally affected is not thinkable. Radiation into or out of the gas can primarily increase or decrease only the internal energy. All other possibilities of changing the energy content (and these are the only ones we shall consider in what follows) must at first produce a change in the motion of the center of mass of the molecules and, accordingly, bring about a change in the kinetic energy of translation (E_a). If, for example, we assume an adiabatic compression produced by thrusting a piston into a thermally insulated flask, at first only the momenta of the molecules striking the piston are increased. It is only after a time that through collisions with other molecules the increase in external energy is transformed in part into inner energy and the equilibrium state is reestablished. Since only extraordinarily short time intervals are in question here it is only for very rapid changes of state that one may expect a modification of the thermodynamical statistical laws as a consequence of the delayed reestablishment of energy equilibrium.

There is a strong suggestion that one should apply these considerations to the propagation of sound.[1] Every volume element of a gas mass in which a sound wave is being propa-

---

[1] H. A. Lorentz has already done this in his classic work *Les Equations du mouvement des gaz* [*Arch. Neerland.*, **16**, 1 (1881)]; see also K. F. Herzfeld and F. O. Rice, *Phys. Rev.*, **31**, 691 (1928). We go into both works later. Finally, D. G. Bourgin [*Phil. Mag.*, **7**, 821 (1929)] has carried out calculations of similar kinds, whose results, however, are difficult to compare with experiment. A hint in this direction is found in L. Boltzmann, *Vorlesungen über Gastheorie*, Vol. 2, p. 131, Leipzig, 1912.

gated suffers adiabatic pressure and temperature changes with sign alternating in time. One must expect (and this is fully confirmed by experiment) that the transfer of external energy into internal energy takes place the more incompletely, and the E_i lag behind the equilibrium values corresponding to the momentum state of the gas $[E_a : f(T)]$ by a greater amount, the more the period of the sound wave approaches the time interval which is necessary for the establishment of equilibrium. In the limiting case of infinitely small wave period (infinitely high frequency), the variation in the energy amount transformed into inner energy will vanish. This must manifest itself as a decrease in the specific heat for the periodic phenomenon taking place. For equal energy transfer now demands a stronger increase in the external energy and consequently a greater increase in temperature. If C is the molar specific heat at constant volume, we have

$$C = \left(\frac{\partial E}{\partial T}\right)_v = \left(\frac{\partial E_a}{\partial T}\right)_v + \left(\frac{\partial E_i}{\partial T}\right)_v = C_a + C_i$$

Hence one will expect that with increasing frequency of state alteration there will take place an apparent decrease in C to the limiting value C_a. Now C enters the classical expression for the velocity of sound in gases in the following way[2]:

$$V = \sqrt{\frac{p}{\rho}\left(1 + \frac{R}{C}\right)}$$

Here p denotes pressure, ρ density, and R is the molar gas constant. With increasing frequency V must also increase asymptotically to the value

$$V = \sqrt{\frac{p}{\rho}\left(1 + \frac{R}{C_a}\right)}$$

The question now arises: On what molecular kinetic quantities does the velocity depend and with which is the energy equilibrium established? Since the energy exchange between outer and inner degrees of freedom is quantized, the elementary act of transition to the inner energy state takes place at a definite time instant, and likewise the return transition. Between these times there is a time interval which one can designate as the lifetime of the energy quantum. It can now be shown that it is in essence this lifetime which determines the time needed to establish equilibrium. One can perhaps describe the situation schematically as follows. The lifetime plays for the energy equilibrium the role of a minimum time quantum. The energy distribution will not be appreciably influenced, if at all, by phenomena that take place in shorter times than this. The incomplete energy exchange, and thereby the increase in sound velocity, should accordingly make themselves evident when the period of the sound wave is comparable with or smaller than the lifetime.

The following analysis attempts to provide the theoretical foundation which these

---

[2] The standard form for V is, of course,

$$V = \sqrt{\frac{p}{\rho}\frac{C_p}{C_v}}$$

Above we have eliminated C_p by means of the relation (Mayer) $R = C_p - C_v$.

considerations demand. This analysis completely neglects the influence of viscosity and heat conduction in the gas on the sound velocity. It is only in this way that one can arrive at an explicit dispersion formula.

The meaning of the symbols is as follows:
p and ρ denote the equilibrium pressure and density of the gas, respectively.
$(p+\pi)$ and $\rho(1+s)$ denote the same quantities in the presence of a plane sound wave. Here we assume that $\pi \ll p$ and $s \ll 1$.
x is the coordinate in the direction of the wave normal.
X is the displacement of a mass particle in the x direction from its equilibrium position.
v is the molar volume of the gas.
The Eulerian equation of motion of the gas becomes

$$-\frac{\partial \pi}{\partial x} = \rho \frac{\partial^2 X}{\partial t^2} \qquad (1)$$

The equation for the conservation of matter is

$$-s = \frac{\partial X}{\partial x} \qquad (2)$$

and π and s for a plane, harmonic sound wave become

$$\begin{aligned}
\pi &= \pi_0 \cos\left[\omega\left(t-\frac{x}{V}\right)+\psi\right] \\
&= \pi_0 \exp\left\{i\left[\omega\left(t-\frac{x}{V}\right)+\psi\right]\right\} \\
s &= s_0 \cos\left[\omega\left(t-\frac{x}{V}\right)\right] \\
&= s_0 \exp\left\{i\left[\omega\left(t-\frac{x}{V}\right)+\psi\right]\right\}
\end{aligned} \qquad (3)$$

Here V denotes the sound velocity, ω is the angular frequency of the harmonic wave, $i=\sqrt{-1}$, and ψ is an arbitrary phase constant. Since all influences producing absorption are to remain unconsidered, a dissipation factor does not need to be included.

Equations (1) and (2) lead to the wave equation

$$\frac{1}{\rho} \frac{\partial^2 \pi}{\partial x^2} = \frac{\partial^2 s}{\partial t^2}$$

If we now use Eq. (3), recalling that

$$\frac{\partial^2 \pi}{\partial x^2} = \frac{-\omega^2}{V^2} \pi \qquad \frac{\partial^2 s}{\partial t^2} = -\omega^2 s$$

the result is

$$V^2 = \frac{1}{\rho} \frac{\pi}{s} = \frac{1}{\rho} \frac{\pi_0}{s_0} e^{i\psi} \qquad (4)$$

If, as is the case in the following analysis, $(1/\rho)(\pi/s)$ turns out to be a complex quantity, this means clearly that $\psi \neq 0$; that is, there exists a phase difference between pressure and density waves. The absolute value of the complex quantity $(1/\rho)(\pi/s)$ is then the square of the sound velocity, and $\tan \psi$ is the ratio of the real and imaginary parts.

Since $v\rho =$ constant and $dv/d\rho = -v/\rho$, we can further write[3]

$$\frac{1}{\rho}\frac{\pi}{s} = \frac{dp}{d\rho} = \frac{\partial p}{\partial v}\frac{dv}{d\rho} = \frac{1}{v}\left(\frac{d(pv)}{dv} - p\right)\left(\frac{-v}{\rho}\right) = \frac{p}{\rho}\left[1 - \frac{1}{p}\frac{d(pv)}{dv}\right] \tag{5}$$

For adiabatic processes the statement of conservation of energy takes the form

$$C\, dT + p\, dv = 0 \tag{6}$$

where C denotes the molar specific heat at constant volume and T is the absolute temperature. As has already been remarked, the first term can be split into two parts, the first of which refers to the kinetic energy of the motion of the center of mass of the molecules, $(dE_a = C_a\, dT)$, and the second refers to the energy content of the molecules in the form of rotation, vibration, and electron energies $(dE_i = C_i\, dT)$. In the equilibrium state, and for slow changes of state, E_i is purely a function of the temperature; but for rapid, periodic change of state, on the other hand, E_i will also depend on the frequency, because the portion of the energy corresponding to temperature change can no longer be completely taken up by the internal degrees of freedom.

We carry through the analysis by writing

$$dE_i = C_i'\, dT = C_i F(\omega)\, dT \tag{7}$$

in which, to begin with, no further assumption is made about $F(\omega)$ than that

$$F(0) = 1 > F(\omega) > 0 = F(\infty)$$

Equation (6) now becomes

$$C_a\, dT + C_i F(\omega)\, dT + p\, dv = 0 \tag{6'}$$

Combined with the equation of state of an ideal gas, $d(pv) = R\, dT$ (all taken for one mole), Eq. (6') yields, when substituted into (5),[4]

$$\frac{1}{\rho}\frac{\pi}{s} = \frac{p}{\rho}\left[1 + \frac{R}{C_a + C_i F(\omega)}\right] \tag{8}$$

The problem now is to fix the function $F(\omega)$ appropriately. $E_i = v\Sigma n_k \varepsilon_k$, where n_k is the number of molecules per unit volume, each of which possesses the internal energy ε_k. For simplicity we replace the sum by the simple term $n\varepsilon$. This means that we assume there is only one energy state (ε), with an occupation number n that is markedly different from zero. This assumption must, of course, be tested from case to case.

We then have

---

[3] The derivation here is closely connected with that used by A. Einstein, *Ber. Berlin Akad.*, p. 380, 1920.
[4] See Appendix Note 1 at the end of the article.

$$E_i = vn\varepsilon = \int C_i' \, dT \tag{9}$$

It follows that n and the total internal energy will change only if, in the collision of two molecules, translational energy is changed into internal energy, that is, one of the collision partners is excited by a collision of the first kind. On the other hand, n and E_i remain unaltered if, in the collision of an excited molecule with an unexcited molecule the energy quantum goes over to the unexcited one, that is, only the possessor changes. The number of collisions of the first kind per unit volume per unit time is proportional to $\mathcal{N}(\mathcal{N}-n)$, where $\mathcal{N}$ denotes the total number of molecules per unit volume and $(\mathcal{N}-n)$ is the number of unexcited ones. The number of reverse processes in which internal energy is transformed into translational energy is thus proportional to $n\mathcal{N}$.

In this way we arrive at a "reaction" equation,

$$\frac{dn}{dt} = \frac{1}{a}(\mathcal{N}-n)\mathcal{N} - \frac{1}{b}n\mathcal{N} \tag{10}$$

The constants $1/a$ and $1/b$ represent, respectively, the probability that a collision will lead to a transfer of translational energy with internal energy, or conversely. We recognize in the quantity $b/\mathcal{N}$ the "decay" time of the internal energy,[5] which is formally completely identical to the lifetime of the energy quantum, that is, the time during which on the average a quantum is retained in the form of internal energy. This time is to be distinguished from the lifetime of the excited state, which indicates how long on the average the molecule remains in this state. If transfer of internal energy as such does not take place, the former time is equal to the latter; otherwise it is greater than the latter.

Equation (10) is now to be applied to a time-periodic disturbance of the equilibrium. We set

$$\mathcal{N} = \mathcal{N}_0(1 + Qe^{i\omega t}) \tag{11}$$

On the assumption, already used in Eqs. (1) and (2), that only small disturbances are in question ($Q^2 \ll 1$), the stationary-state solution has the form

$$n = n_0[1 + qe^{i(\omega t - \phi)}] \tag{12}$$

If we insert these expressions for $\mathcal{N}$ and n in Eq. (10), the result is

$$i\omega n_0 \, qe^{i(\omega t - \phi)} = \frac{1}{a}\mathcal{N}_0^2(1 + 2Qe^{i\omega t}) - \left(\frac{1}{a} + \frac{1}{b}\right)\mathcal{N}_0 n_0[1 + Qe^{i\omega t} + qe^{i(\omega t - \phi)}]$$

If we introduce the abbreviation

$$\frac{1}{\beta} = \left(\frac{1}{a} + \frac{1}{b}\right)\mathcal{N}_0 \tag{13}$$

we get

---

[5] If $1/a = 0$, the solution of the differential equation becomes $n = n_0 \, e^{-(N/b)t}$, and for the decay of the internal energy we get $E_i = vn\varepsilon = E_{i_0} e^{-(N/b)t}$.

Dispersion Theory of Sound

$$\frac{1}{a}\mathcal{N}_0^2 - \frac{1}{\beta}n_0 = \left[\frac{2}{a}\mathcal{N}_0^2 Q - \frac{1}{\beta}n_0(Q + qe^{-i\phi}) - i\omega n_0 qe^{-i\phi}\right]e^{i\omega t}$$

Both the left-hand side and the factor multiplying $e^{i\omega t}$ on the right must equal zero. Hence we have

$$n_0 = \frac{\beta}{a}\mathcal{N}_0^2$$

or
(14)

$$\frac{n_0}{\mathcal{N}_0} = \frac{\beta}{a}\mathcal{N}_0 = \frac{b}{a+b}$$

This fixes the equilibrium state. From the factor of $e^{i\omega t}$ we get

$$i\omega + \frac{1}{\beta} = \frac{Q}{q}\left(\frac{2}{a}\frac{\mathcal{N}_0^2}{n_0} - \frac{1}{\beta}\right)e^{i\phi} = \frac{Q}{q}\frac{1}{\beta}e^{i\phi} \quad (15)$$

By equating the absolute values of the complex expression on each side, we arrive at

$$q = \frac{Q}{\sqrt{1 + \omega^2\beta^2}} \quad (16)$$

and

$$\tan \phi = \omega\beta \quad (17)$$

This says that the relative oscillation of n about its equilibrium value take place with amplitude q, which from value $q_0 = Q$ for vanishing frequency decreases to the value $q_\infty = 0$ for infinitely great frequency. There is also a phase lag ϕ, which increases from the value zero at vanishingly small frequency to the value $\pi/2$ for very large frequency (Fig. 1).

The relation already derived between q and Q (or n and $\mathcal{N}$) is not sufficient to fix the frequency dependence of the specific heat since it says nothing about the connection between n and T. A further assumption is needed, one which is indeed suggested by what has gone before.

We assume that the internal energy per unit volume (εn to a first approximation) varies linearly with both pressure (or $\mathcal{N}$) as well as temperature:

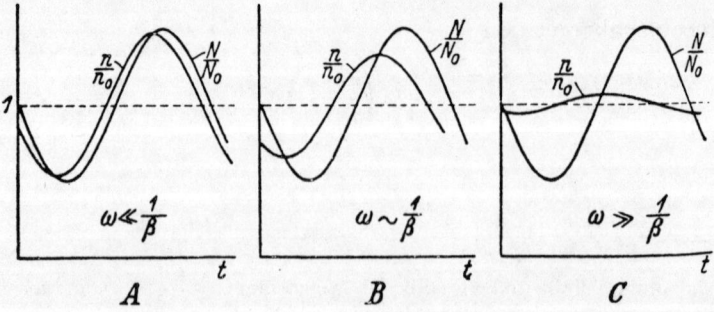

Figure 1. Time dependence of $\mathcal{N}$ and n for different frequencies.

$$\frac{\partial n}{\partial \mathcal{N}} = P \frac{\partial n}{\partial T}$$

in which P denotes a constant that must be independent of the velocity of change of state.[6] This assumption says nothing other than that in rapid change of the number of gas kinetic collisions, the turnover in internal energy remains just as much under the value which holds for slow processes as in an equally rapid change of the energy of the collisions. Using (11) and (12) we find

$$\frac{\partial n}{\partial \mathcal{N}} = \frac{n_0}{\mathcal{N}_0} \frac{q}{Q} e^{-i\phi} = \frac{n_0}{\mathcal{N}_0} \frac{1}{1+i\omega\beta}$$

From this

$$\frac{\partial n}{\partial T} = \frac{1}{P} \frac{n_0}{\mathcal{N}_0} \frac{1}{1+i\omega\beta}$$

and following (7),

$$C_i' = \frac{\partial(n\varepsilon v)}{\partial T}\bigg)_v = \varepsilon v \frac{1}{P} \frac{n_0}{\mathcal{N}_0} \frac{1}{1+i\omega\beta}$$

Through this we have won a useful aid for the establishment of the frequency dependence of the specific heat. According to definition (7) we have

$$F(\omega) = \frac{1}{1+i\omega\beta} \tag{18}$$

a function which fulfills the conditions laid down.

We introduce this function into (8) and get

$$\frac{1}{\rho}\frac{\pi}{s} = \frac{p}{\rho}\left(1 + \frac{R}{C_a + \dfrac{C_i}{1+i\omega\beta}}\right) \tag{8'}$$

The absolute value of this expression yields the square of the sound velocity. Instead of this, for the sake of simplicity in the analysis we shall use the real part, which, as will be demon-

---

[6] This assumption certainly holds for an ideal gas. For the energy content per mole is independent of v and p (or $\mathcal{N}$), which is also per unit volume proportional to $1/v$ or p (or therefore $\mathcal{N}$), so that we have

$$\frac{\partial \varepsilon n}{\partial p} \sim s \frac{\partial n}{\partial \mathcal{N}} = \text{constant}$$

On the other hand, we have that C is a constant and also

$$C_i = sv \frac{\partial n}{\partial T}\bigg)_v = \text{constant}$$

If one takes into account the deviation from the ideal gas state, one gets a dependence of the sound velocity on the amplitude. For this, however, there is no evidence either in earlier measurements (save for the case of shock waves) or in those to be described later in this paper.

Dispersion Theory of Sound

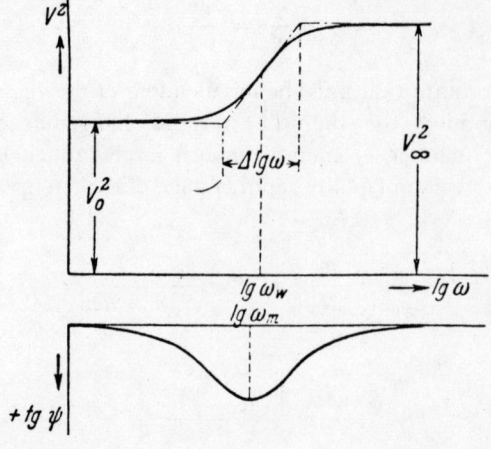

Figure 2

strated, does not differ essentially from the absolute value. We thus get the dispersion formula[7]

$$V^2 = \frac{p}{\rho}\left(1 + R\frac{C + \omega^2\beta^2 C_a}{C^2 + \omega^2\beta^2 C_a^2}\right) \quad (19)$$

in which $C_a + C_i$ is again replaced with C (molar specific heat at constant volume).

From (19) we calculate

$$\frac{dV^2}{d(lg\omega)} = \frac{p}{\rho}\frac{2RC_i C_a C\omega^2\beta^2}{(C^2 + \omega^2\beta^2 C_a^2)^2}$$

$$\frac{d^2V^2}{d(lg\omega)^2} = \frac{p}{\rho}\frac{4RC_i C_a C\omega^2\beta^2(C^2 - C_a^2\omega^2\beta^2)}{(C^2 + \omega^2\beta^2 C_a^2)^3}$$

The curve which represents V^2 as a function of $lg\omega$ (shown in the upper part of Fig. 2) is parallel to the axis of abscissas for $\omega = 0$ and for $\omega = \infty$. The corresponding ordinates are

$$V_0^2 = \frac{p}{\rho}\left(1 + \frac{R}{C}\right)$$

$$V_\infty^2 = \frac{p}{\rho}\left(1 + \frac{R}{C_a}\right) \quad (20)$$

with

$$V_\infty^2 - V_0^2 = \Delta V^2 = \frac{p}{\rho}\frac{RC_i}{CC_a}$$

---

[7] See Appendix Note 2.

The curve has a point of inflection for $\omega = \omega_w = (1/\beta)(C/C_a)$. The slope at this point is

$$\left.\frac{dV^2}{d\lg\omega}\right)_{\omega_w} = \frac{p}{\rho}\frac{RC_i}{2CC_a} = \frac{1}{2}\Delta V^2$$

This is a measure of the maximum dispersion. The rise in the sound velocity takes place primarily in the frequency domain,

$$\Delta\lg\omega = \frac{\Delta V^2}{dV^2/d\lg\omega} = 2$$

that is, within approximately three octaves.

We notice that the quantity $1/\beta$ determines the position of the dispersion region. It plays a role similar to the resonance frequency in the theory of optical dispersion.[8] The significance of β will become clear if a is eliminated between (13) and (14). Thus

$$\beta = \frac{b}{\mathcal{N}_0}\left(1 - \frac{n_0}{\mathcal{N}_0}\right) \tag{21}$$

Insofar as s is known, $n_0/\mathcal{N}_0$ is given on the basis of the Maxwell–Boltzmann distribution law. Therefore, β is proportional to the lifetime of the energy quantum $b/\mathcal{N}_0$ and in general is not very different from it.

We proceed, further, to calculate $\tan\psi$ as the ratio of the imaginary and the real parts of the right-hand side of (8'). Thus

$$\tan\psi = \frac{\omega\beta RC_i}{C^2 + \omega^2\beta^2 C_a^2 + RC + \omega^2\beta^2 RC_a}$$

$$= \frac{\Delta V^2 \omega_w \omega}{V_0^2 \omega_w^2 - V_\infty^2 \omega^2}$$

as one may readily verify with the use of (20). For $\omega = 0$ and $\omega = \infty$, $\tan\psi$ vanishes. It reaches a maximum for

$$\omega_m = \frac{V_0}{V_\infty}\omega_w$$

where it has the value

$$\tan\psi_m = \frac{\Delta V^2}{2V_0 V_\infty}$$

The order of magnitude of $\tan\psi$ is only a few percent. Hence the real part of (8') differs

---

[8] The characteristic difference with the theory of optical dispersion is that here we do not have to do with a resonance effect and, therefore, without the introduction of a damping term the index of refraction is never pure imaginary. A completely analogous analysis has been carried out by P. Debye, *Polar Molecules*, Leipzig, 1929.

from the absolute value by very little, even in the most unfavorable case. This confirms the validity of the approximation made in the calculation of the dispersion curve.

The half-width of the $\tan \psi$ curve turns out to be

$$lg \frac{2+\sqrt{3}}{2-\sqrt{3}}$$

which corresponds to an interval of almost four octaves.

The quantity ψ gives the phase difference between the pressure wave and the density wave.[9] It is of importance in the computation of the sound intensity L, that is, of the energy that is transmitted through unit area per unit time in the direction of propagation. If τ denotes the period of the wave, we have

$$L = \frac{1}{\tau} \int_{t=0}^{t=\tau} \pi \, dx = \frac{1}{\tau} \int_0^\tau \pi \frac{\partial X}{\partial t} \, dt$$

$$= \frac{V}{\tau} \int_c^\tau \pi s \, dt$$

$$= \frac{1}{2} V \pi_0 s_0 \cos \psi$$

A sound wave whose frequency lies in the dispersion zone, where $\tan \psi$ approximates its maximum value, will, therefore, transport less energy than in a nondispersion medium, other things being equal.

Appendix Note 1

The phenomenological derivation of formula (8) may be compared with a more deductive one resulting from an extension of the analysis of H. A. Lorentz, mentioned in an earlier footnote.

From the laws of conservation of matter, momentum, and energy, Lorentz derives three equations of motion, a_4, b_4, and c_4. If we restrict these to the one-dimensional case and omit the terms that correspond to viscosity and heat conduction, these equations take the form

$$\frac{\partial u}{\partial x} = -\frac{\partial s}{\partial t} \tag{a_3}$$

$$\frac{1}{3} \frac{\partial}{\partial x}(h + h_0 s) + \frac{\partial u}{\partial t} = 0 \tag{b_3}$$

$$\frac{1}{3} h_0 \frac{\partial u}{\partial x} + \frac{1}{2}[1 + 2\theta(h_0)] \frac{\partial h}{\partial t} + \frac{\nu}{\rho} \frac{\partial^2 u}{\partial x \, \partial t} = 0 \tag{c_3}$$

---

[9] $\psi = 0$ means, therefore, that the Boyle–Mariotte law does not hold for the instantaneous state quantities.

Here

$$u = \frac{\partial X}{\partial t}$$

$h =$ mean value of $(\xi^2 + \eta^2 + \zeta^2)$, where $\xi, \eta,$ and ζ are the velocity components of a molecule and are averaged over all molecules in a mass element
$h_0 =$ equilibrium value of h
$\theta(h) =$ "intramolecular" energy per mass element

The term ν, formed analogously to the coefficients of viscosity and heat conduction, enters in the analysis because of a special expression for the energy content of the gas per mass element. We have for this quantity

$$\frac{1}{2}u^2 + \frac{1}{2}h + \theta(h) + \frac{\nu}{\rho}\frac{\partial u}{\partial x} \qquad (d)$$

The first term corresponds to the external work, the second and third terms to the heat content. The last term corresponds to the fact that during a change of state the internal energy lags behind the value that it would have under equivalent circumstances in the equilibrium state.[1] For example, in compression of the gas the value of u will always be positive in the x direction and will decrease as x increases. Hence $\partial u/\partial x$ will be negative and the energy content will be smaller than in the equilibrium state, other state quantities remaining the same.

Objection may be raised to the use of Eq. (d) on the ground that for a sufficiently fast compression $\partial u/\partial x$ can take on arbitrarily large negative values, and hence the energy content itself can become negative. Apparently the expression with ν treated as constant may not be used for the calculation of the velocity of arbitrarily high-frequency sound waves. As a matter of fact, if one uses (a_3), (b_3), and (c_3) to calculate the sound velocity V (a calculation carried through by Lorentz only for the case where $\nu = 0$) one gets

$$V = \frac{5 + 6\theta'(h_0) + (6/\rho h_0)i\omega\nu}{9[1 + 2\theta'(h_0)]}$$

From this it results that $V_{\omega=\infty} = \infty$, a result that cannot agree with experience at all.

Calculation of the delayed establishment of equilibrium will now be undertaken using a different energy expression, corresponding to the one used earlier:

$$\tfrac{1}{2}u^2 + \tfrac{1}{2}h + \theta(h)F(\omega)$$

In the place of the term $(p/\rho)(\partial u/\partial x)$ we now have $\theta(h)[F(\omega) - 1]$, with the same restriction on $F(\omega)$ as before. Then from (c_3) we obtain

$$\frac{1}{3}h_0\frac{\partial u}{\partial x} + \frac{1}{2}[1 + 2\theta'(h_0)]\frac{\partial h}{\partial t} + [F(\omega) - 1]\theta'(h_0)\frac{\partial h}{\partial t} = 0.$$

With the help of (a_3) we eliminate $\partial u/\partial x$ and obtain by rearrangement of terms,

---

[1] See the work of Lorentz previously referred to: Eq. (3), p. 78; also pp. 111, 168.

Dispersion Theory of Sound

$$\frac{3}{h_0}\left[\frac{1}{2}+F(\omega)\theta'(h_0)\right]\frac{\partial h}{\partial t}=\frac{\partial s}{\partial t} \tag{e}$$

If for h and s we make the same assumptions earlier made for π and s, we obtain, after differentiating twice with respect to x,

$$\frac{3}{h_0}\left[\frac{1}{2}+F(\omega)\theta'(h_0)\right]\frac{\partial^2 h}{\partial x^2}=\frac{\partial^2 s}{\partial x^2} \tag{f}$$

On the other hand (b_3) yields, after single differentiation with respect to x,

$$\frac{1}{3}\frac{\partial^2 h}{\partial x^2}+\frac{h_0}{3}\frac{\partial^2 s}{\partial x^2}=\frac{\partial^2 u}{\partial x\,\partial t}=\frac{\partial^2 s}{\partial t^2} \tag{g}$$

We eliminate $\partial^2 h/\partial x^2$ between (f) and (g) and get

$$\frac{h_0}{9[1/2+F(\omega)\theta'(h_0)]}=\frac{\partial^2 s/\partial t^2}{\partial^2 s/\partial x^2}-\frac{h_0}{3} \tag{h}$$

For a plane wave the ratio of the second derivative of u with respect to t and x is equal to the square of the propagation velocity. Hence

$$V^2=\frac{h_0}{3}\left[1+\frac{1}{\frac{3}{2}+3F(\omega)\theta'(h)}\right] \tag{i}$$

In the language of thermodynamics h represents a measure of the pressure. Thus

$$p=\tfrac{1}{3}MN(\xi^2+\eta^2+\zeta^2)=\tfrac{1}{3}\rho h$$

in the notation used previously. Further, by definition we have $\theta(h)=E_i/m$, where m is the molecular weight and

$$\theta'(h_0)=\frac{1}{m}\frac{\partial E}{\partial T}\frac{\partial T}{\partial \rho}\frac{dp}{dh}$$

$$=\frac{1}{m}C_i\frac{m}{\rho R}\frac{1}{3}p=\frac{1}{3}\frac{C_i}{R}$$

since because of the equation of state $(m/\rho)p=RT$. Finally we get, from (i),

$$V^2=\frac{p}{\rho}\left[1+\frac{R}{\tfrac{3}{2}R+C_i F(\omega)}\right]$$

This agrees with Eq. (8), in which the left side [from (4)] can be replaced by V^2, while C_a, the specific heat for the external degrees of freedom, is like that of a monatomic gas, $\tfrac{3}{2}R$.

Appendix Note 2

One can derive a similar dispersion formula by the analysis of Nerzfeld and Rice mentioned earlier. They define a temperature T associated with internal degrees of freedom. They also define a heat conductivity for the internal degrees of freedom in addition to the usual one. They represent the energy contribution (per mole and unit time) that is transferred from the external to the internal degrees of freedom by the expression

$$\frac{C_a dT}{dt} = C_a \frac{T - T'}{\tau}$$

in which τ denotes a time "which measures the extent of the exchange between external and internal degrees of freedom." The equations of motion and continuity are set up in the usual form. The energy equation, however, is split into two parts, one for the external degrees of freedom

$$\frac{C_a dT}{dt} = \frac{pM}{\rho^2} \frac{\partial \rho}{dt} + (\cdots) - C_a \frac{T - T'}{\tau}$$

and the other for the internal degrees of freedom

$$\frac{C_i dT}{dt} = (\cdots) + C_a \frac{T - T'}{\tau}$$

The terms in $(\cdots)$ involve the internal viscosity and heat conduction. With the help of the assumption of equal space and time periodicity, from the above they obtain four equations, from which the state variables are eliminated. A complicated equation between V^2 and ω^2 results, whose solution carried out to a first approximation yields a dispersion formula of the type

$$V = V_0 [1 + \omega^2 (K_0 + K_1 \tau + K_2 \tau^2)]$$

If, however, from the beginning one omits the influence of viscosity and heat conduction and carries on further in the spirit of the present paper, one obtains

$$V^2 = \frac{p}{\rho} \left[1 + R \frac{1 + i\omega\tau(C_i/C_0)}{C + i\omega\tau C_i} \right]$$

which gives a similar dispersion curve to that of Eq. (19). The point of inflection at $\omega = \omega_w = (C/C_i)(1/\tau)$. The analogy here rests on the fact that the equation defining τ is of the same type as (10).

Summary

A dispersion formula (19) for the propagation of sound in gases is derived on the basis of simple assumptions concerning the exchange of quantized excitation energy in the collision between molecules. This formula contains only one quantity not determined thermodynamically, namely the lifetime of the excitation energy.

The author acknowledges with thanks the criticism and advice received from E. Grüneison, G. Jaffe, and H. Kneser.

The energy exchanges between molecules. By P. S. H. HENRY, Ph.D., Trinity College, 1851 Senior Student. (The Colloid Science Laboratory, Cambridge.)

[*Received and read* 8 February 1932.]

The marked disagreement between the specific heats at high temperatures of the diatomic gases as found by sound velocity measurements, and those calculated from theoretical considerations, has for long been a source of perplexity to statistical mechanists; a perplexity all the greater in view of the simple nature of the theoretical calculations which appeared to form as direct a test of Boltzmann's hypothesis and the theory of band spectra as was possible to large scale measurements. One way out of the difficulty is to attribute the differences to the experimental difficulties inherent in the sound velocity method of measurement at high temperatures[1], but it cannot be denied that there is a certain measure of agreement between the separate velocity determinations. It consequently becomes necessary to examine the theory.

The only apparent assumptions made in the derivation of the Planck formula for the vibrational specific heat are:

(1) that we are dealing with a state of equilibrium;

(2) that the distribution of energy in this state is that given by Boltzmann's hypothesis;

(3) that the vibrational energy levels and statistical weights have been correctly determined by the spectroscopists.

The spectra of the simple diatomic gases are now so well known that there is little possibility of error in the third assumption. Boltzmann's hypothesis, too, while not yet proved with complete logic from first principles, agrees so well with experiment in other branches of science that it can almost be regarded as an experimental fact. There remains, then, the assumption that the gas is always in a steady state when the sound is passing through it, and this would seem to be easiest of the three to attack.

In two recent papers[2] Kneser has shown, by means of sound velocity experiments over a wide range of frequencies in carbon dioxide, that there is a lag in the transfer of energy between the transverse vibrations of the CO_2 molecule and the other degrees of freedom much greater than that which would be expected from classical mechanics. In fact he found that the "mean life of a quantum" of the vibration concerned, which can also be regarded as the period of relaxation for the vibrational energy, was of the order of 10^{-5} of a second at room temperature in place of the usually assumed period of the order of 10^{-10} of a second. He obtained an expression for the sound velocity which, when expressed in terms of the specific heat, becomes

$$C_{app} = \frac{C_v^2 + 4\pi^2 n^2 \beta^2 C_a^2}{C_v + 4\pi^2 n^2 \beta^2 C_a} = \frac{C_v^2 + \alpha^2 C_a^2}{C_v + \alpha^2 C_a},$$

where C_v is the true specific heat at constant volume, C_a is that part of the specific heat which is contributed by the translations and rotations (i.e. $\tfrac{5}{2}R$ for a diatomic gas), n is the frequency of the sound, β is the period of relaxation of the vibrational energy, and $\alpha = 2\pi n\beta$.

Now if one admits a period of relaxation of the order of 10^{-5} seconds for CO_2, why should one not suppose that the vibrations of oxygen and nitrogen might have a relaxation period of the order of 10^{-4} of a second? Unless we are to disregard the sound velocity results in toto, we must make some fundamental alteration to the theory, and this seems to be the easiest to make. The results of Shilling and Partington can be accounted for if we assume the following periods of relaxation:

Temperature	Relaxation Period in Secs.		Transparency Coefficient	
	Oxygen	Nitrogen	Oxygen	Nitrogen
0° C.	—	—	370	8600
300° C.	1.0×10^{-4}	—	19.5	58
600° C.	1.2	0.88×10^{-4}	11.1	20
900° C.	1.1	0.92	9.4	12.5
1200° C.	1.0	—	7.9	10.3

At room temperatures the vibration comes in so little that the expected effect is smaller than the probable errors of the sound velocity determinations, and I have not included values for the relaxation period for these temperatures. The figures in the last two columns are explained below.

In confirmation of this theory it may be noted here that the experimental results available for oxygen above room temperatures lie in the order of the quickness with which the gas was heated. Thus the sound velocity determinations[3], using a frequency of about 3000 vibrations per second, gave the lowest results, the adiabatic expansion experiments of Eucken and Lüde[4] gave somewhat higher results, while the author's experiments[5], in which the heating took place in about a quarter of a second, gave much higher results, which were, in fact, a little higher than the theoretical.

It may be pointed out that similar theories have been propounded before, by O. K. Rice[6] in connection with unimolecular reactions, and by N. Semenoff and A. Schechter[7] in connection with the dissociation of hydrogen molecules by ions.

There are various ways in which the problem might be attacked experimentally. The most obvious is the determination of the sound velocity over a wide range of frequencies in oxygen and nitrogen at temperatures considerably above that of the atmosphere. The various experiments which have been carried out at room temperatures in air are of little use for this purpose, since the predicted effect is too small at these temperatures to be verified or disproved with certainty and those experiments which have been done at higher temperatures used only one frequency.

Another method would be to investigate the absorption of the sound energy by the gases at elevated temperatures. At those frequencies at which the velocity changes, there should be considerable absorption owing to the irreversible heat transfer. The gas should, in fact, show a diffuse absorption band for each mode of vibration of the molecule, the frequencies of the maxima depending upon the relaxation period for the vibration in question. Thus, as Kneser remarks, Pierce[8] found that CO_2 was nearly opaque to sound at just that frequency where Kneser found that the velocity changed most rapidly.

Calculations of the effect of such a lag in the energy transfer have been made by K. F. Herzfeld and F. O. Rice[9], and more elaborate ones by D. G. Bourgin[10]. The case which we are considering, however, is not so complicated as theirs, and it may be worth while to give a simpler derivation of the effect, especially as the former authors assumed that the relaxation period was small compared with the period of a sound vibration.

Let C_v, C_a and C_x be the total specific heat per gm. mol., the specific heat of the translation and rotations, and that of the vibrations respectively, so that

$$C_v = C_a + C_x. \tag{1}$$

Let E_x be the actual energy per gm. mol. of the vibrations at any instant, and E_T the energy which the vibrations would have at equilibrium at the temperature T.

We shall suppose that

$$\frac{dE_x}{dt} = \frac{1}{\beta}(E_T - E_x), \tag{2}$$

so that we may call β the period of relaxation of the vibrational energy. This equation will be obeyed with sufficient exactness where there is only one vibrational level above the ground level appreciably in evidence; for small deviations from the equilibrium state such as occur in sound waves of average intensity it is likely enough to hold in any case. Where there is only one excited state concerned, β is equal to the "mean life of a quantum" of Kneser.

Now suppose that the gas is subject to small sinusoidal adiabatic variations of temperature of frequency $\frac{\omega}{2\pi}$. We can write
$$T = T_0 + T_1 e^{i\omega t},$$
$$E_T = E_0 + C_x T_1 e^{i\omega t}.$$
Substituting this in (2) and solving, we get
$$E_x = E_0 + \frac{C_x T_1}{1 + i\omega\beta} e^{i\omega t}. \qquad (3)$$
We can thus regard the gas as having a complex specific heat given by
$$C_v' = C_a + \frac{C_x}{1 + i\omega\beta}. \qquad (4)$$
In the usual way it can now be shown that the adiabatic elasticity of the gas is given by $H = p\gamma'$, where $\gamma' = 1 + R/C_v'$.

Substituting this in the equation of motion for plane waves,
$$\rho \frac{\partial^2 \xi}{\partial t^2} = H \frac{\partial^2 \xi}{\partial x^2},$$
and trying the solution $\xi = \Xi e^{i\omega t + (l+im)x}$, we get
$$-\rho\omega^2 = \gamma' p (l + im)^2.$$
If in this we write $\gamma' = \gamma_1 + i\gamma_2$, equate real and imaginary parts, and put $\mu = l/m$, we get
$$\left.\begin{array}{l} \mu^2 + \dfrac{2\gamma_1}{\gamma_2} \mu - 1 = 0, \\[1ex] \mu^2 - \dfrac{2\gamma_2}{\gamma_1} \mu - 1 = -\dfrac{\rho}{p} \dfrac{\omega^2}{\gamma_1 m^2} = -\dfrac{\gamma_{\mathrm{app}}}{\gamma}, \end{array}\right\} \qquad (5)$$
where γ_{app} is the value for γ which would be deduced from the velocity measurements if the lag effect was ignored.

From these we get
$$\left.\begin{array}{l} \mu = -\dfrac{\gamma_1}{\gamma_2} + \sqrt{\dfrac{\gamma_1^2}{\gamma_2^2} + 1}, \\[1ex] \gamma_{\mathrm{app}} = 2\gamma_1 \left(\dfrac{\gamma_1}{\gamma_2} + \dfrac{\gamma_2}{\gamma_1}\right) \mu. \end{array}\right\} \qquad (6)$$

If μ is small, as it must be if the velocity can be measured with any accuracy, these become
$$\left.\begin{array}{l} \mu = \gamma_2/2\gamma_1, \\ \gamma_{\mathrm{app}} = \gamma_1, \end{array}\right\} \qquad (7)$$

and from these we get, using equation (4) and the relation
$$\gamma' = 1 + R/C_v',$$

$$\text{apparent specific heat} = \frac{C_v^2 + \omega^2\beta^2 C_a^2}{C_v + \omega^2\beta^2 C_a}, \qquad (8)$$

this expression being identical with that given on page 250, since $\omega = 2\pi n$, and equivalent to that given for the sound velocity by Kneser.

The absorption of the sound is most simply expressed in terms of the number of wave-lengths which the sound travels before its amplitude is diminished to $1/e$ times its initial value. It will be seen that this number is

$$N = \frac{1}{2\pi\mu} = \frac{C_v(C_v + R) + a^2 C_a(C_a + R)}{a\pi R(C_v - C_a)}, \qquad (9)$$

where a, as before, is equal to $\omega\beta$ or $2\pi n\beta$.

The frequency at which the absorption, reckoned in this way, is a maximum is given by

$$n = \frac{1}{2\pi\beta}\sqrt{\frac{C_a(C_a + R)}{C_v(C_v + R)}}; \qquad (10)$$

and the value of N at this frequency is

$$N_m = \frac{2\sqrt{C_v C_a(C_v + R)(C_a + R)}}{\pi R(C_v - C_a)}. \qquad (11)$$

It will be seen that this latter is independent of the relaxation period, depending only on the relative values of the vibrational and other specific heats. This is only true for the transparency as defined above, for the ordinary absorption coefficient $(-2l)$ has no maximum at a finite frequency, but tends to an upper limit as the frequency tends towards infinity. Values of N_m calculated for oxygen and nitrogen are given in the table on page 250. The frequency at which the gases should show the maximum absorption is about 1400 vibrations per second in each case. It will be seen that the absorption is negligible at room temperatures, but should become easily measurable at somewhat higher temperatures. The only absorption experiments which have been done, so far as the author is aware, are those at room temperatures[11]. The calculated energy absorption coefficient for air arising from the effect under discussion is of the order of 2×10^{-4} per cm., while that arising from viscosity and heat conduction at the high frequencies used in the experiments is of the order of 10^{-1} or 10^{-2}, so that these experiments do not help us. It may be noted here that when Herzfeld and Rice deduced from these experiments a relaxation period of about 10^{-8} seconds, they were referring mainly to the rotational energy.

A more direct method of investigating the matter would be to photograph the ultra-violet absorption spectrum of oxygen as it is escaping from a cylinder at a high pressure (and possibly also high temperature). The gas would be subject to a sudden cooling on account of the expansion, and it might be possible to observe bands corresponding to transitions from the first excited vibrational state when the distribution of intensity in a given band shows that the rotations correspond to a temperature so low that there would normally be hardly any of the molecules in the excited vibrational state. The molecules would, in fact, have two temperatures, the rotational and the vibrational. An attempt to carry out this experiment will shortly be made in this laboratory, and more will be said about it later.

It is interesting to speculate about the manner of the energy exchanges if this theory be true. The slight decrease of the relaxation period of oxygen as the temperature rises would seem to indicate that vibrational energy can only be turned into translational energy in a collision of considerable relative kinetic energy, for the total number of collisions per molecule in a gas at constant pressure varies inversely as the square root of the temperature, and only the high-energy collisions increase in number as the temperature rises. This would mean, according to the principle of detailed balancing, that a vibration can only be started in a collision where the relative kinetic energy is considerably in excess of that required. The data are not sufficiently certain yet, however, for such considerations to have much weight.

In a recent paper Zener[12] has made tentative calculations of the probability that energy of vibration will be transferred from one molecule to another during a collision, using partly wave-mechanical principles and partly classical conceptions. Such a transfer would not, of course, affect the total vibrational energy of the gas, and what we are concerned with is transformation of vibrational energy into other forms. Such a transformation would have a smaller probability than the transfer which Zener considered; and it is interesting to note that he gave for the probability of transfer during a head-on collision between two diatomic molecules at $0°$ C., where one of them possessed one vibrational quantum, the value of 2×10^{-5}. The probability of transformation of the vibrational energy in a given collision between oxygen molecules at $600°$ C. calculated from the relaxation period given above is about 0.5×10^{-5}.

REFERENCES.

(1) *Proc. Phys. Soc.*, 43, p. 340 (1931). See also (2).

(2) *Ann. d. Phys.*, 11, pp. 761, 777 (1931).

(3) *Phil. Mag.*, 3, p. 273 (1927); 6, p. 920 (1928); 9, p. 1020 (1930).

(4) *Zeit. f. Phys. Chem.* B, 5, p. 413 (1929).

(5) *Proc. Roy. Soc.* A, 133, p. 492 (1931).

(6) *Zeit. f. Phys. Chem.* B, 7, p. 226 (1930).

(7) *Nature*, 126, p. 436 (1930).

(8) *Am. Acad. Sci.*, 60, p. 271 (1925).

(9) *Phys. Rev.*, 31, p. 691 (1928).

(10) *Phil. Mag.*, 7, p. 821 (1929).

(11) Neklepajew, *Ann. d. Phys.*, 35, p. 175 (1911); Rich and Pielemeier, *Phys. Rev.*, 25, p. 117 (1925); and Abello, *Phys. Rev.*, 31, p. 1083 (1928).

(12) *Phys. Rev.*, 38, p. 277 (1931). See also *Phys. Rev.*, 37, p. 556 (1931), and O. K. Rice, *Phys. Rev.*, 38, p. 1943 (1931).

THE EFFECT OF HUMIDITY UPON THE ABSORPTION OF SOUND IN A ROOM, AND A DETERMINATION OF THE COEFFICIENTS OF ABSORPTION OF SOUND IN AIR

By Vern O. Knudsen

University of California at Los Angeles

1. *Introductory.* In making reverberation measurements in a room, one soon observes that the condition of the air in the room, especially with respect to humidity, affects the reverberation time for frequencies above 1000 d.v. P. E. Sabine[1] noted that for frequencies above about 2000 d.v. the rate of absorption of sound in his reverberation room increased markedly when the relative humidity in the room was low. E. Meyer[2] obtained a somewhat similar effect for frequencies of 3200 and 6400 d.v., but observed no perceptible change of reverberation time with humidity at frequencies below 1600 d.v. Meyer's results are shown in Fig. 1. These curves give the measured times of reverberation in his reverberation room for different values of relative humidity between about 47 percent and 100 percent, and at a temperature of about 17°C.

About a year ago the writer undertook an investigation of the effect of humidity and temperature upon the rate of absorption of sound in a room. The results obtained to date, on the effect of humidity, have confirmed the general findings of Sabine and Meyer, and, in addition, have led to some results which not only are important in connection with problems in architectural acoustics and sound signaling but also are of theoretical interest in connection with the nature of the absorption of sound vibrations in gases.

2. *Theoretical Considerations.* It is a general characteristic of a wave motion propagated through a ponderable medium that part of the energy of the wave motion will be converted into other forms of energy. Thus, the intensity of a plane wave after traveling a distance x in a homogeneous medium is, provided the displacements be not too large, $I_0 e^{-mx}$, where I_0 is the intensity of the wave at the position $x=0$, and m is the attentuation or absorption coefficient for the plane wave in the medium. The attentuation constant m depends upon viscosity, heat conduction, heat radiation,[3] and possibly a number of other factors, such

---

[1] "The Measurement of Sound-Absorption Coefficients," Jour. Frank. Inst., March, 1929.

[2] "Ein neues automatisches Verfahren der Nachhallmessung," Zeits. f. Technische Physik. 7, 253, 1930.

[3] Rayleigh, "Theory of Sound" II, pp. 312–323.

as wave distortion and molecular absorption.[4] This aspect of sound-absorption will be considered in a later paper.

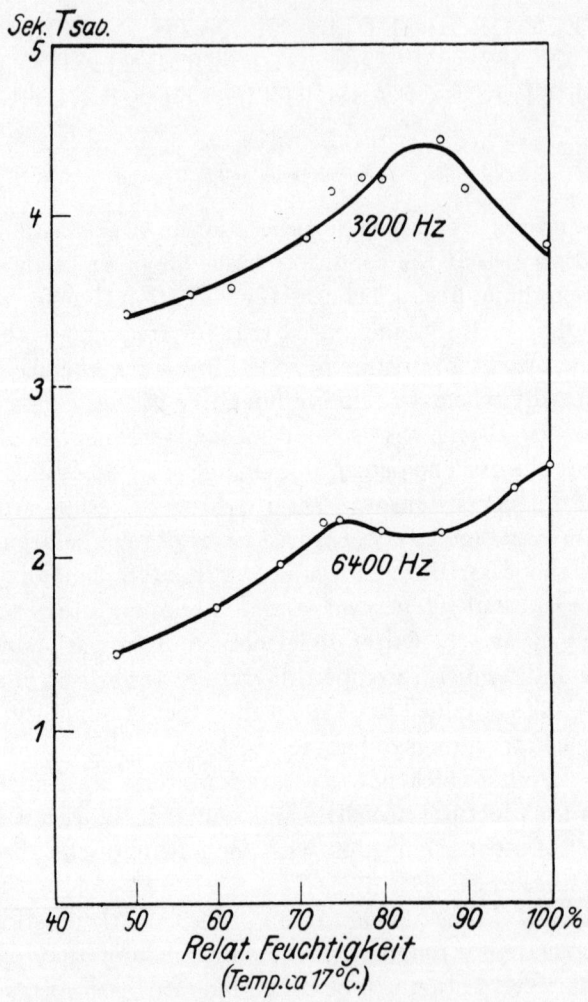

FIG. 1. *Curves obtained by Meyer showing the relation between the relative humidity and the time of reverberation in Meyer's reverberation chamber.*

In the existing theories of reverberation, it has been assumed that the absorption in the air of audible vibrations is so small as to be negligible, and therefore that the absorption of sound in a room is wholly attributable to the surface absorption by the boundaries of the room. It

[4] Herzfeld and Rice, Phys. Rev., 31, 691, April, 1928. Bourgin, Phil. Mag., 7, 821, 1929; Phys. Rev., 34, 521, August 1, 1929. Rocard, Jour. de Physique, 1, No. 12, December, 1930.

will be seen presently that this assumption is justified for frequencies below 512 d.v., but that the absorption in the air may become a significant factor for higher frequencies.

The equation for the rate of decay of sound in a room, assuming that all of the absorption takes place at the boundaries, is, using the recent formula proposed by Eyring,[5]

$$I = I_0 e^{\frac{S \log_e (1-\alpha) ct}{4V}} \qquad (1)$$

where S is the interior surface of the room, α is the average coefficient of absorption of the surface S, c is the velocity of sound, t the time, and V the volume of the room.

If the effect of absorption in the air be introduced in this equation, it becomes

$$I = I_0 \left(e^{\frac{S \log_e (1-\alpha) ct}{4V}} \right) e^{-mx}. \qquad (2)$$

And since $x = ct$,

$$I = I_0 e^{\left[\frac{S \log_e (1-\alpha)}{4V} - m \right] ct}. \qquad (3)$$

If we solve this equation for t, when $I_0/I = 10^6$, we obtain the reverberation formula. In British units, for a rectangular room, and for a temperature of 21° C, (3) becomes

$$t = \frac{.049V}{-S \log_e (1-\alpha) + 4mV}. \qquad (4)$$

It will be noted that the second term in the denominator, $4mV$, represents the effective absorption in the room contributed by losses in the air. When m is negligibly small, (4) reduces to the well known reverberation equation which does not take account of the absorption in the air.

The usual reverberation measurements in a room do not distinguish between the absorption which takes place at the boundaries of the room and that which takes place in the air. However, if reverberation measurements be made in two rooms which have the same boundary material, but different mean free paths, it is possible to determine both the absorption coefficient m in the air, and the absorptivity α of the boun-

---

[5] "Reverberation Time in Dead Rooms," Jour. Acous. Soc. of Amer., 1, 2, 217–241, January 1930.

dary material. Thus, suppose two such rooms have volumes V_1 and V_2, and interior surfaces S_1 and S_2. Then, if t_1 and t_2 be the reverberation times for a particular frequency in the two rooms, under the same conditions of temperature and humidity,

$$t_1 = \frac{.049 V_1}{-S_1 \log_e (1-\alpha) + 4mV_1}, \tag{5}$$

$$t_2 = \frac{.049 V_2}{-S_2 \log_e (1-\alpha) + 4mV_2}. \tag{6}$$

Solving for m and $\log_e (1-\alpha)$, there results

$$m = \frac{.0122 \left(\dfrac{S_2 V_1}{S_1 V_2 t_1} - \dfrac{1}{t_2} \right)}{\left(\dfrac{S_2 V_1}{S_1 V_2} - 1 \right)}, \tag{7}$$

$$\log (1_e - \alpha) = \frac{.0122 \left(\dfrac{1}{t_1} - \dfrac{1}{t_2} \right)}{\left(\dfrac{S_2}{4V_2} - \dfrac{S_1}{4V_1} \right)}. \tag{8}$$

3. *Experimental Findings.* Experimental studies on the absorption of sound in the air have been conducted in the reverberation chamber and in the two test chambers in the Acoustical Laboratory at the University of California at Los Angeles.[6] Each room is enclosed by two separate walls of 12″ thick reenforced concrete, and the floors of these rooms rest upon a 6″ fill of sand and a 2″ slab of cork. The rooms are therefore well insulated against external vibration, sound, and heat. Owing to the small temperature changes throughout the seasons of the year, the temperature in the room is never lower than 20° C and never higher than 22° C, and usually differs by only two or three tenths of a degree from 21° C. The rooms therefore are well suited for constant temperature measurements. The relative humidity, on the other hand, follows closely the changes in the outside humidity, and it is not uncommon to have relative humidity changes from 14 percent to 70 percent within a period of forty-eight hours. Opportunity is thus afforded to investigate the effects of relative humidity upon sound-absorption throughout a wide

---

[6] Plan and sectional drawings of these rooms can be found in the author's article on "The Measurement and Calculation of Sound-Insulation," Jour. Acous. Soc. of Amer., 2, 129–140, July, 1930.

range of humidities, without resorting to artificial means of controlling the humidity. Humidities above 70 percent, up to saturation, were readily obtained by evaporating water in the room.

The reverberation times were measured (1) by obtaining oscillograms of the decay of sound in the rooms, (2) by measuring the times

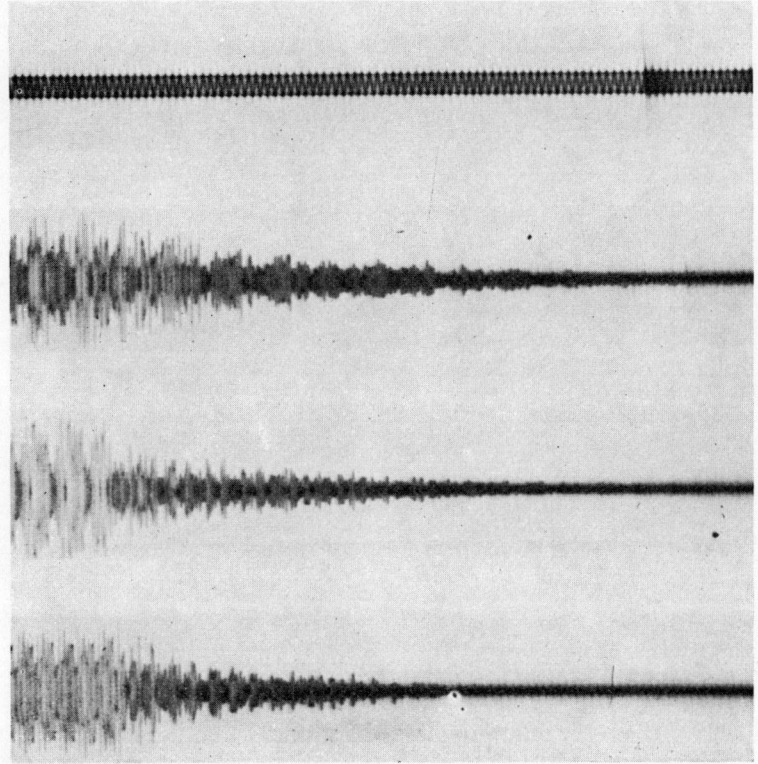

FIG. 2. *Oscillograms of the decay of a warble tone 4096 ± 100 d.v., for different conditions of humidity in the reverberation room at the University of California at Los Angeles. The upper record is a time curve 50 cycles per second; the next record shows the decay in the room at a relative humidity of 74 per cent; the next record shows the decay at a relative humidity of 56 per cent; and the lower record shows the decay at a relative humidity of 26 per cent.*

for specified amounts of decay of sound in the rooms by means of a microphone pick-up, a five stage amplifier, and a relay controlled chronograph, and (3) by making reverberation measurements with the ear and a recording chronograph. In the oscillograph and relay methods, "warble" tones were used having a band width of 200 cycles and a frequency of warble of five to seven cycles per second. Thus, at 2048 d.v., the warble tone consisted of 2048 ± 100 d.v. The reverberation measurements made with the ear and chronograph were all made with single-frequency pure tones. Because of the speed and reliability of measurements made with the relay-controlled meter, this method was used for obtaining most of the data, but check measurements were made by both the oscillograph method and the reverberation method at representative frequencies and conditions of humidity. The times of reverberation determined by the relay method and the oscillograph method, under the same conditions of temperature and humidity, and on the same day, did not differ by more than 5 percent; and the times of reverberation de-

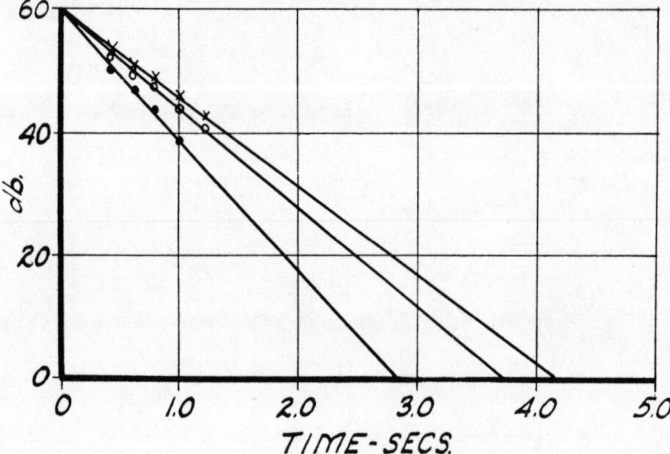

FIG. 3. *Logarithmic decay curves obtained from a series of fifteen oscillograms similar to those shown in Fig. 2.*

termined by the relay and reverberation methods did not differ by more than 3 percent. Each time of reverberation, as determined by the relay method, was based upon 100 separate measurements of the times required for decays of 20, 30, 40, or 50 db. The observed times for decays of 20, 30, 40, and 50 db were very closely in the ratio of 2:3:4:5, indicating that the decay of sound in the room was logarithmic, and that measurements made for a drop of 20 or 30 db can be extrapolated to give the time required for a drop of 60 db. The average departure from the mean time of reverberation, as determined by this method, was 1.4 percent at 2048 d.v. and 1.7 percent at 4096 d.v. Each determination of the time of reverberation with the oscillograph method is based upon measurements of five ocsillograms taken under identical conditions. The over-all characteristics of the oscillograph and its associated amplifier were determined, so that the amplitudes shown on the oscillograms could be converted to pressure variations in the decadent sound. The amplitudes on the oscillograms are measured, corrected for distortion, and then the rate of decay in decibels is plotted against the time. The reverberation time is then given by extrapolating the decay curve for a drop of 60 db.

Typical oscillograms for 4096±100 d.v., obtained in the large reverberation room at relative humidities of 26 per cent, 56 per cent, and 74 per cent (at a temperature of 21.4° C), are shown in Fig. 2. It can be seen from these three oscillograms that the rate of decay of sound in the room decreased as the relative humidity increased. The data given in

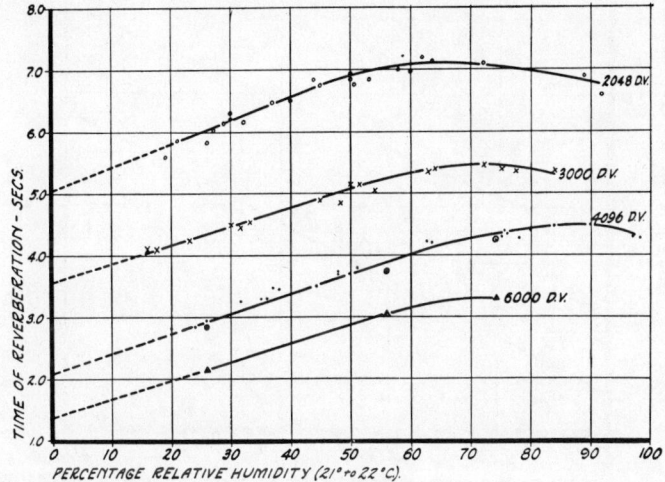

FIG. 4. *Curves showing the dependence of the time of reverberation upon relative humidity in a large room 19'×20'×16'. Measurements made when wall temperature was lower than the temperature of the air in the room.*

Fig. 3 were obtained from measurements on fifteen such oscillograms—five oscillograms for each condition of humidity. The times of reverberation are given by the intercepts on the time axis, namely 2.85 seconds at 26 per cent relative humidity, 3.75 seconds at 56 per cent, and 4.25 seconds at 74 per cent.

In Fig. 4 are shown the results of reverberation measurements in the large room (19'×20'×16') for different conditions of air humidity. The data for 2048 d.v., 3000 d.v. and 4096 d.v. were obtained with the relay-controlled meter; and the few data for 6000 d.v. were obtained with the oscillograph. The time of reverberation is plotted as a function of the percentage relative humidity. Since the temperature was maintained at 21.4°±0.6° C, the absolute humidity can be readily computed from the relative humidity. It will be noted that the time of reverberation increases almost uniformly with an increase in relative humidity, up to 60 per cent relative humidity. For higher values of relative humidity, up to about 80 per cent, the reverberation time increases more slowly, and above about 80 percent the reverberation time decreases. In the series of tests shown in Fig. 4, the temperature of the walls surrounding the room was always slightly lower than the temperature of the air in the room, and varied from 20.8° to 21.6° C. Consequently, there always would be some condensation on the walls before the air became com-

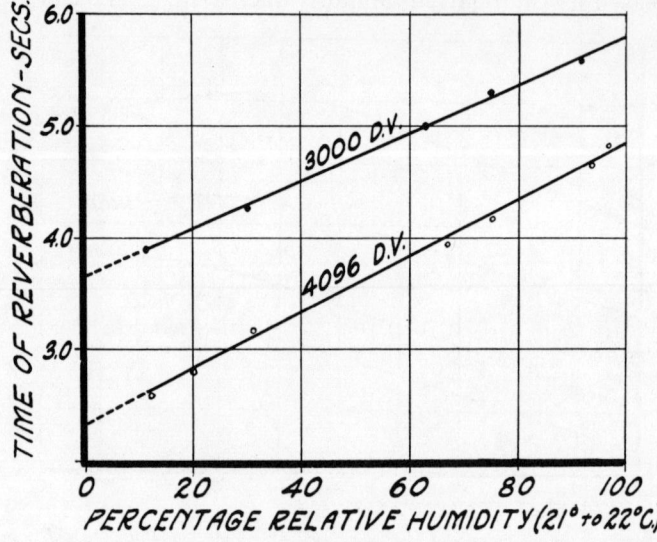

FIG. 5. *Curves showing the dependence of reverberation time upon relative humidity. These measurements were obtained when the temperature of the wall was higher than the temperature of the air in the room.*

pletely saturated. It was repeatedly observed that condensation on the walls began at a humidity of about 70 to 80 per cent, and the detection of condensation on the walls was always coincident with the change in slope of the reverberation-humidity curve. It appears therefore that the bending down of the curves for humidities above about 80 per cent is attributable to a surface effect, that is, to the condensation of moisture on the walls. This is suggested by some humidity measurements which were taken in the summer of 1930, at a time when the temperature of the walls was probably higher than the temperature in the air.[7] Under

---

[7] No measurements of the wall temperature were taken at that time, but the assumption seems reasonable since the air supplied to the building in the summer time is somewhat cooler than the average outside temperature. This point will be investigated in the summer of 1931.

these conditions there was no condensation on the walls, and the time of reverberation increased almost uniformly as the relative humidity was increased to about 100 per cent. This is shown in Fig. 5, which gives the results of measurements for 2048 and 4096 d.v. If there had been no condensation on the walls during the winter of 1930–31, it is probable that the curves shown in Fig. 4 would have been approximately straight lines similar to those shown in Fig. 5.

When the relative humidity was maintained at 100 per cent, and fog appeared in the room, the reverberation times became markedly lower for all frequencies. This is shown in the following table:

Frequency	Time of Reverberation at 80% Relative Humidity	Time of Reverberation with Air Saturated and Fog in Room
128 d.v.	16.15 seconds	10.35 seconds
256 d.v.	14.30 "	8.91 "
512 d.v.	12.65 "	6.52 "
1024 d.v.	10.80 "	6.29 "
2048 d.v.	7.00 "	5.00 "
496 d.v.	4.42 "	2.93 "

It is probable that most of this effect of increased absorption at saturation is attributable to condensation on the walls, although there is a possibility that the fog itself absorbs some sound.

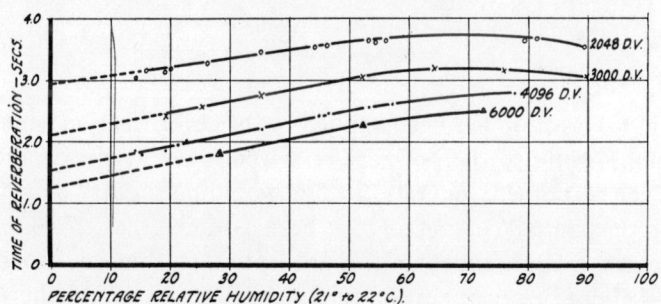

FIG. 6. *Curves showing the dependence of reverberation time upon the relative humidity in a room $8' \times 8' \times 9'\text{-}6''$.*

The curves shown in Fig. 6 give the reverberation time as a function of the relative humidity in a small test room, $8' \times 8' \times 9'-6''$. The walls, floor and ceiling of this room (all reenforced concrete) were painted and varnished with the same kind of paint and varnish which had been used for the interior surfaces of the large reverberation room. The data given in Figs. 4 and 6 therefore are for two rooms which have the same boundary material, namely painted and varnished concrete, but different mean free paths. The reverberation data for these two rooms, as given in Figs. 4 and 6, consequently satisfy the requirements for Eqs. (5) and (6), and hence, by substituting the appropriate values of V_1, V_2, S_1, S_2, t_1 and t_2 in Eqs. (7) and (8), it is possible to solve for m, the absorption coefficient in the air, and α, the absorptivity of the concrete, for different frequencies and humidities. The two-room experiment therefore furnishes a critical experiment for determining whether the effect of humidity upon the absorption of sound in a room is attributable to a change in the surface of the boundaries, to a change in the attenuation constant in the air, or to both.

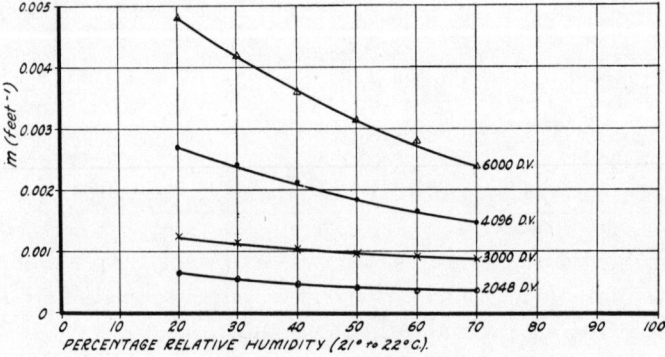

FIG. 7. Curves giving the value of m for different conditions of humidity, for tones of 2048 to 6000 d.v.— temperature of air 21° to 22°C.

A slight correction has been applied to the reverberation times in the small room, owing to the presence of a wood door (painted and varnished as the rest of the concrete), which would be slightly more absorptive than the painted concrete. It has been assumed that this door would add an extra .5 square foot unit of absorption to the small room.[8] The measured times of reverberation in the small room would therefore be about 4 per cent to 6 per cent shorter than they would be if the boundary materials were identical in the two rooms. The times indicated by the curves in Fig. 6 are 4 per cent to 6 per cent above the measured times, in

[8] There is some uncertainty in applying this correction but it is highly probable that the correction is of the right order of magnitude. Thus the coefficient of absorption of painted and varnished concrete is of the order of .017 at frequencies of 2000 to 4000 d.v., and the coefficient of painted and varnished wood is of the order of .037 at these frequencies. Since the area of the wood door was 25 sq. ft., the excess absorption supplied by the door would be $25(.037 - .017) = .50$ unit.

order to make the times more nearly what they would have been if the entire interior surface of the small room had been the same as that in the large room.

From the data given in Figs. 4 and 6 and from the dimensions of the two rooms, the values of m and α have been calculated for different frequencies and different conditions of humidity. The volumes and interior surfaces of the two rooms are as follows (subscript 1 refers to the large room and subscript 2 refers to the small room): $V_1 = 6080$ cu. ft., $V_2 = 608$ cu. ft. $S_1 = 2010$ sq. ft., and $S_2 = 432$ sq. ft. The values of m for different amounts of water vapor in the air, and for frequencies of 2048, 3000, 4096 and 6000 d.v., as calculated by Eq. (7), are given in the curves in Fig. 7. It will be noted that at any one frequency m decreases as the relative humidity increases, and that at any one condition of humidity m increases approximately with, but not quite so rapidly as, the square of the frequency. But since the value of m depends upon the difference of the reciprocals of the times of reverberation in the two rooms, and since there is a possible error of approximately 2 per cent to 3 per cent in the measured times of reverberation (including the uncertainty of the excess absorption in the small room owing to the wood door), there are probable errors in the values of m as large as 8 per cent to 10 per cent at 4096 d.v., and as large as 20 per cent to 30 per cent at 2048 d.v.

In the following table are given the values of α obtained by sub-

Relative Humidity	Coefficients of Sound-Absorption α		
	2048 d.v.	3000 d.v.	4096 d.v.
20%	.018	.020	.020
30%	.018	.019	.019
40%	.017	.018	.018
50%	.017	.017	.017
60%	.016	.017	.016
70%	.016	.017	.016
80%	.017	.016	.019

stituting the appropriate values of V_1, V_2, S_1, S_2, t_1 and t_2 in Eq. (8). It will be noted from these values of α that the coefficient of absorption of painted and varnished concrete, at these frequencies, does not depend appreciably upon the frequency, and, within the limits of accuracy of the measurements, the absorptivity of the painted concrete is nearly independent of the humidity of the air, although the calculated values of α show a tendency to increase slightly as the humidity decreases. More precise experiments must be made to verify this finding.

From reverberation measurements at 512 d.v. in the large room, the coefficient of absorption of the painted concrete is .012 at this frequency. It will be noted that the coefficients at 2048 and even 4096 d.v. are only slightly higher than the coefficient at 512 d.v., whereas if the absorption in the air had been neglected, the calculated coefficient of absorption at 4096 d.v. would have been .054 at 20 per cent relative humidity and .032 at 70 per cent relative humidity. If these values of α be compared with .020 at 20 per cent relative humidity and .016 at 70 per cent relative humidity, that is with the coefficients which are obtained when correction is made for the absorption of sound in the air, it appears that the presence of water vapor in the air in a room does not sensibly effect the absorptivity of the boundaries (painted concrete) of the room, at least until condensation commences on the surface, and that the principal effect of humidity is to change the rate of absorption in the air. Further, it appears that the absorption of high frequency sound in the air places an upper limit upon the time of reverberation which can be attained in a reverberation chamber, and that but little is to be gained in an attempt to obtain long times of reverberation at high frequencies by using more reflective surfaces than painted concrete for the interior of a reverberation chamber.

The results obtained in these experiments indicate that the absorption of sound in the air is an important factor which must be taken into consideration in problems in sound signaling and in architectural acoustics. Thus, if a tone of 4096 d.v., in the form of a plane parallel beam, were used for long range signaling there would be, at a temperature of 21° C. and a relative humidity of 44 per cent, an attenuation of 9.8 db per second, or about 46 db per mile.[9] On the other hand, the attenuation would be less than 1 db per mile for a frequency of 512 d.v.

Again, consider the reverberation time for a tone of 4096 d.v. in a rectangular room $80' \times 100' \times 30'$. Suppose the relative humidity to be only 20 per cent at 21° C., and the interior boundaries of the room to have an average absorptivity of .20. Then $V = 240,000$ cu. ft., and $S = 26,800$ sq. ft. Neglecting the air absorption.

$$t = \frac{.049 \times 240,000}{-26,800 \log_e (1 - .20)} = 1.97 \text{ seconds}.$$

[9] This follows from the manner in which m enters in Eq. (2). The reciprocal of m will give the distance in feet a plane wave must travel in order to have the intensity reduced to $1/e$. Since $e = 2.718$ the intensity of the wave is reduced 4.34 db for each $1/m$ feet the wave advances. Since $m = .002$ ft.$^{-1}$ at a temperature of 21°C, and a relative humidity of 44 per cent, for a sound wave of 4096 d.v., the intensity will be reduced 4.34 db for each 500 feet the plane

And, including the absorption in the air ($m = .0027$),

$$t = \frac{.049 \times 240{,}000}{-26{,}800 \log_e (1 - .20) - 4 \times .0027 \times 240{,}000} = 1.37 \text{ seconds.}$$

At a relative humidity of 70 per cent, $m = .0015$, and

$$t = \frac{.049 \times 240{,}000}{-26{,}800 \log_e (1 - .20) - 4 \times .0015 \times 240{,}000} = 1.58 \text{ seconds.}$$

The effect of the absorption in the air is thus seen to be a very appreciable factor in calculating the reverberation time or absorption in a room at frequencies above 4000 d.v. Further, owing to this absorption in the air, the high frequency components of speech and music suffer an appreciable attenuation in large auditoriums and in open-air theaters, especially when the air is dry. Finally, in order to compensate for this increased absorption in the air at high frequencies, materials which are to be used for the control of reverberation in auditoriums should have a decreasing absorptive characteristic for frequencies above about 2000 d.v.

The writer gratefully acknowledges the assistance he has received from Messrs. Leo P. and Lewis A. Delsasso. Both have contributed largely to the design of the apparatus and to the technique and labor required in conducting the experiments.

---

wave advances. Assuming a velocity of 1125 feet per second, the attenuation would amount to 9.8 db per second.

The Absorption of Sound in Air, in Oxygen, and in Nitrogen—Effects of Humidity and Temperature

VERN O. KNUDSEN, *University of California at Los Angeles*
(Received August 2, 1933)

THE results of an investigation on the absorption of sound in air, published by the author two years ago,[1] indicated that, at least in the audible range of frequencies, the absorption coefficient (1) was of the order of 10 to 25 times greater than the coefficient calculated by classical theory, and (2) increased rather markedly as the humidity of the air decreased. The more recent work of Hopper[2] and of Chrisler and Miller[3] has led to similar results, although their experiments were not designed to determine the absolute values of the absorption coefficients. In addition, the work of Chrisler and Miller indicates that the absorption increases with a rise of temperature. The work reported in this paper extends the author's earlier measurements to dry air and to a range of temperature between $-15°C$ and $55°C$. The new results give a satisfactory confirmation of the former ones and also reveal regions of selective absorption at low humidities which are of interest both practically and theoretically.

I. Apparatus and Technique of Measurement

The experimental procedure is patterned after the two-chamber method previously devised by the author,[1] but both the apparatus and the technique of measurement have been improved. The two reverberation chambers are made of the same grade of cold rolled blue steel boiler plate, 3/16″ thick. One chamber is a six-foot cube; the other a two-foot cube. The two-foot cube is rigidly reinforced around its edges with angle irons, spot-welded to the steel walls; the six-foot cube is reinforced similarly with angle irons spaced 2′ on centers, thus dividing the walls into $2' \times 2'$ panels and supported in the same manner as the walls of the smaller chamber. The inner walls of both chambers are coated with shellac. These precautions insure a high degree of uniformity of the absorptivity of the boundaries of the two chambers—an essential requirement in the method of measurement. Access to the interiors of the chambers is made by removing one wall, which is fastened with bolts and wing nuts. A rubber gasket is used to insure a tight seal. Each chamber is equipped with a motor-driven paddle, as shown in Fig. 1. Both paddles are of the same shape, but the large one (having outside dimensions of $4' \times 4'$) is three times as large as the small one ($1'-4'' \times 1'-4''$). The "mean free paths" of the two chambers are

FIG. 1. Interior view of six-foot chamber showing motor driven paddle.

---

[1] V. O. Knudsen, *The Effect of Humidity upon the Absorption of Sound in a Room, and a Determination of the Coefficients of Absorption of Sound in Air*, J. Acous. Soc. Am. **3**, 126–138 (1931).

[2] F. L. Hopper, *The Determination of Absorption Coefficients for Frequencies up to 8000 Cycles*, J. Acous. Soc. Am. **3**, 415–427 (1932).

[3] V. L. Chrisler and Catherine E. Miller, *Some of the Factors Which Affect the Measurement of Sound-Absorption*, Bur. Standards J. Research **9**, 175–186 (1932).

therefore in the ratio of 3 to 1. The paddle in the large chamber is driven at a speed of 35 r.p.m.; the one in the small chamber at a speed of 88 r.p.m. This provides thorough mixing of the sound, an indispensable requirement for obtaining satisfactory decay curves. Without the paddles, or with the paddles stationary or in slow motion, the decay curves are irregular and unreliable.

By measuring the rates of decay of pure tones in the two chambers, keeping the air or gas mixture in both chambers under the same conditions, it is possible to determine both the surface absorption of the boundaries of the two chambers and the absorption coefficient of the air or any gas mixture within the chambers.[1]

The formulas for calculating the absorption coefficient m of the air or gas in the chambers and the coefficient α of surface absorption of the boundaries are readily obtained, as follows. The equations here given are essentially the same as those given two years ago, but are more convenient for calculations. The decay of sound intensity in the large chamber is given by

$$I_1 = I_1' \exp\{[\ln(1-\alpha)/l_1 - m]ct\}, \quad (1)$$

where I_1' is the average value of the initial or steady state intensity in the large chamber, l_1 the mean free path in the large chamber, c the velocity of sound, and t the time. Similarly, for the small chamber,

$$I_2 = I_2' \exp\{[\ln(1-\alpha)/l_2 - m]ct\}. \quad (2)$$

By solving for m and $\ln(1-\alpha)$

$$m = (2.303/20c)(3\delta_1 - \delta_2), \quad (3)$$

$$\ln(1-\alpha) = \frac{2.303}{10c} \frac{\delta_2 - \delta_1}{1/l_1 - 1/l_2}, \quad (4)$$

where $\delta_1 = (10 \ln I_1'/I_1)/2.303t$, i.e., the rate of decay, in db/sec., in the large chamber; and $\delta_2 = (10 \ln I_2'/I_2)/2.303t$, i.e., the rate of decay, in db/sec., in the small chamber. When both α and either l_1 or l_2 are known, (1) or (2) can be solved directly for m. It then becomes possible to determine m, for air or other gases under various conditions of humidity, temperature, pressure, or contamination by a foreign gas, by measurements of the rate of decay of sound in only one chamber. By solving (2) for m (since this chamber was used alone for many of the measurements), there results

$$m = \ln(1-\alpha)/l_2 + 2.303 \delta_2/c. \quad (5)$$

Since a large part of this investigation consisted of determining the absorption coefficients of air containing different amounts of water vapor, it was necessary to provide a convenient and reliable means for controlling and measuring the humidity. This was done (1) by forcing the air (or gas) from the reverberation chamber through a system of glass tubes in which were sealed a wet and a dry bulb thermometer; and (2) by combining two streams of air—one dry and the other saturated with water vapor at a known temperature. Dry air was obtained by forcing air through two bottles of concentrated H_2SO_4, a calcium chloride tube, and a long tube filled with mineral wool. Saturated air was obtained by bubbling air through a long tube containing water at a known temperature. The two streams of air were mixed in known proportions by metering the air through two Venturi flow meters, which had been previously calibrated against a Sargent gas meter. This means of controlling humidity, by combining in known proportions dry air and air saturated at a known temperature, and a method which involved the direct weighing of the loss of water by the saturation tube, were used principally as a means of calibrating the wet and dry bulb thermometer method. The air which ventilated the wet bulb was flowing at a rate of approximately three meters per second, so that the conditions for the use of wet and dry bulb thermometers as recommended by the United States Weather Bureau were fulfilled. The values of relative humidity obtained by this method, when using the tables prepared by the Weather Bureau, agreed within one to two percent with the values of humidity obtained by either method 2 or by direct weighing.[4] The values of

---

[4] In the work which the author reported two years ago (reference 1) the relative humidity of the air was determined by wet and dry bulb thermometers, when using an old set of tables published by the Smithsonian Institute in 1868. These tables are in error, especially at low humidities,

humidity obtained from the wet and dry bulb thermometer were thus regarded as sufficiently accurate for the purpose of these experiments, and therefore this more convenient method was used for determining the amount of water vapor in the air; although frequent check measurements were made by using either method 2 or by direct weighing. The wet and dry bulb method for determining the relative humidity of air also was found to be a satisfactory method for determining the relative humidity of either oxygen or nitrogen.

Practically all of the commonly used methods for measuring the rate of decay of sound in a reverberation chamber were tested to determine the most precise and dependable method. The method finally chosen is a modification of the one described by Norris and Andree,[5] although by the method here devised it is possible to measure the rate of decay over a range of 70 to 80 db for frequencies up to 6000 cycles, 50 db at 8000 cycles, and 30 db at 10,000 cycles. The apparatus and method therefore appear to possess advantages for determining the absorption coefficients of small specimens of acoustical materials, especially at high frequencies.

The general arrangement of the electro-acoustical apparatus is shown in Fig. 2. The

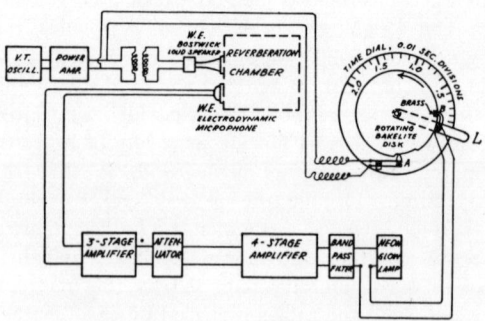

Fig. 2. Block diagram of electro-acoustical circuit.

where they give values of relative humidity approximately 7 percent too low. The error in these tables diminishes from 7 percent at low humidities to practically no error at nearly saturated air. The data in the earlier publication are subject to this correction.

[5] R. F. Norris and C. A. Andree, *An Instrumental Method of Reverberation Measurement*, J. Acous. Soc. Am. **1**, 366–372 (1930).

alternating current generated by the vacuum tube oscillator does not contain more than one percent of harmonics above the fundamental frequency. This alternating current operates a Western Electric Bostwick loudspeaker, which serves as the source of tone. This loudspeaker will provide an abundant supply of sound energy for frequencies between 1000 and 11,000 cycles. It is fastened to an opening in one wall of the reverberation chamber, and fits flush with the inner wall of the chamber, so that it does not offer any obstruction to the flow of sound in the chamber. The microphone is a Western Electric electrodynamic type which is also fitted in the wall of the reverberation chamber in a manner similar to the mounting of the loudspeaker. The tone picked up by the microphone is amplified by a seven-stage amplifier and then passes through a band-pass filter to a neon glow lamp. If the voltage impressed upon the glow lamp is above the critical voltage, about 80 volts, the lamp glows.

The Bakelite disk on the commutator shown at the right in Fig. 2 rotates once each four seconds. The telephone key A, with broad platinum points, is closed at all times except for 0.45 second while the slotted portion of the Bakelite disk is passing the protuberant finger of key A. During this 0.45 second, while key A is open, the loudspeaker is sounding; during the remaining 3.55 seconds of each rotation it is silent. The contact noises (transients) of key A—especially when closing, which stops the tone—are practically inaudible, and do not affect the rate of decay of the pure tones generated by the loudspeaker, as is borne out by the linearity of the decay curves for tones of either low or high frequency. The lever L, which can be set manually at any position on the time dial, has mounted on it a pair of sliding contacts shown at B. These contacts are closed, by means of the brass inlay, for 0.02 second during each rotation of the Bakelite disk. The glow lamp is connected in the circuit only when key B is closed. If key B is closed at a sufficiently short interval of time after the tone stops, the glow lamp will flash. If it is closed at a much later instant, that is after the reverberant sound in the chamber has been reduced below a certain critical value, there is no flash. By means of the manually operated lever L

THE ABSORPTION OF SOUND 115

the key B can be made to close at any chosen time, up to 2.0 seconds, after the tone in the reverberation chamber is stopped. In making a decay curve the attenuator (connected between the third and fourth stages of the amplifier) is first set at say 60 db, which means that 60 db of attenuation is introduced in the amplifier circuit. (The attenuator was calibrated for all frequencies at which measurements were made. Small corrections, especially at 6000 cycles, have been applied to the readings of the attenuator.) The lever L is then moved over the time scale until the position is found at which the glow lamp just doesn't flash each time the contacts at B are closed. The time which has elapsed between the cessation of the tone and the closing of the key B is then indicated by the position of the lever L. This time interval is just sufficient for the decadent sound to reach such a level that, when detected by the microphone and amplified through the seven-stage amplifier, the voltage thus developed will just attain the critical voltage required to flash the glow lamp. The time interval at which this critical voltage is reached will depend, among other things, upon the particular setting of the attenuator. By obtaining a set of readings of time intervals on the time dial corresponding to different settings of the attenuator, it is possible to obtain data for plotting the decay, in db, as a function of the time. Such a decay curve is obtained simply by plotting the readings of the attenuator against the corresponding time intervals indicated on the time dial at which the glow lamp just does not flash.

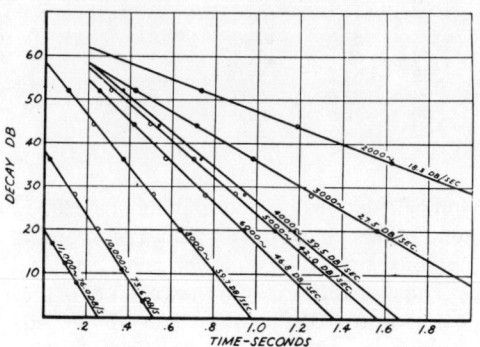

FIG. 3. Decay curves in large chamber filled with air at a temperature of 20°C and a relative humidity of 71 percent.

The setting of the lever L is found to be quite critical, and when the paddles in the chamber are rotating, repeated settings of the time lever will agree to about 0.01 second or less for frequencies above 3000 cycles. In general, readings of the times for the critical flashing potential of the glow lamp are obtained for attenuator settings of 60, 52, 44, 32, 20, and 12 db. In Fig. 3 are shown a number of decay curves obtained in the large chamber filled with air at a temperature of 20°C and a relative humidity of 71 percent. The frequency at which each decay curve was obtained is indicated above each curve, as is also the rate of decay in db per second.

II. ABSORPTION COEFFICIENTS OF SOUND IN AIR AT 20°C CONTAINING DIFFERENT AMOUNTS OF WATER VAPOR

The curves in Fig. 4 are decay curves for a tone of 6000 cycles dying away in the six-foot chamber

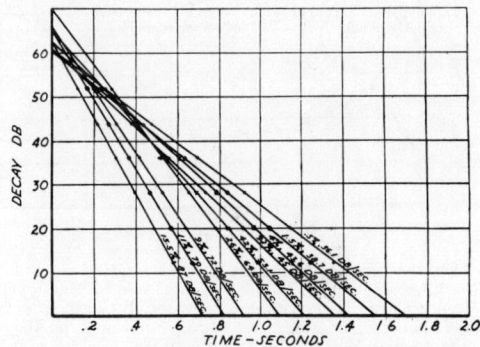

FIG. 4. Decay curves for a tone of 6000 cycles in six-foot chamber filled with air at 20°C. Note the large increase in the rate of decay as the humidity approaches 13 percent from either higher or lower humidities.

filled with air at 20°C. The relative humidity at which each decay was obtained is indicated above each decay curve. It will be seen that the rate of decay was slowest for the most nearly dry air; that it reached a maximum at a relative humidity of about 13 percent; and that it decreased again for higher humidities. Similar series of decay curves were obtained for frequencies of 500, 1000, 2000, 3000, 4000, 8000, 10,000, and 11,000 cycles. In each case the rate of decay was slowest for the dry air; the rate of decay reached a maximum at a certain humidity, between 5 and 20 percent; and diminished for higher humidities. Similar decay curves were then obtained, under the same conditions of temperature and humidity, in the two-foot chamber.

The two lower curves in each of Figs. 5, 6 and 7 summarize the data obtained on the rates of decay in the two chambers for tones of 3000, 6000 and 10,000 cycles. Each point in these curves was obtained from the slope of a decay curve similar to those shown in Fig. 4. The relative humidity has been converted into the absolute concentration of water vapor molecules, and is expressed as a percentage. Thus, a concentration of 1.0 percent of H_2O molecules corresponds to a relative humidity of 43.2 percent at 20°C.

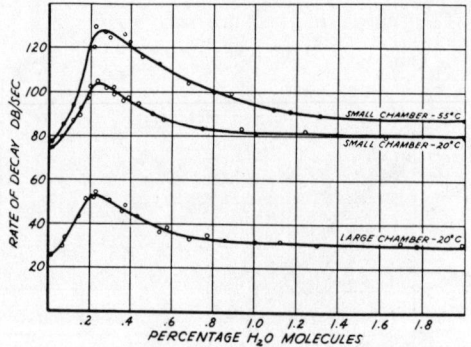

FIG. 5. Curves showing the rates of decay in air, in both the large and small chambers, for a tone of 3000 cycles. Note the nearly constant separation between the two lower curves. The uppermost curve shows how the absorption increases with a rise of temperature.

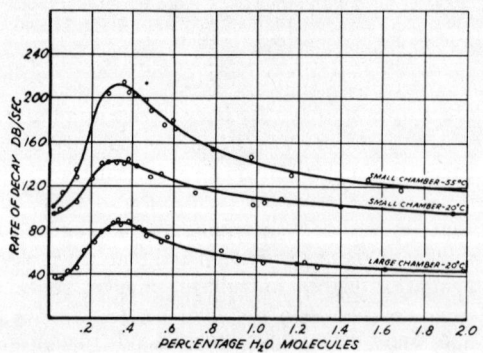

FIG. 6. Curves showing the rates of decay in air for a tone of 6000 cycles

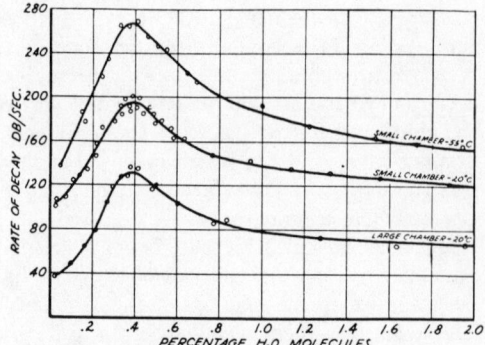

FIG. 7. Curves showing the rates of decay in air for a tone of 10,000 cycles.

Although there is some divergence among these experimental points—the magnitude of which is greater than can be ascribed to the errors of measurement, and possibly is attributable to some impurity or some peculiar condition of the air—they reveal a definite dependence upon humidity which must conform very closely to the curves as drawn in these three figures.

These curves furnish the data required by Eq. (3) for calculating the values of the absorption coefficients m. In Table I is given a set of data

TABLE I. *Values of absorption coefficients at 3000 cycles and 20°C.*

Percentage H_2O molecules	δ_1	δ_2	$3\delta_1 - \delta_2$	m ($\times 10^{-4}$ cm^{-1})	$\delta_2 - \delta_1$
0.01	25.5	73.5	3.0	0.10	48.0
.05	29.0	77.5	9.5	.32	48.5
.10	35.5	84.5	22.0	.74	49.0
.20	51.5	100.3	54.2	1.82	48.8
.25	54.0	104.0	58.0	1.95	50.0
.30	50.5	101.4	50.1	1.68	50.9
.50	37.8	89.0	24.4	0.82	51.2
1.00	32.0	82.0	14.0	.47	50.0
2.00	30.0	80.0	10.0	.33	50.0

and calculations for determining m at 20°C for a tone of 3000 cycles. The values of δ_1 and δ_2 are obtained from the two lower curves in Fig. 5.

The probable errors in the values of m given in Table I are not greater than 10 percent for concentrations of water vapor between 0.10 and 1.00 percent, but the probable errors become increasingly larger at lower and higher humidities

THE ABSORPTION OF SOUND

because of the dependence of m upon the difference $3\delta_1-\delta_2$. If the surface absorption be independent of the relative humidity, and this certainly seems reasonable, the difference $\delta_2-\delta_1$ should be constant. It will be noted from the last column in Table I that this difference is nearly constant. The average value of $\delta_2-\delta_1$ is used for calculating $\ln(1-\alpha)$ by the help of Eq. (4).

The lowest curve in Fig. 8 was obtained from the values of m in Table I. The two other curves for m at the temperature of 20°C were obtained from similar tables based on the data given by the two lower curves in each of Figs. 6 and 7. As these curves show, the absorption coefficient at each frequency has its lowest value for dry air; then, as the amount of water vapor in the air increases the absorption increases, reaching a maximum at a certain concentration, the position of the maximum depending upon the frequency; and then, as the concentration of water vapor further increases the absorption decreases. The maximal values of the absorption coefficients for frequencies of 3000, 6000, and 10,000 cycles are 0.000195, 0.000392, and 0.000655 cm^{-1}, respectively. These coefficients are much more nearly proportional to the first power of the frequency than to the second power, as is required by classical theory. They are in the ratio of 3.00 : 6.03 : 10.08, which is almost identical with the ratio of the frequencies.

III. Absorption Coefficients in Air and Water Vapor at a Temperature of 55°C

To test the effect of temperature on the absorption of sound in air and water vapor, the two-foot reverberation chamber was enclosed in an insulation chamber constructed of Celotex. By means of four electrical heating coils the temperature inside this chamber was maintained at an average temperature of 55°C. A large rotating paddle was used to stir the air in the outer chamber. This helped to maintain a uniform temperature within the chamber—a thermometer near the top of the chamber read about 3°C higher than a thermometer near the bottom. The mean reading of these two thermometers was taken as the average temperature of the air inside the reverberation chamber.

The uppermost curves in each of Figs. 5, 6 and 7 give the observed values of the rates of decay in the small chamber filled with air and water vapor at a temperature of 55°C, for frequencies of 3000, 6000 and 10,000 cycles, respectively. From these three curves, the velocity of sound at 55°C, and the known values of α and l_2, the coefficients of absorption m at 55°C were calculated by means of Eq. (5). (The value of the mean free path l_2 was obtained by direct measurement, by using an optical method.[6] The value of l_2 was 34 cm which was in agreement with the value based upon measurements of rates of decay in the chamber with and without the paddle.)

The calculated values of m are plotted in Fig. 8. It will be observed (1) that the maximal values

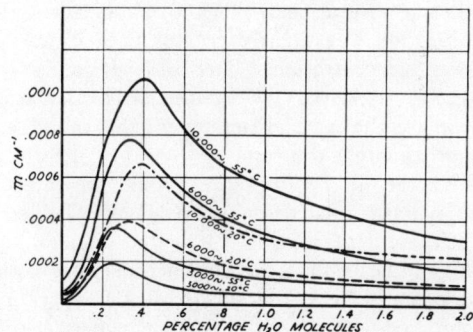

Fig. 8. Curves showing the values of the absorption coefficient m for different concentrations of water vapor, and at temperatures of 20°C and 55°C, for tones of 3000, 6000 and 10,000 cycles.

of the absorption coefficients occur at practically the same concentrations of water vapor as they do at 20°C, and (2) that the coefficients at these maxima are approximately twice as large as the corresponding maxima at 20°C. The maxima are in the ratio of 3.00 : 6.83 : 9.17, which, within the limits of experimental errors, agree with the ratio of the frequencies, namely 3 : 6 : 10.

IV. Absorption in Air and Water Vapor at Temperatures of −15°C and 2°C

The small chamber was maintained at a temperature of −15°C by means of CO_2 ice inside the insulation chamber. At this tempera-

---

[6] V. O. Knudsen, *Architectural Acoustics*, pp. 133–135, John Wiley and Sons (1932).

ture there was no conclusive evidence that the absorption coefficient of air was altered as water vapor was added to dry air, although the data presented in Table II show that with the highest

TABLE II. *Rates of decay of sound in small chamber filled with air and H_2O vapor at $-15°C$.*

Percentage H_2O	Rate of decay, db/sec.	
	3000$\sim$	6000$\sim$
0.02	76	91
0.05	78	92
0.09	78	91
0.11	80	94

concentration of water vapor, 0.11 percent (which was near saturation at this low temperature), the absorption was slightly greater than it was at lower concentrations. The absorption is too small to make possible a dependable calculation of m by the method here used, but m is very approximately the same as it is for dry air at 20°C. (See Fig. 8 for values.) Only the rates of decay in the small chamber are given in Table II, but these data, compared with those for 20 and 55°C, indicate the almost negligible effect of the water molecule at this sub-freezing temperature compared with the very potent effects at higher temperatures.

TABLE III. *Rates of decay in small chamber, air, $2°C$.*

Percentage H_2O	Rate of decay, db/sec.	
	3000$\sim$	6000$\sim$
0.02	75	92
0.11	87	105
0.24	98	125

In Table III some data are presented which were obtained with the small chamber maintained at a temperature of 2°C. The absorption increased as the concentration of water vapor was increased, and the values of m are less than they are for corresponding humidities at a temperature of 20°. The actual values of m at this temperature are not calculated because it is felt that more precise data are necessary before reliable values of m can be determined at these low temperatures.

V. THE ABSORPTION OF SOUND IN OXYGEN AND IN NITROGEN CONTAINING DIFFERENT AMOUNTS OF WATER VAPOR

The nature of the absorption of sound in air suggests that at least a portion of the absorption may result from the transfer of energy between translational and either rotational or vibrational forms of energy during molecular collisions, as has been suggested by the theories of Jeans,[7] Einstein,[8] Herzfeld and Rice,[9] Bourgin,[10] and Kneser,[11] and also by the experiments of Kneser.[12] For this reason, it appeared desirable to determine the effect of water vapor on the absorption of sound in both oxygen and nitrogen. The need for such experiments became urgent after some of the preliminary results on air and water vapor, described in the preceding sections, were investigated during the course of these experiments by Dr. H. O. Kneser. Dr. Kneser was able to propose a highly satisfactory explanation of many of the experimental results by assuming that a large part of the absorption is attributable to collisions between H_2O and O_2 molecules.[13] In a paper which Dr. Kneser is preparing to accompany this one, his theory will be developed; and the degree to which it accords with the present experimental findings will be considered.

The following experiments with oxygen and nitrogen show quite conclusively that the abnormally high absorption in air is almost wholly attributable to an interaction between O_2 and H_2O molecules. The two-foot chamber was filled with commercial oxygen, and decay curves were obtained at 3000 and 6000 cycles with different amounts of water vapor admitted to the chamber, first when the temperature was at 20°C, and then when the temperature was maintained at 55°C.

---

[7] J. Jeans, *Dynamical Theory of Gases*, Second Edition.
[8] A. Einstein, Ber. d. Berl. Akad., p. 380 (1920).
[9] K. F. Herzfeld and F. O. Rice, Phys. Rev. **31**, 691 (1928).
[10] D. G. Bourgin, *Quasi-Standing Waves in a Dispersive Gas*, J. Acous. Soc. Am. **4**, 108–111 (1932), and earlier articles in Phil. Mag. and Phys. Rev.
[11] H. O. Kneser, Ann. d. Physik **5**, 761 (1931).
[12] H. O. Kneser, Ann. d. Physik **5**, 777 (1931).
[13] H. O. Kneser, *The Transfer of Vibrational Energy between Molecules*. Presented at the Washington Meeting of the Am. Phys. Soc., April, 1933. Phys. Rev. **43**, 1051A (1933).

THE ABSORPTION OF SOUND

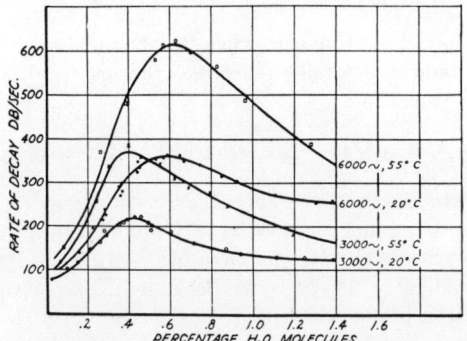

FIG. 9. Decay curves for oxygen and water vapor at frequencies of 3000 and 6000 cycles, and at temperatures of 20°C and 55°C.

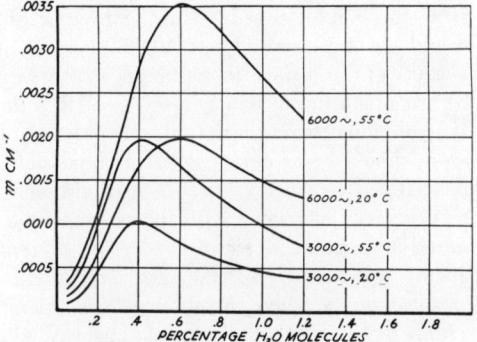

FIG. 10. Curves showing values of m in oxygen and water vapor at temperatures of 20°C and 55°C. Note that the maximal values of m are approximately five times larger than the corresponding maximal values for air, as shown in Fig. 8.

A summary of the results of these decay measurements is given by the curves in Fig. 9. From these curves, using the same values for α and l_2 as were used for air at 55°C, and the approximate value for the velocity of sound in oxygen given by $(\gamma p/\rho)^{\frac{1}{2}}$, the absorption coefficients for oxygen containing different amounts of water vapor were calculated. The results are plotted in Fig. 10. It will be seen that the absorption is much greater than it is in air and water vapor. Thus, the maximal values of the coefficients m at 20°C are 0.001030 cm^{-1} at 3000 cycles and 0.001975 cm^{-1} at 6000 cycles, with a probable error of about 5 percent. These values are 5.28 and 5.04 times the corresponding maximal values of m for air at 20°C. At 55°C, the maximal values of m for

oxygen are 0.001975 cm^{-1} at 3000 cycles, and 0.003525 at 6000 cycles, with a probable error of 10 percent. These are 5.67 and 4.51 times the corresponding values for air at 55°C. Further, the maximal values of the absorption coefficients are, as in the case of air, approximately proportional to the frequency. Also, the maxima in the curves for O_2 occur at greater concentrations of water vapor than they do in the case of air. Thus, at 3000 cycles, the maximal absorption for air occurs at 0.25 percent of H_2O molecules compared with 0.41 percent for O_2; at 6000 cycles the maximum for air occurs at 0.33 percent compared with 0.59 percent for O_2.

The very large magnitude of the absorption coefficient in O_2 is of interest. If we lived in an atmosphere of oxygen having a temperature of 20°C and a relative humidity of 26 percent we should be able to hear high pitched sounds at only a small fraction of the distance we hear them in air. The high notes of the violin and piccolo, for example, would be completely inaudible at a distance of about 50 meters from their source.

Similar measurements with nitrogen and water vapor reveal the interesting result that the absorption coefficient in nitrogen, in the frequency range here investigated, is almost independent of the presence of the water molecule, and does not greatly exceed the classical value. In Table IV are given the results of the rates of

TABLE IV. *Rates of decay in small chamber filled with nitrogen and water vapor, at 20°C and 55°C.*

Percentage H_2O	Rates of decay, db/sec.					
	3000~		6000~		10,000~	
	20°C	55°C	20°C	55°C	20°C	55°C
0.02	76	75	85	93	101	110
0.10	77	77	86	94	101	115
0.28	77	76	86	95	102	116
0.70	78	76	86	95	101	115
1.25	78	76	86	94	101	117
1.60	79	77	87	95	102	118

decay in the two-foot chamber filled with nitrogen and water vapor at temperatures of first 20° and then 55°C, for tones having frequencies of 3000, 6000, and 10,000 cycles. It will be seen that the rate of decay is almost independent of the humidity or the temperature at any one frequency for the entire range of humidity, temperature, and frequency here investigated. For the

6000 and 10,000 cycle tones there is evidence of increased absorption when the temperature is increased to 55°C, and the absorption increases slightly as the humidity is increased.

By referring to Figs. 5, 6 and 7, which give the rates of decay in air at these frequencies and temperatures, it will be seen that the rates of decay in nitrogen and water vapor are very approximately the same as those for dry air; and therefore since the velocities in air and N_2 are nearly the same, the absorption coefficients in N_2 are of the same order of magnitude as are the coefficients for dry air. As noted above, at the higher frequencies, m increases slightly with an increase of either temperature or humidity, but the highest possible value of m in nitrogen and water vapor is only about 0.0001 cm^{-1} at 10,000 cycles and 55°C, compared with 0.0010 cm^{-1} for air and water vapor.

It is reasonable to assume that since air is one-fifth oxygen and four-fifths nitrogen, since the absorption coefficient in oxygen and water vapor is about five times as large as it is in air and water vapor (at least in the region of greatest absorption), and since the absorption in nitrogen does not change appreciably as the amount of water vapor is changed, especially at the lower frequencies of 3000 and 6000 cycles, a large portion of the absorption of audible vibrations in air must be attributable to an interaction between the O_2 and H_2O molecules. Dr. Kneser is able to show that this is consistent with theoretical considerations of the collisions between molecules.

The data on the absorption of sound in air lead to some interesting conclusions which have a practical bearing on the problems of architectural acoustics and sound signaling. In large auditoriums, the reverberation of the high frequency components of speech and music is affected more by the condition of the air in the room than it is by the nature of the materials which form the boundaries of the room. Thus, at a frequency of 10,000 cycles—which is now generally regarded as lying within the range of frequencies necessary for high quality of sound—the value of m at 70°F and 18 percent relative humidity is 0.000655 cm^{-1}, or 0.020 ft.$^{-1}$. The formula for the reverberation time in a room, taking account of the absorption in the air, is

$$t = 0.05 V / [-s \ln (1-\alpha) + 4mV].[1]$$

In the limiting case where the boundaries of the room are totally reflective, that is $\alpha = 0$, this reduces to

$$t = 0.05/4m = 0.05/4 \times 0.020 = 0.625 \text{ second},$$

which is the longest possible time of reverberation any room can have at a frequency of 10,000 cycles when the temperature is 70°F and the humidity 18 percent. Because of the surface absorption, which is always present in a room, the actual reverberation time may be considerably less than this. Of course, this low humidity is rarely, if ever, realized in a room occupied by people, but even at a relative humidity of 50 percent with the temperature at 70°F the longest possible time of reverberation would be 1.30 seconds at 10,000 cycles; and because of the surface absorption probably would be not much more than two-thirds of this time. It is obvious therefore that the boundaries of a room should be as non-absorptive as possible at these high frequencies. Further, it would seem to be necessary to reckon with the phenomenon of sound-absorption in air in the design of reproducing equipment, especially for large theaters, or out of doors, where the absorption in the air is a large factor. Suitable electrical networks, which compensate for the excessive absorption of the high frequencies, should be added to the equipment if high quality sound is to be attained. Proper control of the humidity, as well as the temperature, is likewise necessary in order to secure the highest acoustical quality in all auditoriums.

Sound-signaling in the air by means of high frequency beams encounters very serious limitations because of the absorption in the air. It is of interest to calculate the possible effects of this absorption in the experiments described by Dr. H. C. Hayes at the May, 1933, meeting of the Acoustical Society of America. In these experiments an approximately parallel beam of sound, having a frequency of about 6000 cycles, is used for determining the course and altitude of aircraft. At a temperature of 70°F and a relative humidity of 14 percent, the value of m at 6000 cycles is 0.00039 cm^{-1} or 0.012 ft.$^{-1}$. In the case of a plane parallel beam, the attenuation would amount to 0.051 db per foot. Hence at an altitude of 500 feet, such a beam would suffer an attenuation of 51 db in going from the aircraft to the reflecting surface below and back again to the aircraft. In view of other limitations, such as (1) the unavoidable losses owing to imperfect reflection and to divergence of the reflected beam, (2) the masking effect of any aircraft noise on the sound to be detected, and (3) the finite intensity of the source, it seems probable that about 500 feet is the maximal altitude (for a temperature of 70°F and a relative humidity of 14 percent) at which a 6000 cycle tone could be heard after reflection from either a water or an earth surface. At humidities more commonly encountered, namely 50 percent or more, the absorption in the air would be very much less, so that under such conditions a 6000 cycle signal might be heard up to altitudes of 1000 or 1500 feet. The results of the present investigation would seem to warrant the use of a frequency considerably lower than 6000 cycles for acoustical altimeters and for all long range signaling in air. The data on the absorption of sound in air, together with other considerations, such as are mentioned earlier in this paragraph, should make it possible to choose the optimal frequency for long range acoustical signaling. It is apparent that the optimal frequency will not be the same for different conditions of temperature and humidity; but for given weather conditions it should be possible to select the most favorable signal frequency.

In summary, the results of the present investigation not only provide coefficients which are useful for calculating the attenuation of audible sounds as they are propagated through a room or through a free atmosphere of known humidity and temperature, but, what may prove to be of far greater interest and value, they reveal a means of gaining new information concerning the structure—and possibly important reactions—of gaseous molecules. Some of these molecular phenomena will be considered in the paper by Dr. Kneser.

The author wishes to acknowledge the valued assistance of Messrs. Lewis A. Delsasso and Leonard Obert. Mr. Delsasso has assisted in the design and construction of the apparatus, and Mr. Obert has assisted in making many of the experimental observations. Finally, the author is indebted to Dr. H. O. Kneser of Marburg, Germany, whose careful study of this problem has given more specific and meaningful purpose not only to the experiments described in this paper but to subsequent ones which will be reported later.

Copyright © 1951 by the American Physical Society
Reprinted from *Rev. Mod. Phys.*, **23**, 353–411 (1951)

Absorption of Sound in Fluids*

JORDAN J. MARKHAM

Applied Physics Laboratory, The Johns Hopkins University, Silver Spring, Maryland

AND

ROBERT T. BEYER AND R. B. LINDSAY

Brown University, Providence, Rhode Island

TABLE OF CONTENTS

Chapter I. The Equation of State and Its Relation to Acoustic Propagation
1. Introduction—Hydrodynamic Equations
2. Maxwell's Equation of State
3. Stokes's Viscosity Equation
4. Equation of State for Heat Conduction
5. Equation of State for Heat Radiation
6. Equation of State for Thermal Relaxation
    A. The Method of Irreversible Thermodynamics
7. Equation of State for a General Internal Lag
    A. The Method of Statistical Thermodynamics
        (a) Thermal Relaxation
        (b) Structural Relaxation
        (c) General Linear Case
    B. Section 7 Appendix
8. Various Relaxation Methods
    (a) The Method of Kinetic Theory
    (b) The Method of Irreversible Thermodynamics
    (c) The Method of Statistical Thermodynamics
    (d) The Phenomenological Approach
9. Summary of Results
    (a) Viscosity Mechanism
    (b) Relaxation Mechanism
    (c) Temperature Dependence
10. Temporal Absorption *Versus* Spatial Absorption
11. Combined Effects
    (a) Heat Conduction and Viscosity
    (b) Relaxation and Viscosity
    (c) Double Relaxation
12. Conclusion

Chapter II. The Effect of Viscosity on Sound Absorption
13. Stokes's Relation between the Viscosity Coefficients
    (a) Stokes's Relation for a Gas
    (b) Stokes's Relation for a Liquid
14. Viscosity and Sound Absorption
15. Second-Order Effects

Chapter III. Experimental Methods of Sound Absorption Measurements
16. Mechanical Methods
17. Optical Methods
18. Electrical Methods
    (a) Interferometric Methods
    (b) The Direct Method
    (c) Pulse Methods
    (d) Reverberation Methods
19. Summary of Methods
    (a) Gases
    (b) Liquids

Chapter IV. Experimental Results in Gases
20. Introduction
21. Monatomic Gases
    (a) Helium
    (b) Argon
22. Diatomic Gases
    (a) Hydrogen
    (b) Nitrogen, Oxygen
23. Triatomic Gases

Chapter V. Experimental Results in Liquids
24. Monatomic and Diatomic Liquids
25. Water
26. Liquids with Pronounced Relaxational Effects
    (a) Acetates
    (b) Formates
    (c) Toluene ($C_6H_5CH_3$)
27. Distinction in Behavior between Associated and Non-Associated Liquids
    (a) Benzene (C_6H_6)
    (b) Carbon Bisulfide (CS_2)
    (c) Carbon Tetrachloride (CCl_4)
28. Liquids of High Viscosity
29. Summary

Chapter VI. Experimental Results in Solutions
30. Sea Water
31. Acetate and Formate Electrolytes
32. Various Electrolytes

---

* J. J. M. made his contribution while at Brown University and at A.P.L. (R. T. B. and R. B. L. did their work at Brown University). The contribution from Brown University was supported by Navy Contract N7 onr-35808 while the contribution from A.P.L. was supported by the Bureau of Ordnance, U. S. Navy, under Contract NOrd 7386.

THE object of this paper is to review the absorption of sound in fluids. We shall be more interested in basic principles and concepts rather than detailed calculations which are both complicated and in many cases highly tentative. No attempt has been made to include all the complications in this field; rather we have tried to lay the foundation for a basic understanding of the underlying principles governing this field of physics.

The first two chapters give the basic theory. We have, here, reviewed the older, as well as the more recent theories. This is done to show the connection between the developments of the last century and present day developments. In doing this, the authors hope to show the unity of this field. In the later chapters a review of the data is made. All the data are not listed, but only those which appear to be reliable. On the whole, we have limited the data to the simpler phenomena.[1]

Chapter I. The Equation of State and Its Relation to Acoustic Propagation

1. INTRODUCTION—HYDRODYNAMIC EQUATIONS

Three equations are needed to study the propagation of an elastic disturbance in a fluid medium. Two of these, i.e., the equation of motion and the equation of continuity, are generally accepted in the classical form given by Euler. The third equation is the equation of state or the relation between stress and strain in the medium. It is vital for the study of the dissipative absorption of the disturbance. The variations in the theories of absorption can indeed be traced back to the equation of state used, though this is not always apparent. The equation of state which we shall talk about is not the general thermodynamic one but is along a specific path. Further, time appears in this equation. We shall call the relation between the excess pressure and the excess density the acoustical equation of state, sometimes omitting the word acoustical.

It will therefore be desirable to devote some time to developing various acoustical equations of state and examining their consequences. It should be emphasized at the outset that we must consider more than the static situation which presents no major difficulties but which leads to no absorption. The dynamic state equation, involving first and second time derivatives of the pressure and density, leads to the heart of the problem. Burgers (B36)[2] discusses various state equations for fluids, and we shall base some of our ideas on his development. However, his interest is confined to shearing while we shall introduce more general distortions. Though the expressions for velocity and absorption coefficients deduced by the following analysis are not for the most part new, we have attempted to present a more unified treatment.

A general infinitesimal distortion in a material medium, fluid or solid, can be described in terms of six strain components, which are defined as follows: (J3; p. 152)

where $\mathbf{q} = \mathbf{i}q_x + \mathbf{j}q_y + \mathbf{k}q_z$ is the displacement vector in the medium. The stress components are $P_{xx}, P_{yy}, P_{zz}, P_{xy} = P_{yx}, P_{xz} = P_{zx}, P_{yz} = P_{zy}$ which have their usual meaning. For isotropic bodies, to which we confine our attention, the general linear static relations between the stresses and the strains reduce to

$$P_{jj} = \lambda_s \operatorname{div} \mathbf{q} + 2\theta_s e_{jj},$$
$$P_{jk} = \theta_s e_{jk}, \quad (1.2)$$

where λ_s and θ_s are the usual Lamé constants. λ_s and θ_s depend on the thermodynamic path taken by the e's. The subscript implies that it is isentropic. If volume and total pressure are denoted by V and p, respectively, the bulk modulus, $-V(\partial p/\partial V)_s = K_s$, is related to λ_s and θ_s as follows:

$$K_s = \lambda_s + 2\theta_s/3. \quad (1.3)$$

It should be emphasized that Eq. (1.2) implies static equilibrium and makes no allowance for the dynamical features of deformation. For this review we shall be interested only in those fluids which cannot support a shear; we therefore have $\theta_s = 0$, leading finally to

$$P_{jj} = -p_e = K_s \operatorname{div} \mathbf{q} = -K_s s, \quad (1.4)$$

where p_e is the excess pressure and s is the condensation. Joos shows that

$$s = -\operatorname{div} \mathbf{q} = -\delta V/V = \delta \rho/\rho, \quad (1.5)$$

where ρ is the density. We have indeed utilized the further fact that

$$P_{jj} = -p_e \quad \text{for} \quad j = 1, 2, 3.[4]$$

In what follows p_e and ρ_e will always denote excess pressure and excess density in contrast to the total pressure p and the total density ρ. p_0 and ρ_0 are the equilibrium pressure and the equilibrium density.

Equation (1.4) is the conventional equation of state used in acoustics, and the first-order theory of elastic

$$e_{xx} = \frac{\partial q_x}{\partial x}, \quad e_{yz} = e_{zy} = \frac{\partial q_y}{\partial z} + \frac{\partial q_z}{\partial y}, \quad e_{yy} = \frac{\partial q_y}{\partial y}, \quad e_{zx} = e_{xz} = \frac{\partial q_x}{\partial z} + \frac{\partial q_z}{\partial x}, \quad e_{zz} = \frac{\partial q_z}{\partial z}, \quad e_{xy} = e_{yx} = \frac{\partial q_x}{\partial y} + \frac{\partial q_y}{\partial x}, \quad (1.1)[3]$$

---

[1] Throughout this review, considerable use has been made of the compilations of data and references made by L. Bergmann (B15), C. Kittel (K8), H. O. Kneser (K11), W. T. Richards (R4), and D. Sette (S5). The authors wish to acknowledge the assistance provided by these writings in the preparation of the present paper.
[2] See bibliography at end of article.
[3] A list of symbols is given in Appendix I.
[4] $P_{jj} = -p_e$ follows from the theorem that states (L1; p. 1) "If the stress exerted across any small area situated at a point is wholly normal, the pressure is the same in all directions." The sign appears because in the theory of elasticity a tension is positive while for fluids, a tension is considered as a negative pressure.

waves in fluids is based on it. However, it cannot account for the absorption of sound, and modification is necessary. This can be done in a large variety of ways, and indeed the successful modification seems to depend on the type of fluid in question, so that successful generalization has up to now eluded search.

We shall now consider certain changes that have been made in Eq. (1.4) to include processes which are affected by rates of change, i.e., the first and higher derivatives of the pressure or the volume with respect to time. The two principal types of modifications of historical interest are due to Maxwell and Stokes, respectively. In view of the confusion which seems to exist in published literature it will be worthwhile to review the early work before passing on to more recent developments.

2. MAXWELL'S EQUATION OF STATE

Maxwell (M5) modified Eq. (1.4) by incorporating a term depending on the integral $\int_0^t P_{jj} dt$. If we confine our attention once more to the case in which P_{jj} reduces to $-p_e$, his equation takes the form

$$\frac{p_e}{\rho_0} = \frac{1}{K_s} p_e + \frac{\gamma_s}{K_s} \int_0^t p_e dt, \qquad (2.1)$$

where γ_s is a constant depending on the medium.

In differential form (2.1) becomes

$$\frac{dp_e}{dt} + \gamma_s p_e = K_s \frac{1}{\rho_0} \frac{d\rho_e}{dt}. \qquad (2.2)$$

To understand its physical significance let us consider a special case. If ρ_e is held constant, i.e., constant strain, the equation reduces to

$$dp_e/dt = -\gamma_s p_e, \qquad (2.3)$$

with a solution in the form

$$p_e = p_{e0} \exp[-\gamma_s t], \qquad (2.4)$$

if p_{e0} is the initial excess pressure. If γ_s is positive, p_e decreases with time and in time $1/\gamma_s$ is reduced to $1/e$th of its initial value. The excess pressure is said to relax and to have a "relaxation time" of $1/\gamma_s$. The general equation (2.2) of which (2.3) is a special case appears in the paper of Maxwell (M5) in 1867 on the dynamical theory of gases. It is usually referred to as Maxwell's relaxation equation. A model whose behavior follows this equation has been suggested by J. M. Burgers (B36). It is illustrated in Fig. I-1a. The top element is a perfect spring obeying Hooke's law, whereas the lower element is a piston in a tank full of viscous liquid.[5] The two elements are here combined in series in the sense that the same force F acts on both. In the spring it produces a displacement x_1 directly proportional to the force, while in the piston it produces a velocity dx_2/dt directly proportional to the force. Since the total elongation of the spring is $x = x_1 + x_2$, it follows that the differential equation for x is

$$\frac{dF}{dt} + (a/b)F = a\frac{dx}{dt}, \qquad (2.5)$$

where $x_1 = F/a$, $dx_2/dt = F/b$. This has the same mathematical form as Maxwell's equation (2.2) and becomes identical with it if we let $F = p_e$ and $x = \rho_e/\rho_0$. Then a becomes the bulk modulus K_s and $\gamma_s = a/b$.

The physical significance of (2.5) becomes apparent if F is a constant force which acts only for a finite time interval, i.e., if F is a *force pulse*. Thus we assume that

$$\begin{aligned} F &= 0, & t < 0 \\ F &= F_0, & 0 \leq t \leq t_1 \\ F &= 0, & t_1 < t, \end{aligned} \qquad (2.6)$$

a state of affairs illustrated in Fig. I-1b. We can integrate (2.5) in formal fashion to

$$x = F/a + 1/b \cdot \int_0^t F dt. \qquad (2.7)$$

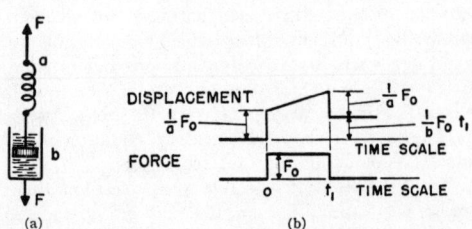

Fig. I-1 (a, b). Mechanical model corresponding to Maxwell's equation (after Burgers).

When $F = 0$, $x = 0$. At the instant F becomes F_0, x immediately rises to F_0/a. As t runs from 0 to t_1, x increases at the constant rate F_0/b and hence at any time t between 0 and t_1

$$x = F_0/a + (F_0/b) \cdot t. \qquad (2.8)$$

When F is reduced to zero at $t = t_1$, the spring at once contracts to its original length but the piston retains the displacement it has gained and hence for $t > t_1$

$$x = (F_0/b) \cdot t_1.$$

The course of x as a function of t is also shown in Fig. I-1b.

Suppose now we shift our attention to the behavior of F with the passage of time as x changes in some presupposed fashion. A simple case is when the spring is initially displaced x_0 and x is maintained constant at this value. The solution of Eq. (2.5) is then

$$F = F_0 \exp[-(a/b) \cdot t], \qquad (2.9)$$

where F_0 is the initial value of F, i.e., $x_0 a$. We see that the force relaxes to zero in infinite time. The time it takes it to decrease to $1/e$th of the initial value can be

---

[5] The reader will readily recognize the electrical circuit analog in which a capacitance C is arranged in parallel with a resistance R across a common electromotive force E.

described as a relaxation time, b/a. It is important to note that the relaxation described by Eqs. (2.2) or (2.5) is one of force (or excess pressure in the acoustical case) and not one of displacement (or condensation).

It will now be of interest to note the effect of assuming the Maxwell relaxational equation of state (2.2) on the propagation of a plane compressional wave in a fluid. For simplicity we confine our attention to harmonic propagation in the x direction with frequency $\omega/2\pi$ and write the equation of motion and the equation of continuity respectively in the form

$$\rho_0 \dot{u}_x = -\partial p_e/\partial x \tag{2.10}$$

$$\rho_0 (\partial u_x/\partial x) = -\dot{\rho}_e, \tag{2.11}$$

where u_x is the particle velocity in the x direction. Here we shall use the dot to mean differentiation with respect to time. For this chapter it is not necessary to distinguish between the total and the partial time derivatives. The difference leads to higher order terms

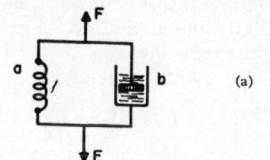
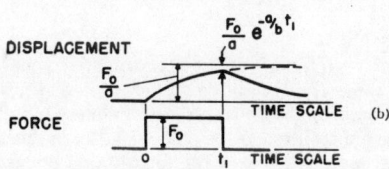

FIG. I-2 (a, b). Mechanical model corresponding to Stokes's equations (after Burgers).

which become important only when second-order effects are considered. Usually in the equation of state the dot means total (hydrodynamic) derivatives. Actually, however, it is not an easy problem to distinguish between total and partial time derivatives in this case. In some sections it will be convenient to use partials and in others totals. When obtaining expressions for the velocity and the absorption only the partial time derivative is needed.

It is now assumed that p_e, ρ_e, and u_x are propagated in accordance with the expressions

$$p_e = P_e \exp[i(\omega t - kx)], \tag{2.12}$$

$$u_x = U_x \exp[i(\omega t - kx)], \tag{2.13}$$

$$\rho_e = R_e \exp[i(\omega t - kx)], \tag{2.14}$$

where P_e, U_x, R_e are respectively complex amplitudes and k is a complex propagation parameter, i.e.,

$$\begin{aligned} k &= k_r + i k_i \\ &= k_r - i\alpha. \end{aligned} \tag{2.15}$$

α is the linear amplitude absorption coefficient of the wave. Substitution into Eqs. (2.2), (2.10), and (2.11) yields the relations

$$P_e(1 - i\gamma_s/\omega) - K_s R_e/\rho_0 = 0, \tag{2.16}$$

$$k P_e - \omega \rho_0 U_x = 0, \tag{2.17}$$

$$\omega R_e - k \rho_0 U_x = 0. \tag{2.18}$$

In order that these may have a nonvanishing solution, the determinant of the coefficients must vanish, yielding the equation

$$\omega^2 (1 - i\gamma_s/\omega) = k^2 K_s/\rho_0. \tag{2.19}$$

In the majority of cases it is satisfactory to assume $|k_i| \ll |k_r|$. Equation (2.19) then yields at once for the velocity of propagation

$$c = \omega/k_r = (K_s/\rho_0)^{\frac{1}{2}}, \tag{2.20}$$

and for the absorption coefficient

$$\alpha = +\gamma_s/2c. \tag{2.21}$$

For ordinary fluids under standard conditions the dispersion of sound is very small, and hence the dependence of c on the frequency may probably be ignored in Eq. (2.21). If γ_s is a genuine constant of the medium, α will not depend on frequency. On this basis the sound absorption coefficient for a fluid resulting from Maxwell's equation is frequency independent. Since this contradicts experience we are tempted to conclude that this particular equation does not play a role in sound absorption in fluids. Actually this conclusion must be accepted as purely tentative since it is conceivable that a mechanism may be developed in which γ_s may turn out to be a function of frequency. So far it does not appear that this has been realized.

3. STOKES'S VISCOSITY EQUATION

An alternative modification of Eq. (1.4) is obtained by adding a term proportional to the rate of change of strain, giving for the linear case

$$p_e = \rho_e \frac{K_s}{\rho_0} + \frac{\zeta}{\rho_0} \dot{\rho}_e. \tag{3.1}$$

The association of ζ with viscosity of the medium is rather suggestive (see Sec. 14). We shall examine this with greater care later.

We have simplified the problem slightly by replacing a tensor by a scalar. Actually p_e is the pressure on a surface at right angles to the axis of wave propagation. This is the only pressure which enters into Eq. (2.10).

For the moment let us consider Burgers, (B36) model of this equation, shown in Fig. I-2a. Here the elements of the system in Fig. I-1 are combined in parallel. The forces on spring and piston are now respectively

$$F_1 = ax \quad \text{and} \quad F_2 = b\dot{x}, \tag{3.2}$$

since the displacement is the same for both elements.

The resultant force F is the sum of F_1 and F_2, the equation for the system becoming

$$F = ax + b\dot{x}, \quad (3.3)$$

which is identical with (3.1) if x becomes ρ_e/ρ_0, F becomes p_r, $K_s = a$, and $\zeta = b$.

To contrast (3.3) with (2.5) we consider once more the square force pulse (2.6). The general solution of (3.3) satisfying the boundary condition $x = x_0$ for $t = 0$ is (S12; p. 284)

$$x = \frac{1}{b} e^{-(a/b)t} \int_0^t F(t') e^{(a/b)t'} dt' + x_0 e^{-(a/b)t}. \quad (3.4)$$

If for simplicity we set $x_0 = 0$, the solutions fitting the square pulse become

$$x = \frac{1}{a} F_0 [1 - e^{-a/b \cdot t}], \quad 0 \leq t \leq t_1,$$

$$x = \frac{1}{a} F_0 [1 - e^{-a/b \cdot t_1}] \cdot e^{-a/b \cdot (t - t_1)}, \quad t_1 \leq t. \quad (3.5)$$

The time b/a is called the retardation time by Burgers. It may in fact be considered a kind of relaxation time. The behavior of x as a function of t is shown graphically in Fig. I-2b.

We next examine the use of Stokes's equation for a plane harmonic compressional wave. We again employ the expressions (2.12), (2.13), and (2.14) and substitute into (3.1), (2.10), and (2.11), respectively. Equations (2.17) and (2.18) remain unchanged but (2.16) is now replaced by

$$p_e = (R_e/\rho_0)(K_s + i\omega\zeta). \quad (3.6)$$

Application of the standard condition on the secular determinant yields the relation

$$\omega^2 = (k^2/\rho_0)(K_s + i\omega\zeta). \quad (3.7)$$

We call K_s/ζ the angular frequency ω_v and then have

$$k^2 = \frac{\rho_0 \omega^2}{\zeta} \cdot \frac{\omega_v - i\omega}{\omega_v^2 + \omega^2}. \quad (3.8)$$

Utilizing (2.15) and separating into real and imaginary parts results in

$$k_r^2 - k_i^2 = \frac{\rho_0}{\zeta} \frac{\omega^2 \omega_v}{\omega_v^2 + \omega^2},$$

$$2 k_r k_i = -\frac{\rho_0}{\zeta} \frac{\omega^3}{\omega_v^2 + \omega^2}. \quad (3.9)$$

The solutions are:

$$k_r^2 = \frac{1}{2} \frac{\rho_0}{\zeta} \frac{\omega^2}{\omega_v^2 + \omega^2} \omega_v \left\{ 1 \pm \left(1 + \frac{\omega^2}{\omega_v^2}\right)^{\frac{1}{2}} \right\} \quad (3.10)$$

$$k_i^2 = -\frac{1}{2} \frac{\rho_0}{\zeta} \frac{\omega^2}{\omega_v^2 + \omega^2} \omega_v \left\{ 1 \mp \left(1 + \frac{\omega^2}{\omega_v^2}\right)^{\frac{1}{2}} \right\}. \quad (3.11)$$

The signs are selected so that the original two equations are satisfied. Since it is desirable that the k's be real, we select the upper signs. If the lower signs are selected, k_i and k_r merely reverse their roles. The phase velocity in general is given by

$$c^2 = \frac{\omega^2}{k_r^2} = \frac{2\zeta}{\rho_0 \omega_v}(\omega_v^2 + \omega^2)\frac{1}{1 + (1 + \omega^2/\omega_v^2)^{\frac{1}{2}}}. \quad (3.12)$$

The solutions can also be put in the following forms, (L14)

$$k_r = \frac{\omega^2/\omega_v}{(2\zeta\omega_v/\rho_0)^{\frac{1}{2}}}\{(1+\omega^2/\omega_v^2)[(1+\omega^2/\omega_v^2)^{\frac{1}{2}} - 1]\}^{-\frac{1}{2}}, \quad (3.13)$$

$$\alpha = \left(\frac{\rho_0}{2\zeta\omega_v}\right)^{\frac{1}{2}} \omega \left[\frac{(1+\omega^2/\omega_v^2)^{\frac{1}{2}} - 1}{(1+\omega^2/\omega_v^2)}\right]^{\frac{1}{2}}. \quad (3.14)$$

All the above expressions are exact. If we are interested in low frequencies for which $\omega \ll \omega_v$, the expressions for velocity and absorption reduce respectively to the following approximations:

$$c = \left(\frac{K_s}{\rho_0}\right)^{\frac{1}{2}} \cdot (1 + \tfrac{3}{8}\omega^2/\omega_v^2), \quad (3.15)$$

$$\alpha = \frac{\omega^2 \zeta}{2\rho_0 c_0^3} \cdot (1 - \tfrac{5}{8}\omega^2/\omega_v^2), \quad (3.16)$$

where $c_0 = (K_s/\rho_0)^{\frac{1}{2}}$ and is approximately equal to c. In this range the absorption coefficient is very nearly proportional to the square of the frequency, and to the same approximation the phase velocity is constant and equal to c_0. If $\zeta = 4\eta/3$ Eq. (3.16) takes the usual form for the viscous absorption η is the shear viscosity coefficient.

On the other hand if $\omega \gg \omega_v$, the expressions (3.11) and (3.12) yield

$$c = c_0(2\omega/\omega_v)^{\frac{1}{2}}, \quad (3.17)$$

$$\alpha = \frac{1}{c_0}\left(\frac{\omega\omega_v}{2}\right)^{\frac{1}{2}} = \left(\frac{\rho_0}{2\zeta}\omega\right)^{\frac{1}{2}}. \quad (3.18)$$

Note that here $k_r = -k_i$. Both velocity and absorption coefficient vary with the square root of the frequency.

Experience indicates that in most fluid media, neither the Maxwell equation (2.1) nor the Stokes equation (3.1) is sufficient by itself to account for acoustic absorption. A combination of the two might be expected to be more successful. This has been shown to be true in the case of water by the theory of L. H. Hall (H1). A derivation of Hall's combined Maxwell-Stokes equation (assumed by him but not derived) will be presented in Sec. 7 where a more general viewpoint will be discussed. Moreover, we must not neglect a review of heat conduction as an origin of sound absorption.

4. EQUATION OF STATE FOR HEAT CONDUCTION

It was pointed out by Kirchhoff (K3) in 1868 that it is not proper to consider the effect of viscosity of a fluid on sound absorption without also accounting for the comparable effect of heat conduction. Applying the first law of thermodynamics we have for the rate at which heat enters the fluid

$$\frac{\partial Q}{\partial t} = \frac{\partial U}{\partial t} + p \frac{\partial V}{\partial t}, \quad (4.1)$$

where U is the total internal energy, Q is the heat, and V is the molar volume.

In this and the following sections we shall be forced to use a considerable number of thermodynamic equations. For the convenience of the reader, we have listed them in Appendix II. All the relations can be proven in an elementary manner as shown in many texts of thermodynamics, e.g., Slater (S8). We shall assume that the process is quasistatic so that thermodynamics can be used. For the one-dimensional case, using Eq. (A-10) of Appendix II, we can write Eq. (4.1) for one mole as

$$\frac{M\kappa}{\rho_0} \frac{\partial^2 T}{\partial x^2} = \frac{C_v}{K\beta} \cdot \frac{\partial p}{\partial t} + \frac{C_p}{V\beta} \cdot \frac{\partial V}{\partial t}, \quad (4.2)$$

in which K is the isothermal bulk modulus, κ the coefficient of thermal conductivity, β the coefficient of volume expansion at constant pressure, and C_p and C_v the usual molar heats. M is the molecular weight. Equation (4.2) thus serves as an equivalent equation of state or equation for time rate of change of state under the influence of heat conduction. We may rewrite (4.2) in the form

$$\frac{\partial p}{\partial t} - \frac{K_s}{\rho_0} \frac{\partial \rho}{\partial t} = \frac{M\kappa}{\rho_0} \frac{\beta K}{C_v} \frac{\partial^2 T}{\partial x^2}, \quad (4.3)$$

where $K_s = (C_p/C_v)K = \gamma K$. We eliminate T with the help of (A-8), and obtain,

$$\frac{\partial p}{\partial t} - \frac{K_s}{\rho_0} \cdot \frac{\partial \rho}{\partial t} = \frac{M\kappa}{\rho_0} \left[\frac{1}{C_v} \frac{\partial^2 p}{\partial x^2} - \frac{K}{\rho_0 C_v} \cdot \frac{\partial^2 \rho}{\partial x^2} \right]. \quad (4.4)$$

On the right-hand side we have thrown away the nonlinear term involving $(\partial \rho/\partial x)^2$ since the changes in p and ρ involved in the passage of the sound wave are very small. In Eq. (4.4) we may replace p and ρ by p_e and ρ_e respectively in order to make comparison with the earlier Eqs. (2.2) and (3.1).

We now examine the use of (4.4) for a plane wave of the form (2.12) to (2.14). Equations (2.17) and (2.18) remain unchanged but (2.16) becomes

$$\left(i\omega + \frac{M\kappa k^2}{\rho_0 C_v} \right) P_e - \left(\frac{i\omega K_s}{\rho_0} + \frac{M\kappa K k^2}{\rho_0^2 C_v} \right) R_e = 0. \quad (4.5)$$

Combined with (2.17) and (2.18) this yields

$$k^2 \left(\frac{i\omega K_s}{\rho_0} + \frac{M\kappa}{\rho_0^2} \frac{K k^2}{C_v} \right) - \left(i\omega + \frac{M\kappa k^2}{\rho_0 C_v} \right) \omega^2 = 0. \quad (4.6)$$

This may be rewritten in the form

$$k^2 = \omega^2 \frac{\rho_0}{K} \cdot \frac{C_v - iWk^2}{C_p - iWk^2}, \quad (4.7)$$

where $W = M\kappa/\rho_0\omega$. This leads to a quadratic for the complex k^2. Fortunately, inspection shows that $k_r \gg k_i$. It will be sufficient, therefore, to replace k^2 in the term Wk^2 by $k_r^2 = \omega^2/c^2$. Moreover on the left-hand side of (4.7) k^2 becomes to a good approximation $k_r^2 + 2ik_r k_i$. This is true as long as we neglect viscosity; when considering viscosity and heat conduction are combined, this approximation is questionable. Hence by equating real and imaginary parts of (4.7) we arrive at

$$\alpha = -k_i = -\frac{1}{2} \frac{c}{c_0^2} (\gamma - 1) \frac{\omega_c}{\omega_c^2 + \omega^2} \omega^2 \quad (4.8a)$$

for the absorption coefficient. We have here set

$$\omega_c^2 = \frac{C_p^2 \rho_0^2 c^4}{M^2 \kappa^2} \quad (4.9)$$

which appears as the square of a kind of relaxation frequency, itself a function of frequency through c.

For comparison with other mechanisms we write (4.8a) as

$$\alpha = -\frac{1}{2} \frac{c}{c_0^2} \frac{K_s^0 - K_s^\infty}{K_s^\infty} \frac{\omega_c}{\omega_c^2 + \omega^2} \omega^2, \quad (4.8b)$$

where in the case of heat conduction

$$K_s^0 = \frac{C_p}{C_v} K$$

$$K_s^\infty = K.$$

Similarly to the indicated approximation, the phase velocity is

$$c^2 = \frac{1}{\rho_0} \frac{\omega_c^2 + \omega^2}{(\omega_c^2/K_s^0) + (\omega^2/K_s^\infty)}. \quad (4.10)$$

The fundamental difference between this mechanism and those described in subsequent sections should be clearly noted, even though the frequency dependence of k_r and α turns out to be approximately the same. The space gradient appears here while in the later equations of state only derivatives with respect to time appear.

5. EQUATION OF STATE FOR HEAT RADIATION

For the sake of completeness we present a review of the effect of heat radiation on sound transmission, following the general method of the preceding sections. We again use the first law in the form (4.1), but now

$$\frac{\partial Q}{\partial t} = -qC_v(T - T_0), \quad (5.1)$$

where we are following Stokes (S18) in using effectively Newton's law of cooling for the rate of radiation from the region traversed by the sound where the temperature is T to the surrounding fluid at temperature T_0.

If we put $V=M/\rho$, Eq. (4.2) now assumes the form, after differentiation with respect to the time

$$\frac{C_v}{K\beta}\frac{\partial^2 p}{\partial t^2} - \frac{MC_p}{\beta V \rho_0^2}\frac{\partial^2 \rho}{\partial t^2} = -qC_v\frac{\partial T}{\partial t}. \quad (5.2)$$

Once more we write, using (A-8)

$$\dot{T} = \frac{1}{K\beta}\dot{p} - \frac{M}{\beta V \rho_0^2}\dot{\rho}, \quad (5.3)$$

and find for (5.2) after some reduction

$$\frac{\partial^2 p}{\partial t^2} + q\frac{\partial p}{\partial t} - \frac{K_s}{\rho_0}\frac{\partial^2 \rho}{\partial t^2} - \frac{qK}{\rho_0}\frac{\partial \rho}{\partial t} = 0. \quad (5.4)$$

Integrating with respect to the time, setting the constant of integration equal to zero and using excess pressure and density p_e and ρ_e, respectively, we arrive at

$$\frac{\dot{\rho}_e}{\rho_0} - \frac{1}{K_s^\infty}\dot{p}_e + \omega_r\left(\frac{\rho_e}{\rho_0} - \frac{1}{K_s^0}p_e\right) = 0. \quad (5.5)$$

where $K_s^\infty = \gamma K$, $K_s^0 = K$, and $\omega_r = q/\gamma$. The K's defined here do not agree with those defined in the last section. They really denote the bulk moduli for a slow or a fast process. In the case of radiation, the fast bulk modulus is adiabatic while in heat conduction, the fast bulk modulus is isothermal. In the case of thermal or structural relaxation, we have a fast and a slow adiabatic bulk modulus, and this is the reason for the notation of K_s^0 and K_s^∞. The model for this acoustical equation of state will be considered in the next section.

For a plane wave of the form (2.12) to (2.14) we obtain in place of (2.16) and (4.5)

$$(\omega - i\omega_r)\frac{R_e}{\rho_0} - \left(\frac{\omega}{K_s^\infty} - \frac{i\omega_r}{K_s^0}\right)P_e = 0. \quad (5.6)$$

The secular equation connecting ω and k now becomes

$$\omega^2\rho_0\left(\frac{\omega}{K_s^\infty} - \frac{i\omega_r}{K_s^0}\right) = k^2(\omega - i\omega_r). \quad (5.7)$$

On the presumption that $|k_r| \gg |k_i|$ we can solve for k_r and k_i and get for the absorption coefficient

$$\alpha = \frac{1}{2}\frac{c}{c_0^2}\frac{\omega^2 \omega_r}{\omega^2 + \omega_r^2}\frac{K_s^\infty - K_s^0}{K_s^\infty}, \quad (5.8)$$

while the phase velocity (and associated dispersion law) is given by

$$c^2 = \frac{1}{\rho_0}\frac{\omega^2 + \omega_r^2}{(\omega^2/K_s^\infty) + (\omega_r^2/K_s^0)}. \quad (5.9)$$

At low frequency

$$c_0^2 = K_s^0/\rho_0, \quad (5.10a)$$

while at high frequency

$$c_\infty^2 = K_s^\infty/\rho_0. \quad (5.10b)$$

Stokes and Rayleigh made only empirical estimates of the magnitude of q, but it would seem possible to study it independently by applying radiation theory. One attempt is made here as a suggestion. Further work in evaluating q is certainly required. From the Stefan-Boltzmann law applied to a sphere of surface area A, with the temperature difference $T-T_0$ between the surface and the surrounding environment, we have

$$4A\sigma T_0^3(T-T_0) = qC_v(T-T_0), \quad (5.11)$$

where σ is the Stefan-Boltzmann constant. Let us apply this to a sphere of diameter $\lambda/2$, which we assume to be at the same temperature. Then

$$q = \frac{48T_0^3\sigma}{\rho_0 c_v \lambda}, \quad (5.12)$$

where c_v is the specific heat capacity per gram. From radiation measurements

$$\sigma = 5.673 \times 10^{-5} \text{ erg/sec cm}^2 \text{ °A}^4. \quad (5.13)$$

An estimate of q may readily be obtained, for example, argon, where

$\rho_0 = 1.784 \times 10^{-3}$ g/cm³   at $T_0 = 273$°A,
$c_v = 0.075$ cal/g °A   and $c = 3.08 \times 10^4$ cm/sec.

The result is $q = 1.34 \times 10^{10}$ sec^{-1}, at 1 mc. The value of q in (5.12) is directly proportional to frequency, whereas Rayleigh (referred to by Rocard (R9)) estimated it would be independent of frequency. At low frequencies the value given by (5.12) is not far out of line with Rocard's estimate. It is not intended that our result should be taken seriously, but it does reinforce the conclusion that radiation absorption plays little role in gases, except possibly at very low pressures or very high temperatures.

6. EQUATION OF STATE FOR THERMAL RELAXATION

A. The Method of Irreversible Thermodynamics

In the analysis of Herzfeld and Rice (H10) sound absorption in a fluid is attributed to the lag in the adjustment between external and internal degrees of freedom of the constituent molecules during the passage of the sound wave. These authors apply thermodynamics to a nonequilibrium problem. The nonequilibrium effects are so important that one cannot even approximate them by equilibrium thermodynamics. In essence, their treatment is one of the first treatments of irreversible thermodynamics. The summarization presented here is in a form permitting close comparison with the analysis of the preceding sections.

The total energy U per mole is divided into a part U^e referring to the translational energy of the molecules as a whole and a part U^i characterizing the relative motions of the constituent parts of the molecules. For convenience two corresponding temperatures T^e and T^i are associated with these energies, respectively. These temperatures assume partial equilibrium of the internal and external energies. These postulates undoubtedly need further investigation. U^i is a function of T^i only. It is assumed that the only way in which U^i may change is through a change in the temperature T^e. We may express the above assumption analytically in the form

$$\dot{T}^i = (T^e - T^i)/\tau. \quad (6.1)$$

The quantity τ plays the role of a relaxation time.

The process by which the energy of the system is changed is assumed to be adiabatic. Under these conditions the first principle of thermodynamics takes the form

$$dU^e + dU^i + pdV = 0, \quad (6.2)$$

where we are dealing with fluids in which it is possible to assign unambiguous meaning to p and V. Now from the definition of specific heat at constant volume

$$dU^i = C_v{}^i dT^i, \quad (6.3)$$

where $C_v{}^i$ is the internal molar heat at constant volume. Moreover, from Appendix II (Eq. A-11) we can write at once, so far as the external energy is concerned,

$$dU^e + pdV = C_v{}^e dT^e + T^e K\beta dV = 0. \quad (6.4)$$

The use of this equation assumes that conventional thermodynamics holds for the relations among U^e, T^e, V, and p, i.e., for a given U^e and T^e, p and V are known. Equation (6.2) then becomes, if we apply time variation,

$$C_v{}^i \dot{T}^i + C_v{}^e \dot{T}^e + T^e K\beta \dot{V} = 0. \quad (6.5)$$

To eliminate T^i from (6.5) we transform it to the form

$$C_v{}^i \left(\frac{d^2 T^i}{dt^2} + \frac{1}{\tau} \frac{dT^i}{dt} \right) + C_v{}^e \left(\frac{d^2 T^e}{dt^2} + \frac{1}{\tau} \frac{dT^e}{dt} \right)$$

$$+ T^e \beta K \left(\frac{d^2 V}{dt^2} + \frac{1}{\tau} \frac{dV}{dt} \right) + \beta K \frac{dT^e}{dt} \frac{dV}{dt} = 0. \quad (6.6)$$

By using (6.1) and (6.5), we obtain

$$C_v{}^e \frac{d^2 T^e}{dt^2} + K\beta T^e \frac{d^2 V}{dt^2} + \frac{1}{\tau}(C_v{}^i + C_v{}^e) \frac{dT^e}{dt}$$

$$+ K\beta \frac{dV}{dt} \left(\frac{dT^e}{dt} + \frac{1}{\tau} T^e \right) = 0. \quad (6.7)$$

To compare this equation of state with those for viscosity, heat conduction, and heat radiation, we must express it in terms of the changes in p and V. For this purpose we must use the thermodynamic Eq. (A-8) of Appendix II and write

$$\dot{T}^e = \dot{p}/K\beta + \dot{V}/V\beta, \quad (6.8)$$

$$\frac{d^2 T^e}{dt^2} = \frac{1}{K\beta} \frac{d^2 p}{dt^2} + \frac{1}{V\beta} \frac{d^2 V}{dt^2} - \frac{1}{V^2 \beta} \left(\frac{dV}{dt} \right)^2. \quad (6.9)$$

On substitution back into (6.7) there results

$$(C_v{}^e/K\beta) \frac{d^2 p}{dt^2} + [(C_v{}^i + C_v{}^e)/\tau K\beta] \frac{dp}{dt} + \left(\frac{dp}{dt} \right) \left(\frac{dV}{dt} \right)$$

$$+ (C_v{}^e/\beta V + T^e K\beta) \frac{d^2 V}{dt^2} + \left(\frac{K}{V} - \frac{C_v{}^e}{\beta V^2} \right) \left(\frac{dV}{dt} \right)^2$$

$$+ [(C_v{}^e + C_v{}^i)/\tau \beta V + T^e \beta K/\tau] \frac{dV}{dt} = 0. \quad (6.10)$$

Writing $\dot{V} = -M/\rho_0{}^2$, $\dot{\rho}_e$, etc. and neglecting the nonlinear terms as small compared with the rest, we finally secure the equation of state in the form (with $\dot{p} = \dot{p}_e$, as before)

$$C_v{}^e \frac{d^2 p_e}{dt^2} + \frac{C_v{}^t}{\tau} \frac{dp_e}{dt} - \frac{K}{\rho_0}(C_v{}^e + T^e \beta^2 KV) \frac{d^2 \rho_e}{dt^2}$$

$$- \left(\frac{K}{\rho_0 \tau} \right) (C_v{}^t + T^e \beta^2 KV) \frac{d\rho_e}{dt} = 0. \quad (6.11)$$

We have here set $C_v{}^t = C_v{}^e + C_v{}^i$ the total specific heat at constant volume for a slow process where $T^e = T^i$. Examination of the order of magnitude of the various terms in (6.10) for harmonic time dependence with angular frequency ω indicates the validity of the neglect of the nonlinear terms. We now find it convenient to make the following definitions:

$$K_s{}^0 = K(1 + T^e \beta^2 KV/C_v{}^t), \quad (6.12)$$

and

$$K_s{}^\infty = K(1 + T^e \beta^2 KV/C_v{}^e). \quad (6.13)$$

A very fast process does not affect U^i and hence $C_v{}^e$ is the net specific heat. Thus (6.13) comes naturally from the general definition of the adiabatic bulk modulus (A-14)(Z2; p. 229). For a slow process $T^i = T^e$ and the net specific heat is $C_v{}^t$, hence again (6.12). This transforms (6.11)

$$\frac{\dot{\rho}_e}{\rho_0} - \frac{\dot{p}_e}{K_s{}^\infty} + \omega_0 \left(\frac{\rho_e}{\rho_0} - \frac{p_e}{K_s{}^0} \right) = 0. \quad (6.14)$$

We have also defined

$$\omega_0 = \frac{1}{\tau}(C_p{}^t/C_p{}^e). \quad (6.15)$$

A comparison with Eq. (5.5) shows that the acoustical equation of state for thermal relaxation is identical in form with the one for heat radiation.

A mechanical model governed by an equation of the form (6.14) has been suggested by Frenkel (F5; p. 208) and is shown schematically in Fig. I-3a (compare Figs. I-1a and I-2a).

Since (6.14) is identical in form with (5.5) the equations for the velocity and the absorption in this case are the same as before, i.e., (5.9) and (5.8) with ω_0 replacing ω_r. The meaning of the K's have changed. The equation for α is given by

$$\alpha = -\frac{1}{2}\frac{c}{c_0^2}\omega^2\frac{\omega_0}{\omega^2+\omega_0^2}\frac{K_s^\infty - K_s^0}{K_s^\infty}$$

$$= -\frac{1}{2}\frac{c}{c_0^2}\omega^2\frac{\omega_0}{\omega^2+\omega_0^2}\left(\frac{C_p{}^i - C_v{}^i}{C_v{}^i C_p{}^e}\right)C^i. \quad (6.16)$$

To obtain the last form we have used Eqs. (A-12) and (A-13) as well as the fact that on this model β and K are independent of frequency.

7. EQUATION OF STATE FOR A GENERAL INTERNAL LAG

A. The Method of Statistical Thermodynamics

In this section we shall employ a statistical model to derive an equation of state for a fluid for use in sound transmission calculations. This method has been used by Kneser (K11) and others but will be presented here in somewhat more general form, applicable to both gases and liquids (M2). We assume that the pressure changes associated with dilatations of the fluid are connected with changes in the energy states of the constituent molecules. Transitions between the various states are governed by finite probabilities, leading to finite relaxation times and hence to absorption and dispersion of acoustical radiation as suggested by the earlier parts of this review.

To make the model more concrete, several cases will now be examined in detail. First, consider gaseous hydrogen (F1; p. 84), which is complicated by the fact that two types exist, *ortho* and *para*. We shall, for simplicity, confine our attention to only one type, say *para*. A vibrational excitation requires high energy, so that below several thousand degrees, the average fraction of molecules in the excited vibrational state is very small. A useful concept is the characteristic temperature θ, i.e., $\Delta\epsilon/k'$, where $\Delta\epsilon$ is the energy difference of the transition, while k' is Boltzmann's constant. The characteristic temperature of vibration for hydrogen is about 6100°A.

The rotational levels are therefore the ones of interest here. The energy levels are given approximately by (H8).

$$Bc_l h(J+1)J,$$

where B is a constant which can be determined optically, c_l is the velocity of light and h is Planck's constant. In general, $J = 0, 1, 2, 3$. However, for *para*hydrogen, the values of J are 0, 2, 4. The statistical weight of the levels is $2J+1$. Since $B = 61$ cm^{-1} for H_2, we find that the fraction of molecules in state $J=0$ is 0.28, in state $J=2$ is 0.58 and in state $J=4$ is 0.13 for $T=600°A$. Therefore, in this case one must consider at least the three lower states. The two-state model to be developed in this section is not completely adequate for H_2. Rhodes (R3) has considered a more elaborate model which applies here.

In other molecules θ_{rot} is much lower than in hydrogen. One may assume that it is possible to assign a statistical weight to each vibrational level. Using the above information as to the weight of the rotational level and the energy distribution, the statistical weight of a vibrational level[6] is

$$\sum (2J+1)\exp[-Bc_l h J(J+1)/k'T].$$

B has only a slight dependence on the vibrational state. There is no reason to believe that a high rotational level of a lower vibrational level has less energy than a lower rotational level of an excited vibrational band. In general, as a matter of fact, this is not the case

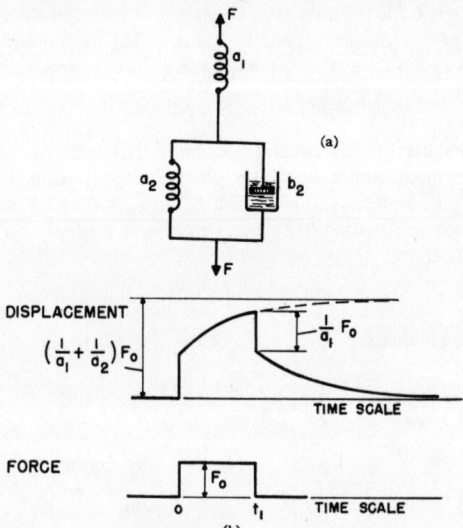

FIG. I-3 (a, b). Mechanical model corresponding to the relaxational equation (after Frenkel).

(H8; p. 107). By the proper definition of a vibrational statistical weight given above, however, we may consider the vibrational molecule as having evenly spaced energy levels. The characteristic temperatures for vibration range from about 6100°A for H_2 to 310°A for I_2 (S8; p. 142). For I_2 at 620°A, the fraction of molecules in the ground state, 1st, 2nd, 3rd, and 4th excited states are 0.39, 0.24, 0.14, 0.09, and 0.05; and we cannot expect a two-state model to apply in this case. θ_v for Cl_2 is 810°A so that at 300°A the fractions of states in the lowest and first excited state are 0.94 and 0.06. This gas can therefore be approximated by a two-state model. For vibrational states, Landau and Teller have been able to work out a model for an n-state gas (L3).

---

[6] For simplicity, we shall not define the summation over J. This depends on the type of molecules considered.

The lumping of the rotational levels into one is permissible only as long as there is a rapid readjustment of these levels. One might expect that as one goes up in frequency the readjustment in the vibrational levels will lag and produce an absorption region. Later, at higher frequency, another region of absorption will appear due to lags caused by readjustments of the rotational levels. Such effects have recently been observed by Zmuda[7] in N_2 at room temperature.

One may also consider the problem of dissociation of a diatomic gas for which there exists two states. One would expect that the general theory to be presented here can be adapted to this case, though the details have not been carried through. Another method has been used by Einstein (E4) and by Luck (L18) to solve this problem. (See also the paper of Kneser and Gauler (K10).)

At present we are unable to give as complete a physical theory of a liquid as of a gas or of a crystalline solid. For our purposes it will be necessary to distinguish between a normal liquid and an associated liquid. A normal liquid is a monomer. The basic unit is the individual atom; crudely speaking, the liquid is a dense imperfect gas. One would expect that the internal vibrations of normal liquids would be nearly the same as in the gaseous phase (we ignore the possibility of lower frequency interlattice modes), and such is the case for many fluids.

The associated liquid (H9; p. 534) on the other hand, has a group of molecules which bond together to form multi-molecular groups, or polymers. Formic acid in the liquid state shows a vibrational bond at 3080 cm^{-1}, whereas at high temperatures in the gaseous phase, the characteristic vibration is at 3570 cm^{-1} due to a vibration between H and O atoms. This has been interpreted by assuming that at lower temperatures molecular groups appear, such as

$$
\begin{array}{c}
O-H\cdots O \\
H-C C-H \\
O\cdots H-O
\end{array}
$$

The new bond arises because of links characterized by $H\cdots O$. Such binding is known as a hydrogen bond because it usually requires a hydrogen atom. The actual binding may take the form of rings or chains. It is to be stressed that the idea of the association is not based on the simple physical illustration but is required from basic chemical considerations. Associated liquids appear when there are hydroxyl (OH) and amino (NH_2) units (W4; p. 46).

A dilatation may cause two effects on the fluid. It may in the normal fluid cause transitions within the molecules, or in the associated liquid it may cause changes in the association. Both Hall (H1) and Ghosh (G2) assume that part of the absorption in water is due to the rearrangements of the polymer structure. In ice, each molecule has four nearest neighbors. In water, the number of nearest neighbors ranges from three to two. Hall postulates two states, of which the lower is more ice-like, while in the upper state the molecules are more random. We cannot, at present, make the picture more specific nor show that there are just two states, nor define just what the unit, which will be loosely called a molecule, is. Even the detailed Hall and Ghosh calculations do not answer these important questions.

In acetic acid, Lamb and Pinkerton (L2) suggested that part of the absorption is due to the breaking down of a dimer to a monomer. A difficulty of the concept is that there is no independent way of finding out just what the process is. It takes 16 kcal/mole at 25°C to dissociate the double molecule in the gaseous phase, but the acoustical processes suggest a much smaller value. We are, therefore, forced to conclude that here, as in the case of water, the exact model of what is broken up remains unclear. Equally uncertain is the model suggested by Liebermann (L12) for the absorption in sea water. This model is connected with the dissociation of $MgSO_4$ and will be discussed further in Sec. 30 below.

Liebermann has been able to get a numerical check of some of the *para*meters, but his method appears to be semi-empirical. For instance, Liebermann (L10) previously had been able to explain the absorption in sea water ignoring the presence of $MgSO_4$ altogether. This suggests that a more fundamental investigation of the problem should be made.

In spite of the many limitations of the two-state model, we shall develop it, since it shows the nature of the problem and is the most general model (known to the authors) using statistical thermodynamics, which would apply to gases and liquids.

For the sake of simplicity we suppose the two states of each molecule are alike: an unexcited state has an energy E_1 (per mole) and an excited state has an energy E_2 (per mole). It is assumed that the process of excitation alters the intermolecular forces so as to distort the molecular structure and change the average volume per molecule. The average volumes per molecule in the unexcited and excited states are denoted by v_1 and v_2, respectively, and the corresponding values per mole for V_1 and V_2. The change in volume V_2-V_1 may be expected to play a greater role in liquids than in gases.[8]

The instantaneous numbers of molecules per mole in the unexcited and excited states are N_1, and N_2, respectively, and $N_1+N_2=N$, where N is the total number of molecules per mole. It is assumed that for equilibrium the distribution is canonical and therefore at temperature T the average number of molecules in

---

[7] A. J. Zmuda, private communication and (Z3).

[8] The association of v_1 and v_2 with states may seem artificial. Actually we shall only be interested in V_2-V_1 associated with the increase or decrease of the population in state 2.

state 1 is
$$N_1^0 = Cw_1 \exp[-E_1/RT], \quad (7.1)$$
where w_1 is a weighting factor, C is a constant, and $R = k'N$, where k' is Boltzmann's constant. Similarly
$$N_2^0 = Cw_2 \exp[-E_2/RT]. \quad (7.2)$$
We find it convenient to assume $w_1 = w_2$. Let A_{12} be the transition probability from state 1 to state 2 (i.e., the average number of transitions per second per molecule) and A_{21} the corresponding probability from state 2 to state 1. The fundamental equation governing transitions between the two states is
$$\dot{N}_1 = N_2 A_{21} - N_1 A_{12}. \quad (7.3)$$
For equilibrium, $\dot{N}_1 = 0$, and therefore
$$A_{12}^0/A_{21}^0 = N_2^0/N_1^0 = \exp[\Delta E/RT], \quad (7.4)$$
where $\Delta E = E_1 - E_2$.

Now the dilatation produces a disturbance which alters N_1, N_2, A_{21}, A_{12} as well as $\dot{N}_1$. Using ΔN_1, etc., to denote the change in value, by neglecting higher order terms we have from (7.3)
$$\Delta \dot{N}_1 = N_2^0 \Delta A_{21} + A_{21}^0 \Delta N_2 - N_1^0 \Delta A_{12} - A_{12}^0 \Delta N_1. \quad (7.5)$$
From (7.4) and the fact that $\Delta N_1 = -\Delta N_2$, it finally follows that
$$\Delta \dot{N}_1 = -(A_{12}^0 + A_{21}^0) \Delta N_1 - N_2^0 A_{21}^0 \Delta W \quad (7.6)$$
with
$$W = \Delta E/RT. \quad (7.7)$$
We now set for convenience
$$A_{12}^0 + A_{21}^0 = 1/\tau, \quad (7.8)$$
where τ has the dimensions of time and indeed will be referred to as the relaxation time for the particular mechanism. W is treated as a function of the independent macroscopic variables p and T. We select T and p as the macroscopic independent variables—T, because of its unique position in statistics and p because it seems related to the external forces on a molecule. This selection is not unique and it might be advisable to use T and V as has been done by Mandelstam and Leontovich (M1). Since the process we are dealing with is irreversible, the selection of independent variables has an important effect on the detailed development of the theory. However, the final results should be independent of the choice of macroscopic variables. The existence of macroscopic variables assumes an intermolecular equilibrium.

We can now write (7.6) in the form
$$\Delta \dot{N}_1 = -\Delta N_1/\tau - N_2^0 A_{21}^0 \left[\frac{\partial(W)}{\partial p} \Delta p + \frac{\partial(W)}{\partial T} \Delta T \right]. \quad (7.9)$$
Next we employ (7.4) again to replace $N_2^0 A_{21}^0$ with a more convenient expression and obtain for (7.9)[9]
$$\Delta \dot{N}_1 = -\Delta N_1/\tau - (N/\tau)(2 + A_{21}^0/A_{12}^0 + A_{12}^0/A_{21}^0)^{-1}$$
$$\times \left[\frac{\partial(W)}{\partial p} \Delta p + \frac{\partial(W)}{\partial T} \Delta T \right]. \quad (7.10)$$
Finally it will be convenient to set
$$N_1 = N_1^0 + n, \quad N_2 = N_2^0 - n, \quad (7.11)$$
so that
$$\dot{N}_1 = \dot{n}, \text{ etc.,}$$
and to write
$$2 + A_{21}^0/A_{12}^0 + A_{12}^0/A_{21}^0 = 2(1 + \cosh W). \quad (7.12)$$
Equation (7.10) can then be given the form of the differential equation
$$\tau \frac{d^2 n}{dt^2} + \frac{dn}{dt} = B_p \frac{dp}{dt} + B_T \frac{dT}{dt}, \quad (7.13)$$
if we set
$$B_p = -\frac{N}{2(1 + \cosh W)} \cdot \frac{\partial W}{\partial p},$$
$$B_T = -\frac{N}{2(1 + \cosh W)} \cdot \frac{\partial W}{\partial T}. \quad (7.14)$$

It is necessary to introduce thermodynamical considerations, in particular the first law. For an adiabatic process this will now appear in the form
$$\Delta U^e + \Delta U^i + p \Delta V^e + p \Delta V^i = 0. \quad (7.15)$$
The total energy of the system is
$$U = U^e + U^i, \quad (7.16)$$
where U^e is the "external" energy or that associated with the motion of the molecules as a whole, while U^i is "internal" energy or that associated with intramolecular energy states, such as those referred to at the beginning of this section. It is necessary to introduce two types of volume, V^e relating to the *macroscopic* volume per mole, and ΔV^i denoting effective change in volume per mole associated with the transition from state 1 to state 2 or vice versa. Thus
$$\Delta V^i = (v_1 - v_2) n = \Delta v \cdot n, \quad (7.17)$$
and
$$\Delta U^i = (\epsilon_1 - \epsilon_2) n = \Delta \epsilon \cdot n, \quad (7.18)$$
where $\epsilon_1 = E_1/N$ and $\epsilon_2 = E_2/N$. Introducing time derivatives and continuing to treat T and p as the independent variables, we can express the content of the

---

[9] Equation (7.10) follows from (7.9) since
$$\frac{N_2^0 A_{21}^0}{A_{12}^0 + A_{21}^0} = \frac{N_1^0 A_{12}^0}{A_{12}^0 + A_{21}^0} = \frac{N_1^0 A_{21}^0}{A_{21}^0 + (A_{21}^0)^2/A_{12}^0}$$
$$= \frac{N_1^0 A_{21}^0 + N_2^0 A_{21}^0}{A_{12}^0 + A_{21}^0 + A_{21}^0 + (A_{21}^0)^2/A_{12}^0} = \frac{N}{2 + (A_{21}^0/A_{12}^0) + (A_{12}^0/A_{21}^0)}.$$

first law for an adiabatic process in the form

$$\left[\left(\frac{\partial U^e}{\partial T}\right)_p + p\left(\frac{\partial V^e}{\partial T}\right)_p\right]\dot{T}$$
$$+ \left[\left(\frac{\partial U^e}{\partial p}\right)_T + p\left(\frac{\partial V^e}{\partial p}\right)_T\right]\dot{p} + \dot{n}(\Delta\epsilon + p\Delta v) = 0. \quad (7.19)$$

For the sake of simplicity we shall write $V^e = V$ in what follows, since V^i enters only through the term in Δv. From the principles of thermodynamics we have

$$(\partial U^e/\partial T)_p + p(\partial V^e/\partial T)_p = C_p^e, \quad (7.20)$$

the external molar heat capacity at constant pressure. The usual (reversible) relations are assumed to hold between U^e, T, and p. dU^e does not include effects due to changes in population. The use of the second law Eq. (A-9) yields

$$dU^e = [C_p^e - pV\beta^e]dT - [TV\beta^e - pV/K^e]dp, \quad (7.21)$$

so that

$$\left(\frac{\partial U^e}{\partial p}\right)_T = -T\beta^e V + pV/K^e, \quad (7.22)$$

and from Eq. (A-2),

$$p(\partial V/\partial p)_T = -pV/K^e. \quad (7.23)$$

In Eqs. (7.22) and (7.23)

$$K^e = -\frac{1}{(1/V)(\partial V/\partial p)_T}, \quad (7.24)$$

which is the external isothermal bulk modulus, and

$$\beta^e = (1/V)(\partial V/\partial T)_p, \quad (7.25)$$

which is the external coefficient of expansion at constant pressure.

If we use (7.20), (7.22), and (7.23) we can put (7.19) in the form

$$C_p^e \dot{T} - T\beta^e V\dot{p} + \dot{n}[\Delta\epsilon + p\Delta v] = 0. \quad (7.26)$$

For convenience later we differentiate (7.26) with respect to time, neglecting nonlinear terms like $\dot{p}\dot{T}$ and $\dot{p}\dot{n}$ as small compared with $(d^2p/dt^2)T$, etc., multiply by τ and add to (7.26). The somewhat artificial result is

$$C_p^e\left[\frac{dT}{dt} + \tau\frac{d^2T}{dt^2}\right] - T\beta^e V\left[\frac{dp}{dt} + \tau\frac{d^2p}{dt^2}\right]$$
$$+ [\Delta\epsilon + p\Delta v]\left[\frac{dn}{dt} + \tau\frac{d^2n}{dt^2}\right] = 0. \quad (7.27)$$

This immediately enables us to use (7.13) and hence to write (7.27) in the form

$$C_p^e \frac{d^2T}{dt^2} - T\beta^e V\frac{d^2p}{dt^2}$$
$$+ \frac{1}{\tau}\left\{[C_p^e + B_T\Delta\epsilon + B_T p\Delta v]\frac{dT}{dt}\right.$$
$$\left. - [T\beta^e V - B_p\Delta\epsilon - B_p p\Delta v]\frac{dp}{dt}\right\} = 0. \quad (7.28)$$

We may simplify (7.28) by making the following definitions:

$$C_p^i = \Delta\epsilon B_T + B_T p\Delta v, \quad (7.29)$$
$$\beta^i = -(1/TV)(\Delta\epsilon B_p + p\Delta v B_p), \quad (7.30)$$
$$1/K^i = -(1/V)\Delta v B_p. \quad (7.31)$$

These definitions require justification.

For a slow process, the first two terms of (7.28) will be very small. If the process is slow enough, we may also assume that the process is quasistatic. This means that the right-hand side of (7.3) is always very small. Such a process can be considered reversible as well as adiabatic and, therefore, from thermodynamic arguments it follows that

$$C_p^t \dot{T} - T\beta^t V\dot{p} = 0. \quad (7.32)$$

The superscript refers to the total, or static, values of C_p and β. Definitions (7.29) and (7.30) assure that

$$C_p^t = C_p^i + C_p^e \quad (7.33)$$

and

$$\beta^t = \beta^e + \beta^i. \quad (7.34)$$

We see from (7.32), (7.33), and (7.34) that C_p^i is the additional molar heat and β^i is the additional thermal coefficient of expansion. This justifies definitions (7.29) and (7.30).

Another way to satisfy ourselves about these definitions is to return to (7.10) and consider equilibrium states, i.e., $\dot{n} = 0$. Then, with the help of (7.10) and (7.14), we have

$$n = B_p\Delta p + B_t\Delta T, \quad (7.35)$$

where by (7.11) n is the change in population. The specific heat at constant pressure is defined as

$$C_p = (\partial U/\partial T)_p + p(\partial V/\partial T)_p. \quad (7.36)$$

By (7.35) it follows that the shift in the populations per unit temperature at constant pressure is B_T; hence, the first term on the right of Eq. (7.29) corresponds to the first term on the right of Eq. (7.36). Likewise the same holds for the second term.

We can also justify this by referring to the customary definition of a complex specific heat. This definition is (K11, R4, and E6).

$$C^{\text{eff}} = C^i/(1 + i\omega\tau), \quad (7.37)$$

based on the assumption that the disturbance is harmonic. C^{eff} is the effective value of the specific heat of vibration (or rotation) at frequency $\omega/2\pi$. From (7.13) it follows that

$$(\partial n/\partial T)_p = B_T/(1+i\omega\tau). \quad (7.38)$$

The complex number in (7.37) and (7.38) implies a phase lag between changes of T and n. The effective internal specific heat is

$$C_p^{\text{eff}} = (\Delta\epsilon + p\Delta v)(\partial n/\partial T)_p$$
$$= \{\Delta\epsilon B_T + p\Delta v B_T\}/(1+i\omega\tau). \quad (7.39)$$

Comparison of (7.37) and (7.39) provides justification of (7.29). Similar justification can be made for (7.30) and (7.31). The usual theory does not distinguish between C_p^i and C_v^i. Our system is more general, and these quantities are not equivalent. C_p^i is the more basic (to our theory) because our independent variables are p and T.

An alternative definition of β^i is possible. Following (7.39) we may write that

$$\beta^{\text{eff}} = (\Delta v/V)(\partial n/\partial T)_p = \Delta v B_T/[V(1+i\omega\tau)], \quad (7.40)$$

or

$$\beta^i = \Delta v B_T/V. \quad (7.41)$$

One can also obtain (7.41) by arguments similar to those connected with Eqs. (7.35) and (7.36). Comparing (7.41) and (7.30), one must conclude that

$$\frac{\Delta v}{V}B_T = -\frac{1}{TV}(\Delta\epsilon B_p + p\Delta v B_p). \quad (7.42)$$

It seems that this requirement must be imposed on this model if it is to obey reversible thermodynamics at frequencies far below and far above $1/\tau$.

Using the definitions just made, we can simplify (7.28) to the form

$$C_p^e\frac{d^2T}{dt^2} - T\beta^e V\frac{d^2p}{dt^2} + \frac{1}{\tau}\left\{C_p^t\frac{dT}{dt} - TV\beta^t\frac{dp}{dt}\right\} = 0. \quad (7.43)$$

This is the adiabatic equation of state for our model. To compare it with the previous equations of state we must eliminate T. Further, we have four variables—u_x, ρ_e, p_e, and T—but only three equations, (2.10), (2.11), and (7.43). The density is not an independent variable in the usual sense, and V is not a function of T and p alone, but also depends on time. To find dV which will depend on dp and dT as well as on the change in volume produced by changes in the population, we use (7.17), (7.24), and (7.25). The result is

$$dV = -(V/K^e)dp + V\beta^e dT + \Delta v \cdot n. \quad (7.44)$$

If we obtain $\dot{V}$ and d^2V/dt^2 from (7.44), divide the former by τ and add to the latter, we obtain (neglecting as usual terms like $\dot{V}p/K^e$, etc.)

$$\frac{d^2V}{dt^2} + \frac{1}{\tau}\frac{dV}{dt} = -\frac{V}{K^e}\left(\frac{d^2p}{dt^2} + \frac{1}{\tau}\frac{dp}{dt}\right)$$
$$+ V\beta^e\left(\frac{d^2T}{dt^2} + \frac{1}{\tau}\frac{dT}{dt}\right) + \frac{1}{\tau}\Delta v\left(B_p\frac{dp}{dt} + B_T\frac{dT}{dt}\right), \quad (7.45)$$

in which (7.13) accounts for the last term on the right. Utilizing (7.31) and (7.41), and transposing a few terms, we can rewrite (7.45) as

$$\beta^e\frac{d^2T}{dt^2} + \beta^t\frac{1}{\tau}\frac{dT}{dt}$$
$$= -\frac{1}{K^e}\frac{d^2p}{dt^2} + \frac{1}{\tau K^t}\frac{dp}{dt} + \frac{1}{V}\left(\frac{d^2V}{dt^2} + \frac{1}{\tau}\frac{dV}{dt}\right), \quad (7.46)$$

where

$$1/K^t = 1/K^e + 1/K^i. \quad (7.47)$$

We now differentiate with respect to time, keeping only first-order terms, and multiply through by β^e. We then multiply (7.43) by β^t/τ and add the two. The result is:

$$C_p^e\left(\beta^e\frac{d^3T}{dt^3} + \frac{1}{\tau}\beta^t\frac{d^2T}{dt^2}\right) - T\beta^e V\left(\beta^e\frac{d^3p}{dt^3} + \frac{1}{\tau}\beta^t\frac{d^2p}{dt^2}\right)$$
$$+ \frac{1}{\tau}C_p^t\left(\beta^e\frac{d^2T}{dt^2} + \frac{1}{\tau}\beta^t\frac{dT}{dt}\right)$$
$$- \frac{TV\beta^t}{\tau}\left(\beta^e\frac{d^2p}{dt^2} + \frac{1}{\tau}\beta^t\frac{dp}{dt}\right) = 0. \quad (7.48)$$

Changing $\dot{v}/v$ to $-\dot{\rho}_e/\rho_0$ and using (7.46), the general acoustical equation of state is obtained:

$$\frac{C_v^e}{K^e}\frac{d^2p_e}{dt^2} - \frac{C_p^e}{\rho_0}\frac{d^2\rho_e}{dt^2} + \frac{1}{\tau}\left\{\frac{C_p^e}{K^t} + \frac{C_p^t}{K^e} - 2TV\beta^e\beta^t\right\}\frac{dp_e}{dt}$$
$$- \frac{1}{\tau}(C_p^e + C_p^t)\frac{1}{\rho_0}\frac{d\rho_e}{dt} + \frac{1}{\tau^2}\frac{C_v^t}{K^t}p_e - \frac{1}{\tau^2}C_p^t\frac{\rho_e}{\rho_0} = 0.[10] \quad (7.49)$$

We have used Eq. (A-13) to define C_v^e and C_v^t:

$$C_v^t = C_p^t - TV(\beta^t)^2 K^t \quad (7.50)$$

$$C_v^e = C_p^e - TV(\beta^e)^2 K^e. \quad (7.51)$$

Further, the definition is made

$$C_v^i = C_v^t - C_v^e. \quad (7.52)$$

We shall not attempt to examine the acoustical behavior of this equation of state in this general form. Three simplifying assumptions will now be made and discussed. They are:

[10] On integrating, we assume an appropriate constant, so that $\rho_e = \rho - \rho_0$ and $p_e = p - p_0$ appear.

(a) Thermal Relaxation:

$$B_T \neq 0,$$
$$\Delta\epsilon \neq 0,$$

but

$$B_p = \Delta v = 0.$$

(b) Structural Relaxation:

$$B_p \neq 0,$$
$$\Delta v \neq 0,$$

but

$$B_T = 0.$$

(c) General Linear Case:

Here the assumption is made that the products of any of the terms defined by (7.29), (7.30), and (7.31) are negligible.[11]

(a) Thermal Relaxation

This case applies to gases one would expect $\Delta\epsilon$ to be independent of temperature, and W is a function of T only. Since we neglect Δv, $\beta^i = 0$, and $\beta^e = \beta^t$. Further, $K^e = K^t = K$. From these relations and (A-13), it follows that $C_p{}^t - C_p{}^e = C_v{}^t - C_v{}^e = C_p{}^i = C_v{}^i$. Consequently, in Eq. (7.49) the coefficient of $\dot{p}_e$ becomes

$$\frac{C_p{}^e}{K} + \frac{C_p{}^t}{K} - 2TV\beta^e\beta^t = \frac{C_v{}^e}{K} + \frac{C_v{}^t}{K}, \quad (7.53)$$

and we may write (7.49) in the following operational form:

$$\left(\frac{d}{dt} + \frac{1}{\tau}\right)\left\{\frac{C_v{}^e}{K}\dot{p}_e + \frac{C_v{}^t}{\tau K}p_e - C_p{}^e\frac{\dot{\rho}_e}{\rho_0} - \frac{C_p{}^t}{\tau}\frac{\rho_e}{\rho_0}\right\} = 0. \quad (7.54)$$

The operator on the left does not equal zero; hence

$$\frac{\dot{\rho}_e}{\rho_0} - \frac{1}{K_s{}^\infty}\dot{p}_e + \omega_0\left(\frac{\rho_e}{\rho_0} - \frac{1}{K_s{}^0}p_e\right) = 0. \quad (7.55)[12]$$

Here we have defined $\omega_0 = C_p{}^t/\tau C_p{}^e$, which is the relaxation frequency for this process. Since K is frequency independent for this case, $K_s{}^0$ and $K_s{}^\infty$ are given (6.12) and (6.13). Equation (7.55) is equivalent to the acoustical equation of state obtained by the method of Herzfeld and Rice, i.e., Eq. (6.14).

(b) Structural Relaxation

Here, $B_T = 0$. Hence $C_p{}^i = 0$ and from (7.41) $\beta^i = 0$. Now from (7.30) we see that

$$\Delta\epsilon B_p + p\Delta v B_p = 0, \quad (7.56)$$

---

[11] One might assume a relaxation process where $B_T \neq 0$ and $\Delta v \neq 0$, but $\Delta\epsilon = B_p = 0$. Such a model leads to a contradiction because of (7.30) and (7.41). The same holds if one assumes that $B_p \neq 0$ and $\Delta\epsilon \neq 0$, but $B_T = \Delta v = 0$. These are thermodynamic limitations on the model.

[12] One may obtain (7.54) more simply by returning to (7.44) since, for this case, $V = V(p, T)$.

or

$$\Delta\epsilon + p\Delta v = 0. \quad (7.57)$$

Equation (7.57) looks like a statement of the first law for an adiabatic process and may not be an unreasonable requirement. Since $C_p{}^i = 0$, $C_p{}^e = C_p{}^tC_p$. From (7.52) we know that $C_v{}^i \neq 0$. Equation (7.49) takes the form

$$\left(\frac{d}{dt} + \frac{1}{\tau}\right)\left\{\frac{C_v{}^e}{K^e}\dot{p}_e - C_p\frac{\dot{\rho}_e}{\rho_0} + \frac{1}{\tau}\frac{C_v{}^t}{K^t}p_e - \frac{C_p}{\tau}\frac{\rho_e}{\rho_0}\right\} = 0. \quad (7.58)$$

Thus the acoustical equation of state for structural relaxation is

$$\frac{\dot{\rho}_e}{\rho_0} - \frac{1}{K_s{}^\infty}\dot{p}_e + \omega_0\left(\frac{1}{\rho_0}\rho_e - \frac{1}{K_s{}^0}p_e\right) = 0, \quad (7.59)[13]$$

where

$$K_s{}^\infty = K^eC_p/C_v{}^e \quad (7.60)$$

$$K_s{}^0 = K^tC_p/C_v{}^t \quad (7.61)$$

$$\omega_0 = 1/\tau. \quad (7.62)$$

Though Eq. (7.59) is formally similar to (7.55), $K_s{}^0$, $K_s{}^\infty$, and ω_0 have different definitions.

As far as the authors know, Eqs. (7.55) and (7.59) first appear in a paper by Frenkel and Obraztsov (F4). The connection shown here between the equation of state and the statistical model, as well as the linear case, is more recent (M2). It was also derived in an independent manner by Hoff Lu (H13) (see Sec. 8). Equation (7.59) was used by Hall in his calculation of the absorption of sound in water. Since, in this case, C_p is nearly equal to C_v, and C^i is very small, Hall neglected the adiabatic correction. For many liquids $C_p - C_v$ is not small, and consideration of these corrections may be of value.

From Eq. (7.47) we note that $K^t < K^e$ and because of (7.50) and (7.51) $C_v{}^e < C_v{}^t$, and hence $K_s{}^e > K_s{}^t$, an assumption made by Hall. Since Eq. (7.59) is identical in form with (5.5) the absorption and velocity dispersion are given by (5.8) and (5.9) except that the K's and the ω_0 have different meaning.

(c) General Linear Case

We shall make the special assumption that quotients and products such as $(C_p{}^i/K^i)$ and $C_p{}^iC_v{}^i$ are small. It is to be stressed that this is not generally true. Dutta (D4) has calculated C^i (he assumed thermal relaxation, where $C_p{}^i = C_v{}^i$) for several liquids. His calculations are based on Einstein's specific heat equation (S8; p. 142), the internal vibration of the molecules having been obtained from observed Raman spectra. Under the assumption that $(C_p{}^i)^2$, etc., are small, the third term on

---

[13] This equation can also be obtained without a double differentiation by returning to (7.44).

the left of (7.49) can be transformed as follows:

$$\frac{C_p{}^e}{K^t} + \frac{C_p{}^t}{K^e} - 2TV\beta^e\beta^t = \frac{C_p{}^e}{K^e} + \frac{C_p{}^t}{K^t} - TV(\beta^e)^2$$

$$-TV(\beta^t)^2 = \frac{C_v{}^e}{K^e} + \frac{C_v{}^t}{K^t}, \quad (7.63)$$

which gives for the general linear equation of state

$$\frac{C_v{}^e}{K^e} \frac{d^2 p_e}{dt^2} - \frac{C_p{}^e}{\rho_0} \frac{d^2 \rho_e}{dt^2}$$

$$+ \frac{1}{\tau}\left\{\left(\frac{C_v{}^e}{K^e} + \frac{C_v{}^t}{K^t}\right)\frac{dp_e}{dt} - (C_p{}^e + C_p{}^t)\frac{1}{\rho_0}\frac{d\rho_e}{dt}\right\}$$

$$+ \frac{1}{\tau^2}\left\{\frac{C_v{}^t}{K^t}p_e - C_p{}^t\frac{\rho_e}{\rho_0}\right\} = 0. \quad (7.64)$$

The difference between this case and the two previous cases for a harmonic disturbance is in the region where $\omega \sim 1/\tau$. At low frequencies there is a well-defined set of constants, $C_v{}^t$, K^t, and $C_p{}^t$ as there are at high frequency, $C_v{}^e$, K^e, and $C_p{}^e$.

There is no essential difference in the behavior of an acoustic wave in this case, however. Equations (2.17) and (2.18) still apply but (2.16) is replaced by

$$\left(i\omega + \frac{1}{\tau}\right)\left[i\omega\frac{C_v{}^e}{K^e}P_e + \frac{C_v{}^t}{\tau K^t}P_e - i\omega\frac{C_p{}^e}{\rho_0}R_e - \frac{1}{\tau}\frac{C_p{}^t}{\rho_0}R_e\right] = 0. \quad (7.65)$$

To obtain (7.65) we have used (2.12) and (2.14). Since the expression in the first parenthesis does not effect the evaluation of k, (7.65) can be replaced by

$$\left(\frac{i\omega}{K_s{}^\infty} + \frac{\omega_0}{K_s{}^0}\right)P_e - (i\omega + \omega_0)\frac{R_e}{\rho_0} = 0, \quad (7.66)$$

where in this case

$$K_s{}^\infty = K^e(C_p{}^e/C_v{}^e), \quad (7.67)$$

$$K_s{}^0 = K^t(C_p{}^t/C_v{}^t) \quad (7.68)$$

and

$$\omega_0 = C_p{}^t/\tau C_p{}^e. \quad (7.69)$$

These relations are generalizations of (6.12), (6.13), and (6.15) as well as (7.60) and (7.62). The absorption and dispersion are again given by (5.8) and (5.9) with appropriate interpretations of the constants. An alternative expression for α is

$$\alpha = \frac{1}{2}\frac{\omega^2\omega_0}{\omega^2 + \omega_0^2}\frac{c}{c_0^2}\left[K^e\left(\frac{1}{K^t} - \frac{1}{K^e}\right) + \frac{C_v{}^i}{C_v{}^e} - \frac{C_p{}^i}{C_p{}^e}\right]. \quad (7.70)$$

One cannot but raise the question as to what happens if $K_s{}^\infty \leqslant K_s{}^0$. In this system, energy will be taken from the body and transformed into mechanical work, a clear violation of the second law. One must, therefore, conclude that $K_s{}^\infty \geqslant K_s{}^0$ from this argument. We shall not examine the definitions of K^i, $C_p{}^i$, and β^i to show this is true, and, as far as we know, it has not been attempted. It could be done but would require some thermostatistical arguments along the line followed by Hall.

B. Section 7 Appendix

One may object to Eq. (7.3) on the grounds that it is too specialized and that many possible reactions cannot be characterized that simply. Actually the form of (7.3) is not a fundamental assumption to the equations derived in this section. Let us examine a more complex case. Consider the reaction

$$2Y_1 \rightleftarrows Y_2. \quad (7.71)$$

Let us assume that the backward reaction is of first order with respect to Y_2 while the forward reaction is of second order with respect to Y_1, i.e.,

$$\dot{n}_2 = k_{12}n_1^2 - k_{21}n_2, \quad (7.72)$$

where the n_i's are the numbers of particles per unit volume. We may rewrite (7.72) in the form

$$\dot{N}_2 = A_{12}N_1^2 - A_{21}N_2, \quad (7.73)$$

where

$$A_{21} = k_{21}, \quad (7.74a)$$

$$A_{12} = k_{12}/V. \quad (7.74b)$$

V is the volume which contains a mole of atoms some of which are associated and some of which are disassociated. V, of course, is a function of the equilibrium thermodynamic variables.

We introduce the Helmholtz free energy for a system which obeys classical statistics. It can be defined as follows (F1; p. 67)

$$F = -RT\{\ln f - \ln N + 1\}. \quad (7.75)$$

Here,

$f =$ the partition function, namely

$$= \sum_i \exp\{-E_i/RT\}.$$

The free energy for a mole of associated molecules can be defined as

$$F_2 = -RT\{\ln f_2 - \ln N + 1\}, \quad (7.76)$$

where for f_2 the sum extends only over the associated states. From (7.76) it follows that

$$f_2 = \frac{N}{e}\exp(-F_2/RT). \quad (7.77)$$

Likewise for a mole of disassociated atoms

$$f_1 = \frac{N}{e}\exp(-F_1/RT). \quad (7.78)$$

By standard statistics (F1; p. 158) we know that

$$N_1^2/N_2 = f_1^2/f_2 = \exp\{\Delta F/RT + \delta\} \quad (7.79)$$

where

$$\Delta F = F_2 - 2F_1,$$
$$e^\delta = N/e, \text{ a constant.}$$

Returning to (7.73), we consider a small perturbation from the equilibrium value

$$\Delta \dot{N}_2 = (A_{12}^0 + \Delta A_{12})(N_1^0 + \Delta N_1)^2 - (A_{21}^0 + \Delta A_{21})(N_2^0 + \Delta N_2). \quad (7.80)$$

At equilibrium

$$\frac{(N_1^0)^2}{N_2^0} = \frac{A_{21}^0}{A_{12}^0} = \exp(\Delta F^0/RT + \delta) \quad (7.81)$$

and

$$\Delta A_{21} = \frac{(N_1^0)^2}{N_2^0}\Delta A_{12} + A^0{}_{12}\frac{(N_1^0)^2}{N_2^0}\Delta W, \quad (7.82)$$

where ΔW is the variation of $\Delta F/RT$. Since $\Delta N_2 = -\frac{1}{2}\Delta N_1$ (7.80) gives, with the help of (7.81) and (7.82),

$$\Delta \dot{N}_2 = -(A_{21}^0 + 4A_{12}^0 N_1^0)\Delta N_2 - A_{21}^0 N_2^0 \Delta W$$
$$= -(1/\tau)\Delta N_2 - A_{12}^0(N_1^0)^2 \Delta W, \quad (7.83)$$

where now

$$1/\tau = (A_{21}^0 + 4N_1^0 A_{12}^0). \quad (7.84)$$

Consider two extreme cases. For a substance which is almost completely dissociated, i.e., $N_1 \gg N_2$,

$$A_{12}^0(N_1^0)^2 = A_{21}^0 N_2^0,$$
$$A_{12}^0 N_1^0 \ll A_{21}^0 \quad (7.85)$$

and

$$1/\tau \approx A_{21}^0 \quad (7.86)$$

which means that $1/\tau$ is independent of concentration. If, however, $N_1 \ll N_2$—the substance is almost completely associated, then

$$A_{12}^0 N_1^0 \gg A_{21}^0. \quad (7.87)$$

Here

$$1/\tau \approx 4A_{12}^0 N_1^0 = 4k_{12}n_1^0. \quad (7.88)$$

That is $1/\tau$ is proportional to the concentration of nonassociated molecules. Equation (7.83) is identical in form with Eq. (7.6) if we consider (7.8). (7.84) suggests that $1/\tau$ may be a complicated function of the thermodynamic variables and may depend on the concentration. It would seem that the basic assumption made in going from (7.3) to (7.6) is that only linear terms are kept. One would expect, however, that the definition of τ and the coefficients of $(\partial W/\partial p)$ and $(\partial W/\partial T)$ in Eq. (7.14) would depend on (7.3). The general shape of the absorption curves is independent of the form of (7.3), but the detailed behavior may depend on it.

8. VARIOUS RELAXATION METHODS

In addition to the methods presented in Sections 6 and 7, many approaches to the relaxation problem, i.e., (6.14), (7.55), (7.59), and (7.69), are possible and have been made in the past 50 years. The occurrence of several distinct approaches to the same problem is, of course, nothing new. Classical mechanics has several different formulations—i.e., Newton's second law, D'Alembert's principle, Hamilton's principle, Lagrange's equations, etc. In such a case the various methods complement each other and lead to a deeper understanding of the field. The object of this section is to review briefly the various approaches that have been made to the relaxation problem. A complete analysis of all the theories is clearly beyond the scope of any one paper. We shall attempt only to classify the various approaches and, in some cases, describe the basic steps.

It seems useful to classify the various relaxation theories into four groups according to the methods on which the theories are based, namely: (a) kinetic theory, (b) irreversible thermodynamics, (c) statistical thermodynamics, and (d) phenomenological approach. The equivalence of these various approaches is not always demonstrated, although many papers assume that the methods are essentially equivalent. We have seen above, that for gases a formulation based on statistical thermodynamics leads to the same acoustical equation of state as one based on irreversible thermodynamics. In this section we shall show that the kinetic theory approach and the phenomenological approach lead to the same acoustical behavior.

(a) The Method of Kinetic Theory

There are at least three treatments of relaxation which use kinetic theory. The oldest one is that of Jeans (J2) who in 1904 considered a gas of rough nonsymmetric molecules. Here it takes time to establish equilibrium between rotational and translational energy. From his development it is possible to define a translational and a rotational temperature, indeed it is possible to arrive at an equation corresponding to (6.1). Later, Bourgin developed a general theory of absorption in gases. Another variation of this method has been given by Saxton, (S1a). Saxton's treatment includes heat conduction and viscosity.[14] Although Bourgin's theory (B26 to B32) is applicable to far more complicated cases, a simplified version will be presented here which applies to the two state gas. In the case of gases his treatment can be applied to very complicated systems.

[14] An interesting point appears in Saxton's theory. At low frequency, i.e., below the relaxation maximum the contribution of viscosity to the amplitude absorption per wavelength is given by

$$\alpha\lambda = 4\pi\omega\eta/3\rho_0 c_\infty^2. \quad (8.1)$$

Here c_∞^2 is the square of the velocity at high frequency (beyond the absorption maximum) which seems surprising since one might expect c_0^2 to appear as it does in Sec. 11, where we have treated a similar problem by a different method. For H_2 at room temperature $C_v{}^i$ is nearly equal to the gas constant (for further details see Sec. 22). At this temperature the vibration degrees of freedom do not enter in, and $C_v{}^i$ is due entirely to rotation. Hence

$$c_\infty^2/c_e^2 = (C_p^\infty/C_v^\infty)(C_v^0/C_p^0) = 25/21 = 1.19.$$

This means a decrease of 20 percent in the viscous absorption term if Saxton's result is correct.

We start by considering a gas with the internal states 1, 2, 3, $\cdots$. Let n_i be the number of molecules per unit volume in state i and $n=\sum_i n_i$, where n is the total number per unit volume. The n of this section is not related to the n of Sec. 7. The equations of continuity and of motion can be found in Jeans (J1; Eqs. (286), (290), and (298)). For the one-dimensional case, where second-order terms have been neglected, these are:

equation of continuity:

$$\partial n/\partial t = -\partial nv/\partial x; \qquad (8.2)$$

equation of motion:

$$mn\partial v/\partial t = -\partial mn\langle u^2\rangle/\partial x = -(2/3)\partial(nE_k)/\partial x; \qquad (8.3)$$

equation of the conservation of kinetic energy:

$$\partial(nE_k)/\partial t = -(5/3)\partial nE_k v/\partial x + \Delta E_k, \qquad (8.4)^{15}$$

where $v=$ the average particle velocity along the x axis, i.e., the drift velocity (particle velocity); $m=$ the mass of the molecule; $E_k=$ the kinetic energy of translation $=3/2k'T$; $\langle u^2\rangle=$ the square of the velocities along the x axis; and $k'=$ Boltzmann's constant. Kinetic theory shows that $\langle u^2\rangle$ equals $k'T/m$.

We require a fourth equation; this will describe the transition of the molecules from one state to another. By state we mean an energy grouping within which there is some kind of an equilibrium, i.e., the time to establish equilibrium within the state is short relative to the time to establish equilibrium between the states. We select an arbitrary particle in state i. The probability that this particle is transformed in unit time to state j due to a collision with another particle also in state i, is

$$\bar{f}_{ij}(n_i-1) \approx \bar{f}_{ij}n_i.$$

The above expression defines $\bar{f}_{ij}$. The net number of transitions from state i to j due to collisions with molecules in state i is

$$n_i^2 \bar{f}_{ij}.$$

Similarly we have to consider collisions between molecules in state i and state j. The corresponding expression is

$$n_i(n-n_i)f_{ij}.$$

The net change of the number in the ith state, due to collision, is

$$\Delta n_i = -\sum_j\{[n_i^2(\bar{f}_{ij}-f_{ij})+nn_if_{ij}] \\ -[n_j^2(\bar{f}_{ji}-f_{ji})+nn_jf_{ji}]\}. \quad (8.5)$$

The first terms in the square bracket give the number of particles that leave the ith state and the second bracket is for the number of particles that enter the ith state.

One now assumes that n_i can be calculated from an expansion about the equilibrium value. This results in

$$\Delta n_i = -\sum_j\{\delta n(n_if_{ij}-n_jf_{ji}) \\ + \delta n_i[2n_i(\bar{f}_{ij}-f_{ij})+nf_{ij}] \\ - \delta n_j[2n_j(\bar{f}_{ji}-f_{ji})+nf_{ji}] \\ + [n_i^2\delta(\bar{f}_{ij}-f_{ij})+nn_i\delta f_{ij} \\ - n_j^2\delta(\bar{f}_{ji}-f_{ji})-nn_j\delta f_{ji}]\}. \quad (8.6)$$

Bourgin, borrowing a concept from the thermodynamical development, here assumes that a change of the population in the ith state is given by

$$\delta n_i = \frac{n_i}{n}\delta n + \beta_i n_i'\delta E_k \qquad (8.7)$$

where $n_i' = (dn_i/dE_k)$.[16] Implicitly this equation assumes that the disturbance which makes $\Delta n_i \neq 0$ is harmonic. The first term of (8.7) arises because of a change in density. The second is caused by a change of the populations in each of the levels due to changes in temperature (since $E_k=3k'T/2$). β_i is a complex function of the frequency implicit in (8.7) and describes the phase lag between the translational energy and the energy in the ith state, as well as the fraction of the translational energy δE_k which affects this state.

For a slow process, where equilibrium is approached, $\beta_i \to 1$ and $\Delta n_i \to 0$. Substituting (8.7) into (8.6) and recalling (8.5), we get for a slow process,

$$\delta E_k n \sum_j(n_i'R_{ij}-n_j'R_{ji}) \\ = -\sum_j[n_i^2\delta(\bar{f}_{ij}-f_{ij})+nn_i\delta f_{ij} \\ - n_j^2\delta(\bar{f}_{ji}-f_{ji})-nn_j\delta f_{ji}], \quad (8.8)$$

where

$$R_{ij} = (1/n)[2n_i(\bar{f}_{ij}-f_{ij})+nf_{ij}]. \qquad (8.9)$$

Δn_i can be expressed in terms of δn, δE_k, and δf by using (8.6). Further, the terms multiplying δn equal zero because of (8.5). Eliminating the δf terms with the help of (8.8) we finally obtain for $\beta_i \neq 1$,

$$\Delta n_i = -n\delta E_k \sum_j\{n_i'R_{ij}(\beta_i-1)-n_j'R_{ji}(\beta_j-1)\}. \quad (8.10)$$

Equations (8.2), (8.3), (8.4), and (8.10) are the basic equations of Bourgin's treatment. We consider the simplest case where all the β_i's are equal. That is, all the internal modes are in phase with each other. Our next problem is to eliminate ΔE_k from (8.4); for this purpose the equation of continuity for the n_ith state,

$$\partial n_i/\partial t = -(\partial/\partial x)(n_iv) + \Delta n_i, \qquad (8.11)$$

is used. As previously, we assume that our solutions are of the form $\exp\{i(\omega t-kx)\}$, where k is complex. Specifically, for n_i we assume the solution

$$\delta n_i^0 \exp\{i(\omega t-kx)\}.$$

δn_i^0, δv^0 etc. are the amplitudes. Substituting this type

---

[15] Equation (8.4) follows from Jeans, Eq. (298), if we assume that the velocity of a molecule in the x direction is $[u_x+v(x)]$, while the other components are u_y and u_z.

[16] In Bourgin's 1929 paper, β_i is introduced only by the word "evidently," and we have been unable to find out Bourgin's exact interpretation of this step but believe that it can be justified as is done above. More details are given in his later papers.

of solution into (8.2) and (8.11) we express Δn_i in terms of δn^0 and $\delta n_i{}^0$. Now using (8.7), (8.10), and the assumption that the β_i's are equal, we obtain

$$\beta_i = nA_i/(A_i n + i\omega), \quad (8.12)$$

where

$$A_i = 1/n_i' \sum_j (n_i' R_{ij} - n_j' R_{ji}). \quad (8.12a)$$

The last term of (8.4) arises because of a redistribution of the levels which increases or decreases the internal energy at the expense of the translational energy. We may evaluate it by first writing

$$\Delta E_k = -\sum_i \epsilon_i \Delta n_i, \quad (8.13)$$

where ϵ_i is the energy of the ith state. It follows by means of (8.10) that

$$\Delta E_k = \{\sum_i \epsilon_i n(\beta_i - 1) \sum_j (n_i' R_{ij} - n_j' R_{ji})\} \delta E_k$$
$$= \{(\beta_i - 1) nA_i \sum_i \epsilon_i n_i'\} \delta E_k.$$

Since β_i is independent of the state, A_i must be also, because of (8.12). Further,

$$\Delta E_k = \left\{ -\frac{i\omega A_i n}{nA_i + i\omega} \sum_i \epsilon_i \frac{\partial n_i}{\partial E_k} \right\} \delta E_k$$

$$= -\frac{2}{3k'} \left\{ \frac{i\omega A_i n}{nA_i + i\omega} \sum_i \epsilon_i \frac{\partial n_i}{\partial T} \right\} \delta E_k,$$

$$= -\frac{2}{3k'} \left\{ \frac{i\omega}{(nA_i + i\omega)} A_i n^2 c^i \right\} \delta E_k, \quad (8.14)$$

where

$$c^i = \frac{1}{n} \sum_j \epsilon_j \frac{dn_j}{dT}$$

$$= \frac{1}{n} \left(\frac{3}{2} k' \right) \sum_j \epsilon_j n_j'. \quad (8.15)$$

If we insert (8.14) into (8.4) and recall the type of solutions in which we are interested, we have

$$i\omega[n \delta E_k{}^0 + E_k \delta n^0] = (5/3) ikn E_k \delta v^0$$
$$- \frac{2}{3k'} \left(\frac{i\omega}{nA_i + i\omega} \right) A_i n^2 c^i \delta E_k{}^0. \quad (8.16)$$

The higher order terms have been omitted. Using (8.2), (8.3), and (8.16) the desired relation between k^2 and ω^2 is finally obtained:

$$\frac{\omega^2}{k^2} = \frac{2}{3} \frac{E_k}{m} \left[\frac{5k'/2 + c^i + i5k'\omega\tau/2}{3k'/2 + c^i + i3k'\omega\tau/2} \right], \quad (8.17)$$

where the relaxation time τ is now given by

$$1/\tau = A_i N. \quad (8.18)$$

If we make use of the fact that in kinetic theory the pressure equals $1/3 mn\langle u^2 \rangle = 2nE_k/3$ and employ the definitions

$$c_p{}^e = 5k'/2$$
$$c_v{}^e = 3k'/2$$

(8.17) can be transformed into

$$\frac{\omega^2}{k^2} = \frac{p}{\rho_0} \frac{(1/\tau) C_p{}^t + i C_p{}^e \omega}{(1/\tau) C_v{}^t + i C_v{}^e \omega}, \quad (8.19)$$

where the C's are the molar heats and $C_v{}^t = C_v{}^e + C^i$. Since $K = p$ for an ideal gas, (8.19) is equivalent to

$$\omega^2 \rho_0 \left(\frac{\omega}{K_s{}^\infty} - \frac{i\omega_0}{K_s{}^0} \right) = k^2(\omega - i\omega_0) \quad (8.20)$$

where

$$K_s{}^0 = K C_p{}^t / C_v{}^t = p C_p{}^t / C_v{}^t, \quad (8.21)$$

$$K_s{}^\infty = K C_p{}^e / C_v{}^e = p C_p{}^e / C_v{}^e, \quad (8.22)$$

and

$$\omega_0 = C_p{}^t / \tau C_p{}^e. \quad (8.23)$$

Equation (8.20) is equivalent to Eq. (5.7) and this shows the equivalence of the kinetic method to the other approaches to the subject.

(b) The Method of Irreversible Thermodynamics

The usual methods of thermodynamics consider only equilibrium processes. A treatment of absorption based on irreversible thermodynamics has been given by De-Groot (D2a; p. 51). Although the second law usually is stated as an inequality, use is rarely made of the fact that it is indeed an inequality. This occurs because we most often discuss only reversible processes. It is necessary now to go beyond this limitation and include processes which are irreversible. If we neglect heat conduction and heat radiation relaxation, then we are dealing with an adiabatic but a non-isentropic process. This means that a certain extension of conventional thermodynamics must be made. For an isolated system, a reversible process can be characterized by the statement $\Delta S = 0$, where S is the entropy. For an irreversible process S increases with time, and we require a statement of its rate of increase. In general, however, this equation need not contain the entropy explicitly. We shall call this relation, giving the rate of entropy increase, "the equation of irreversibility."

As we have already seen the method of Herzfeld and Rice states this equation in terms of two temperatures, the internal T^i and the external T^e—i.e.,

$$\dot{T}^i = (T^e - T^i)/\tau. \quad (8.24)$$

In the literature the equation of irreversibility takes many forms; only in recent years has it been related to more basic thermodynamic concepts. Some of the forms of the equation of irreversibility will be reviewed.

Several years after the paper of Herzfeld and Rice, Henry (H6) suggested a variation of (8.24). After making the following definitions: E_T = Total equilibrium energy, E_0 = Total energy without vibration (or rotation), and E_x = Actual energy of system at time t, he states the equation as

$$d(E_x - E_0)/dt = (E_T - E_x)/\tau. \quad (8.25)$$

From (8.25) one may obtain the same results as from (8.24).

Eucken and Becker (E6) use the concept of the internal specific heat in setting up the equation of irreversibility. If one defines C_t^i = the internal specific heat (vibrational) at instant t, and C^i = the internal specific heat for a static process, then

$$\dot{C}_t^i = (C^i - C^t)/\tau. \quad (8.26)$$

In liquids, the process is more complicated than in gases and the above concept must be generalized. One of the most direct means is that recently presented by Hoff Lu (H13) in which he used the volume directly.

By letting s_0 = the static value of $-\Delta V/V$, s_∞ = the value of $-\Delta V/V$ at very high frequencies, and s = the value of $-\Delta V/V$ at instant t, Hoff Lu's equation of irreversibility becomes:

$$d(s - s_\infty)/dt = (s_0 - s)/\tau. \quad (8.27)$$

Using (8.27) one can obtain the acoustical equation of state for a single relaxation process. The assumption is made that the equation holds for liquids and gases. In general, however, (8.27) depends on the thermodynamic path selected.

The approach of Mandelstam and Leontovich (M1) shows an advance in that they relate the problem more closely to thermodynamics.[17] Thermodynamics enters in Eqs. (8.24) to (8.27) only explicitly and these equations do not seem to give a deep insight into the problem. Mandelstam and Leontovich, on the other hand, consider a system with three (in the simplest case considered) independent variables. They select T, ρ, and ξ. ξ is not defined precisely. It may be related to the population of the states or to an internal temperature. Since T and ρ are being used, the thermodynamic function chosen is the Helmholtz free energy F.[18] F is a function of T, ρ, and ξ—i.e., $F(p, \rho, \xi)$. Conventional thermodynamics applies when

$$\partial F/\partial \xi = 0. \quad (8.28)$$

Since (8.28) does not always hold, our problem is to find a relation between F and $\dot{\xi}$. The one chosen by Mandelstam and Leontovich, is

$$\dot{\xi} = K' \partial F/\partial \xi, \quad (8.29)$$

where K' is a constant. This leads to the usual expression for absorption and velocity. K' is related to τ of Herzfeld and Rice as follows:

$$1/\tau = K' \partial^2 F/\partial \xi^2. \quad (8.30)$$

---

[17] J. Meixner recently presented a similar theory at the 1951 Ultrasonics Conference in Brussels.

[18] The use of the letter F to denote the Helmholtz free energy is not universal since students of G. N. Lewis' school denote the Gibbs free energy by F. Since the reaction rate theory has been developed to a large extent by American chemists, the use of F for the Gibbs function appears in a few papers referred to in this review. At times, the distinction between the Gibbs and the Helmholtz function is not made in these papers. We use F for the Helmholtz free energy, the notation followed by Guggenheim. (G8).

The difference between this method and the ones previously mentioned is that the terms here can be defined with greater care, and, in principle, the development is not confined too closely to a particular problem or a specific experiment.

One would like to connect the equation of irreversibility with the recent developments of irreversible thermodynamics. At least three such attempts have been made, namely: Damköhler's (D1), Meixner's (M6), and Eckart's (E3a). We shall discuss only the paper of Eckart, which considers a simple case. In setting up the equation of irreversibility, he uses the concept of chemical potential (G8; p. 17) or (Z2; p. 322) (also known as partial potential).

Consider a reaction which can be characterized by the relation $Y_1 \rightleftarrows Y_2$. Eckart's equation of irreversibility is

$$\dot{N}_2 = -g(\mu_2 - \mu_1), \quad (8.31)$$

where g is a constant which can be related to the law of mass action. The quantity μ_1 is the chemical potential due to molecules in state 1, i.e., $(\partial F/\partial N_1)_{T,V,N_2}$, and μ_2 is the chemical potential corresponding to state 2. In this theory,

$$1/\tau = (g \partial/\partial N_2)(\mu_2 - \mu_1). \quad (8.32)$$

The importance of Eckart's paper is that Eq. (8.31) has not been set up *ad hoc* to explain the absorption of sound but is part of a more general theory of irreversible processes (E1). The problem of sound absorption in Eckart's formulation becomes just one phase of a more basic field—i.e., "thermodynamics" as contrasted with "thermostatics."[19]

We have listed some of the equations of irreversibility that have been used in the past. We may deduce (8.25) from (8.24) by assuming that E_0 is in equilibrium with the macroscopic variables; for a gas this is characterized by T^e. The nonvibrational part of E_x is E_0. The vibrational part of E_x will be characterized by an internal temperature T^i. It seems logical to expect the following relations to hold:

$$E_x - E_0 = a(T^i - \bar{T}) + b, \quad (8.33)$$

$$E_T - T_0 = a(T^e - \bar{T}) + b. \quad (8.34)$$

The parameters a, b, and $\bar{T}$ have the following meaning: $\bar{T}$ corresponds to some average temperature; b is the internal (vibrational) energy at $\bar{T}$; and a is the rate of change of the internal energy with temperature. By substituting into (8.25) we obtain

$$A\dot{T}^i = a(T^e - T^i)/\tau, \quad (8.35)$$

which is the same as (8.24). An argument such as this

---

[19] There is a slight difference between Eckart's development as contrasted with the others. His theory defines $\omega_0 = K_s^0/\tau K_s^\infty$ instead of $C_p^e/\tau C_p$. This must be the result of the method used in proceeding from the equation of irreversibility. The definition of ω_0 in the paper of Mandelstam and Leontovich is the same as that of Herzfeld and Rice in the case of thermal relaxation. The authors would like to thank Dr. Eckart for the use of his notes.

can also be used to show that (8.26) is equivalent to (8.24).

By equating the ξ of Mandelstam and Leontovich to the N_2 of Eckert's development, one may show that they are equivalent. On the other hand, Mandelstam and Leontovich have defined an internal temperature by assuming that a given distribution of level population can be characterized by a given value of ξ. For this ξ and macroscopic temperature T^e,

$$\partial F\{T^e, \rho, \xi\}/\partial \xi \neq 0. \qquad (8.36)$$

But an internal temperature T^i can be defined such that

$$\partial F\{T^i, \rho, \xi\}/\partial \xi = 0. \qquad (8.37)$$

Using these equations, Mandelstamm and Leontovich have shown that

$$\dot{T}^i - A\dot{\rho} = (1/\tau)(T^e - T^i), \qquad (8.38)$$

where A is a complicated thermodynamic function. For a gas, the pressure is independent of the level population and $A=0$; hence, (8.38) is the equation of Herzfeld and Rice, with a more precise meaning of T^i.

(c) The Method of Statistical Thermodynamics

This method has already been given essentially in Sec. 7, and we shall not elaborate on it here. Two interesting modifications of the theory will be given here— the development of Landau and Teller (L3) for a many-state vibrational gas and a means of calculating collision efficiencies for gases.

Landau and Teller consider a gas with a single nondegenerate vibrational mode of frequency ν_M. The states of the gas can be listed as $l=0, 1, 2 \cdots$. It is assumed that transitions occur because of binary collisions and that A_{ij} is the transition probability from the ith to the jth state. One assumes further that the effectiveness of a collision depends on the translational energy and not on the vibrational state of the hitting molecule. The following relations between the A's are assumed.

$$A_{01}: A_{12}: A_{23} \cdots = A_{10}: A_{21}: A_{32} \cdots = 1:2:3 \cdots$$
$$A_{ij} = 0 \quad \text{if} \quad i-j \neq \pm 1. \qquad (8.39)$$

These assumptions may be justified by considering quantum-mechanical definitions of the transition probability. The A's should depend on the square of the matrix element

$$\int \psi_i V_p \psi_j d\tau, \qquad (8.40)$$

where the ψ's are the wave functions of the states, and V_p is the perturbed part of the Hamiltonian. If one assumes that V_p is directly proportional to the normal vibrational coordinate of the molecule (D3 and R10; p. 343) then (8.40) results. From (8.40) we see that the temperature affects A_{01} in exactly the same way as it does A_{12}, etc.

Consider $\dot{N}_0$, which is given by

$$\dot{N}_0 = A_{10}N_1 - A_{01}N_0 \qquad (8.41)$$

or, at equilibrium, as before,

$$\frac{A_{01}}{A_{10}} = \frac{N_1^0}{N_0^0} = \exp(-h\nu_M/k'T) = e^{-x}, \qquad (8.42)$$

where h is Planck's constant and x is defined by (8.42). We now define

$$Z = \sum_l N_l l, \quad l = 0, 1, 2, \cdots, \qquad (8.43)$$

and it follows that

$$\dot{Z} = \sum_l [(l+1)N_l A_{l,l+1} - lN_l A_{l,l+1} - lN_l A_{l,l-1} + (l-1)N_l A_{l,l-1}] = \sum_l [N_l A_{l,l+1} - N_l A_{l,l-1}].[20] \qquad (8.44)$$

Using (8.39) we obtain

$$\dot{Z} = A_{01}\sum_l (l+1)N_l - A_{10}\sum_l l N_l$$
$$= A_{01}N - (A_{10} - A_{01})Z = A_{01}\{N - Z(e^x - 1)\}. \qquad (8.45)$$

If we expand (8.45) about the equilibrium value and neglect higher order terms, we get

$$\dot{Z} = [A_{01} + \Delta A_{01}]\left[N - (Z^0 + \Delta Z)\left(e^x - 1 - \frac{x}{T}e^x \Delta T\right)\right].$$

Since at equilibrium $\dot{Z} = 0$, we know that

$$N - Z^0(e^x - 1) = 0,$$

or

$$\dot{Z} = (A_{01} - A_{10})\Delta Z + A_{01}Z^0(x/T)e^x \Delta T. \qquad (8.46)$$

Here A_{01}, A_{10}, Z, and T are evaluated at the equilibrium value. If we assume that

$$\Delta Z = \Delta Z^0 e^{i\omega t},$$
$$\Delta T = \Delta T^0 e^{i\omega t},$$

then (8.46) gives

$$\Delta Z^0 [i\omega + (A_{10} - A_{01})]$$
$$= A_{01}(x/T)Z^0 e^x \Delta T^0 = B\Delta T^0. \qquad (8.47)$$

B is defined by (8.47). Returning to (8.43) we see that $h\nu_M \Delta Z$ is just the change of the oscillator's energy due to the shift in population caused by the temperature change ΔT. Following Landau and Teller, we define

$$C_t{}^i = C^i/(i\omega\tau + 1), \qquad (8.48)$$

where

$$C^i = h\nu_M B/(A_{10} - A_{01}), \qquad (8.49a)$$
$$\tau = 1/(A_{10} - A_{01}). \qquad (8.49b)$$

It should be noted that for a harmonic disturbance, the τ of (8.49b) is the same as the τ of (8.26). Equation (8.49b) shows that the relaxation frequency for a vibrating molecule is not proportional to $A_{10} + A_{01}$, as derived in Sec. 7, but to $A_{10} - A_{01}$. The result obtained in Sec. 7 holds only if $e^{-x} \ll 1$ and, under these conditions, the difference between the two results is very small.

---
[20] For convenience $A_{0(-1)}$ is set equal to zero.

Equation (8.49), of course, does not apply to rotational modes or to liquids.

Before leaving this subject we would like to show the relation between the relaxation time τ and the ratio of collisions which cause transitions to the total number of collisions. Let us assume that for a gas the A's of Sec. 7 are due to binary collisions; then we may write

$$A_{10} = (1/V)k_{10}(N_0+N_1) \approx k_{10}N_0/V, \quad (8.50a)$$

$$A_{01} = (1/V)k_{01}(N_0+N_1) \approx k_{01}N_0/V, \quad (8.50b)$$

if one assumes that $N_0 \gg N_1$. Equations (8.50a) and (8.50b) define the k's. The volume V appears because N_i is the total number of molecules in state i. The actual number of collisions between N_1 and N_0 can be calculated from statistical mechanics (F1; p. 491, Eq. (1201, 18)) and is

$$Z_{ij} = \frac{2}{V} \frac{N_i N_j}{\sigma_{ij}} D_{ij}^2 \left(2\pi \frac{k'T}{m} \right)^{\frac{1}{2}}. \quad (8.51)$$

Here D_{ij} = the average diameter of the jth and ith molecule (distance of nearest approach) and m = the reduced mass $m_i m_j/(m_i+m_j)$.

$$\sigma \begin{cases} =2 \text{ if } i \text{ and } j \text{ are the same molecule.} \\ =1 \text{ if they are not the same molecule.} \end{cases}$$

Since $A_{10}N_1$ is the number of N_0 molecules that are produced by collisions per unit time, the fractional number of collisions which are effective is

$$P_{10} = \frac{A_{10}N_1}{Z_{10}} = \frac{k_{10}}{2D_{10}^2 [4\pi(RT/M)]^{\frac{1}{2}}}, \quad (8.52)$$

where M is the molar mass. Because the masses in states 0 and 1 are equal, $m = M/2N$. Since $A_{01} \ll A_{10}$, for $N_1 \ll N_0$ the experimental value of ω_0 effectively gives A_{10}, and one may calculate P_{10}. Actually, a slight correction [known as the Sutherland correction (J1; p. 176)] should be applied to Eqs. (8.51) and (8.52).

Further modifications of the statistical model are needed for more complex gases such as CO_2 or COS. CO_2 is a linear molecule and has four internal degrees of vibration, one of which is doubly degenerate. The question now arises as to whether one can treat the molecules as ones having a single internal temperature corresponding to a single ω_0, or several, corresponding to the different degrees of freedom. Results seem to show, at least to first approximation, that one needs only a single internal temperature. The static value of the specific heat is due mainly to the three lower vibrational levels. Since C_p/C_v enters into the velocity, one may obtain C_v by acoustical means. Measurements of Eucken (E7) and his co-workers indicate that the various vibrations relax essentially at a single frequency [for CO_2 and COS]. Schafer (S2) has indicated how the data on COS can be interpreted in terms of a double relaxation process, but the evidence does not seem decisive.

More complicated vapors, however, may have more than one distinct relaxation process. One such example is given in the work of Alexander and Lambert (A1) on acetaldehyde (CH_3CHO). They have demonstrated the existence of three distinct relaxation times.

(d) The Phenomenological Approach

Each of the preceding approaches is based on some kind of molecular model. This is not necessary, as Stokes's treatment shows. Considering the macroscopic features, one may write down relations between the density and the pressure and then explore the acoustical consequences. An example of this type of approach is found in a paper of Frenkel and Obraztsov (F4). They examine two possible acoustical equations of state, viz.,

$$a p_e + b \dot{p}_e + d \rho_e + e \dot{\rho}_e = 0, \quad (8.53)$$

and

$$a p_e + b \frac{d p_e}{dt} + c \frac{d^2 p_e}{dt^2} + e \frac{d \rho_e}{dt} + f \frac{d^2 \rho_e}{dt^2} = 0. \quad (8.54)$$

Their theory is broad enough to describe a medium which can support a shearing stress. The great advantage of this method is that it is possible to explore many equations of state without constructing complicated molecular models. This has great usefulness, as for example, in treating glass-like substances. The main disadvantage is that the final equations are usually very complicated. Frenkel and Obraztsov unfortunately are able to give only a limited amount of experimental evidence to support their generalized theory.

9. SUMMARY OF RESULTS

Absorption occurs when the density gets out of phase with the pressure. This can be caused by two mechanisms. One is a frictional lag which we call viscosity (Sec. 3). The other is caused by a change in the bulk modulus which has one value for a slow process and another for a fast process. The second mechanism we call a relaxation process (L14). Reserving the term relaxation to the second process is not universal, for some authors speak of a viscous relaxational process. The change of the bulk modulus with frequency occurs when one considers heat radiation (Sec. 5), thermal relaxation (Secs. 6 and 7) structural relaxation (Sec. 7), a combination of thermal and structural relaxation (Sec. 7), and heat conduction (Sec. 4). In general, the dispersion and the absorption are different, depending on the mechanism. In Fig. I-4 we have plotted the sound velocity and the absorption per wavelength as a function of angular frequency for the two types of effects. At high frequency, where there is a lot of to and fro motion per unit length one would expect higher linear absorption simply because of the additional motion; for this reason we take the absorption per wavelength ($\alpha\lambda$).

If it is possible to obtain values of the absorption or velocity over a large range, then one may easily interpret the results and decide which mechanism is the underlying cause. If we have data only over the range where $\omega \ll \omega_0$, then it may be impossible to interpret the results fully. Nonacoustical information regarding the medium can, at times, help our interpretation.

The results of Secs. 3 to 8 may be summarized as follows:

(a) *Viscosity Mechanism.* (*Stokes 1845*)

Phase velocity

$$c^2 = \frac{2\zeta}{\omega_v \rho_0} \frac{\omega_v^2 + \omega^2}{\{1 + [1 + (\omega^2/\omega_v^2)]^{\frac{1}{2}}\}}, \quad (9.1)$$

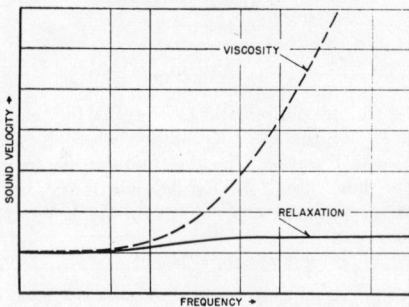

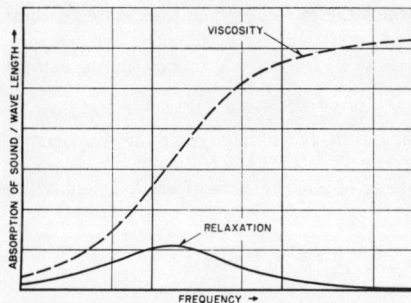

FIG. I-4 (a). Velocity vs frequency for a viscous mechanism and a relaxational mechanism (b). Amplitude absorption per wavelength vs frequency for a viscous mechanism and a relaxational mechanism.

where

$$\omega_v = K_s/\zeta.$$

For $\omega \ll \omega_v$,

$$c^2 \approx \frac{K_s}{\rho_0}\left(1 + \frac{3}{4}\frac{\omega^2}{\omega_v^2}\right); \quad (9.2)$$

for $\omega \gg \omega_v$,

$$c^2 \approx 2\zeta\omega/\rho_0 \text{---viscous wave.} \quad (9.3)$$

Absorption coefficient:

$$\alpha = \left(\frac{\rho_0 \omega_v}{2\zeta}\right)^{\frac{1}{2}} \left\{\frac{\omega^2}{\omega^2 + \omega_v^2}\left[\left(1 + \frac{\omega^2}{\omega_v^2}\right)^{\frac{1}{2}} - 1\right]\right\}^{\frac{1}{2}}. \quad (9.4)$$

For $\omega \ll \omega_v$,

$$\alpha = \frac{1}{2}\frac{1}{c_0^3}\frac{\zeta}{\rho_0}\omega^2\left(1 - \frac{5}{8}\frac{\omega^2}{\omega_v^2}\right) \approx \frac{1}{2}\frac{1}{c_0^3}\frac{\zeta}{\rho_0}\omega^2; \quad (9.5)$$

for $\omega \gg \omega_v$,

$$\alpha \approx (\rho_0 \omega/2\zeta)^{\frac{1}{2}}\text{---viscous wave.} \quad (9.6)$$

Absorption per wavelength:

$$\alpha\lambda = 2\pi\alpha c/\omega. \quad (9.7)$$

For $\omega \ll \omega_v$,

$$\alpha\lambda = \pi \frac{\zeta}{\rho}\frac{\omega}{c_0^2}\left(1 - \frac{1}{4}\frac{\omega^2}{\omega_v^2}\right); \quad (9.8)$$

for $\omega \gg \omega_v$,

$$\alpha\lambda = 2\pi. \quad (9.9)$$

Setting $\zeta = 4\eta/3$ we have,

for air $\quad \omega_v = 6000$ mc at normal pressure
for water $\quad \omega_v = 2 \times 10^6$ mc.

(b) *Relaxation Mechanism*

Phase velocity

$$c^2 = \frac{1}{\rho_0}\frac{\omega^2 + \omega_0^2}{[(\omega^2/K_s^\infty) + (\omega_0^2/K_s^0)]}. \quad (9.10)$$

Absorption coefficient:

$$\alpha = \frac{c\omega_0}{2c_0^2}\frac{\Delta K_s}{K_s^\infty}\frac{\omega^2}{\omega_0^2 + \omega^2}, \quad (9.11)$$

where

$$\Delta K_s = K_s^\infty - K_s^0.$$

Absorption per wavelength,

$$\alpha\lambda = \pi\frac{\Delta K_s}{(K_s^\infty K_s^0)^{\frac{1}{2}}}\omega_m\frac{\omega}{\omega^2 + \omega_m^2}. \quad (9.12)$$

Here

$$\omega_m^2 = \omega_0^2/K_s^\infty/K_s^0, \quad (9.13)$$

for which $\alpha\lambda$ has its maximum value. At $\omega = \omega_m$ we obtain

$$\alpha\lambda = \frac{\pi}{2}\frac{\Delta K_s}{(K_s^\infty K_s^0)^{\frac{1}{2}}}. \quad (9.14)$$

Types of relaxation
(i) Heat radiation[21] (Stokes 1851)

$$\omega_0 = \omega_r = q/\gamma, \quad (9.15)$$

$$\Delta K_s = (\gamma - 1)K. \quad (9.16)$$

For air $\omega_r = 0.002$ cycle, based on a value of q estimated

---

[21] We have tried to list the names of the authors who developed the theory. At places several authors are listed when each developed his own point of view. Several attempts to develop a generalization of (ii) and (iii) which are in process of being published are not listed.

roughly by Rocard (R9).[22] He assumes it is independent of frequency.

(ii) Thermal relaxation (Jeans 1904, Bourgin 1928, Herzfeld and Rice 1928, Kneser 1931).

$$\omega_0 = C_p{}^t/C_p{}^e \tau \qquad (9.17)$$

$$\Delta K_s = TV\beta^2 K^2(C^i/C_v{}^i C_v{}^e). \qquad (9.18)$$

(iii) Structural relaxation (Hall 1948, B. B. Ghosh 1950).

$$\omega_0 = 1/\tau. \qquad (9.19)$$

(iv) General case (Mandelstam and Leontovich 1937, Liebermann 1949, Markham 1950).

$$\omega_0 = C_p{}^t/\tau C_p{}^e. \qquad (9.20)$$

The actual expression of ΔK_s in the text for (iii) and (iv) are not reproduced here.

(v) Thermal conduction. (Kirchhoff 1868).

$$\omega_c = C_p \rho_0 c^2 / M\kappa. \qquad (9.21)$$

In this case,

$$\Delta K_s = K_s{}^0 - K_s{}^\infty = (\gamma - 1)K \qquad (9.22)$$

for air $\omega_c = 6 \times 10^9$ cycles, while for water $\omega_c = 10^{13}$ cycles. One might classify thermal conduction as a third mechanism because it is based on a different equation of state: ΔK_s is $K_s{}^0 - K_s{}^\infty$ instead of $K_s{}^\infty - K_s{}^0$. Here $K_s{}^0 > K_s{}^\infty$, and ω_c is a function of the frequency through c.

(c) *Temperature Dependence*

(i) Viscosity

For a gas, the viscosity coefficient depends on the mean velocity; hence, one would expect the coefficient to be a function of the square root of the absolute temperature (J1; p. 170)[23] ρ_0 at constant pressure depends inversely on the temperature, while K is independent of the temperature. Therefore, at low frequencies, the velocity is given by

$$c = \text{const } T^{\frac{1}{2}},$$

while at high frequencies

$$c = \text{const } T^{\frac{1}{4}}.$$

On the other hand at low frequencies the absorption is given by

$$\alpha = \text{constant, independent of temperature,}$$

while at high frequencies,

$$\alpha = \text{const } T^{-\frac{1}{2}}.$$

This means that in gases the absorption and velocity

[22] Rocard (R9) has calculated α for air. He gets for the radiation part of α at 6 kc, 1.5×10^{-8}. Assuming that $\omega_r \ll \omega$ at this frequency one obtains $\alpha = c_\infty \Delta K \omega_r / 2 c_0^2 K_s{}^\infty$, where c_∞ is the high frequency sound velocity i.e., the adiabatic value. The above equation gives $\omega_r = 2\alpha c_\infty/(\gamma - 1)$.
[23] Actually the temperature exponent is not exactly $\frac{1}{2}$. For He the viscosity varies as $T^{0.65}$.

have only a small dependence on the temperature for the viscous case.

For many liquids, the viscosity coefficient depends exponentially on the reciprocal of the temperature—i.e., $\zeta = Ae^{B/T}$. The factor A may have a slight temperature dependence, which we shall ignore (G2a; p. 477). We shall ignore the temperature dependence of K and ρ since ζ depends so critically on T. Hence, for low frequencies one obtains

$$c = \text{const}$$
$$\alpha = \text{const } e^{B/T},$$

and, on the other hand, at high frequencies

$$c = \text{const } e^{\frac{1}{2}B/T}$$
$$\alpha = \text{const } e^{-\frac{1}{2}B/T}.$$

(ii) Relaxation

We shall limit our remarks to cases (ii), (iii), and (iv) of (b), and to the special condition where $N_2{}^0 \ll N_1{}^0$. First, we shall consider the absorption at low frequencies. For these cases $K_s{}^\infty - K_s{}^0$ depends on $N_2{}^0$, which[24] in turn depends on $\exp[-(E_2-E_1)/RT]$. We may, therefore, write

$$K_s{}^\infty - K_s{}^0 = ae^{-b/T}. \qquad (9.23)$$

The form of ω_0 depends on $A_{12}{}^0 + A_{21}{}^0$. Since $N_2{}^0 \ll N_1{}^0$, $A_{12}{}^0 \ll A_{21}{}^0$ and ω_0 are proportional to $A_{21}{}^0$ or the rate with which the molecules drop from state 2 to state 1, $A_{21}{}^0$ and ω_0 should have the same temperature dependence as do the reaction rate constants, namely

$$A_{21}{}^0 \sim \omega_0 \sim a' \exp(-b'/T). \qquad (9.24)$$

For our estimates we shall ignore the temperature dependence of the a's. At low frequencies the temperature dependence of α should arise through ω_0 and ΔK, or

$$\alpha \approx \frac{1}{\omega_0}(K_s{}^\infty - K_s{}^0) \approx \exp[-(b-b')/T]. \qquad (9.25)$$

α may decrease or increase with temperature depending on the values of the b's.

At high frequencies α depends on the temperature through $\Delta K \omega_0$ or

$$\alpha \approx \Delta K \omega_0 \approx \exp[-(b+b')/T]. \qquad (9.26)$$

[24] To justify this statement we return to Eq. (7.10) and show that

$$N\left(2 + \frac{A_{21}{}^0}{A_{12}{}^0} + \frac{A_{12}{}^0}{A_{21}{}^0}\right)^{-1} = \frac{N_1{}^0 N_2{}^0}{N} \approx N_2{}^0.$$

The last step holds if $N_1{}^0 \gg N_2{}^0$. If we use the equilibrium condition,

$$N_1{}^0 A_{12}{}^0 = N_2{}^0 A_{21}{}^0 = (N - N_1{}^0) A_{21}{}^0$$

or

$$\frac{A_{21}{}^0}{A_{21}{}^0 + A_{12}{}^0} = \frac{N_1{}^0}{N}.$$

Using this relation, Eq. (7.12) and the footnote after Eq. (7.9) it follows that

$$\frac{N}{2(1+\cosh W)} = \frac{N_2{}^0 A_{21}{}^0}{A_{12}{}^0 + A_{21}{}^0} = \frac{N_2{}^0 N_1{}^0}{N}.$$

Hence for $N_2{}^0 \ll N_1{}^0$, B_p, and B_T depend on $N_2{}^0$ or e^W.

Hence one would expect α to increase with temperature depending on the sizes of the b's.

The important point is that the temperature dependence does not depend on the type of relaxation, i.e., thermal, structural, or general. This is certainly true if $N_2{}^0 \ll N_1{}^0$. If $N_2{}^0$ is not much smaller than $N_1{}^0$, then the temperature dependence may be more complicated and, perhaps a difference between the various relaxation mechanisms may appear. This argument seems to indicate that one must be very careful in arriving at conclusions from the temperature dependence of absorption.

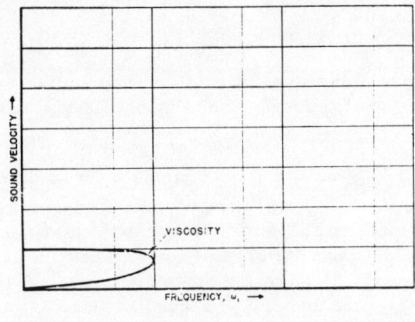

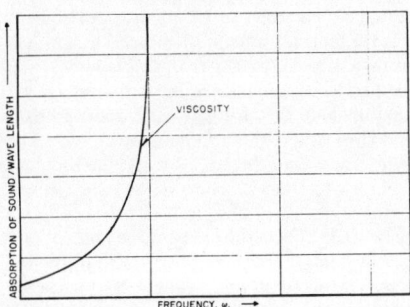

FIG. I-5 (a). Velocity *vs* frequency for a viscosity mechanism in the case of temporal absorption. (b). Amplitude absorption per wavelength *vs* frequency for a viscosity mechanism in the case of temporal absorption.

10. TEMPORAL ABSORPTION *VERSUS* SPATIAL ABSORPTION

We pause here, to introduce a consideration with respect to absorption and dispersion of sound not usually emphasized (L14a). Going back to Eq. (2.13), we note that if we set $k = k_r + ik_i$ we get

$$p_e = P_e \exp(k_i x) \exp[i(\omega t - k_r x)]. \quad (10.1)$$

Here $\alpha = -k_i$ appears as the coefficient measuring attenuation of the wave in space. We may call it the *spatial* absorption coefficient. It is of course the commonly used coefficient in ordinary progressive wave propagation studies. However, it is equally possible to set

$$\omega = \omega_1 + i\omega_2 \quad (10.2)$$

and obtain

$$p_e = P_e \exp(-\omega_2 t) \exp[i(\omega_1 t - kx)]. \quad (10.3)$$

Stokes (S17) in his early study of sound attenuation in viscous media used (10.3). Here ω_2 gives the coefficient which measures attenuation in time. We may call it the *temporal* absorption coefficient. It has been used extensively in studies of the attenuation of sound in finite solid rods. It is clear, that it applies appropriately to the temporal decay of a wave train or standing wave in a medium, whereas α applies more appropriately to a progressive wave due to a constant source. An interesting point is that the dispersion may be different in the two cases and indeed, for the case of viscous absorption the velocity corresponding to (10.1) is always greater than c_0, while that corresponding to (10.3) is always less than c_0.

The temporal absorption due to viscosity alone is interesting, since in the case of liquids, heat conduction plays a much smaller role. The relation between ω and k given by (3.7) yields the results,

$$c = \frac{\omega_1}{k} = c_0 \left(1 - \frac{1}{4} \zeta^2 \frac{k^2}{\rho_0{}^2 c_0{}^2} \right)^{\frac{1}{2}} \quad (10.4)$$

and

$$\omega_2 = \frac{1}{2} \zeta \frac{k^2}{\rho_0}. \quad (10.5)$$

Equation (10.5) gives the temporal absorption coefficient precisely for a fluid in which viscosity provides the only attenuating mechanism. It, of course, agrees in form with Stokes's original calculation, if ζ is properly reinterpreted. The dispersion equation (10.4) is interesting because of its prediction of a frequency cutoff.

We have plotted c against ω_1 in Fig. I-5a. ω_2/c should correspond to α so that in Fig. I-5b, $2\pi\omega_2/kc$ is plotted against ω_1.

To explore temporal absorption in the case of a relaxation mechanism we return to Eq. (5.7) and use assumption (10.2). We write (5.7) in the form

$$\frac{\omega^2}{k^2} = \frac{1}{\rho_0} \frac{1}{[(\omega/K_s{}^\infty)^2 + (\omega_0/K_s{}^0)^2]}$$

$$\times \left\{ \frac{\omega^2}{K_s{}^\infty} + \frac{\omega_0{}^2}{K_s{}^0} + i\omega\omega_0 \left(\frac{1}{K_s{}^0} - \frac{1}{K_s{}^\infty} \right) \right\}, \quad (10.6)$$

where ω_0 has replaced ω_r, since we are interested in the general relaxational case. Assuming that $\omega_1 \gg \omega_2$ we have

$$\frac{\omega_1{}^2}{k^2} = \frac{1}{\rho_0} \frac{(\omega_1{}^2/K_s{}^\infty) + (\omega_0{}^2/K_s{}^0)}{[(\omega_1/K_s{}^\infty)^2 + (\omega_0/K_s{}^0)^2]} \quad (10.7)$$

and

$$2\frac{\omega_1 \omega_2}{k^2} = \frac{1}{\rho_0} \frac{\omega_1 \omega_0}{[(\omega_1/K_s{}^\infty)^2 + (\omega_0/K_s{}^0)^2]} \left(\frac{1}{K_s{}^\infty} - \frac{1}{K_s{}^0} \right). \quad (10.8)$$

Several approximations have been made in arriving at (10.7) and (10.8). They can be justified if $\omega_1 \gg \omega_2$, a condition which always seems to be true. Solving (10.8) for ω_2 we obtain,

$$\omega_2 = \frac{1}{2\rho_0} \frac{\omega_0}{[(\omega_1/K_s^\infty)^2 + (\omega_0/K_s^0)^2]} \frac{\omega_1^2}{c^2}\left(\frac{1}{K_s^\infty} - \frac{1}{K_s^0}\right). \quad (10.9)$$

The velocity given by (10.7) is slightly different from that given by (5.9). However, again at low frequencies,

$$c_0^2 = \frac{K_s^0}{\rho_0}, \quad (10.10)$$

and at high frequencies

$$c_\infty^2 = \frac{K_s^\infty}{\rho_0}. \quad (10.11)$$

One would expect that α might equal ω/c. This actually holds to a good approximation, since we may obtain from (10.9)

$$\alpha \approx \frac{\omega_2}{c} = \frac{1}{2\rho_0} \frac{\omega_t}{\omega^2 + \omega_t^2} \frac{\omega_1^2}{c^3} K_s^0 K_s^\infty \left(\frac{1}{K_s^\infty} - \frac{1}{K_s^0}\right), \quad (10.12)$$

where

$$\omega_t = \omega_0 \frac{K_s^\infty}{K_s^0}. \quad (10.13)$$

If we make a further slight approximation,

$$\alpha \approx \frac{1}{2c} \frac{\omega_t}{\omega_1^2 + \omega_t^2} \omega_1^2 \left(\frac{K_s^\infty - K_s^0}{K_s^\infty}\right). \quad (10.14)$$

This means that there is no essential difference in the expressions for c and α in the case of temporal and spatial absorption. As we have seen this is not the case when viscosity is considered. The reason for this difference seems to be that in the viscous case k_i is not always much smaller than k_r. Indeed at some frequencies they are almost equal. However, in the relaxation case where K_s^∞ is not much larger than K_s^0, k_i is always much smaller than k_r, a fact we have used all along and will now prove. If

$$k_r \gg k_i, \quad (10.15)$$

then

$$1 \gg \alpha\lambda/2\pi, \quad (10.16)$$

From Sec. 9 Eq. (9.14) we know that the maximum value of $\alpha\lambda$ for relaxation is given by

$$\frac{\pi}{2} \frac{\Delta K_s}{(K_s^\infty K_s^0)^{\frac{1}{2}}}.$$

By using (6.12), (6.13), (10.16), and (A-13) of Appendix II we may show that for gases our approximation (10.15) requires

$$1 \gg \frac{1}{4} \frac{C_v{}^i(C_p{}^0 - C_v{}^0)}{[C_p{}^\infty C_p{}^0 C_v{}^\infty C_v{}^0]^{\frac{1}{2}}}. \quad (10.17)$$

For gaseous hydrogen at room temperature where $C_v{}^i = R$ (see Sec. 22) $C_p{}^0 = 7R/2$, $C_p{}^\infty = 5R/2$, $C_v{}^0 = 5R/2$, and $C_v{}^\infty = 3R/2$, the right-hand side of (10.17) gives approximately 0.04. Actually in spatial absorption we assume that $k_r{}^2 \gg k_i{}^2$ which is certainly true for gases since the maximum correction is less than 0.1 percent.

We have to proceed slightly differently in the case of liquids. Let us assume that $\omega_0 \gg \omega$ and write (5.8) in the form

$$\alpha = B\nu^2, \quad (10.18)$$

where now B is an experimental constant and ν is the frequency, i.e., $\omega/2\pi$. Equation (10.16) takes the form

$$1 \gg (1/2\pi)B\nu c. \quad (10.19)$$

The most absorbing liquid known is CS_2 where $c = 1.2 \times 10^5$ cm/sec (B15 (1949 p. 279)) and B ranges from 6×10^{-14} sec^2/cm at 3 mc to 1.4×10^{-14} sec^2/cm at 75 mc (see Sec. 27). Since B and c are known we may calculate the frequency at which the right-hand side of (10.19) equals 0.1. This frequency is about 100 mc. Our treatment should hold at least to 100 mc in an extreme case. CS_2 however, relaxes above 70 mc so that at higher frequencies $\alpha\lambda/2\pi$ is smaller than the value calculated here; further it does not increase with frequency. We may thus conclude that for CS_2 our formulas hold and we should thus expect them to hold for all cases known at present. Since $\alpha \approx \omega_2/c$ and $k_\gamma = \omega_1/c$ the above arguments hold for temporal absorption as well.

For high values of $\alpha\lambda$ the form of the solution, i.e., (10.1) and (10.3) affects the value of the absorption and velocity. Present experimental results seem to fit Eq. (10.1). The possibility is, however, that some future experiment may give results which agree with (10.3). In this review we have found solutions to a set of equations but we have *not* considered initial or boundary conditions. These conditions would dictate the form of the solution and may in some extreme cases influence the definition of the velocity and the absorption. Combinations of (10.1) and (10.3) are quite possible.

11. COMBINED EFFECTS

In any fluid, the problem of sound propagation is far more complicated than the treatment given so far, since one must correct for all the effects acting at the same time. We therefore, desire an equation which includes heat radiation, heat conduction, viscosity, and the various relaxational phenomena. One should even include effects of diffusion of various gases in gas mixtures, such as air.

During the last century, effects of both viscosity and thermal conductivity have been considered by Kirchhoff (K3) and his work is reproduced in Rayleigh

(R2; Vol. II, p. 319) and Lamb (L1; p. 648). He considered the propagation of a plane wave and a wave in a cylindrical tube. For the plane wave at low frequencies ($\omega \ll \omega_v$ or ω_c) the absorption due to heat conduction can simply be added to the absorption due to viscosity.

Herzfeld and Rice have considered the combined effects of heat conduction, viscosity, and thermal relaxation. If $\omega \ll \omega_c$, ω_v, and ω_0 then the net absorption is the sum of the individual effects. As mentioned in Sec. 8, Saxton has considered heat conduction, viscosity, and thermal relaxation. So has Sakadi (S1) in a more recent paper.

We shall now consider the combined effect of: heat conduction and viscosity; viscosity and relaxation; and two relaxation processes.

(a) Heat Conduction and Viscosity

The equation of state for the classical problem of Kirchhoff, combined effect of heat conduction and viscosity, can be obtained most simply by going back to Eq. (4.4) for heat conduction alone,

$$\dot{p}_e - \frac{K_s}{\rho_0}\dot{\rho}_e = \frac{M\kappa}{\rho_0}\left[\frac{1}{C_v}\cdot\frac{\partial^2 p_e}{\partial x^2} - \frac{K}{\rho_0 C_v}\frac{\partial^2 \rho_e}{\partial x^2}\right]. \quad (11.1)$$

Now the equation of state for viscosity alone is Eq. (3.1). Since we shall shortly apply this development to a monatomic gas we set $\zeta = 4\eta/3$. This step will be discussed more fully in Sec. 13. By differentiating the viscosity equation with respect to the time, we get

$$\frac{\partial p_e}{\partial t} = \frac{K_s}{\rho_0}\frac{\partial \rho_e}{\partial t} + \frac{4}{3}\frac{\eta}{\rho_0}\frac{\partial^2 \rho}{\partial t^2}. \quad (11.2)$$

One's intuition suggests, that one may modify (11.1) to include viscosity by adding to $K_s\dot{\rho}_e/\rho_0$ of the left-hand side, the term $(4\eta/3\rho_0)\partial^2\rho_e/\partial t$ and by adding to $K\rho_e/\rho_0$ of the right-hand side, the term $(4\eta/3)\dot{\rho}_e/\rho_0$. This leads to the equation

$$\frac{\partial p_e}{\partial t} - \frac{K_s}{\rho_0}\frac{\partial \rho_e}{\partial t} = \frac{4}{3}\frac{\eta}{\rho_0}\frac{\partial^2 \rho_e}{\partial t^2} + \frac{M\kappa}{\rho_0 C_v}\frac{\partial^2 p_e}{\partial x^2}$$

$$-\frac{M\kappa}{\rho_0 C_v}\frac{\partial^2}{\partial x^2}\left[\frac{K\rho_e}{\rho_0} + \frac{4}{3}\frac{\eta}{\rho_0}\frac{\partial \rho_e}{\partial t}\right]. \quad (11.3)$$

(We use partial time derivatives here for convenience (see Sec. 2).) Equation (11.3) is really only one term of a tensor equation, as is Eq. (3.1), and applies only to a plane wave in this form. p_e is the negative of the stress at right angles to the wave vector.

We introduce the kinematic viscosity, namely,

$$\eta' = \eta/\rho_0, \quad (11.4)$$

and the thermometric conductivity which will be defined as

$$\kappa' = \frac{M}{\rho_0}\frac{\kappa}{C_v}. \quad (11.5)$$

Using solutions (2.12) and (2.14), (11.3) gives

$$P_e(i\omega + k^2\kappa')$$

$$-R_e\left[ic_0^2\omega - \frac{4}{3}\eta'\omega^2 + k^2\kappa'\left(\frac{c_0^2}{\gamma} + \frac{4}{3}i\omega\eta'\right)\right] = 0. \quad (11.6)$$

Proceeding as usual with (2.17) and (2.18) we obtain ultimately,

$$\omega^3 - k^2\omega\left[c_0^2 + i\omega\left(\frac{4}{3}\eta' + \kappa'\right)\right]$$

$$+ik^4\left(\frac{c_0^2}{\gamma} + i\omega\frac{4}{3}\eta'\right) = 0. \quad (11.7)$$

Equation (11.7) was first derived by Kirchhoff (K3) by another method; this justifies, in part, the use of (11.3).

We shall consider spatial absorption only here. For every value of ω there are two independent values of k. At low frequencies one k will give a very high value of α and is important, only near the source of sound. We shall call this solution k' to distinguish it from the usual acoustical solution. As Dr. H. Grad has pointed out, in a private communication, α increases with frequency and α' decreases so that in some region the primed solution may be important.

At present, the most interesting application of this theory is to rarefied monatomic gases. The most detailed experimental work has been done in helium by Greenspan (G3, G4) so that we shall specialize (11.7) to that gas. This reduces the constants in the equation.

The kinetic theory of transport phenomena in gases leads to the general relation

$$\kappa = \epsilon\eta c_v, \quad (11.8)$$

where c_v is the specific heat per unit mass at constant volume, and ϵ is a constant. The evaluation of ϵ, which depends on the assumed intermolecular force, has been summarized by Loeb (L16; p. 240). The most reliable value is that of Chapman and Enskog who used a repulsive force law of the form Gr^{-n} for effectively monatomic molecules, i.e., those in which the energy is translational only. Enskog showed that

$$\epsilon = \frac{5}{2}\cdot\frac{1+(n-5)^2/4(n-1)(11n-13)+\cdots}{1+3(n-5)^2/2(n-1)(101n-113)+\cdots}. \quad (11.9)$$

For $n=5$, $\epsilon=5/2$. This is the celebrated result of Maxwell. Actually ϵ varies little with n. Thus for $n=2$, $\epsilon=2.71$ and for $n\to\infty$, $\epsilon\to 2.52$. We therefore, choose the factor $5/2$ for ϵ. Hence,

$$\kappa' = 5\eta'/2. \quad (11.10)$$

A word or two must be said about the limiting velocity, which depends on the adiabatic bulk modulus K_s. The latter in turn comes from the conventional thermodynamic (as distinct from the acoustical) equation of state, through the isothermal bulk modulus (Eq. (A-2) of Appendix II.) If we use the Holborn and

Otto equation for a real gas (see Zemansky (Z2; p. 94)) we may write

$$pV = A + Bp + Cp^2 + Dp^3 + \cdots, \quad (11.11)$$

where $A, B, C, D, \cdots$ are the virial coefficients, which are functions of the temperature and the mass of gas. Using (11.11) we obtain,

$$K = p[1 + (Bp/A)] \approx p. \quad (11.12)$$

The term Bp/A is a correction term of the order of 0.05 percent under standard conditions for a gas like helium. We shall ignore this correction. We have thus reduced the constants in (11.7).

Instead of changing the frequency the experimenter usually varies the pressure and obtains curves (velocity and absorption), as a function of the pressure. The pressure enters into (11.7) through η' and κ'. Thus the imaginary parts of the coefficients of k^2 and ik^4 are proportional to ω/ρ_0. This suggests plotting the two velocities (ω/k_r and ω/k_r') and the two absorptions (α and α') against $1/p_0$, since $1/p_0$ plays a role similar to ω. This is done in Fig. I-6.[25] The graph figures indicate that even at 0.2 mm of Hg the "prime" wave is attenuated much faster. Some comparison of these curves with experiments will be made in Sec. 21.

General expressions which can be used for various frequencies and pressures have been developed by Greenspan (G4) and by Tsien and Schamberg (T4). For low enough frequencies and high enough pressures Wang Chang (W2) (see also Wang Chang and Uhlenbeck (W1)) has developed the following approximate expression for α

$$\alpha = \frac{7}{10} \frac{\omega}{c_0} \frac{\omega}{p} \eta \left[1 - 1.114 \left(\frac{\omega\eta}{p} \right)^2 \right], \quad (11.13)$$

which holds for monatomic gases since in this development they have assumed (11.8) with $\epsilon = 2.5$ and $\gamma = 5/3$.

(b) Relaxation and Viscosity

We shall now consider the combined effect of viscosity and relaxation. The equation of state for a simple

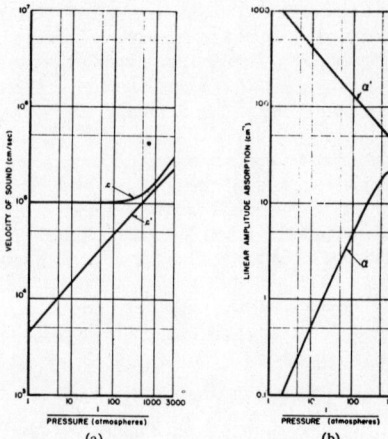

Fig. I-6 (a). Velocity vs 1/pressure for helium at room temperature. (b). Amplitude absorption vs 1/pressure for helium at room temperature.

relaxation process, Eq. (5.5), (6.14), or (7.59), is

$$\frac{\dot{\rho}_e}{\rho_0} - \frac{1}{K_s^\infty}\dot{p}_e + \omega_0\left(\frac{\rho_e}{\rho_0} - \frac{1}{K_s^0}p_e\right) = 0. \quad (11.14)$$

To add viscosity, which will give a term in $(1/\rho_0)\partial^2\rho/\partial t^2$ at high frequencies, and a term in $\dot{\rho}_e/\rho_0$ at low frequencies, we write

$$\frac{1}{\rho_0}\frac{\partial \rho_e}{\partial t} + \left(\frac{\zeta}{\rho_0 K_s^\infty}\right)\frac{\partial^2 \rho_e}{\partial t^2} - \frac{1}{K_s^\infty}\frac{\partial p_e}{\partial t}$$

$$+ \omega_0\left[\frac{\rho_e}{\rho_0} + \left(\frac{\zeta}{\rho_0 K_s^\infty}\right)\frac{\partial \rho_e}{\partial t} - \frac{1}{K_s^0}p_e\right] = 0. \quad (11.15)$$

This equation is "written down by inspection" as was (11.2) and is a combination of (11.14) and (3.1). For simplicity, ζ is assumed to be independent of frequency. As we shall show in Sec. 14, one may derive Eq. (11.15) for a plane wave from Stokes's basic postulates of viscosity. Using (11.15) one may, by the method used before, obtain the following relations,

$$c^2 = \frac{1}{\rho_0}\frac{\omega_0^2 + \omega^2 + 2\omega^2\omega_0\zeta[(1/K_s^0) - (1/K_s^\infty)] + \omega^2\zeta^2[(\omega/K_s^\infty)^2 + (\omega_0/K_s^0)^2]}{(\omega_0^2/K_s^0) + (\omega^2/K_s^\infty)}, \quad (11.16)$$

and

$$\alpha = \frac{\frac{1}{2}\rho_0\omega^2 c\{\omega_0(1/K_s^0 - 1/K_s^\infty) + \zeta[(\omega/K_s^\infty)^2 + (\omega_0/K_s^0)^2]\}}{\omega_0^2 + \omega^2 + 2\omega^2\omega_0\zeta[(1/K_s^0) - (1/K_s^\infty)] + \omega^2\zeta^2[(\omega/K_s^\infty)^2 + (\omega_0/K_s^0)^2]}. \quad (11.17)$$

To simplify the above expressions we assume that the dispersion due to viscosity occurs at a frequency

[25] The following constants were used: $\omega = 2\pi \times 10^6$, $c_0 = 1.02 \times 10^5$ cm/sec, $\eta = 1.97 \times 10^{-4}$ poise, $\gamma = 5/3$, and $\rho_0 = 2.35 \times 10^{-7} p_0$ (in mm of Hg). These values correspond approximately to room temperature. If these values were extrapolated to exactly the same temperature, the difference between our curves and the corrected graphs would not show up in Fig. I-6.

well beyond ω_0. Below the frequency where one gets viscous dispersion (we recall that $\omega_v = K_s/\zeta$, Sec. 3), the velocity is given by the usual expression

$$c^2 = \frac{1}{\rho_0}\frac{\omega_0^2 + \omega^2}{(\omega_0^2/K_s^0) + (\omega^2/K_s^\infty)} \quad \text{for } \omega \ll \omega_v$$

$$\text{and } \omega_0 \ll \omega_v, \quad (11.18)$$

while for the region well beyond ω_0, the velocity expression becomes

$$c^2 = (K_s^\infty/\rho_0)[1 + \omega^2(\zeta/K_s^\infty)^2] \quad \text{for} \quad \omega_0 \ll \omega \ll \omega_v \quad (11.19)$$

and

$$\omega_0/\omega \ll \omega/\omega_v.$$

This equation is not identical with Eq. (3.15) because we have made a different approximation here: namely, we have set $k_r^2 \approx k_r^2 - k_i^2$.

If the absorption is small, so that one can neglect the terms in ζ^2 and $\zeta(1/K_s^\infty - 1/K_s^0)$, one obtains in place of Eq. (11.17)

$$\alpha = -\frac{1}{2}\frac{1}{c_0^2}c\frac{\omega^2}{\omega_0^2 + \omega^2}\omega_0\frac{K_s^\infty - K_s^0}{K_s^\infty}$$

$$+ \frac{1}{2}\rho_0 c\zeta\frac{\omega^2}{\omega_0^2 + \omega^2}\left[\left(\frac{\omega}{K_s^\infty}\right)^2 + \left(\frac{\omega_0}{K_s^0}\right)^2\right]. \quad (11.20)$$

This can be simplified still more by assuming that for the correction in the second term $K_s^0 = K_s^\infty$ and $c_0 = c$, or

$$\alpha = -\frac{1}{2}\frac{1}{c_0^2}c\omega_0\frac{\omega^2}{\omega_0^2 + \omega^2}\frac{K_s^\infty - K_s^0}{K_s^\infty} + \frac{1}{2}\frac{\omega^2}{c^3}\zeta/\rho_0 \quad (11.21)$$

for $\omega \ll \omega_v$ and $\omega_0 \ll \omega_v$, i.e., a simple combination of (5.8) with ω_0 in place of ω_r and (3.16).

For a frequency well beyond ω_0, the absorption becomes

$$\alpha = \frac{1}{2}\rho_0 c(K_s^\infty)^2(1/K_s^0 - 1/K_s^\infty)\frac{\omega_0}{(K_s^\infty)^2 + \omega^2 \zeta^2}$$

$$+ \frac{1}{2}\rho_0 c\zeta\frac{\omega^2}{(K_s^\infty)^2 + \zeta^2 \omega^2} \quad (11.22)$$

for $\omega_0 \ll \omega \ll \omega_v$ and $\omega_0/\omega \ll \omega/\omega_v$. Equation (11.22) makes the basic assumption that $k_r \gg k_i$. If this is not true (11.19) and (11.22) are not valid. The method used in Sec. 3 has to be employed.

Our conclusion is that at most frequencies of interest in ordinary fluids one would expect the effects of relaxation and viscosity to be additive. The dispersive effect of viscosity can be disregarded below ω_0. This agrees with Herzfeld and Rice in the range considered.

(c) Double Relaxation

Let us consider a model with two internal energies $U^{(1)}$ and $U^{(2)}$. We postulate no direct interaction between the two internal parts. This may occur in a fluid with a complicated molecule, which has two means of getting excited. If the activation energy for a direct transition from one type of excitation to another is very large compared to the activation energy from the unexcited to the excited levels, we may assume two separate U^{i}'s which do not interact. Other models are possible.

A simple way to proceed is to expand the treatment of Herzfeld and Rice. Schafer (S2) has considered two ways of doing this. We adopt his first treatment to our method and write

$$C_v^{(1)}dT_1 + C_v^{(2)}dT_2 + C_v^e dT^e + T^e\beta K dV = 0, \quad (11.23)$$

$$\dot{T}_1 = -\frac{1}{\tau_1}(T^e - T_1), \quad (11.24)$$

and

$$\dot{T}_2 = -\frac{1}{\tau_2}(T^e - T_2). \quad (11.25)$$

This set of equations corresponds to Eqs. (6.1), (6.2), (6.3), and (6.4) of Sec. 6. $C_v^{(1)}$ is related to $U^{(1)}$ and $C_v^{(2)}$ is related to $U^{(2)}$. By differentiation and some relatively simple algebra, we obtain

$$C_v^e \frac{d^3 T^e}{dt^3} + T^e \beta K \frac{d^3 V}{dt^3} + \frac{1}{\tau_2}\left\{C_{v2}^e \frac{d^2 T^e}{dt^2} + T^e \beta K \frac{d^2 V}{dt^2}\right\}$$

$$+ \frac{1}{\tau_1}\left\{C_{v1}^e \frac{d^2 T^e}{dt^2} + T^e \beta K \frac{d^2 V}{dt^2}\right\}$$

$$+ \frac{1}{\tau_1 \tau_2}\left\{C_v^t \frac{dT^e}{dt} + T^e \beta K \frac{dV}{dt}\right\} = 0, \quad (11.26)$$

where we have defined

$$C_{v1}^e = C_v^e + C_v^{(1)}, \quad (11.27)$$

$$C_{v2}^e = C_v^e + C_v^{(2)}, \quad (11.28)$$

and

$$C_v^t = C_v^e + C_v^{(1)} + C_v^{(2)}. \quad (11.29)$$

Making use of (A-8) and (A-13) of Appendix II the acoustical equation of state becomes

$$\frac{C_v^e}{K}\frac{d^3 p_e}{dt^3} - \frac{C_p^e}{\rho_0}\frac{d^3 \rho_e}{dt^3} + \frac{1}{\tau_2}\left\{\frac{C_{v2}^e}{K}\frac{d^2 p_e}{dt^2} - \frac{C_{p2}^e}{\rho_0}\frac{d^2 \rho_e}{dt^2}\right\}$$

$$+ \frac{1}{\tau_1}\left\{\frac{C_{v1}^e}{K}\frac{d^2 p_e}{dt^2} - \frac{C_{p1}^e}{\rho_0}\frac{d^2 \rho_e}{dt^2}\right\}$$

$$+ \frac{1}{\tau_1 \tau_2}\left\{\frac{C_v^t}{K}\frac{dp_e}{dt} - \frac{C_p^t}{\rho_0}\frac{d\rho_e}{dt}\right\} = 0, \quad (11.30)$$

where

$$C_{p1}^e = C_{v1}^e + T^e V \beta^2 K, \text{ etc.} \quad (11.31)$$

This equation further reduces to

$$\frac{1}{K_s^\infty}\frac{d^3 p}{dt^3} - \frac{1}{\rho_0}\frac{d^3 \rho_e}{dt^3} + \omega_2\left(\frac{1}{K_s^{(2)}}\frac{d^2 p_e}{dt^2} - \frac{1}{\rho_0}\frac{d^2 \rho_e}{dt^2}\right)$$

$$+ \omega_1\left(\frac{1}{K_s^{(1)}}\frac{d^2 p_e}{dt^2} - \frac{1}{\rho_0}\frac{d^2 \rho_e}{dt^2}\right)$$

$$+ \omega_1 \omega_2\left(\frac{1}{K_s^0}\frac{dp_e}{dt} - \frac{1}{\rho_0}\frac{d\rho_e}{dt}\right) = 0, \quad (11.32)$$

ABSORPTION OF SOUND IN FLUIDS

where

$$\omega_1 = C_{p1}^e/\tau_1 C_p^t, \quad (11.33)$$

$$\omega_2 = C_{p2}^e/\tau_2 C_p^t, \quad (11.34)$$

$$K_s^0 = KC_p^0/C_v^0, \quad (11.35)$$

$$K_s^{(1)} = KC_{p1}^e/C_{v1}^e, \quad (11.36)$$

$$K_s^{(2)} = KC_{p2}^e/C_{v2}^e, \quad (11.37)$$

$$K_s^\infty = KC_p^e/C_v^e. \quad (11.38)$$

To arrive at (11.32) we have assumed that $C_v^{(1)}$ and $C_v^{(2)}$ are small. We have, therefore, written

$$C_p^e C_p^t = C_{p1}^e C_{p2}^e, \quad (11.39)$$

which neglects the term $C_v^{(1)} C_v^{(2)}$. Using (2.17), (2.18), and (11.32), we arrive at the following expression for the velocity

$$c^2 = \frac{1}{\rho_0} \frac{\omega^4 + \omega^2(\omega_1^2 + \omega_2^2) + \omega_1^2 \omega_2^2}{(\omega^4/K_s^\infty) + \omega^2[(\omega_1^2/K_s^{(1)}) + (\omega_2^2/K_s^{(2)})] + \omega_1^2 \omega_2^2/K_s^0} \quad (11.40)$$

while for the absorption

$$\alpha = \frac{1}{2} \frac{c\rho_0 \omega^2}{\omega^4 + \omega^2(\omega_1^2 + \omega_2^2) + \omega_1^2 \omega_2^2}$$

$$\times \left\{ \omega^2 \omega_1 \left(\frac{1}{K_s^{(1)}} - \frac{1}{K_s^\infty} \right) + \omega^2 \omega_2 \left(\frac{1}{K_s^{(2)}} - \frac{1}{K_s^\infty} \right) \right.$$

$$\left. + \omega_1^2 \omega_2 \left(\frac{1}{K_s^0} - \frac{1}{K_s^{(1)}} \right) + \omega_2^2 \omega_1 \left(\frac{1}{K_s^0} - \frac{1}{K_s^{(2)}} \right) \right\}. \quad (11.41)$$

We have assumed that $C_v^{(1)}$ and $C_v^{(2)}$ are small compared to C_v^e.

For simplicity we shall consider only the case where $\omega_2 \gg \omega_1$. Then we see that there are two dispersive ranges. Let us first explore the region around ω_1. Since in this case ω approximates ω_1, we have the further inequality $\omega \ll \omega_2$. Equations (11.40) and (11.41) reduce to

$$c^2 = \frac{1}{\rho_0} \frac{\omega^2 + \omega_1^2}{(\omega^2/K_s^{(2)}) + (\omega_1^2/K_s^0)} \quad (11.42)$$

for $\omega \ll \omega_2$ and $\omega_1 \ll \omega_2$ and

$$\alpha = \frac{1}{2} c\rho_0 \omega^2 \left\{ \frac{\omega_1}{\omega^2 + \omega_1^2} \left(\frac{1}{K_s^0} - \frac{1}{K_0^{(2)}} \right) \right.$$

$$\left. + \frac{1}{\omega_2} \left(\frac{1}{K_s^{(2)}} - \frac{1}{K_s^\infty} \right) \right\} \quad (11.43)$$

for $\omega \ll \omega_2$ and $\omega_1 \ll \omega_2$. Again, we have assumed that $C_v^{(1)}$ and $C_v^{(2)}$ are small compared to C_v^0.

The second dispersive regions occur when ω is in the region of ω_2. Now $\omega \gg \omega_1$ and the equations reduce to

$$c^2 = \frac{1}{\rho_0} \frac{\omega^2 + \omega_2^2}{(\omega^2/K_s^\infty) + (\omega_2^2/K_s^{(2)})} \quad (11.44)$$

for $\omega_1 \ll \omega$ and ω_2, and

$$\alpha = \frac{1}{2} c\rho_0 \left\{ \frac{\omega^2}{\omega^2 + \omega_2^2} \omega_2 \left(\frac{1}{K_s^{(2)}} - \frac{1}{K_s^\infty} \right) \right.$$

$$\left. + \omega_1 \left(\frac{1}{K_s^{(1)}} - \frac{1}{K_s^\infty} \right) \right\} \quad (11.45)$$

for $\omega_1 \ll \omega$ and ω_2.

Equations (11.42) to (11.45) are exactly what one might expect by simply adding two relaxation effects together. The method of derivation, of course, has a serious limitation in that we have made rather restricting assumptions. If the ω_i's are about equal then one cannot split up the effects and one is required to use (11.40) and (11.41). In such a case, the simple conclusions arrived at here do not hold.

Schafer has carried out some numerical calculations on this problem. He has also considered a variation of Eqs. (11.23), (11.24), and (11.25).

Some authors (Korn (K15), and Alfrey (A1a)) have suggested summing over a large number of relaxation effects and even integrating over a range of values of τ. Their method is slightly different from ours in that they simply divide the pressure into p_1, p_2, p_3, etc., the first p being related to ρ_e/ρ_0 by a static equation—i.e., Eq. (1.4)—while the p_i (for $i \neq 1$) are related to ρ_e/ρ_0 by means of a Maxwellian equation—i.e., Eq. (2.2). This second approach has the great advantage of mathematical simplicity. However, it seems to depart a great deal from the simple physical picture presented here. It is essentially like a mathematical way of expressing data which does not give much physical insight into the problem.

12. CONCLUSION

The summary of the individual effects is given in Sec. 9. Actually the theory presented in this chapter is rather formal in that it does not attempt to evaluate some of the basic parameters in the equations. In Sec. 3 we talk about ζ without attempting to evaluate it. If Stokes's relation is assumed, η still is not evaluated because we have to know the shear viscosity which is usually obtained from experiments. In Sec. 6 we have two unknowns τ and C_v^i. They appear as parameters in the theory. This is also true of the A's of Eq. (7.3) and the B's of Eq. (7.14). The theory also assumes partial equilibrium without considering the importance and consequences of this hypothesis. For instance, can one really write Eq. (7.3) in the form (7.9)? That is, is one allowed to assume that the A's are functions of the temperature and the pressure? These assumptions are certainly a weakness in the theory and offer an opportunity for further study. Attempts to evaluate τ of

Sec. 6 have been made in the past and work is continuing along this line. The evaluation of ζ and η is a problem in kinetic theory of gases and liquids. For gases, $C_v{}^i$ can be calculated from optical data and attempts to evaluate the B's of (7.14) are also in progress. We shall make no attempt to review these problems here. They are mentioned simply to point out that they do exist and imply that the theory of sound absorption is far from complete. One may hope, however, that the framework of the theory is correct and that the problem is to discover what combination of the various effects, viscosity, heat conduction, and relaxation accounts for the experimental results. Of course one cannot be sure of this until all the data have been analyzed and until better methods of computing the various parameters have been found.

From Sec. 11 we can conclude that the contribution of the various mechanisms can usually be added together. This has been proven, only for relatively simple models neglecting higher order terms. In many cases the specific heat need not be small, and higher order terms may well be considered. Consideration of more elaborate models than have been made here or by Schafer (S2) would be useful.

Chapter II. The Effects of Viscosity on Sound Absorption

Usually, the absorption of sound due to viscosity is treated very briefly, since it is generally believed to be well understood. This idea stems from the acceptance of Stokes's relation between the viscosity coefficients. The use of this relation has been questioned recently for liquids and polyatomic gases by Tisza (T3) as well as by Liebermann. Further, experiments of Liebermann (L11) indicate that Stokes's assumption may be incorrect for many liquids. If this interpretation is true, then relaxation (Sec. 6 and Sec. 7) may not play a role in some liquids.

In view of the renewed interest in viscosity, it seems advisable to return to the basic assumption made by Stokes over a century ago and consider the limitation of his proof, then bring the subject up to date by considering the special case of gases, the recent suggestions of Tisza, and discuss some of the problems of second-order acoustic fields. Some of the developments are found in standard texts, but we hope that by bringing all of them together the reader's attention may be focused on the basic problems in this field.

13. STOKES'S RELATION BETWEEN THE VISCOSITY COEFFICIENTS

We shall here follow closely the derivation of Stokes's relation given by Lamb (L1; p. 571). As pointed out in the first chapter, a general stress can be characterized by three stresses along the principal axes of stress, and a general strain is characterized by three strains along the principal axes of strain. A definition of an isotropic body is that the principal axes are in the same direction. By simple generalization, we replace the six components of strain (these reduce to three along the principal axes) by six components of the rate of change of strain. Analogous to Eq. (1.1) we write:

$$\dot{e}_{xx} = \frac{\partial u_x}{\partial x}, \quad \dot{e}_{yz} = \dot{e}_{xy} = \frac{\partial u_y}{\partial z} + \frac{\partial u_z}{\partial y},$$

$$\dot{e}_{yy} = \frac{\partial u_y}{\partial y}, \quad \dot{e}_{zx} = \dot{e}_{xz} = \frac{\partial u_z}{\partial x} + \frac{\partial u_x}{\partial z}, \quad (13.1)$$

$$\dot{e}_{zz} = \frac{\partial u_z}{\partial z}, \quad \dot{e}_{xy} = \dot{e}_{yx} = \frac{\partial u_x}{\partial y} + \frac{\partial u_y}{\partial x}.$$

The u's are the components of velocity along the x, y, and z, axes. By rotating the axes, one may reduce these quantities to three—i.e., $\dot{e}_1$, $\dot{e}_2$, and $\dot{e}_3$. Here $\dot{e}_1$ is defined as $\partial u_1/\partial x_1$, where u_1 is the component of velocity along the first principal axis.

Analogous to the static case, we define an isotropic viscous body as one in which the principal axes of rate of strain are in the same direction as the principal axes of stress. Therefore,

$$p_i = -p_0 + \eta' \sum_j \dot{e}_j + 2\eta \dot{e}_i. \qquad (13.2)$$

Here η and η' are the viscosity coefficients, η being the shear viscosity and $\eta' + \tfrac{2}{3}\eta$ the bulk viscosity. The reader will recall that in elasticity a tensor is positive, while pressure is negative. The p's with subscripts are components of the stress matrix, and the negative sign appears because of the difference between the ordinary pressure and tension. In the usual development p_0 is the static pressure.[26] Shortly we shall attempt the generalization of (13.2) and p_0 will be the nonviscous pressure which may not equal the static value.

By rotating the axes, one finds that for arbitrary Cartesian coordinates,

$$\begin{aligned} p_{jj} &= -p_0 + \eta' \nabla \cdot \mathbf{u} + 2\eta \dot{e}_{jj}, \\ p_{ji} &= \eta \dot{e}_{ji} \quad \text{for} \quad i \neq j, \end{aligned} \qquad (13.3)$$

where

$$\nabla \cdot \mathbf{u} = \dot{e}_{xx} + \dot{e}_{yy} + \dot{e}_{zz}. \qquad (13.4)$$

If we have a static fluid or uniform motion, the $\dot{e}$'s are zero and

$$p_1 = p_2 = p_3 = -p_0.$$

Let us now find the relations between $\sum_i p_{ii}$ along the principal axes and along any arbitrary Cartesian system. One may show from geometrical arguments that the pressure along the x axis is related as follows to the pressure along the principal axes:

$$p_{xx} = p_1 l_1{}^2 + p_2 l_2{}^2 + p_3 l_3{}^2, \qquad (13.5)$$

where the l's are the direction cosines. From similar ex-

---

[26] In this section, as well as the next, p_0 is not the equilibrium pressure. In a fluid where both the $\dot{e}_i$'s and $\dot{e}_j$'s are not zero, p_0 will not equal the equilibrium value.

pressions for p_{yy} and p_{zz}, we obtain the equality

$$p_{xx}+p_{yy}+p_{zz}=p_1+p_2+p_3. \quad (13.6)$$

The x, y, z system is completely arbitrary, and Eq. (13.6) makes it sensible to define the average pressure, p_{av}, as

$$p_{av}=-\tfrac{1}{3}(p_{xx}+p_{yy}+p_{zz})=-\tfrac{1}{3}(p_1+p_2+p_3). \quad (13.7)$$

For the static case, we see that

$$p_{av}=p_0,$$

while for the dynamic case,

$$p_{av}=-p_0+\tfrac{1}{3}(3\eta'+2\eta)\sum_i \dot{e}_{ii}. \quad (13.8)$$

Equation (13.8) is the core of the lengthy argument regarding Stokes's viscosity relation. Is p_{av} equal to p_0, or are they not equal? Stokes's original proof in 1845 (S17) can hardly be considered rigorous. Stokes himself admits this in his original paper.

The original assumption is that

$$p_{av}=p_0, \quad (13.9)$$

and

$$3\eta'+2\eta=0. \quad (13.10)$$

General thermodynamic arguments may be presented to show that the following equality or inequality must hold:

$$\eta'+2\eta/3 \geqslant 0. \quad (13.11)$$

The proof of this relation may be found in an elegant paper by J. H. C. Thompson (T2a).

Basset (B3), in his treatment, derives Stokes's relation for gases by assuming that Eq. (13.9) holds. This he claims is a third assumption required to obtain Stokes's viscosity equations, i.e., Eqs. (13.3) and (13.10). He considers that liquids are incompressible and that $\rho_e=0$.

To find a more rigorous proof of Eq. (13.10) we must turn to models of the fluid used, and the subject separates itself into two parts—the gas and the liquid.

(a) Stokes's Relation for a Gas

At present, it seems fair to state that the general basis of the kinetic theory is accepted for real monatomic gases, (J1). From the rigorous development of the theory, one may show that Stokes's relationship holds. The only assumption as to the nature of the gas is that one has central forces between molecules. A further assumption is usually made that the process occurs at constant temperature. This, of course, is not true, and would lead to a term arising from thermal diffusion. The effect of thermal diffusion is a separate one which one would expect in the first-order approximation to be superimposed on viscosity. Polyatomic gases have additional degrees of freedom which are not considered in the usual development. Recently Grad (G2b and G2c) developed a new kinetic theory of gases. For monatomic gases Stokes's relation is not changed.

Wang Chang and Uhlenbeck (W2) (see also Wang Chang (W1)) recently developed equations for the absorption and velocity dispersion, using terms (nonlinear) beyond those which were considered by Stokes. These terms arise from an expansion of the kinetic theory. Assuming that ϵ of (11.8) equals 2.5, they find that α of (11.13) for the combined effect of heat conduction and viscosity should be replaced by

$$\alpha=\frac{7}{10}\frac{\omega}{c_0}\frac{\omega}{p}\eta\left[1-3.68\left(\frac{\omega\eta}{p}\right)^2\right]. \quad (13.12)$$

One would expect, therefore, that one is not allowed to expand Stokes's equation (for a gas) and obtain corrections to the conventional viscosity absorption term, i.e., Eq. (9.5).

Experimental work by Greenspan (given in Sec. 21 below) indicates, however, that Stokes's approximation gives better results for He gas than do "corrected" expressions. For absorption measurements, the agreement between Stokes's theory and experiment is very good; for velocity dispersion, both Stokes and higher approximations are off, the Stokes's form being slightly inferior.

One must therefore, conclude that at present our knowledge is too limited to do effective work using terms beyond those derived by Stokes. The work of Wang Chang and Uhlenbeck indicates that one should attempt to derive an expression for sound absorption and dispersion from a generalized transport equation rather than the cruder methods presented here. For sound absorption, the transport equation, however, does not seem to give reliable results at present. Further experimental and theoretical work in this field is very necessary. It is to be hoped that Grad's new development will give better results than the previous theories.

(b) Stokes's Relation for a Liquid

The kinetic theory of liquids is at present being developed by Born and Green (B24) and by Kirkwood (K5). A recent paper by Kirkwood, Buff, and Green (K6) indicates that the calculation of the viscosity coefficients is still in a primitive stage. Indeed, they rely on sound absorption measurements to estimate the value of the bulk viscosity coefficient. Turning to experimental data, we find (Sec. 24) an agreement between theory and experiment for several simple liquids. This indicates that Stokes's relation holds for some liquids.

On the basis of this discussion, we must conclude that Stokes's relation for liquids cannot at present be established on theoretical grounds. Since the burden of proof should be on the establishment of the relation, the Stokes's relation cannot be accepted unconditionally. As Basset has pointed out, the viscosity relation is a basic assumption made when setting up the viscosity equation for fluids. This procedure can be justified theoretically for real, dilute monatomic gases. Further, there is good experimental evidence to support this assumption in some gases and liquids, and this assumption leads to no contradiction at present.

14. VISCOSITY AND SOUND ABSORPTION

Let us return to Eq. (13.2) and consider its connection with an acoustical equation of state. First, consider the case where there is no relaxation. p_1, p_2, and p_3 are related to the force on an element of volume. These p's have to be used in the equation of motion (2.10). Also p_0 has to be related to the condensation by Eq. (1.4) or an equivalent relation. If the p_i's were related directly to the condensation, then the viscous term of Eq. (13.2) would not play a role. Using Eq. (1.4), i.e.,

$$(p_0)_e = K_s \rho_e / \rho_0 \qquad (14.1)$$

and Eq. (13.3), we get

$$(p_{jj})_e = -K\rho_e/\rho_0 + \eta' \nabla \cdot \mathbf{u} + 2\eta \dot{e}_{ii} \qquad (14.2)$$
$$\cdot p_{ji} = \eta \dot{e}_{ji}.$$

Equation (14.2) is then the basic acoustical equation of state for a viscous medium. One should emphasize that it is a tensor relation and when substituting into the equation of motion, the term on the right must be replaced by terms involving tensors or dyads.

If we are interested in a plane wave traveling along the x axis, $\mathbf{q}$ becomes $q_x = f(x)$, and

$$p_e = (-p_{11})_e = K_s \frac{\rho_e}{\rho_0} + (2\eta + \eta') \frac{\partial}{\partial t} \frac{\rho_e}{\rho_0}, \qquad (14.3)$$

where

$$\dot{\rho}_e/\rho_0 = -\partial u_x/\partial x$$

by Eq. (2.11). The use of (2.11) means that a slight approximation has been made in (14.3); it is the same type of approximation made when using (2.10) and (2.11). This is equivalent to Eq. (3.1). Here p_e is the excess pressure at right angles to the x axis, which does not equal the pressure in the other directions. By equating Eq. (14.3) to Eq. (3.1), we have

$$\zeta = \eta' + 2\eta. \qquad (14.4)$$

Or, if Stokes's viscosity relation holds,

$$\eta' = -2\eta/3 \qquad (14.5)$$

and

$$\zeta = 4\eta/3, \qquad (14.6)$$

the standard classical expression.[27]

We should like to combine viscosity and relaxation, by assuming that (1.4) is not valid and considering effects within and between molecules. In general, one is not permitted to make this separation. However, for many substances, namely, gases, many liquids, and molecular

[27] The standard derivations (see Lamb (L1; p. 646); Rayleigh (R2; Vol. II, p. 315)) proceed in a slightly different manner. They use Eq. (13.3) and the equation of motion, then in a final step introduced the static equation of state (1.4). We have proceeded differently to try to bring out the equation of state for the viscous process to compare it with other processes.

solids, such a step may be useful. For substances made of complex molecules, the phase lag between pressure and density occurs because of intermolecular processes and processes within the molecules themselves. The intermolecular processes can be accounted for by means of viscosity, while the intramolecular processes can be accounted for by means of a relation between ρ_e and p_0'. When generalizing (14.2), we replaced p_0 by p_0' where p_0' includes both the static pressure and the pressure arising from the relaxation process. (It is necessary to interpret "within the molecules" quite broadly (Sec. 7).) The above approach is probably only a rough approximation to the truth, which holds for gases and probably for some liquids.

It is impossible to justify completely this last paragraph. Indeed, for some substances, this separation may be incorrect and useless, but there are some arguments to support this concept.

For Gases:—As mentioned above, the rigorous theory of real gases requires us to accept Stokes's relation, if intramolecular effects are ignored. The method of statistical thermodynamics gives us a model which associates the internal processes with a relation between ρ_e and p_0'. It seems, therefore, completely logical to make this separation for a gas. The general kinetic theory of gases is preserved, yet the absorption is explained. More elaborate kinetic theories may be able to arrive at a relation between p_e and ρ_e.

For Liquids:—Here we know very little at present. The calculation of Hall (H1) and the model of a general relaxation implies this separation, since relaxation is something that happens within a molecule or a small cluster of molecules. Further, data taken by Liebermann on ethyl formate can be easily explained by this hypothesis (M2a).

We, therefore, assume that p_0' is related to ρ_e by means of a relaxation equation (Eq. (6.14)). Here p_0' is no longer a static pressure. It is only the static pressure for processes slow relative to the internal equilibrium. The acoustic equation for viscosity and relaxation becomes

$$\frac{1}{\rho_0}\frac{\partial \rho_e}{\partial t} + \frac{1}{K_s^\infty}\frac{\zeta}{\rho_0}\frac{\partial^2 \rho_e}{\partial t^2} - \frac{1}{K_s^\infty}\frac{\partial p_e}{\partial t}$$
$$+\omega_0\left(\frac{\rho_e}{\rho_0} + \frac{1}{K_s^0}\frac{\zeta}{\rho_0}\frac{\partial \rho_e}{\partial t} - \frac{1}{K_s^0}p_e\right) = 0. \quad (14.7)$$

This is identical to Eq. (11.15) and is the "derivation" mentioned.

Tisza (T3) has suggested that one should force all the effects into an equation similar to Eq. (14.3). η is the ordinary-shear viscosity, but $3\eta' + 2\eta$ no longer has its traditional meaning. By making this sum complex and frequency dependent, one may obtain the experimental results for gases. In this case, the sum will be dependent on the type of process, i.e., isothermal or adiabatic. To evaluate the sum, Tisza uses the conventional theory of thermal relaxation. This approach, while mathematically correct, seems less useful than the usual approach. If one uses Tisza's method and adopts it to other hydrodynamical problems, one must keep in mind that his evaluation of $3\eta'+2\eta$ is for an adiabatic process, and that it may be zero for an isothermal process.

Frenkel and Obraztsov (F4) have suggested that one may define a low frequency bulk viscosity by returning to (6.14), i.e.,

$$\frac{\dot{p}_e}{\rho_0} - \frac{1}{K_s^{\infty}} \dot{p}_e + \omega_0 \left(\frac{\rho_e}{\rho_0} - \frac{1}{K_s^0} p_e \right) = 0. \quad (14.8)$$

For a slow process $\dot{p}_e = (K_s^0/\rho^0)\dot{\rho}_e$, or we may write (14.8) in the form

$$p_e = \frac{K_s^0}{\rho_0} \rho_e + \frac{K_s^0}{\omega_0} \left(\frac{K_s^{\infty} - K_s^0}{K_s^{\infty}} \right) \frac{\dot{\rho}_e}{\rho_0}. \quad (14.9)$$

For low frequencies,

$$\eta'' = \frac{K_s^0}{\omega_0} \left(\frac{K_s^{\infty} - K_s^0}{K_s^{\infty}} \right) \quad (14.10)$$

can be defined as a viscosity coefficient. To extend η'' to higher frequencies it has to be made complex and frequency dependent. While one can make the mathematical treatment of viscosity and relaxation similar by using Eq. (14.10), the physical difference shown in Fig. I-4 remains.

15. SECOND-ORDER EFFECTS

It is well known that the derivation of the standard acoustical wave equation is based on certain approximations involving the neglect of terms usually considered too small to be retained; e.g., terms like $\mathbf{u}V\cdot\mathbf{u}$ in the hydrodynamic equations of motion and $\mathbf{u}\cdot\nabla\rho$ in the equation of continuity. Such terms must be retained or accounted for in some way when higher order effects are considered. It is interesting to note the breakdown of the analogy between acoustic and electromagnetic radiation in this respect: no approximations are involved in the derivation of the electromagnetic wave equation, since it follows directly from the field equations. It is consequently not permissible to draw conclusions about second-order quantities in acoustics such as average energy and radiation pressure, from electromagnetic analogies.

Though much work on second-order acoustical effects has been published by Eckart (E3), Bergmann (B16), and Westervelt (W3) among others, the present authors believe that the field has not yet been sufficiently explored to justify a thorough review. This section therefore will be confined to a few general remarks.[28]

A convenient method of dealing with second-order effects has been used by Eckart. He expands the density, pressure and particle velocity in series of terms of successively decreasing order of magnitude as in standard perturbation technique. Thus

$$\begin{aligned} p &= p_0 + p_1 + p_2 + \cdots, \\ \rho &= \rho_0 + \rho_1 + \rho_2 + \cdots, \\ \mathbf{u} &= 0 + \mathbf{u}_1 + \mathbf{u}_2 + \cdots. \end{aligned} \quad (15.1)$$

Here p_0[29] and ρ_0 are the zero-order terms, i.e., equilibrium pressure and density. Terms like p_1, ρ_1, and $\mathbf{u}_1$ are first-order terms, etc. A product like $\rho_0\mathbf{u}_1$ is assumed to be of first order while $\rho_1\mathbf{u}_1$ is of second order, i.e., in this respect on a par with p_2, ρ_2, and $\mathbf{u}_2$.

The usual acoustic wave equation is satisfied by terms of the first order. Let us consider what can be done with a problem involving second-order terms, such as the energy density in a sound field (M3). The stored potential energy for compressing a mass of fluid Δm is

$$-\int p\, dV = \Delta m \int_{\rho_0}^{\rho_0+\rho_1+\rho_2} p\, \frac{d\rho}{\rho^2}$$

$$= \Delta m \int \frac{(p_0+p_1)}{(\rho_0+\rho_1)^2} d\rho. \quad (15.2)$$

For a nonabsorbing medium we can use the relation $p_1 = c_0^2 \rho_1$ and obtain for the stored energy

$$\quad\quad\quad (a)\quad\quad (b)\quad\quad (c)\quad\quad\quad (d)$$

$$\frac{\Delta m}{\rho_0^2} \left\{ p_0(\rho_1+\rho_2) + \left(\frac{c_0^2}{2} - \frac{p_0}{\rho_0} \right) \rho_1^2 \right\}. \quad (15.3)$$

In Eq. (15.3), the terms have the following meaning: (a) is a harmonic term, (b) arises because of Eq. (15.1), (c) is the term which is commonly given. We shall call it the Rayleigh term, since it dates back to that period. (d) is a term which, as far as the authors know, has not been included previously. At least it does not appear in the "standard" treatments.

For a liquid where $c \approx 10^5$ cm/sec, $p_0 \approx 10^6$ dynes/cm^2 and $\rho_0 \sim 1$, the ratio of (c) to (d) is about 1 to 10^{-4}, and the omission of (d) is correct. One may likewise show that one can omit (b) for liquids. For a gas, $c \sim 3 \times 10^4$, $\rho \sim 10^{-3}$, and (d) cannot be omitted. To obtain the average of (15.3) for a gas one needs more exact solutions than are usually used.

A very complicated problem exists in setting up the viscosity tensor in the equation of motion for a fluid. The viscosity coefficients are highly dependent on the temperature, and, since one is interested in an adiabatic process, the variation with temperature must be taken into account. Correcting the viscosity coefficient in the tensor, i.e., expanding η and η' in Eq. (13.3) in terms of ρ_1, leads to very complicated equations which have not been studied fully. Finally, it would seem advisable to include relaxation in any development of second-order acoustical effects.

---

[28] Further details will be published elsewhere.

[29] p_0 is the equilibrium pressure and therefore has a different meaning in this section from p_0 of Secs. 13 and 14.

In conclusion we should like to mention Eckart's (E3) theory and Liebermann's experiments (L11). Using these data, one may conclude that Stokes's viscosity relation does not hold, and that its breakdown is sufficient to explain the absorption in liquids. While this theory and experiment are of major importance, they, by themselves, do not resolve the problem. The reason for this conclusion is that when one carries out Eckart's theory a little further, one can show that a relaxation effect combined with viscosity in a manner similar to (14.7) can equally well explain the results (M2a). The recent work of Nyborg (N2) on acoustic streaming adds weight to this conclusion. Further, the viscosity idea alone cannot explain Liebermann's experiment on ethyl formate, or the work of Lamb and Pinkerton (L2) on acetic acid. More work is required here, before one fully understands the phenomenon of streaming and its relation to absorption.

Chapter III. Experimental Methods of Sound Absorption Measurements

Sound absorption measurements can be classified roughly into three groups: (1) mechanical, (2) optical, and (3) electrical. Although some absorption measurements have been made on the basis of thermal effects, notably by Richardson (R5), their reliability is open to some question, and they will not be considered here.[30] We shall give here only a brief outline of the most frequently used methods, referring the reader to particular papers for the experimental details. We shall also endeavor to classify the methods by range of usefulness and accuracy. In this way we shall be better able to judge the validity of the experimental measurements.

16. MECHANICAL METHODS

The mechanical method which is most frequently used is based on radiation pressure. When a rigid wall confronts a sound beam, there is a difference between the pressure at the wall and the pressure in the same medium at rest, behind the wall (i.e., the pressure in the medium in the absence of the beam). This net pressure is known as the Langevin radiation pressure, and is approximately equal in magnitude to twice the energy density of the oncoming sound wave (B18, B34, H7). This net pressure can be employed to measure a quantity proportional to the sound intensity in several ways. For example, if the sound is allowed to rise vertically in a tank of liquid, a cone or plate may be suspended in the sound field, and its apparent weight measured both in the presence and absence of the sound beam (C4, H14). In a variation of this method, a horizontal sound beam may be used to displace a plate or bead, either on the end of a long wire (F3), or mounted as the vane of a torsion balance (A3). Finally, the movable vane has been made one plate of a condenser, and the intensity measured by the change in capacitance of the system (B38).

The successful use of these mechanical methods is confined almost exclusively to liquids. As a class, measurements of this type suffer from five general difficulties:

1. Surface tension or other retarding forces which act on the wire or detector may easily be of a magnitude comparable to that of the forces being measured (which usually are of the order of 10 dynes). These extraneous forces can be reduced in magnitude, but are never entirely eliminated. Their effect is to cause a "sticking" of the detector in the vicinity of its balance point, thus decreasing the accuracy with which the apparent weight changes can be measured.

2. Specular reflection from the walls of the container, if it is too narrow, or from the detector, will result in standing waves, and lead to grossly incorrect results. This was the case in some of the early work, particularly that of Sörensen (S11), and Hartmann and Focke (H5). A rather striking confirmation of this difficulty was given by Claeys, Errera, and Sack in the article previously cited (C4). These observers measured the apparent absorption with tanks of different diameters, and were able to get almost any desired (higher) value for the absorption coefficient by using a sufficiently narrow container. As the diameter of the container was increased, the apparent value of the absorption coefficient fell off to a constant value.

3. It would seem that the first error mentioned above could be minimized by increasing the sound intensity. However, high intensity can cause cavitation, in which case a larger fraction of the energy of the beam is lost than in simple propagation. Cavitation can be avoided if the instantaneous pressure in the medium is not permitted to fall below zero. The average energy density $\bar{E}$ in a plane wave is related to the maximum excess pressure p_e by

$$\bar{E} = p_e^2/2\rho_0 c^2, \quad (16.1)$$

where c is the velocity of propagation and ρ_0 is the mean density of the liquid. If the pressure in the medium is not to fall below zero, the excess pressure should not exceed the hydrostatic pressure.[31] If this latter is the atmospheric value, then for water, $\bar{E} \sim 20$ dynes/cm². Since the mean intensity $\bar{I}$ is given by $\bar{I} = \bar{E}c$, this limits us to intensities no higher than about 0.3 watt per sq cm.

4. As its frequency is lowered, a sound beam diverges more and more, so that a detector of fixed size will

---

[30] A refinement of this technique, involving hot-wire interferometry, has recently been reported by Matta and Richardson (M4). This instrument is believed by its designers to give very accurate measurements of the absorption coefficients in gases.

[31] If a dissolved gas is present, the excess pressure should not exceed $p_A - p_V$ where p_A and p_V are the atmospheric and vapor pressures respectively. This has been pointed out by Boyle and Taylor (B33).

eventually fail to intercept the entire beam. Consequently, the simple formula for plane waves

$$I = I_0 e^{-2\alpha x} \qquad (16.2)$$

cannot correctly be employed. (Here I_0 is the sound intensity at the plane $x=0$ and I is the intensity at $x=x$.)

5. Hydrodynamic flow (a manifestation of quartz wind) can itself exert a net force on the detector, giving rise to an incorrect value of the radiation pressure. This can be reduced by placing a thin Cellophane screen (H14) directly in front of the detector. This serves to stop the liquid flow while it allows the sound to pass through with small loss.

These various difficulties limit the radiation pressure method to the use of moderate sound intensities at high frequencies. For most liquids, careful measurements with the radiation pressure method can be made with a probable error of ±5 or 10 percent at frequencies above 10 megacycles. In the region 3–10 megacycles, the measurements are less trustworthy, except in liquids of fairly high absorption. (In liquids of very high viscosity, the detector cannot move with sufficient freedom, so that an upper limit of usefulness also exists.) Few if any measurements made below 3 megacycles can be relied upon.[32]

17. OPTICAL METHODS

Most optical measurements of the absorption coefficient have been carried out by means of a method developed by Biquard (B20), based on the Debye-Sears (D2) effect. While the number of measurements made recently by this method is not large, it is still a useful method within certain ranges of frequency.[33]

In the Debye-Sears experiment, Fig. III-1, light from a narrow slit traverses a beam of sound which is at right angles to the light. The light is then focused on a screen. The successive compressions and rarefactions of the sound beam alter periodically the refractive index of the liquid. Thus the sound beam acts as a diffraction grating, and a series of parallel diffraction lines are produced on the screen. The greater the intensity of the sound beam, the more the light is diffracted away from the main beam (i.e., the central maximum). In the application of Biquard, the light in the main beam falls on a photocell, and a current is measured which is proportional to the sound intensity. If the intensity of the sound at a distance x cm from the crystal is $I_0 e^{-2\alpha x}$, then the loss in intensity of the light beam at this point is proportional to this quantity, and hence if J_0 is the intensity of the transmitted light in the absence of sound, and J is the intensity in the presence of sound,

$$1 - (J/J_0) = k I_0 e^{-2\alpha x}. \qquad (17.1)^{34}$$

The output θ of the photocell is directly proportional to the incident light intensity. Hence

$$1 - (\theta/\theta_0) = k I_0 e^{-2\alpha x}. \qquad (17.2)$$

In the measurement θ is measured as a function of x. If the equation is put in the form

$$\ln[1 - (\theta/\theta_0)] = \ln k I_0 - 2\alpha x, \qquad (17.3)$$

then a graph of $\ln[1-(\theta/\theta_0)]$ vs x will have a slope of -2α, so that the absorption coefficient can be determined.

This measurement suffers from many of the same limitations as the mechanical method. Thus, it assumes a plane wave with constant intensity across the wave front. As the frequency decreases, the divergence of the beam increases, so that the method is not a practical one at low frequencies.

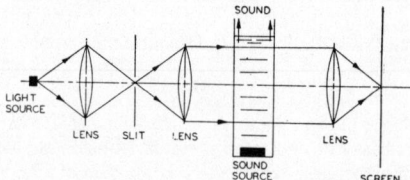

Fig. III-1. Sound absorption measurement by the optical method.

Multiple reflections and high intensity are responsible for the same type of errors here as in the mechanical method. In addition, high intensity will cause appreciable excitation of higher order images, in which case some light may be diffracted back into the main beam. More simply, the loss in light intensity in such a case is no longer directly proportional to the sound intensity. Finally, alignment problems are very important. The light beam must be narrow, and accurately perpendicular to the sound beam; the medium must be sufficiently transparent so that appreciable light can penetrate it.

More recent observers have made improvements in the general technique. Burton (B37) employs a monochromatic light source and an electron-multiplier tube so that much narrower light beams and lower acoustic intensities can be used.

The frequency range over which optical methods can be trusted is similar to that for mechanical methods. Where comparison by method is possible in a given liquid, good agreement is obtained by most observers using optical methods in comparison with other methods. Below 10 megacycles many of the measurements which are available are not in good agreement with those of other methods. Most of these lower fre-

---

[32] Strictly speaking, any such classification of results must take the absorption into account, since as it increases, the errors due to (2) and (3) become proportionally less significant. As a general rule, measurements in which the values of the amplitude absorption coefficient are larger than 0.05 cm^{-1} are quite reliable when the tank is at least 10 cm in diameter; values in the range 0.01 cm^{-1} to 0.05 cm^{-1} are less satisfactory, and values below 0.01 cm^{-1} are usually invalid in a tank of this size.

[33] Within the past year, however, D. Sette (S6a) has published a considerable number of measurements obtained by an optical method. These results appear to be of high accuracy.

[34] k is an experimental constant not related to the previous k's.

quency measurements, such as those of Parthasarathy (P1), are older works, where the techniques were less satisfactorily developed. The recent works of Burton (B37), Willis (W7), and Sette (S6, S6a) indicate that agreement with accepted values in some liquids exists at frequencies as low as 4 mc.

Hydrodynamic flow is also a serious problem with the optical method, since it leads to a sizeable variation in the optical properties of the medium. As mentioned above, a Cellophane screen can be used to decrease the amount of this flow.

18. ELECTRICAL METHODS

In this section are grouped all methods in which a microphone, piezoelectric or otherwise, is used to receive sound signals. There are four general methods used to determine the absorption coefficient with such equipment and each of these will be discussed in turn.

(a) Interferometric Methods

In the basic interferometric method, a plane quartz crystal is used as a transducer. A plane reflector is set accurately parallel to the transducer at a distance which can be varied. If sound waves emanate from the transducer, they will be reflected from the reflecting surface. If the reflected wave returning to the crystal is 180° out of phase with the signal emanating from the crystal, the disturbance at the crystal will be reduced essentially to zero. This will produce a considerable rise in the plate current of the output stage of the driving oscillator (or in the plate current in the tube). Since the problem is essentially one of standing waves, it is clear that a maximum in the plate current will be produced every time the reflector is moved a half-wavelength.[35] The interferometer is therefore of fundamental importance in the measurement of the sound velocity.

The instrument has also been used to measure absorption. The variation in the current reading in the output stage of the driving oscillator was shown by Pielemeier (P5) to be proportional to the excess pressure in the sound beam at the face of the crystal. If now the reflector is moved through a distance x, the path length is increased by $2x$, so that the pressure of the returning wave will be decreased by the factor $e^{-2\alpha x}$. Hence, by registering the current at two maxima, the absorption coefficient can then be computed.

The work of Hubbard and his associates (A2, H15, H16, S14, S16) indicates that the current is a more complicated function of the absorption coefficient, the reflection coefficient of the reflecting surface, and the distance moved by the reflecting plate.

The interferometer has become a standard instrument for absorption measurements in gases. It has also had some use in liquids. Recently, Hunter and Fox (H18, H19) have developed an interferometer using the liquid-air interface as the reflecting surface.

The use of an interferometer is limited by the following difficulties:

1. Imperfect alignment of the crystal and the reflecting surface.
2. Departure of the sound beam from a plane wave. Pumper (P9) has shown that a correction can be made for deviation from plane waves, and that this correction is essentially a constant, experimentally measurable term (at a given frequency) which is to be subtracted from the measured absorption value to give the correct value.
3. Lack of adequate knowledge concerning the reflection coefficient. According to Herzfeld (H11) and Hubbard (H16) the reflection coefficient is modified by heat conduction, so that separate measurements must be made of this quantity.

As a general estimate, interference methods appear to be capable of measuring absorption coefficients accurately, provided that the values of α are greater than about 0.1 cm^{-1}. In general, older values are less reliable, especially in gases, because of deviations of the reflection coefficients from theoretical values, and because of the presence of impurities in the gases under measurement. This latter difficulty will be discussed more fully later (Sec. 22).

(b) The Direct Method

In the so-called direct method, a microphone is located on the axis of an ultrasonic beam, and excess pressure or intensity is measured along the axis. If the beam approximates a plane wave, the decay law $I = I_0 e^{-2\alpha x}$ may be used to evaluate the absorption. If, at the other extreme, the waves are spherical, the law $I = (I_0/r^2)e^{-2\alpha r}$ can be employed.

While this method has been widely used in both liquids and gases, there are many difficulties attendant upon it, and frequently neither of the above formulas can be employed. For example, if the transducer is assumed to operate as a piston-like source, one can show that Fraunhofer diffraction takes place within a distance a^2/λ of the crystal (a = radius of transducer, λ = wavelength of sound) while the beam diverges, and a typical Fresnel pattern is obtained at much larger distances. King (K2) and H. Born (B23) have obtained approximate theoretical expressions for the radiation field and these formulas may be employed to determine the absorption. Corrections must also be included for the finite size of the microphone (W6), if the wavelengths are not very long compared with the dimensions of the microphone.

Standing waves are again a problem here (as they must be for all methods employing continuous waves). Where the beam is well defined, it is possible to tilt the microphone so that it is not quite perpendicular to the

---

[35] A more rigorous treatment by Grossmann (G6) indicates this is not strictly true. Because of the curvature of the wave fronts, the distances between successive maxima near the crystal are slightly larger than $\lambda/2$. A correction term has been calculated by him.

axis of the beam. This will reduce standing waves from the front face of the microphone. To reduce reflection from walls, various absorbent materials have been used. The problem here is more serious in gases, where the reflection coefficients are larger than in liquids.

The frequency range over which successful measurements have been made is not very great, except in water. The method has been used from 1 mc down to 50 kc in measurements in water, where a medium of very great extent is available. In the laboratory, measurements have been made in the range 1 to 4 mc. Some of these measurements are rather inaccurate. The absorption is low and since α is computed as a log of a number which almost equals unity, it does not take a very large error of measurement to make the result meaningless.

In gases, the frequency range over which measurements are made is much lower due both to the higher absorption values in gases and to the shorter wavelengths. Measurements are made mainly in the region 20–150 kc.

A variation of this method has been employed by Knudsen and Fricke (K14). From the measurements of other observers [e.g., van Itterbeek and Thys (V3)] it appears that absorption in nitrogen is almost exactly classical. (See Sec. 22.) The absorption coefficient of nitrogen can then be used as a standard. The intensity of sound is first measured by a microphone with a nitrogen-filled chamber, and again with the chamber filled with the test gas. Corrections are made for wall absorption. While there is a small difference in the results obtained by Fricke (F6) by this method for CO_2 and those obtained by Leonard (L5) for the same gas, it is quite probable that the difference is due entirely to the presence of impurities.

(c) Pulse Methods

Two general objections may be raised against all methods involving continuous waves. In the first place, there is always the possibility of the creation of standing waves, which would then lead to incorrect values for 2α. In the second place, the amount of energy introduced into the medium may change the local temperature somewhat, leading to refractive effects, since the temperature in the center of the beam would be higher than on its edges. In addition, the changed temperature would also bring it about that measurements would actually be made at a higher temperature than that recorded by a thermometer located outside of the beam. Both of these difficulties are avoided by the use of pulses. In the arrangement of Pellam and Galt (P2) a quartz transducer and a polished reflector are set up, parallel to each other, as in the interferometer method. The high frequency voltage applied to the transducer is pulsed, with a repetition rate of the order of 1000 per second. The width of the pulse in this particular case was 1 microsecond, with a frequency of 15 mc. Thus the average power is only 1/1000 that of a continuous signal of the same amplitude.

The signal is reflected from the polished surface and is picked up by the quartz, now acting as a receiver. If the distance between the transducer and the reflector is appropriately chosen, (relative to the repetition rate), the reflected signal will arrive while there is no transmitted pulse so that standing waves are entirely avoided. By the use of suitable electronic equipment, the initial and the reflected pulse may be compared on an oscilloscope. The customary technique (P2) is to pass the initial pulse through a calibrated attenuator. The attenuation required to reduce the size of the initial pulse to that of the reflected one measures the power lost in transmission, plus reflection losses. If the reflector is now moved through a distance x, parallel to the axis of the beam, the path length is increased by $2x$. A graph of attenuation vs $2x$ will then permit a calculation of the absorption coefficient.

The theory of measurement at this point becomes identical to that of the direct method. Regions of Fraunhofer and Fresnel diffraction must be treated differently. An adequate treatment of this problem is given by Pinkerton (P8).

One problem raised by the pulse method is that the use of a narrow pulse increases the spread of frequencies in the Fourier spectrum of the pulse. A simple calculation (P3) shows however that for pulses containing at least 15 cycles (at a frequency of 15 mc), this error is no greater than 1 part in 250.

Inherently, the pulse method is the most accurate method of making absorption measurements—provided that the experimental procedures are sufficiently refined. While observers using pulse techniques estimate errors as being below 15 percent, and frequently of the order of 2–3 percent, large discrepancies have appeared in certain measurements, even though very similar apparatus and procedures have been used.[36]

So far, pulse techniques have been applied mainly to liquids and solids. In the latter field, they form in fact the principal method of measurement. The frequency range of pulse measurements extends from about 1 mc up to 200 mc and even beyond.

(d) Reverberation Methods

The methods discussed so far have been strongly limited in their application to frequencies of the order of a megacycle and higher in liquids, with a lower range possible in gases. The use of the reverberation method in liquids extends greatly the lower limit of frequencies which can be employed. This method was first employed by Knudsen (K13) for gases, and has been extended to liquids by Leonard (L6), Liebermann and Wilson (L9), Mulders (M10), and Moen (M8).

In principle, the method rests on the measurement of

---

[36] An example of this is given by the values which have been obtained for ethyl alcohol. Measurements by four observers, all using the pulse technique, include the following values of $\alpha/\nu^2 \times 10^{17}$ cm^{-1} sec^2: Pellam and Galt (P2)23, Pinkerton (P8)52, Rapuano (R1)54, Teeter (T1)225. ν is the frequency.

TABLE III-I. Range of absorption measurements in liquids.

Method	Range of α in cm^{-1}	Corresponding frequency range in water (in mc)
Mechanical	0.001 (C3) to 1 (B17)	2 to 60
Optical	0.02 (B22) to 5 (B1)	7 to 140
Interferometric	0.05 (H17) to 4 (H17)	15 to 130
Pulse	0.05 (P3) to 310 (R1)	15 to 800
Reverberation	10^{-5} (L7) to 0.003 (M8)	0.15 to 3

the reverberation time in a given enclosure. In the application of Liebermann and Wilson, a brass sphere containing the liquid is excited in a radial mode of vibration by a crystal attached to the surface of the sphere. The transmitter is shut off and the same crystal is used as a detector. During this stage spherical waves are repeatedly reflected from the walls of the container, with the energy of the beam gradually being dissipated by the absorption in the medium and the walls. The rate at which the intensity falls off is then a function of these two quantities. If a radial mode is used, and if the brass sphere has a thickness of an odd number of quarter wavelengths, the energy transmitted through the walls is essentially negligible. Under these conditions, the absorption coefficient α will be given by

$$\alpha = (1/ct) \ln(I_0/I_t), \quad (18.1)$$

where I_0 is the initial sound intensity, I_t is the intensity t seconds later, c is the speed of sound. The theoretical analysis of this problem was given by C. F. Eyring (E9). This method or a slight variation of it has been successfully applied in measurements in the frequency range in liquids from 24 kc to 200 kc. The frequency limitations are the following. At very low frequencies, the absorption coefficients are so small that any error in calculating wall losses will be significant. At high frequency, resonant modes lie very close together so that it is difficult to excite a radial mode only. If nonradial modes are excited, wall transmission will become more complicated and in general its magnitude becomes greater. In addition, the decay time gets smaller and therefore more difficult to measure accurately.

The arrangement by Mulders makes possible measurements in liquids in the somewhat higher frequency range of 500 to 1500 kc. In this arrangement, a frequency modulated source is employed and as many as 10^4 modes of vibration are excited. Thus the sound becomes essentially diffuse. Many corrections must be employed, but the results for water are in substantial agreement with those of other methods. The method of Moen also enables an extension of the frequency range up to 1 megacycle.

The diffuseness of the sound was obtained by Knudsen in gases by employing a motor-driven paddle.

In all low frequency measurements, a particular problem is presented by the formation of bubbles. The presence of bubbles increases the absorption and also produces scattering so that the absorption coefficient which is measured may be considerably larger than that of the pure liquid. Since the value of α in water at 50 kc is of the order of 0.75×10^{-6} neper per centimeter, it does not take the presence of many bubbles to make the results wholly meaningless. Elaborate procedures are required to avoid any dissolved gases in the liquid. This leads to some uncertainty as to the validity of open water measurements at these same low frequencies, since such precautions cannot be taken.

19. SUMMARY OF METHODS

It is clear from the foregoing that no one method will give satisfactory results over the range of frequencies experimentally available. The following should serve as an approximate criterion for measurements.

(a) Gases

In general, it appears that systematic errors in the various methods employed today are small in comparison with the error caused by the presence of minute amounts of impurities. These will not only change the magnitude of the total absorption at a given frequency, but may also have a profound effect on the relaxation frequencies. Thus the presence of only 0.01 percent of H_2O in CO_2 *doubles* the relaxation frequency (K14) (see Sec. 22). In comparing experimental results by different observers, greatest credibility can be given in general to the one obtaining the lowest relaxation frequency. At low frequencies reverberation methods appear to be most satisfactory. From 20-200 kc, the direct method has been quite successful, while for measurements in the highest frequency ranges, interferometry is employed. It should be pointed out that in gases, increase in frequency or decrease in pressure are essentially equivalent. Using an interferometer at low pressures, Zartman (Z1) has obtained consistent results up to 85 mc/atmosphere. At atmospheric pressure, the highest frequency at which measurements have been made are those of Stewart (S14) in hydrogen (up to 6 mc). Both of these sets of measurements were made by interferometric methods.

(b) Liquids

Because the absorption in liquids varies more widely, from one substance to another, it is more convenient to express ranges of usefulness in terms of measured values of α, citing the frequency range in water at room temperature which corresponds to these values. This is done in Table III-I for the principal methods of measurement. It is to be noted that the absorption has not necessarily been measured in water over these frequencies. The values are listed merely for comparison purposes. In addition the range of frequencies measured in open water has been omitted, since this constitutes a rather special case which obviously cannot be repeated for any other liquid. It should be pointed out that *the standard criterion for the accuracy of a given method (in a liquid) is the obtaining of a value of α/ν^2 for water (above 1 megacycle) which agrees with the universally accepted value.*

Chapter IV. Experimental Results in Gases

20. INTRODUCTION

The absorption coefficient attributed to shear viscosity and heat conduction (the so-called classical absorption coefficient) as calculated by Stokes (S17) and Kirchhoff (K3) may be written

$$\frac{\alpha}{\nu^2} = \frac{2\pi^2}{\gamma p_0 c}\left[\frac{4}{3}\eta + \frac{\gamma-1}{c_p}\kappa\right], \quad (20.1)$$

where p_0 = mean pressure, γ = ratio of specific heats, η = coefficient of shear viscosity, κ = thermal conductivity, c_p = specific heat at constant pressure, and ν = frequency. Equation (20.1) is a combination of

TABLE IV-I. Ratio of α heat conduction/α shear viscosity, computed at 20°C.

Gas	$\alpha/\nu^2 \times 10^{13}$ cm^{-1} sec^2		α_{hc}/α_{vis}
	heat conduction	viscosity	
Argon	0.77	1.08	0.71
Helium	0.216	0.309	0.70
Neon	0.75	0.07	0.70
Hydrogen	0.052	0.117	0.44
Oxygen	0.47	1.14	0.41
Nitrogen	0.39	0.96	0.41
Air	0.38	0.99	0.39
Sulfur dioxide	0.27	1.10	0.28
Ammonia	0.110	0.453	0.25
Carbon dioxide	0.31	1.09	0.24

Eq. (3.16) and (4.8a) of Chapter I. Use has been made of the condition that $\zeta = 4/3\eta$ and that $K = p$ (see Sec. 11a).

In general, the viscosity term is somewhat larger than that due to heat conduction. The magnitudes of these two effects have been computed at atmospheric pressure for several gases from standard physical data (I1, L4), and the results are shown in Table IV-I.

In addition to viscosity and heat conduction, a number of other causes of absorption have been discussed in the past, and may be added to the "classical" group. Chief among these are the thermal radiation losses (see Sec. 5, above), and the losses due to diffusion in a gas mixture (C1, R8). The radiative absorption coefficient, as calculated by Rayleigh (R2, v. II, p. 24) and Stokes (S18), is frequency independent. In air the absorption due to the interdiffusion of the nitrogen and oxygen molecules must also be considered. The sizes of these four effects in air are listed in Table IV-II.

Thus, at a frequency of 6 kc, the absorption resulting from radiation is only about 0.5 percent of that which results from shear viscosity. The absorption caused by radiation is even less significant at higher frequencies and may therefore be neglected.

In making absorption measurements in gases, it is generally more convenient to employ a constant frequency of sound and vary the gas pressure, rather than vice versa. It is therefore more meaningful to give the experimental values of $\alpha p/\nu^2$ instead of α/ν^2. In this case, it is desirable to use ν/p as the abscissa of the graph of the absorption, rather than ν. (The reason for this choice will become clearer later when it is shown that for a relaxational effect, an increase in pressure has the same effect as a decrease in the frequency.)

In the case of a monatomic gas, we should expect $\alpha \cdot p/\nu^2$ to be independent of either p or ν, since no relaxational effects are present.

21. MONATOMIC GASES

A considerable number of measurements have been made in argon and helium over a wide range of frequencies and pressures. Here as elsewhere in gases, the presence of impurities, and a lack of accurate knowledge of the radiation field led to many errors in the results of the earlier investigators. Among the values which are quoted, the results of Van Itterbeek and Mariens (V5) for helium are probably in error due to the departure of the character of the sound beam from a plane wave.

(a) Helium

The results of the various experiments in helium have been put in the form of Eq. (20.1) and are plotted in Fig. IV-1.[37] These experimental values are recorded at temperatures in the ranges 15–25°C. No attempt has been made to adjust for this variation. The velocity of sound in helium was taken to be 1.01×10^5 cm/sec. The classical value of $\alpha p/\nu^2$ is computed from standard critical (I1) (L4) data at or near 15°C and equals 0.545×10^{-13} cm^{-1} sec^2 atmosphere.

It can be seen from the graph that the classical absorption value for helium is substantially verified in the

TABLE IV-II. Values of α/ν^2 for various types of classical absorption in air.

(20°C, atmospheric pressure)	$\alpha/\nu^2 \times 10^{15}$ cm^{-1} sec^2
Shear viscosity	99
Thermal conduction	38
Diffusion	7.4
Radiation	0.42 [at 6 kc (R9, p. 57)]

---

[37] Copies of the numerical data on which this and other graphs in this paper are based, can be obtained from the authors.

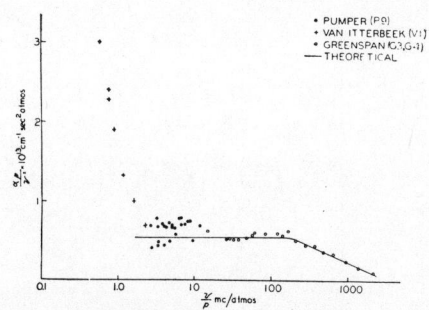

FIG. IV-1. $\alpha p/\nu^2$ vs ν/p for helium.

range above 2 mc/atmos. The rise in the values of $\alpha p/\nu^2$ at the lower end of the frequency scale is in all probability due to experimental errors.[38]

The falling off of the curve at high values of ν/p can also be explained on the basis of classical behavior (see Sec. 11). Greenspan has computed the effect of viscosity and heat conduction when higher order terms are considered. The solid line at the high values of ν/p represents the numerical solution of the hydrodynamic equation. The agreement with experiment is thus shown to be excellent. It should be emphasized that this is obtained from the classical equations, without altering the expression for viscosity. It therefore resembles the behavior found experimentally in highly viscous liquids (see Sec. 28).

(b) Argon

The $\alpha p/\nu^2$ data for argon are plotted in Fig. IV-2. The computed value of the classical $\alpha p/\nu^2 = 1.88 \times 10^{-13}$ cm^{-1} sec^2 atmosphere at 20°C. While there is a considerable spread of values in the figure, they fluctuate about the classical value. The spread is occasioned, at least partly, by the spread in temperatures at which the measurements were recorded. The behavior of $\alpha p/\nu^2$ at large ν/p has been computed as in the helium case, and is plotted on the graph.

On the basis of these measurements one can conclude that the measured absorption in monatomic gases is entirely accounted for by the classical theory.[39]

22. DIATOMIC GASES

The absorption per wavelength for a thermal relaxation may be written approximately (see Secs. 6 and 7):

$$\mu = \alpha\lambda = A\omega_0\omega/(\omega_0^2+\omega^2), \quad (22.1)$$

where

$$A = \pi R C^i/C_v(R+C_v), \quad (22.2)$$

[38] Discussions of these errors may be found in Halpern (H2) and Pumper (P9).

[39] A recent paper by E. Skudrzyk (S7) has advanced the hypothesis that Stokes's assumption (that the dilatational viscosity is equal to zero) fails even for ideal monatomic gases. On this basis he adds 50 percent to the viscosity term in the expression for the absorption coefficient (changing $4/3\eta$ to 2η). The experimental evidence collected here appears to be contrary to such a hypothesis.

C^i = internal molar heat capacity, C_v = molar heat capacity at constant volume, R = molar gas constant, and ω_0 = angular relaxation frequency. We recall the more rigorous expression (6.16) which can be put in the form,

$$\alpha\lambda = \frac{1}{2}\frac{c^2}{c_0^2}\frac{2\pi\epsilon\omega\tau}{1+\omega^2\tau'^2}, \quad (22.3)$$

where

$$\frac{c^2}{c_0^2} = \frac{1+\omega^2(\tau')^2}{1+\omega^2(\tau')^2(1-\epsilon)}, \quad (22.4)$$

$$\frac{1}{\omega_0} = \tau' = \tau\frac{C_p^\infty}{C_p^0} \quad (22.5)$$

and

$$\epsilon = \frac{R}{C_v^0(C_v^\infty+R)}C^i. \quad (22.6)$$

C_v^0 is the low frequency value of C_v and C_v^∞ is the high frequency value. With the exception of hydrogen, the value $C^i = C_v^0 - C_v^\infty$ is quite small, so that $\tau' \approx \tau$, $\epsilon \ll 1$, $C_v^\infty \approx C_v$, and

$$\alpha\lambda = \pi\epsilon\omega\tau/(1+\omega^2\tau^2). \quad (22.7)$$

It has been observed by van Itterbeek and Mariens (V2) and also by Keller (K1) that the relaxation time τ of Sec. 6 is inversely proportional to the pressure. This result is a consequence of the fact that the reaction rate $\omega_0 \approx 1/\tau$ is directly proportional to the number of molecular collisions which occur per second, and this latter quantity is in turn directly proportional to the pressure (see Sec. 8c). We may therefore rewrite Eq. (22.1) in the form

$$\mu = \frac{A'(\nu/p)}{1+B(\nu/p)^2}. \quad (22.8)$$

Hence, both classical (Eq. (20.1)) and relaxational (Eq. (22.8)) effects can be plotted in a graph of μ vs ν/p. The expression for the velocity may be written in a similar form. We now consider specific cases.

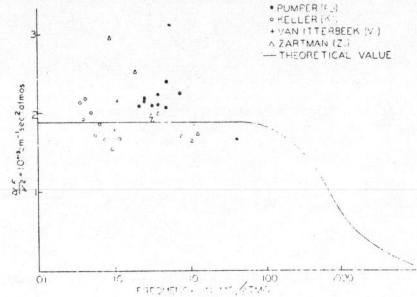

FIG. IV-2. $\alpha p/\nu^2$ vs ν/p for argon.

(a) Hydrogen

The internal specific heat of hydrogen is due to the rotational degrees of freedom. The characteristic temperature for vibration of the hydrogen molecule is 6140°A, which is so high that the vibrational specific heat is entirely negligible at room temperature.[40] The characteristic temperature for rotation on the other hand is 171°A, so that many energy states are excited at room temperature. Epstein gives the molar heat capacity due to rotation as $0.97R$ at 293°A ($R=$ molar gas constant). Therefore, using the exact relations Eq. (22.3) and Eq. (22.4), one should expect

$$c_\infty^2/c_0^2 = 1.19,$$
$$\mu_{max} = (\alpha\lambda)_{max} = 0.273. \quad (22.9)$$

These quantities should be virtually independent of temperature in the range above 0°C, although the relaxation frequency itself may be expected to depend upon the temperature. The principal results are shown in Fig. IV-3 and Fig. IV-4. In Fig. IV-3 the smooth curves have been drawn to give a best fit to the experimental data. The values plotted in Fig. IV-4 are for the excess absorption per unit wavelength, i.e.,

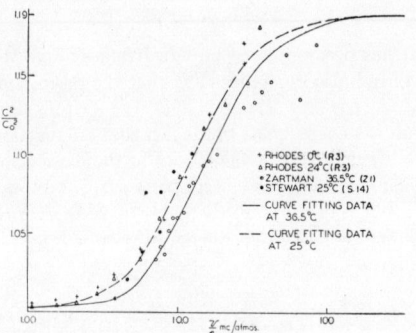

FIG. IV-3. Sound velocity in hydrogen.

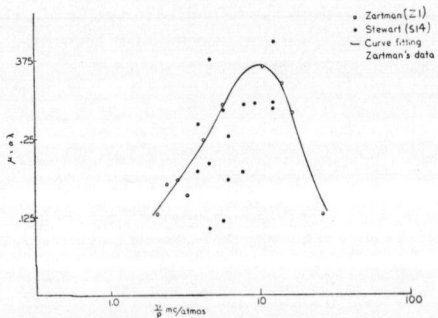

FIG. IV-4. $\alpha\lambda$ vs ν/p for hydrogen.

[40] The texts of Slater (S8) and Epstein (E5) should be consulted for a more detailed discussion of this topic. The reader should also recall Sec. 7 of this review.

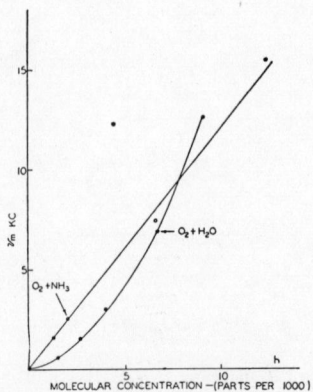

FIG. IV-5. Effect of impurities on sound absorption oxygen. (L. and H. Knötzel (K12).)

above the classical value. The results of Zartmann indicate conclusively that the velocity dispersion takes place over the frequency range predicted theoretically.[41] The temperature dependence is reflected only in the value of the relaxation frequency which is approximately 10.0 mc at 25°C Stewart (S14) and Rhodes (R3) and 13.6 mc at 36.5°C Zartmann (Z1).

The values for the absorption coefficients shown in Fig. IV-4 are somewhat high (compared to the theoretical value) and have a greater spread. Nevertheless, the results, especially those of Zartmann, indicate a relaxation frequency of the order of 10–12 megacycles. Stewart attributes some of her high values to misalignment of the crystal in the interferometer. The data points are too widely scattered to form a definite conclusion on the maximum experimental value of μ.

In summation, the results in hydrogen give satisfactory support to the theory of a relaxation of the rotational degree of freedom.

(b) Nitrogen, Oxygen

The study of relaxation effects in diatomic gases is complicated by the very small values of the internal molar heat capacity C^i and the consequently small values of μ_{max}. In addition, the relaxation frequency is generally a very low one, so low in fact that it might be well below the frequency range in which sound absorption can be accurately studied. A useful technique however has been developed by Kneser and Knudsen (K9) to bring the maximum values of μ within the range of measurement. Their experiments have shown that the presence of an impurity in the gas increases the relaxational frequency ν_m *without affecting the value of* μ_{max}. The absorption is therefore measured at a number of different concentrations of the impurity and the curve

[41] None of the observers measured the velocity at low (dispersion free) frequencies. The curve is drawn with an assumed $V_0 = 1316$ m/sec at 25°C and 1341 m/sec at 36.5°C. An error here could shift the values of the relaxational frequencies somewhat.

TABLE IV-III.

Gas	T°C	μ_{max}(thermal) $\times 10^4$ Exper. Theoret.	ν_m(exper.)	Observer
Nitrogen	20°	4 2.5		Schmidtmüller (S2a)
Oxygen	19°	52	50±10 cps	Knötzel (K12)
Oxygen	20°	54.5 51		Kneser and Knudsen (K9)
Oxygen	22°	54	170 cps	Oberst (O1)
Oxygen	55°	98 102		Kneser and Knudsen (K9)

for ν_m is extrapolated to zero concentration of the impurity. This method is somewhat inaccurate, since the relaxation frequency changes rapidly with concentration, but it does serve to give an upper bound to that frequency. An illustration of this behavior is shown in Fig. IV-5 which is taken from the results of H. and L. Knötzel (K12). Here the relaxation frequency ν_m for oxygen is plotted as a function of the ratio of molecules of impurity to the total number of molecules in parts per mil (h). Both ammonia and water vapor have been used as impurities. While the value of ν_m has been measured at $h=0$, it is somewhat more accurate to obtain a best-fit curve through the various points and solve for $h=0$. This gives $\nu_m = 50 \pm 10$ cps in good agreement with the actual experimental values at $h=0$.

Since μ_{max} is not affected by the impurity, this method can be used to find the value of μ_{max}. A sufficient impurity is introduced to get ν_m in the most convenient frequency range, and μ_{max} can then be measured directly.

Table IV-III summarizes the results. The values of μ_{max} are computed from Eq. (22.3) and Eq. (22.6), using the values of C^i computed from standard data.

The results in Table IV-III indicate substantial agreement between theory and experiment. In addition, the exceedingly low frequencies at which these μ_{max} occur indicate that these gases may safely be used in the calibration of absorption measuring instruments at higher frequencies (where α/ν^2 due to this process may be entirely neglected. See Sec. 18 above).

Some of the principal results obtained in diatomic gases at higher frequencies are shown in Table IV-IV.

The experimental values tabulated here are undoubtedly not more accurate than within ±10 percent, and are perhaps worse in some cases. One may say therefore that a reasonable agreement has been obtained between theory and experiment, although more extensive and more accurate data would be useful here.[42]

23. TRIATOMIC GASES

The relaxation effects in triatomic gases are more pronounced and the relaxational frequencies are considerably higher than for diatomic gases. Among these gases, carbon dixode has been thoroughly studied by a number of investigators. The chief results are plotted in Fig. IV-6.

If we take the results of Fricke (F6) as the more accurate,[43] a relaxation frequency of 20 kc is indicated. The approximate theoretical curves (based on this relaxation frequency) have been drawn in the figure. The combined relaxation-classical (viscosity-heat con-

TABLE IV-IV.

Observer	Range of ν/p in mc/atmosphere	T°C	$\alpha p/\nu^2 \times 10^{13}$ cm^{-1} sec^2 atmosphere Experimental	Classical
	Nitrogen			
Zartmann (Z1)	1–38.5	25.6	1.91	1.35
Keller (K1)	0.35–1.60	18	1.71	1.35
Schmidtmüller (S2a)	0.08–0.115	20	1.85	1.35
Van Itterbeek and Thys (V3)	0.61–1.97	20	1.35	1.35
	Oxygen			
Van Itterbeek and Thys (V4)	0.60–1.69	20	1.57	1.61
Van Itterbeek and Thys (V4)	0.60–1.55	50	2.22	1.70
	Nitric oxide (NO)			
Van Itterbeek and Thys (V4)	0.60–0.97	16.3	1.58	1.48

duction) curve is seen to be in excellent agreement with the results of Fricke, van Itterbeek, and Zartmann over the entire frequency range (8 kc/atmos to 85 mc/atmos).

The chief results in the triatomic gases are summarized in Table IV-V. The data are mainly those of Fricke. The values of μ_{max} were computed by him from thermal data.

Chapter V. Experimental Results in Liquids

24. MONATOMIC AND DIATOMIC LIQUIDS

It is to be expected that the values of the sound absorption coefficient in a monatomic liquid such as mercury should show good experimental agreement with the classical values since the usual types of internal degrees of freedom are lacking. The same should be true of liquefied monatomic or diatomic gases. In the latter case, the internal degrees of freedom of such molecules as O_2, N_2, and H_2 are "frozen" in their liquid state and therefore should not contribute to any relaxational process.

The experimental results for mercury are shown in

---

[42] Several experimental investigations (P1a, T1b, Z3) have recently been reported on the relaxation of rotational degrees of freedom in both oxygen and nitrogen.

[43] Professor Leonard has informed the authors that the CO_2 used by Fricke was probably of somewhat higher purity than that used in his research. The graph suggests that the CO_2 used by Keller was of approximately the same purity as that used by Leonard.

TABLE IV-V.

Gas	T°C	Theoretical μ_{max}	ν_m(exp.) in kc
CO_2	23	0.115	20
COS	23	0.175	28
CS_2	23	0.203	370
N_2O	23	0.148	157
SO_2	23	0.0745	1040

Fig. V-1. The values of Ringo et al. (R7) are the most recent and indicate that the observed absorption lies within about 15 percent of the classical value. This is probably within the margin of experimental error. The case of mercury differs from most other liquids in that heat conduction is the primary source of ultrasonic absorption. It is also to be noticed that no frequency dependence of α/ν^2 has been observed in the range 20–1000 mc.

Pellam and Squire (P4) have measured the absorption coefficient in liquid helium, at 15 mc, as a function of temperature. Their results are reproduced in Fig. V-2, along with a curve showing the classical value of the absorption coefficient. From these data, one may conclude that ultrasonic absorption in liquid helium is a classical phenomenon above the λ-point (2.19°A). In the neighborhood of this point, and below it, there is no agreement between classical and experimental values. The fact is not surprising, in view of the complicated hydrodynamics associated with HeII.

Measurements have also been made by Galt (G1) on liquefied monatomic and diatomic gases. His results are shown in Table V-I. All measurements were made at 44.4 mc. The velocity measurements made by Galt have also been included.

The calculations of the theoretical absorption coefficients (B19) were based on viscosity and density data by van Itterbeek and van Paemel (V6) and Rudenko and L. W. Schubnikov (R11). Experimental values for the thermal conductivity for liquid oxygen and nitrogen are taken from Hamman (H3). Corresponding values for argon or hydrogen were not available, and an empirical formula of Borovik (B25) (which gives at least the order of magnitude) was used in these two cases.

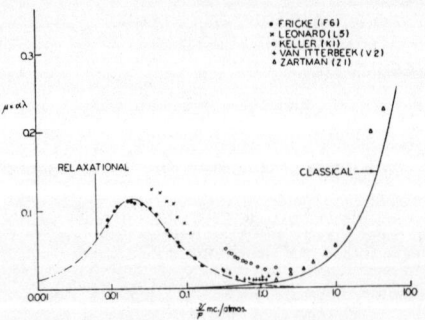

FIG. IV-6. $\alpha\lambda$ vs ν/p for carbon dioxide.

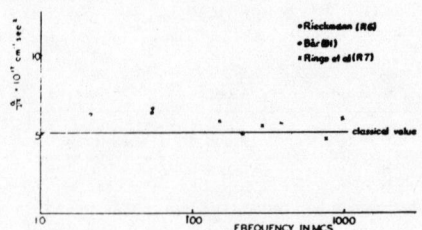

FIG. V-1. α/ν^2 vs frequency for mercury.

In their paper on the viscosity of liquids, Kirkwood, Buff, and Green (K6) made use of Galt's results to estimate an upper bound of $\frac{1}{3}$ for the ratio of bulk to shear viscosity in argon. When thermal conductivity is included, however, it is clear that such a ratio must be far smaller. The bulk viscosity of argon is essentially negligible, so far as ultrasonic absorption is concerned. The general agreement between the experimental and theoretical values in Table V-I warrants the conclusion that the absorption processes in monatomic and diatomic liquids (with the exception of helium below the λ-point) are well accounted for by classical theory.

25. WATER

Of all liquids water has been the one whose sound absorption properties have been studied most frequently (B5, B10, B12, B21, C4, F2, F3, H14, L7, L10, M10, P6, R1, S7, S9, V7). The principal results are shown in Fig. V-3.[44] The temperatures are all in the vicinity of 20°C, but a variation of several degrees for operating temperatures among the different observers increases somewhat the spread of values.

The results indicate a constant value of α/ν^2 at frequencies above 1.0 mc, a value which is about $3\frac{1}{2}$ times the classical value of Stokes. The spread of values below 1.0 mc (which is understandable because of the extremely small value of the absorption coefficient α in

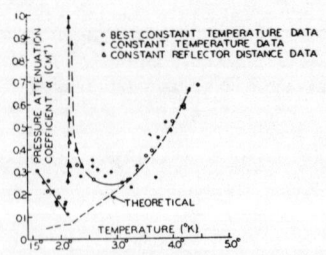

FIG. V-2. α vs absolute temperature for helium (after Pellam and Squire (P4)).

[44] It is to be noted that the values of Sörensen (S11), Hartmann and Focke (H5), and others (B38, O2, R5, R12) are not plotted. These were early values, and have since been shown to have been unreliable (see Sec. 16). In this case, as in all others in this paper, the authors are attempting to plot only those values which are at present believed reliable.

TABLE V-I. Absorption coefficients of liquefied gases.

Liquid	$T(°A)$	Velocity cm/sec² ×10⁻⁴	$\alpha/\nu^2 \times 10^{17}$ cm⁻¹ sec²			
			Shear viscosity	Thermal conductivity	Total theoretical	Experiment
Argon	85.2±0.2	8.53	7.9	2.6	10.5	10.1
Oxygen	87.0±0.2	9.52	5.5	1.8	7.3	8.6
	70 ±1	10.94	5.6	1.1	6.7	8.6
	60 ±5	11.19	7.3	1.0	8.3	8.6
Nitrogen	73.9±0.2	9.62	6.6	2.9	9.5	10.6
Hydrogen	17 ±1	11.87	3.7ᵃ	2.1	5.8	5.6

ᵃ This is considerably larger than the value computed by Galt. The discrepancy lies in his use of a density of liquid hydrogen, taken from Bergmann (B15), of 0.355 g/cm³. The value of 0.075 g/cm³ used here is that given by van Itterbeek and van Paemel (V6) and also appears in the *Handbook for Chemistry and Physics*, thirtieth edition, p. 1703.

this range) makes it difficult to decide whether or not there exists a low frequency relaxation process, but the evidence seems against it, at least at frequencies above 100 kc.[45] The absorption in water was at first thought to be due to a thermal relaxation (Dutta and Ghosh (D4), Herzfeld (H12)) but inconsistencies made this impossible. Thus, as Herzfeld pointed out, the constancy of α/ν^2 up to 200 mc requires, from the relation

$$\frac{\alpha_{\text{excess}}}{\nu^2} = \frac{A\tau}{1+\omega^2\tau^2} \approx 15 \times 10^{-17} \text{ cm}^{-1} \text{ sec}^2,$$

that $(2\pi\tau) \ll 1/(2\times 10^8)$ or $\tau \ll 8\times 10^{-10}$ second. One may calculate A from (6.16) from thermal data giving $A = 2\times 10^{-9}$ cm⁻¹ sec. This figure was obtained from the value of C^i given by Dutta (D3a). The other parameters were taken from Herzfeld's paper (H12). The product $A\tau = 15\times 10^{-17}$ cm⁻¹ sec² at low frequencies yields $\tau \sim 7\times 10^{-8}$ sec which contradicts the above.

In addition to this difficulty, there is a second objection. The thermal term depends on the difference $C_p - C_v$ which vanishes in water at 4°C (since $C_p - C_v$ varies directly as the thermal expansion coefficient). However, the measurement of absorption in the vicinity of 4°C

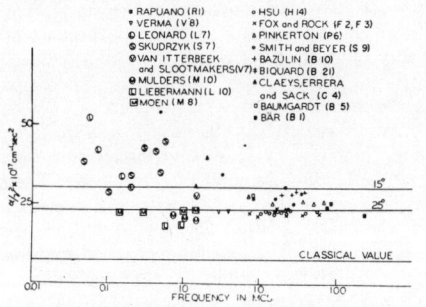

FIG. V-3. α/ν^2 vs frequency for water.

[45] Kneser (K11) has recently (1949) hypothesized the existence of a thermal relaxation in water with a relaxation frequency of about 3 megacycles. This hypothesis was made on the basis of the data obtained by Skudrzyk (S7). The more recent experimental work of Leonard (L7), Liebermann (L10), Mulders (M10), Moen (M8), and Verma (V8), however, indicate that this is not the case.

shows no dip at all in the excess absorption when it is plotted as a function of temperature (see Fig. V-4).

As was observed in Sec. 7, Hall (H1) has attributed the effect to a structural rearrangement in the water molecules: the impact of a sound beam on water first compresses the groups of molecules (which are assumed to possess some degree of order in an ice-like state). Then a structural compression sets in, in which the molecules break their structural bands and move into a close packed arrangement. Since there is a time lag, absorption will appear.

Hall has calculated a theoretical absorption coefficient on the basis of an instantaneous compressibility μ_T^∞. In his theory, this quantity, as well as the relaxation time for the process and the so-called *bulk viscosity* coefficient are evaluated in terms of independently known parameters. The results are shown in Fig. V-4 where Hall's theoretical curve for structural absorption is plotted in the temperature range 0–80°C. The excess absorption found by various investigators, is also plotted.

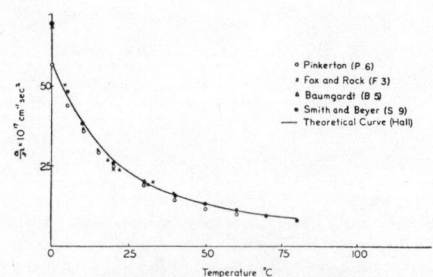

FIG. V-4. Temperature dependence of sound absorption coefficient in water.

While the agreement between Hall's theory and the experimental results is satisfactory, it must be pointed out that the constants from which the final theoretical curve is drawn are known only very approximately. In addition, the ratio of excess to classical absorption is (experimentally) very nearly independent of the temperature, so that any theory which deduces an excess absorption approximately proportional to the *shear viscosity* will take on the same temperature dependence as that shown. Before the theory can be more generally accepted, it should be applied to other associated liquids.

It should also be pointed out that Sette (S4) has recently reported a calculation for ethyl alcohol, based on the Hall theory. He found that the relative change in volume between the two structural arrangements was much smaller than in water, so that the resulting contribution to the absorption coefficient was inadequate to account for the excess above the classical value. Apparently the simple model of a two-state liquid is inadequate in such a case.

As shown in Chapter II a more macroscopic approach to this problem has been made by Liebermann (L11) on the basis of hypotheses set forth by Tisza (T3) and Eckart (E3). Tisza and others (E3, L17, M1) have pointed out that the original Stokes equation for sound propagation contained a term involving a bulk or dilatational viscosity, which Stokes set equal to zero. If one does not make this assumption, then the second viscosity coefficient can be adjusted so as to fit the experimental results.

What was needed was an independent determination of this second viscosity coefficient. An experimental method was first suggested by Eckart (E3) who developed the second-order equation for wave propagation in this case. This method involves the phenomenon of streaming of the fluid (quartz wind or hydrodynamic flow) in which the driving force is dependent on both viscosity coefficients (shear and bulk) but in which the retarding force is dependent only on shear viscosity. Hence the equilibrium velocity of streaming should depend on the ratio of the coefficients. Liebermann has measured the viscosity ratio by this method for a number of liquids at 5 megacycles. In the case of

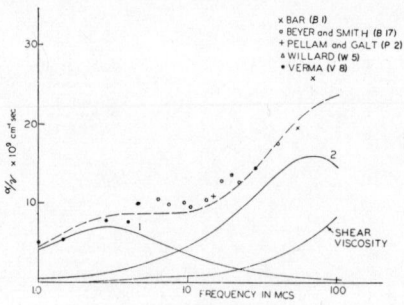

FIG. V-5. α/ν vs frequency for ethyl acetate.

water, the ratio of absorption due to bulk viscosity to that due to shear viscosity was found by him to be 3.3 while the ratio of total absorption to classical absorption at the same temperature is 2.95.

As will be seen later (Table V-VII), rather satisfactory agreement exists between Liebermann's results and data obtained by direct experimental measurement of the absorption. It must be borne in mind however that this identification of excess absorption with a bulk viscosity does not indicate of itself any mechanism by which the second viscosity coefficient can be calculated. Thus Hall's treatment might be reconciled with Liebermann's, in which case the bulk viscosity would be attributed to a structural compression. Unfortunately, Hall and Liebermann have employed somewhat different equations of state, so that a direct equivalence is not immediately possible (see Sec. 15).

26. LIQUIDS WITH PRONOUNCED RELAXATIONAL EFFECTS

The stimulation for the development of a successful theory for sound absorption in gases was provided by

TABLE V-II. Relaxation frequencies in ethyl acetate.

Curve	ν_m (relaxation frequency) in megacycles	$(\alpha'/\nu)_{max}$ (cm^{-1} sec)
I	2.9	7.0×10^{-9}
II	70	16×10^{-9}

the existence of relaxational processes whose relaxation frequencies generally lay within the range of experimental measurement. This is not the usual case for liquids, but some exceptions have been found to exist. It is the study of these exceptions which gives the greatest promise of progress toward understanding absorption mechanisms in liquids. The most important of these liquids are organic compounds containing the acetate (CH_3COO) and the formate ($HCOO$) radicals.

(a) Acetates

The experimental values of α/ν for ethyl acetate ($CH_3COOC_2H_5$) are plotted in Fig. V-5. Most of the values appear to lie in the transition region between two relaxation frequencies. This was first suggested by Kneser (K11). In line with his hypothesis, the curves I and II (along with the curve for shear viscosity absorption) have been fitted to the data. The sum of these three curves is represented by the dashed line. The constants obtained by the curve fitting are given in Table V-II.

While some data exist (B1, B22, C4, P2) for methyl acetate (CH_3COOCH_3) showing at least one relaxational effect, the results are too fragmentary to allow any conclusions to be drawn from them.

The data which are now available for acetic acid (CH_3COOH) (Lamb and Pinkerton (L2)), are quite extensive. It should be pointed out that ultrasonic absorption in this liquid was first studied by Bazulin (B7, B12) in 1936. He was the first to observe a relaxation process in a liquid. His data are largely omitted from what follows only because of the completeness of the work of Lamb and Pinkerton and because of the difficulty of making corrections for small temperature differences. Wherever the two sets can be compared, the experimental agreement is excellent.

Figure V-6 reproduces graphs of $\mu' = (\alpha\lambda)_{excess}$ vs frequency, computed by Lamb and Pinkerton from their data at 20°C, 35°C, and 50°C, respectively. These results demonstrate clearly the existence of a relaxational effect, the characteristic frequency of which increases with temperature. The maximum of the absorption per unit wavelength increases in a similar manner.

The relation between the relaxation time and the temperature is indicated by Lamb and Pinkerton to be of the form

$$\nu_m = AT^n \exp(-\Delta E_a/RT), \quad (26.1)$$

where ν_m is the relaxation frequency, A is a constant, $n =$ a number in the range 0 to 1, and ΔE_a is the apparent

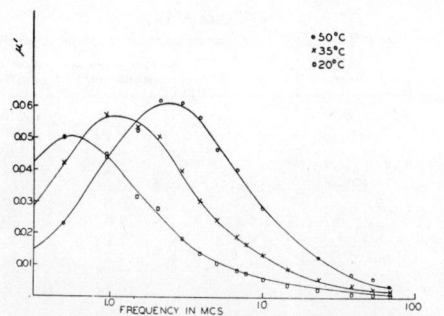

Fig. V-6. Excess absorption per unit wavelength vs frequency for acetic acid (after Lamb and Pinkerton (L2)).

energy of the backward reaction involved in the propagation process.

While A is not known, the form of Eq. (26.1) can be checked against the experimental results. Unfortunately, the experimental values of Lamb and Pinkerton fit the expression equally well for $n=0$ or $n=1$. The value of ΔE_a which is computed is nearly the same in each of these cases, being 8.86 kcal/mole for $n=0$ and 8.46 kcal/mole for $n=1$. It would be very desirable to compute ν_m directly from spectroscopic data, but this is not possible at the present time.

The plot of α/ν^2 vs frequency for acetic acid (see Fig. V-7) indicates that the high frequency value $(132\times10^{-17}$ sec^2 cm^{-1}) is still far above the classical viscosity value of 20×10^{-17} sec^2 cm^{-1}. Thus it seems probable that there is yet another relaxational process, not yet discernible at the highest frequency measured. Lamb and Pinkerton have advanced the hypothesis that there are two relaxation phenomena, each one connected with one of the hydrogen bonds that ordinarily hold the acetic acid in a double molecule.

The temperature dependence of the absorption coefficient can be used to gain considerable information about relaxational processes, as Kittel (K7) has pointed out. The absorption coefficient due to shear viscosity has a negative temperature coefficient. We now consider relaxation processes of the form $M' = \alpha'/\nu^2 = A\tau/(1+\omega^2\tau^2)$.

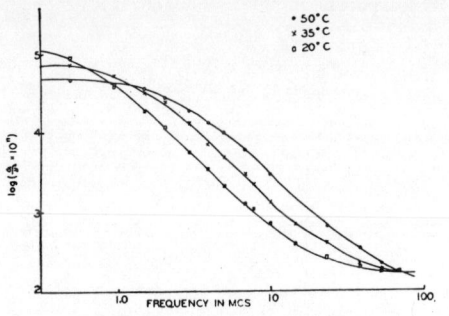

Fig. V-7. $\alpha/\nu 2$ vs frequency for acetic acid (after Lamb and Pinkerton (L2)).

It follows from Eq. (9.24) that τ has a negative temperature coefficient. Then the sign of the temperature coefficient of $M'[=(1/M')\partial M'/\partial T]$ depends on two factors: (i) whether ω is greater or less than $1/\tau$ (i.e., greater or less than $2\pi\nu_m$), and (ii) whether the product $A\tau$ has a positive or negative temperature coefficient. Thus, if $A\tau$ has a positive temperature coefficient, i.e., if A has a large positive coefficient, then $(1/M')\partial M'/\partial T$ will always be positive. On the other hand, if A has a smaller temperature coefficient than τ (be it positive or negative), then $(1/M')\partial M'/\partial T'$ will be negative at frequencies below ν_m and positive at frequencies above ν_m. (See also Sec. 9.)

The process of thermal relaxation may be shown to be one in which $(1/A)\partial A/\partial T$ is positive. The temperature behavior of structural relaxation is not so clear, but the work of Hall indicates a small negative value of $(1/A)\partial A/\partial T$ for water.

In general it appears that the relaxational frequency for a structural or bulk viscosity effect is so high that the resulting temperature coefficient of absorption will always be negative over the frequency range available to investigation. Where this is not the case, one can perhaps distinguish between bulk viscosity and thermal relaxation by the temperature dependence of A.

An example of this temperature dependence is given in Fig. V-7 for acetic acid. At frequencies below the relaxation range ($\sim$1 mc) the absorption coefficient decreases with increasing temperature. At frequencies above this range, the absorption coefficient increases with increasing temperature, at least up to 67.5 mc. Above this frequency, extrapolation seems to indicate that the absorption once more decreases with increasing temperature. However, this is the region in which a second relaxation effect appears to enter the picture. In addition, the relaxation frequency (see Fig. V-6) increases with temperature, while the peak of the $(\alpha\lambda)_{\text{excess}}$ curve (which is proportional to A) also increases.

(b) Formates

Work on absorption in formate compounds has so far been limited to some absorption values for formic acid (HCOOH) obtained by Bazulin (B10) and a few absorption values for ethyl formate (HCOOC$_2$H$_5$) obtained by Parthasarathy (P1). The measurements of the latter investigator, where they can be compared with accurate results by other investigators, tend to be high and also to have large fluctuations, and can be regarded only as very rough or even qualitative information.

Bazulin's results in formic acid indicate that a relaxation process exists with a frequency of the order of 5 megacycles. The values have been recorded over a temperature range of 4°C and for 87 percent concentration. These two limitations, plus the small number of points available, prevent any more detailed analysis of ultrasonic absorption in the liquid.

The main interest in ethyl formate lies in the fact that its second (or bulk) viscosity coefficient has been measured by Liebermann at four different frequencies. His results are given in Table V-III, together with ratio of excess to classical absorption obtained by Parthasarathy's experimental data.

The results of Liebermann indicate a relaxation frequency for the *bulk viscosity* of the order of 2 megacycles, which is at least partially borne out by the direct absorption measurements. A great deal more experimental data, especially in the range 1–10 megacycles, is needed to clarify the picture.

(c) Toluene ($C_6H_5CH_3$)

Recent measurements of Moen (M8) indicate strongly the presence of a relaxational process in toluene at relatively low frequencies. A relatively large number of experimental results exist for this liquid, mostly at higher frequencies. The most reliable of these are listed in Table V-IV. From these it can be observed that a substantially constant value of $\alpha/\nu^2 (\approx 78 \times 10^{-17} \text{ cm}^{-1} \text{ sec}^2)$ is obtained above 1 megacycle. If this value is subtracted from the low frequency data, and the graph constructed of (α'/ν) vs frequency, (where α'/ν^2 is the excess α/ν^2 above the high frequency value) the results are those given in Fig. V-8. The solid curve in the figure has been fitted to the first four points, since the values at the higher frequencies are more in doubt, due to the uncertainty in the value of the subtracted term. The results indicate a relaxation frequency of about 120 kc and a value of $(\alpha'/\nu)_{max}$ of about 3×10^{-10} cm^{-1} sec. This compares with a calculated value of $(\alpha'/\nu)_{max}$ for thermal relaxation of 250×10^{-8} cm^{-1} sec (K11). This low frequency effect would therefore appear to be due to a different phenomenon. On the other hand, the value of α'/ν (above the classical at the highest frequency measured, 75 mc) is 5×10^{-8} cm^{-1} sec. Apparently the relaxation frequency for the thermal effect here is of the order of several hundred megacycles.

27. DISTINCTION IN BEHAVIOR BETWEEN ASSOCIATED AND NON-ASSOCIATED LIQUIDS

Both Hall (H1) and Pinkerton (P7) have pointed out that the absorption results in non-associated liquids exceed the classical value far more greatly than do the results in associated liquids. We have already discussed water in the associated group and toluene among the non-associated liquids. The other more important results follow.

(a) Benzene (C_6H_6)

Benzene has been studied by a very large number of observers over a frequency range from 150 kc to 165 mc. A graph of α/ν^2 vs frequency is shown in Fig. V-9. While these results are taken at various temperatures in the vicinity of 20°C, the spread of values is greater than that which could be attributed to the temperature effect. Moen (M8) points to the possibility that a small impurity could lower the absorption value appreciably. The best evidence of this is furnished by Bazulin (B8, B10) who performed the experiment with benzene of two grades of purity. For the purest sample he obtained a value of $\alpha/\nu^2 = 874 \times 10^{-17}$ cm^{-1} sec^2 while for the less pure he obtained a value of 714×10^{-17} cm^{-1} sec^2.

TABLE V-IV.

Observer	T °C	Frequency in megacycles	$\alpha/\nu^2 \times 10^{17}$ cm^{-1} sec^2
Moen (M8)	27	0.15	276
		0.20	210
		0.25	170
		0.4	111
		0.6	86
		0.8	80
		1.0	79
		1.2	78
Biquard (B22)	20	4.78	83
	19.5	7.96	83.5
Willard (W5)		10.34	90
Bazulin (B10)	19.2	10.69	83.1
	19.4	13.95	80.7
	19.6	17.23	84.2
	21.6	24.39	84.1
	21.3	37.37	79.5
Grobe (G7)	20	30	68
	20	43	74
	20	75	72.5
Verma (V8)	25	1.00	92
		1.46	91
		2.89	93
		4.00	94
Classical			7.8

TABLE V-III.

Freq. mc	$\dfrac{\alpha_{excess}}{\alpha_{classical}}$ (computed from viscosity measurements of Liebermann)	$\dfrac{\alpha_{excess}}{\alpha_{classical}}$ (computed from absorption measurements of Parthasarathy)
2	64	
3	24	17
4	17	
5	12	
7		1.8[a]
16		8.2

[a] In a number of liquids observed by Parthasarathy, the values of the absorption obtained at 7 megacycles are significantly lower than those at either 3 or 16 megacycles and are in disagreement with the results of other observers. This value is therefore open to serious question.

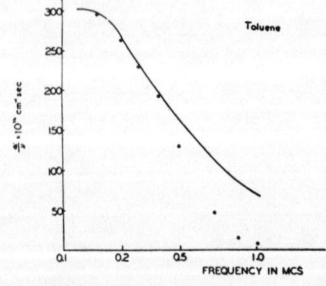

FIG. V-8. Excess absorption in toluene (data from Moen (M8)).

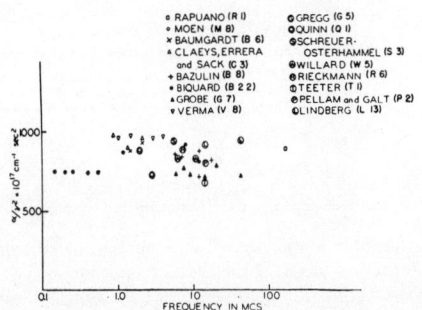

FIG. V-9. α/ν^2 vs frequency for benzene.

With this in mind, it is reasonable to assume that the best value of α/ν^2 is about 900×10^{-17} cm^{-1} sec^2.

The temperature coefficient of excess absorption in benzene is positive in the range 3-15 mc (B8, G5, P2). Since this frequency range is definitely below the relaxation frequency (α/ν^2 is still a constant) it appears from the discussion of the temperature coefficient that the excess absorption here is probably due to thermal relaxation.

(b) Carbon Bisulfide (CS_2)

The absorption values of CS_2 are very high. The most reliable results are shown in Fig. V-10. It is easily seen that the scattering of the points make it difficult to calculate any relaxation frequency. In a recent paper on the theory of sound absorption in unassociated liquids, Bauer (B4) suggested the presence of three vibrational relaxation processes. However, he fitted his calculations to the value of Claeys, Errera, and Sack at 0.87 mc, the value of Willard at 6.57 mc and the values of Rapuano near 100 mc. Since in general, Bazulin's results are more accurate in the range 1-10 mc than those of most other early observers, this selection of points does not appear too satisfactory. The more recent values of Moen (M8) and Verma (V8) make the picture even more uncertain. Because of the very high values of the absorption, it is possible that the results are easily affected by small impurities. It appears that a relaxation frequency exists in the range 8-70 mc, but a more definite appraisal must wait upon additional data.[46]

(c) Carbon Tetrachloride (CCl_4)

In CCl_4, there is no observable decrease in α/ν^2 as the frequency increases.[47] The mean value at room temperature is about 500×10^{-17} cm^{-1} sec^2. At the highest frequency measured, (105 mc (R1)) $\alpha/\nu = 525 \times 10^{-9}$ cm^{-1} sec, as compared with a maximum value of 5050×10^{-9} cm^{-1} sec, computed on the basis of a thermal relaxation (K11). Evidently the maximum occurs at a much higher frequency.

(d) Ethyl Alcohol (C_2H_5OH) and Methyl Alcohol (CH_3OH)

The best available data for ethyl and methyl alcohol are shown in Fig. V-11. No relaxation frequencies are distinguishable in either set of data.

28. LIQUIDS OF HIGH VISCOSITY

One might expect in a general way that highly viscous liquids would come closer to the classical behavior than other liquids. We first recall the Stokes equations for absorption due to viscosity (R2, vol. II, p. 316), which appeared in Sec. 3:

$$k^2 - \alpha^2 = \frac{3\rho_0}{4\eta} \frac{\omega^2 \omega_v}{\omega_v^2 + \omega^2}, \qquad (28.1)$$

$$2k\alpha = \frac{3\rho_0}{4\eta} \frac{\omega^3}{\omega_v^2 + \omega^2}, \qquad (28.2)$$

where $\rho_0 =$ mean density, $c_0 =$ low frequency value of sound velocity, $\eta =$ shear viscosity coefficient, and $\omega_v = 3\rho_0 c_0^2/4\eta$. From these, one can solve explicitly for α^2 and c^2:

$$\alpha^2 = \frac{3\rho_0}{8\eta} \frac{\omega^2 \omega_v}{\omega_v^2 + \omega^2} \left[\left(1 + \frac{\omega^2}{\omega_v^2}\right)^{\frac{1}{2}} - 1 \right], \qquad (28.3)$$

$$c^2 = \frac{8\eta}{3\rho_0} \frac{\omega_v^2 + \omega^2}{\omega_v} \frac{1}{\left(1 + \frac{\omega^2}{\omega_v^2}\right)^{\frac{1}{2}} + 1}. \qquad (28.4)$$

The most interesting measurements in a highly viscous medium are those of Mikhailov and Gurevich (M7) on rosin. They measured the absorption at four fre-

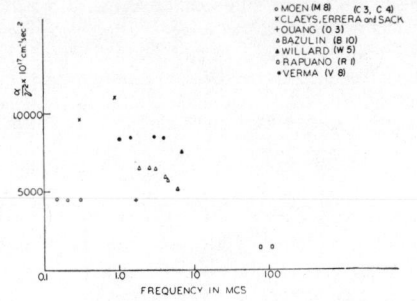

FIG. V-10. α/ν^2 vs frequency for carbon bisulfide.

---

[46] Lamb and Andreae (L2a) have recently found this frequency to be 72 mc at 25°C.

[47] The values of Claeys, Errera, and Sack for CCl_4 have been omitted from consideration here. Their values of α/ν^2 are only about 1/20 of those of other observers. While no experimental reasons can be given for discarding these values (in general their other results are only about 20-30 percent above the commonly accepted values), the great divergence from the results of investigators whose other results are reliable, and who agree in these cases seems to offer sufficient justification. It is quite probable that small amounts of impurities can produce a great effect in these liquids of high absorption.

quencies between 0.5 mc and 5 mc for various temperatures. The shear viscosity was also measured in the same range. Their results indicate that at a fixed frequency the absorption increases with increasing viscosity only for low values of the viscosity. At high values of the viscosity, a maximum absorption coefficient is reached, and thereafter the absorption coefficient *decreases* as the viscosity increases. This is precisely the behavior predicted by Eq. (28.3). A complete check of the theoretical (classical) and experimental values is not possible since the values of the sound velocity in rosin are not available. However, it is possible to check the behavior at the highest values of the viscosity when, according to Eq. (28.3),

$$\alpha = \left(\frac{3\rho_0\omega}{8\eta}\right)^{\frac{1}{2}}, \quad \omega \gg \omega_v, \qquad (28.5)$$

i.e., α is independent of the sound velocity at these frequencies. At 42°C the value of the viscosity of rosin

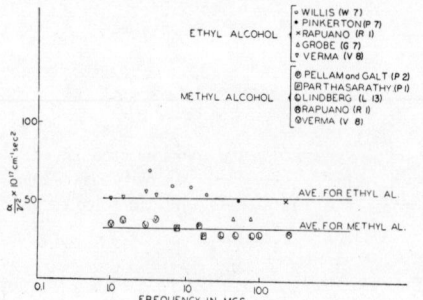

Fig. V-11. α/ν^2 vs frequency for ethyl alcohol and methyl alcohol.

is about 10^8 poise (M7). Assuming a density of 1.07 g/cm^3, we obtain the results shown in Table V-V.

This represents surprisingly good agreement when we recall that η is known only to an order of magnitude.

In the same paper, values of the absorption in methyl-metacrylate at two different frequencies are also reported. Here again, with $\eta > 10^{13}$ poise, the ratio of $\alpha/\nu^{\frac{1}{2}}$ is the same at the two frequencies.

Both of these results seem to point to a perfectly classical behavior for materials of very high viscosity. Unfortunately the only velocity data which are available (for methyl-metacrylate in the same frequency range) show that the velocity is essentially constant over the range 1 to 6.8 mc, whereas the limiting form of the velocity from Eq. (28.4) is

$$c = (8\eta\omega/3\rho_0)^{\frac{1}{2}}. \qquad (28.6)$$

A second objection to this application has been raised by Bazulin and Leontovich (B14), who point out that some approximations had already been made by Stokes in setting up his original equation of motion, and that so

TABLE V-V. Absorption in rosin.

Frequency mc	α_{theor} cm^{-1}	α_{exp}(M7) cm^{-1}
0.66	0.13	0.15
1.52	0.20	0.20
3.04	0.28	0.31
4.67	0.40	0.47

long as these are neglected, it is misleading to employ an "exact" solution of the inexact equation.

Nevertheless it would seem that the excellent agreement between Eq. (28.5) and the experimental results should be more than merely fortuitous. The agreement between classical theory and experiment in helium gas at very low pressures, discussed in Sec. 21 also indicates that the assumptions involved in the Stokes's equation of motion remain valid at what is effectively a very high frequency (i.e., high value of the frequency-pressure ratio). It may be that further corrections affect the velocity equation appreciably, but not the absorption relation.

The results in highly viscous glycerin are also of considerable interest. The frequency dependence of α/ν^2 is shown in Fig. V-12, while the temperature dependence of the ratio $\alpha_{observed}/\alpha_{classical}$ is shown in Fig. V-12.

The run of measurements at room temperature agree reasonably well with the classical values. As the absorption coefficient changes rapidly with temperature, the small variation in operating temperature among the various observers makes this quite understandable.

The recent results of Litovitz (L15) shown in Fig. V-13 indicate the presence of an additional, nonclassical absorption process. In his case, two runs were made with slightly different water contents (and hence different viscosities). In both cases a peak absorption was obtained (−9.2°C, −6.2°C).

These results may be interpreted by assuming that there exists a compressional viscosity which has a characteristic frequency of the same order of magnitude as $\omega_v/2\pi$ in Eq. (28.3). The existence of a maximum absorption (as a function of viscosity) places glycerin in a class with rosin, since, in both cases, the absorption

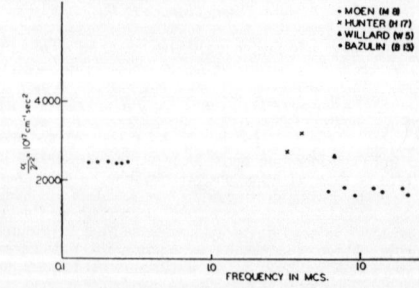

Fig. V-12. α/ν^2 vs frequency for glycerin.

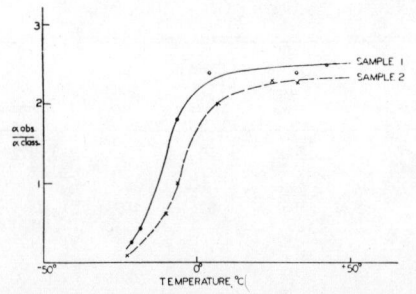

FIG. V-13. Temperature dependence of absorption in glycerin (data from Litovitz (L15)).

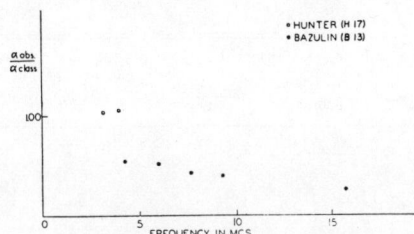

FIG. V-14. Sound absorption in castor oil.

decreases with increasing viscosity (at sufficiently high viscosity).

Another highly viscous liquid is castor oil, the chief experimental results (B13, H17) for which are shown in Fig. V-14.

These results indicate that the experimental value of α/ν^2 lies below the classical value at frequencies above 4 mc. In addition, α/ν^2 falls off with rising frequency. This would be in accord with Eq. (28.3) except that the computed value $\omega_v/2\pi$ for castor oil is about 300 mc, whereas the distribution of points in Fig. V-14 indicates a value of $\omega_v/2\pi$ of about 5 mc.

29. SUMMARY

Pinkerton (P7) has presented a rough classification of liquids according to values of the absorption coefficients. A modified form of this scheme is reproduced in Table V-VII. A detailed listing of specific examples is given in Table V-VIII. The ratio of α as calculated on the basis of Liebermann's experiments to the classical value is also given.

Chapter VI. Experimental Results in Solutions

30. SEA WATER

A great many experimental measurements have been made on the transmission of sound in sea water, especially at frequencies between 10 kc and 100 kc. Most of these measurements have been made in the ocean itself. As an experimental medium, however, the ocean presents many disadvantages. It is not a medium of constant velocity,[48] so that the sound rays are bent away from rectilinear propagation; it is not actually a medium of infinite extent; both surface and bottom reflection must be considered. In addition, there are slight changes in the chemical constitution of sea water from place to place, and finally, air bubbles, seaweed, fish, and similar scattering objects are often present.

Since there are many applications of underwater sound which are concerned with the sound transmitted from point to point, it is well to distinguish between the total attenuation, due to absorption, scattering, etc., and true absorption. It is probable that a failure to distinguish adequately between these quantities accounts for the relatively high "absorption" values sometimes reported at very low frequencies. Fortunately, the perfection of the reverberation tank method has

TABLE V-VII. Classification of liquids.

Classical	$\alpha_{exp}/\alpha_{class}$	Temperature coefficient of α	Type of liquid	Examples
AI anomalous	3–1500	Positive, $(\alpha_{exp}/\alpha_{cl})$ varies with temperature	unassociated, polyatomic	CS_2, C_6H_6, CCl_4
AII anomalous	1.5–3	Negative, $(\alpha_{exp}/\alpha_{cl})$ virtually independent of temperature	associated, polyatomic	water, alcohols
AIII anomalous	5–5000	Depends critically upon frequency	organic acids and esters	acetic and formic acid; ethyl acetate
NI normal	1	Positive	monatomic, diatomic	helium, mercury, liquid oxygen
NII normal	1	Negative, $(\alpha_{exp}/\alpha_{cl})$ varies with temperature and may even become <1	associated polyatomic	glycerin, castor oil, highly viscous liquids

[48] For a general discussion of this problem, see *Physics of Sound in the Sea*, Part I. (N1). A brief, elementary survey has been given by Harnwell (H4).

Table V-VIII.

Class	Liquid	Formula	$T\,°C$	Frequency mc	$\alpha/\nu^2\,\text{exp}$ $\times 10^{-17}$ $\text{cm}^{-1}\,\text{sec}^2$	$\alpha/\nu^2\,\text{Cl}$ $\times 10^{-17}$ $\text{cm}^{-1}\,\text{sec}^2$	$\dfrac{\alpha_{\text{exp}}}{\alpha_{\text{cl}}}$	$\left[\dfrac{\alpha_{\text{total}}}{\alpha_{\text{class}}}\right]$ (Liebermann)	$\dfrac{1}{\alpha}\dfrac{dT}{d\alpha}$	References
AI	Carbon bisulfide	CS_2	20	1–10	6000	5	1200	>150		B10, B22, C4, V8
			21	75–105	1400	5	280	•		R1
	Benzene	C_6H_6	20–25	1–165	900	8.7	103	82	0.006	B7, B8, G7 P7, Q1, V8
	Methyl iodide	CH_3I	25	1–4	820					V8
			20	15	316					P2
			2	15	247	10	24.7		0.010	P2
	Toluene	$C_6H_5CH_3$	27	0.15	205	7.8	26			M8
			20–25	1–75	80	7.8	10.3		0.013	B10, L17, V8
	Acetylene dichloride	$C_2H_2Cl_2$	25	1–10	420	7.7	55			P7, V8, W5
	Methyl bromide	CH_3Br	2	15	304					P2
	Chloroform	$CHCl_3$	20–25	1–10	400	10	40	20		B22, V8, W5
	Carbon tetrachloride	CCl_4	20	1–100	500	20	25	23	0.001	B1, P2, Q1, W5
	Chlorobenzene	C_6H_5Cl	25	1–4	124	8	15.5			V8
	n-butyl chloride	$CH_3(CH_2)_2CH_2Cl$	2	15	108	10	10.8			P2
	Acetone	$(CH_3)_2CO$	25	1–4	70	7	10			V8
			20	5–70	30	7	4.3	5.3		B21, B22
	m-xylene	$C_6H_4(CH_3)_2$	25	1–15	78	8.4	9.3	10		P7, V8, W5
	n-heptane	$CH_3(CH_2)_5CH_3$	22	15	80	10	8			P2
	n-hexane	$CH_3(CH_2)_4CH_3$	21	15	77	10	7.7			P2
	ethyl bromide	CH_3CH_2Br	2	15	61	10	6.1			P2, V8
	nitrobenzene	$C_6H_5NO_2$	25	1–15	80	14	5.7		0.005	P2, V8
	n-propyl chloride	$CH_3CH_2CH_2Cl$	2	15	42	8	5.3			P2
	n-propyl iodide	$CH_3CH_2CH_2I$	2	15	54	14	3.86			P2
	n-butyl bromide	$CH_3(CH_2)_2CH_2Br$	2	15	49	13	3.77			P2
	n-propyl bromide	$CH_3CH_2CH_2Br$	2	15	39	11	3.54			P2
	ethyl iodide	CH_3CH_2I	2	15	40	12	3.25			P2
	n-butyl iodide	$CH(CH_2)_2CH_2I$	2	15	48	17	2.82			P2
AII	water	H_2O	20	7–250	25	8.5	2.95	3.3	−0.031	P6, R1, S9
	methyl alcohol	CH_3OH	20–25	1–250	34	14.5	2.35	2.5	−0.010	P2, R1, V8
	ethyl alcohol	CH_3CH_2OH	20–25	1–220	54	22	2.45	4.4	−0.015	R1, V8
	n-propyl alcohol	$CH_3CH_2CH_2OH$	22–28	15–280	75	36	2.08		−0.008	P2, R1, V8
	n-butyl alcohol	$CH_3(CH_2)_2CH_2OH$	25	1–4	104	50	2.02			P2, R1, V8, W5
	m-amyl alcohol	$CH_3(CH_2)_3CH_2OH$	29	15	106	58	1.83		−0.014	P2
AIII	Acetic acid	CH_3COOH	18	0.5	90000	17	5300		*	L2
			18	67.5	158	17	10.8		−0.010	L2
	Formic acid	$HCOOH$	17.5	4.04	2270	5	454			B10
			20.5	9.83	1170	5	234			B10
	Methyl acetate	CH_2COOCH_3	25	1.00	468	6.8	69			V8
			22	69	34	6.8	5			B1
	Ethyl acetate	$CH_3COOCH_2CH_3$	25	1.00	516	8.3	62			V8
			22	69	37	8.3	4.5			B1
	Ethyl formate	$HCOOCH_2CH_3$	23–28	3	138	7.6	18.1	25		P1
			23–28	16	70	7.6	9.2	13 (at 5 mc)		P1

Class	Liquid	Formula	$T\,°A$	Frequency mc	$\alpha/\nu^2\,\text{exp}$	$\alpha/\nu^2\,\text{Cl}$	$\dfrac{\alpha_{\text{exp}}}{\alpha_{\text{cl}}}$		$\dfrac{1}{\alpha}\dfrac{dT}{d\alpha}$	References
NI	Helium	He	4°A	15	231	204	1.12		0.6	P4
	Argon	A	85°A	44.4	10.1	10.5	0.97			G1
	Hydrogen	H_2	17°A	44.4	5.6	5.8	0.97			G1
	Nitrogen	N_2	73.9°A	44.4	10.6	9.5	1.12			G1
	Oxygen	O_2	87°A	44.4	8.6	7.3	1.18		0.0	G1
	Mercury	Hg	20–25°C	20–50	6	5.05	1.2			B1, R6, R7

Class	Liquid	Formula	$T\,°C$							
NII	Castor oil		21.4	15.72	2100	7980	0.26			B13
			21.5	4.29	4500	7900	0.57			B13
			18.6	3.157	10900	9130	1.20		−0.075	H17
			21.6	3.95	8400	7820	1.17			H17
	Olive oil		21–25	1–4	1250	1100	1.14		−0.038	H17, V8
	Linseed oil		20.5	3.157	1470	1450	1.01		−0.032	H17
NII	Glycerin	$C_3H_8O_3$	20–27	0.15–4	2500				−0.069	M8, W5
			21–23	6–21	1700				−0.056	B13, H17
			32.8	30	1410	590	2.4			H17
			−18.8	30	12500	29100	0.43		0.036	H17

* The value of $(1/\alpha)\partial\alpha/\partial T$ in this case depends critically on both frequency and temperature. It may be either positive or negative. See Lamb and Pinkerton, reference L2.

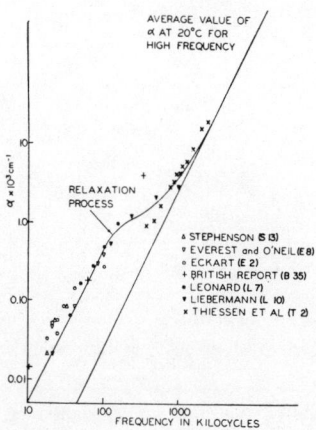

Fig. VI-1. Sound absorption in sea water.

made it possible to compare the open sea results with those obtained under the controlled conditions of the laboratory. A plot of available values of the absorption coefficient in sea water is shown in Fig. VI-1.

It is clear from the characteristic shape of the curve drawn through the experimental points that a relaxational phenomenon exists, with a relaxation frequency of about 145 kc.

The relaxational effect in sea water was at first attributed by Liebermann (L10) to a shift in the chemical equilibrium of ionized sodium chloride. The weight of experimental evidence, however, seems to be against this point of view (B10, B38, C3, R12, T2, V7) although one recent observer (V7) did obtain a large absorption in sodium chloride solution.[49] It is quite probable that

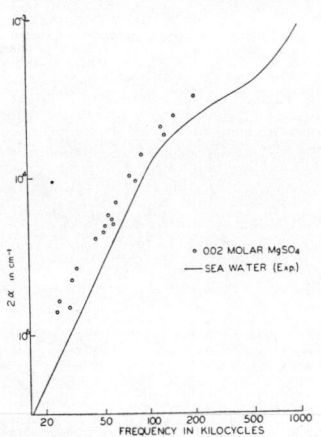

Fig. VI-2. Sound absorption in 0.02 molar magnesium sulfate solution (data from Leonard (L7)).

[49] Professor van Itterbeek has informed the authors that recently repeated measurements with sodium chloride of very high purity gave no excess absorption over the fresh water value.

these values may have been caused by impurities since their experimental methods used appear to be otherwise satisfactory.

Recent experiments of Leonard (L7) and others indicate that the excess absorption in sea water can be explained by the presence of small amounts of $MgSO_4$. This is present in sea water in a concentration of slightly more than 0.02 molar. The available results are plotted in Fig. VI-2. The solid curve represents the best fit of sea water data. Thus the absorption in 0.02 molar $MgSO_4$ more than accounts for the excess absorption.

As was pointed out in Sec. 7 above, Liebermann has advanced a second theory of sound absorption by electrolytes in which the absorption is attributed to a shift in the ionization equilibrium by the passage of the sound wave. He defines a static isothermal partial compressibility

$$\mu_T{}^i = -\frac{1}{V}\left(\frac{\partial V}{\partial p}\right)_T \qquad (30.1)$$

and a static partial molar heat capacity C^i, and introduces the dynamic quantities $\mu_T{}^{\text{eff}}$, C^{eff}, defined by the relations

$$\mu_T{}^{\text{eff}} = \frac{\mu_T{}^i}{1+i\omega\tau}; \quad C^{\text{eff}} = \frac{C^i}{1+i\omega\tau}. \qquad (30.2)$$

TABLE VI-I. Values of A and τ for sea water.

	A	τ
Sea water (Liebermann, L10)	1.45×10^{-10} sec/cm	1.1×10^{-6} sec
0.02M $MgSO_4$ (Leonard, L7)	2.3×10^{-10} sec/cm	1.3×10^{-6} sec
0.02m $MgSO_4$ (Liebermann) theory, L10	6.5×10^{-10} sec/cm	...

From these relations, Liebermann obtained an approximate expression for the absorption:

$$\alpha = \frac{1}{2c}\left[\frac{C^i}{C^2}(C_p-C_v) + \frac{\mu_T{}^i}{\mu_T{}^0}\right]\frac{\omega^2\tau}{1+\omega^2\tau^2}, \qquad (30.3)$$

where C and $\mu_T{}^0$ are the limiting values (at low frequency) of C_p and μ_T respectively.

Liebermann's evaluation of the constants in Eq. (30.3) indicated that the thermal effect is very small in comparison with the compressibility effect. In addition, it appears that the compressibility term is sufficient to account for the absorption present, at least at frequencies below 1 megacycle.

If the total absorption in the solution is written in the form

$$\alpha = \frac{A\tau\omega^2}{1+\omega^2\tau^2} + \text{const } \omega^2, \qquad (30.4)$$

the values of A and τ obtained in sea water and 0.02m $MgSO_4$ and the value of A obtained by Liebermann's theory are given in Table VI-I.

From Table VI-I one may draw the following conclusions:

(1) The experimentally measured absorption in MgSO$_4$, as has already been pointed out, is larger than the total absorption in actual sea water. This indicates that a more complicated reaction among the various ions present in sea water must take place, so as to modify the results which exist for the simpler magnesium sulfate-water solution.[50]

(2) The theoretical value obtained by Liebermann is at least of the same order of magnitude as the experimental value. The discrepancy in numerical values is not surprising in view of the roughness of available data. Some criticism can be made in regard to the use of data for evaluating $\mu_T{}^i$ which were obtained by Bachem at 5 mc, at which frequency this absorption effect is no longer of consequence. One should also emphasize that Liebermann's treatment is not necessarily unique. The application of the method to other solutions or the direct calculation of the relaxation time of the process would be highly desirable.

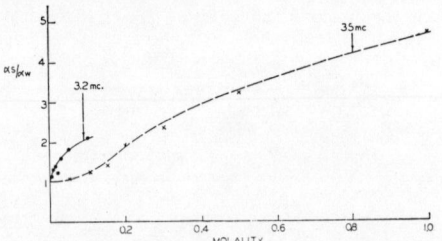

FIG. VI-3. Sound absorption in magnesium sulfate solution as a function of concentration (data from Smith, Barrett and Beyer (S10)).

The sound absorption in magnesium sulfate solutions has more recently been investigated at higher frequencies (S10), above those at which the process just described is significant. These results show that there is an additional absorption, not attributable to any of the classical effects or to the compressibility effect of the chemical reaction. In addition, the temperature dependence of the absorption in such solutions has been measured in the range 0°–30°, and no anomalous effect exists in the neighborhood of 4°C, thus ruling out the thermal relaxation (which being proportional to $C_p - C_v$ must vanish near 4°C).

The variation of absorption with concentration obtained in these experiments is shown in Fig. VI-3. The results indicate that at lower frequencies, absorption is proportional to the square root of the concentration, but that at higher frequencies, there is a more complicated dependence, particularly at low concentrations. In addition, Wilson and Leonard (W8) have recently reported that the relaxation frequency of an aqueous

[50] Tamm and Kurtze (O1a, T1a) have found that the excess absorption in a solution of MgSO$_4$ above the value for water can be virtually eliminated by the addition of large amounts of NaCl.

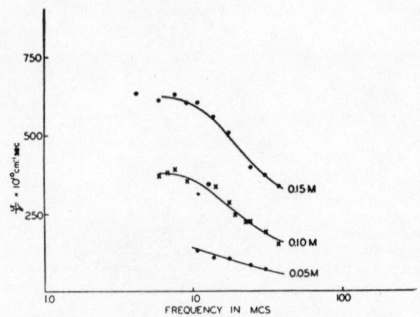

FIG. VI-4. Sound absorption in zinc acetate solution (data from Bazulin (B11)).

MgSO$_4$ solution is virtually a constant over the concentration range 0.003 to 0.02 molal. Thus there is still a great need of further experimental and theoretical developments.

Mention should be made here of an earlier theory for absorption in strong electrolytes, due to Leontovich (L8). He has investigated the effect of the sound wave on the "ionic atmosphere" of the ions, from the point of view of the Debye theory of electrolytes. While his estimates of the magnitude of the absorption to be expected fall far short of what is obtained experimentally, it is of interest to note that the theory predicts that α should be proportional to the square root of the concentration at low frequencies and proportional to the square of the concentration at high frequencies, which is just the dependence shown in Fig. VI-3 (at low concentrations). A further study of this approach would be desirable.

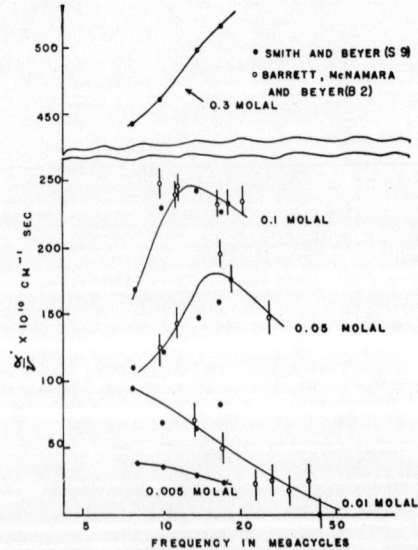

FIG. VI-5. Sound absorption in copper acetate solutions.

Table VI-II.

Concentration of copper acetate (molality)	Relaxation frequency in megacycles
0.005	≪ 5
0.01	~ 5
0.05	15
0.10	13
0.30	~22

It may well be that two or more general types of absorption processes exist in electrolytic solutions, namely, (1) that due to increased dissociation on the part of the solute in the presence of sound (Liebermann theory) and (2) that due to distortions produced in an ionic atmosphere of a wholly dissociated solute (Leontovich theory). Finally, the question may be raised[51] as to whether one is justified in assuming a simple additivity (total absorption = solute absorption + solvent absorption) for the various processes. The presence of solute ions in the neighborhood of water molecules certainly produces some change in the liquid structure, and may

Table VI-III.

Type	Solute	Mol. wt.	Observer	Ref.	Molarity	T °C	Freq. mc	α_s/α_w	α_s/α_w corrected to unit molarity
1-1	NaCl	58.5	Rüfer	R12	0.96	18	7.45	0.98	0.98
	NaNO$_3$	85.0	Bazulin	B9	1.00	20	30.5	0.97	0.97
	KCl	74.6	Bazulin	B9	1.00	24	20.1	0.97	0.97
	KNO$_3$	101.1	Bazulin	B9	1.00	17	30.5	0.97	0.97
	AgNO$_3$	169.9	Bazulin	B9	0.94	19–20.5	30.5	1.11	1.17
	NaBr	102.9	Rüfer	R12	1.63	18	7.45	1.02	1.01
			Rüfer		1.63	18	8.55	0.93	0.96
	KBr	119.0	Bazulin	B10	1.00	16–17	30.5	1.00	1.00
			Rüfer	R12	1.95	18	7.45	1.09	1.04
			Rüfer		1.95	18	8.55	1.00	1.00
	NaClO$_3$	106.5	Rüfer	R12	2.79	18	8.55	1.60	1.22
	NaClO$_4$	122.5	Rüfer	R12	2.06	18	7.45	2.04	1.51
			Rüfer		2.06	18	8.55	1.85	1.41
2-1	MgCl$_2$	95.2	Rüfer	R12	1.05	18	6.3	1.08	1.07
			Rüfer		1.05	18	8.55	1.39	1.37
	SrCl$_2$	158.5	Bazulin	B9	0.43	18	30.5	1.11	1.26
	Cu(NO$_3$)$_2$	187.6	Bazulin	B9	0.5	19–20	30.5	1.25	1.50
			Rüfer	R12	0.47	18	7.45	1.22	1.47
	Cd(NO$_3$)$_2$	236.4	Bazulin	B9	0.45	18–19	30.5	1.32	1.64
	Pb(NO$_3$)$_2$	331.2	Bazulin	B9	0.5	18–19	30.5	1.75	2.50
	UO$_2$(NO$_3$)$_2$	394.2	Rüfer	R12	0.32	18	8.55	1.77	
1-2	(NH$_4$)$_2$SO$_4$	80.1	Bazulin	B9	0.5	17.5	30.5	1.43	1.86
	Na$_2$SO$_4$	142.1	Bazulin	B9	0.5	17–18	30.5	1.50	2.00
	K$_2$SO$_4$	174.3	Bazulin	B9	0.5	17–17.5	30.5	1.89	2.78
2-2	MgSO$_4$	120.4	Bazulin	B9	0.974	22.5	20.64	4.48	4.6
			Rüfer	R12	0.99	18	6.3	5.10	5.1
			Smith, Beyer Barrett	S10	0.99	20	35.44	4.80	4.8
	MnSO$_4$	151.0	Bazulin	B9	0.05	21–24	30.5	1.39	8.8
			Bazulin		0.125	21–23	30.5	2.04	9.2
			Bazulin		0.250	21–24	30.5	2.93	8.7
	NiSO$_4$	154.8	Bazulin	B9	0.125	19–21	30.5	1.75	7.0
	CuSO$_4$	223.2	Bazulin	B9	0.125	20–22	30.5	2.04	9.3
			Rüfer	R12	0.56	18	6.3	4.28	6.9
	CdSO$_4$	208.5	Bazulin	B9	0.125	22–23.5	30.5	2.46	12.7
3-1	La(NO$_3$)$_3$	324.9	Bazulin	B9	0.1	16	12.45	4.6	37
			Bazulin		0.1	16	15.60	4.1	32
			Bazulin		0.1	17	17.35	3.9	30
			Bazulin		0.1	20	20.74	3.3	24
			Bazulin		0.1	20	30.56	2.8	19
			Bazulin		0.1	20–21	37.12	2.4	15
			Bazulin		0.25	17	9.24	10.7	40
			Bazulin		0.25	17	10.74	12.1	45
			Bazulin		0.25	17	12.40	12.0	44
			Bazulin		0.25	17.5	14	10.4	39
			Bazulin		0.25	18	20.64	9.9	37
			Bazulin		0.25	17–18	25.60	8.2	30
3-2	Al$_2$(SO$_4$)$_3$	342.12	Bazulin	B9	0.041	21–24	30.5	2.18	30
			Bazulin		0.084	21–24	30.5	3.14	27
			Bazulin		0.165	21–24	30.5	3.09	26
			Bazulin	B11	0.1	18	15.7	4.94	40
			Bazulin		0.1	17–18	17.35	4.87	40
			Bazulin		0.1	17–18	24	4.55	37
			Bazulin		0.1	20–21	30.5	3.54	26

---

[51] This was suggested by Dr. E. B. Yeager. Some recent measurements (B2a) indicate that the presence of a solute may lower the absorption coefficient below that of pure water.

31. ACETATE AND FORMATE ELECTROLYTES

The anomalies present in acetate and formate liquids have prompted some investigation of aqueous solutions of salts containing these radicals. While the results are only fragmentary, they show some interesting trends, which merit further attention.

The most complete data available are for water solutions of zinc acetate (B11). In order to isolate the relaxation process, the values of α/ν for pure water have been subtracted from the experimentally determined values for the solution and the results have been plotted in Fig. VI-4. The corresponding results for solutions of copper acetate (B2, B17) have been similarly plotted in Fig. VI-5. (The vertical lines indicate the approximate uncertainty range in the more recent values.)

For zinc acetate the relaxation frequency at 0.10 and 0.15M has a value of about 8 megacycles. There is not sufficient evidence to indicate any relation between the concentration and the relaxation frequency. Liebermann's treatment of absorption in magnesium sulfate presumed that this frequency was directly proportional to the concentration.

Some rough information on the relation of the relaxation frequency to concentration may be obtained from the data on copper acetate solutions. The approximate relaxation frequencies are given as a function of concentration in Table VI-II.

It must be remembered that these numbers are based on only a very small number of points and are therefore a very rough estimate. Nevertheless they do point toward an increase in the relaxation frequency with increasing concentration, although not in a linear fashion.

32. VARIOUS ELECTROLYTES

The ratios of α_{solution} to α_{water} for aqueous solutions of various electrolytes are given in Table VI-III. The results are classified according to valence combinations. Molality values have been converted to molarity and a linear correction has been applied to give α_s/α_w for unit molarity in each case. Inasmuch as the actual dependence of α_s on concentration is not clearly known (compare discussion of MgSO$_4$ solutions above), this may be taken only as a rough guide, especially in those cases in which the experimental measurements have been made at very low concentrations.

The following conclusions may be drawn from the table:

(1) For 1–1 solutes, the excess absorption resulting from the solute is negligible, except in the cases of NaClO$_3$ and NaClO$_4$. In these two cases, there appears to be a small but real excess.

(2) In a general way, the ratio α_s/α_w increases with increase in valence of either ion.

(3) For a given negative radical, the absorption increases with the atomic weight of the positive ion. This is most clearly brought out in the series of nitrate compounds of the (2–1) type and to a lesser extent in the series of sulfate compounds of the (2–2) type.

ACKNOWLEDGMENTS

The authors would like to thank: Professor R. W. Morse for editing several of the sections; Professor G. Heller, Dr. F. T. McClure, and Professor A. O. Williams, Jr. for valuable discussions; Mrs. Rachel T. Lindsay and Mrs. Helen Sherwood for their assistance in calculating the curves; Mrs. Marie Gammell and Mrs. Mildred Gordon for typing the manuscript; and Mrs. Betty Grisamore for her assistance in proofreading.

APPENDIX I. LIST OF SYMBOLS

Below are listed the most important symbols:

a—See Eq. (2.5), reciprocal of the compliance in Burgers's model.
b—See Eq. (2.5), fluid resistance in Burgers's model.
c—Sound velocity at frequency $(1/2\pi)\omega$.
c_0—Limiting velocity of sound at low frequencies.
c_∞—Limiting sound velocity for high frequencies.
f_{ij}—See Eq. (8.5)
k—See Eqs. (2.12) to (2.14).
k_r—Real part of k.
k_i—Imaginary part of k.
h—Planck's constant.
k'—Boltzmann's constant.
m—Mass of a molecule.
n—Sec. 7; see Eq. (7.11).
n—Sec. 8, number of molecules per unit volume.
n_i—Number of molecules per unit volume in state i.
p—Total pressure.
p_e—Excess pressure.
p_0—Equilibrium pressure (except in Sec. 13 and 14).
q—Displacement vector.
t—Time.
u—Particle velocity.
u_x—x component of the particle velocity.
v_1—Volume per molecule associated with state 1.
v_2—Volume per molecule associated with state 2.
A_i—See Eq. (8.12a).
A_{12}—Transition probability from state 1 to state 2.
A_{21}—Transition probability from state 2 to state 1.
A_{12}^0—Equilibrium value of the transition probability from state 1 to state 2.
A_{21}^0—Equilibrium value of the transition probability from state 2 to state 1.
B_p—See Eq. (7.14).
B_T—See Eq. (7.14).
C_v—Molar heat at constant volume.
C^{eff}—Dynamic partial molar heat capacity.
C^i—Internal molar heat capacity.
$C_p^e = C_p^\infty$—Molar heat at constant pressure related to intermolecular processes.
$C_v^e = C_v^\infty$—Molar heat at constant volume for intermolecular processes.

$C_p{}^i$—Molar heat at constant pressure for intramolecular processes.
$C_v{}^i$—Molar heat at constant volume for intramolecular processes.
$C_p{}^t = C_p{}^0{}^- = C_p{}^e + C_p{}^i$, see Eq. (7.33).
$C_v{}^t = C_v{}^0{}^- = C_v{}^i + C_v{}^e$.
E_1—Energy of an unexcited state (per mole).
E_2—Energy of an excited state (per mole).
E_k—See Eqs. (8.2) to (8.4), kinetic energy of translation.
F—Force in Burgers's model.
K—Isothermal bulk modulus.
K_s—Adiabatic bulk modulus.
K^e—Isothermal bulk modulus related to intermolecular effects.
$K^t = K^0$—Net isothermal bulk modulus.
$K_s{}^0 = K_s{}^t$—Low frequency adiabatic bulk modulus.
$K_s{}^\infty = K_s{}^e$—High frequency adiabatic bulk modulus.
M—Molecular weight.
N—Number of molecules in given volume usually taken as a mole.
N_1—Number of molecules in state 1.
N_2—Number of molecules in state 2.
$N_1{}^0$—Number of molecules in state 1 at equilibrium.
$N_2{}^0$—Number of molecules in state 2 at equilibrium.
P_e—See Eq. (2.12), pressure amplitude.
Q—Heat.
R—Gas constant.
R_e—See Eq. (2.14), excess density amplitude.
T^e—Temperature related to intermolecular equilibrium.
T^i—Temperature related to intramolecular equilibrium.
U_x—See Eq. (2.13), particle velocity amplitude.
U^e—Internal intermolecular energy.
U^i—Internal intramolecular energy.
V—Molar volume.
α—Amplitude absorption coefficient.
α_s—Amplitude absorption coefficient for solution.
α_w—Amplitude absorption coefficient for H_2O.
α'—Excess α.
β—Volume coefficient of thermal expansion.
β^e—Volume coefficient of thermal expansion related to the intermolecular forces.
β^i—See Eq. (7.30).
β^t—See Eq. (7.34).
γ—Ratio of specific heats.
γ_s—Parameter, related to Maxwell's equation.
$\epsilon_1{}^- = E_1/N$.
$\epsilon_2{}^- = E_2/N$.
ζ—See (Eq. (3.1)), viscosity coefficient.
η—Shear viscosity.
η'—See Eq. (13.2), viscosity coefficient.
θ_s—See Eq. (1.2), elastic constant.
κ—Coefficient of thermal conductivity.
λ—Sound wavelength.
λ_s—See Eq. (1.2), elastic parameter.
$\mu = \alpha\lambda$—Amplitude absorption coefficient per wavelength.
μ_i—Electrochemical potential.

μ'—Excess value of μ.
ρ—Total density.
ρ_e—Excess density.
ρ_0—Equilibrium density.
τ—See Eq. (6.1), relaxation time.
ν—Sound frequency.
ν_m—Relaxation frequency.
ω—Angular frequency.
ω_c—Angular relaxation frequency for heat conduction.
ω_m—See Eq. (9.13).
ω_0—Angular relaxation frequency.
$\omega_r{}^- = q/\gamma$, see Eq. (5.5).
$\omega_v{}^- = C_p\rho_0 c^2/M\kappa$, see Eq. (4.9).

APPENDIX II. SUMMARY OF SOME THERMODYNAMIC EQUATIONS

We are interested in irreversible processes, and we must, therefore, be careful in our use of thermodynamics. In general, we shall assume that there is a functional relation between the four variables, pressure p, volume V, temperature T, and the time t. Namely, the equation

$$\Delta V = \left(\frac{\partial V}{\partial p}\right)_{T,t} \Delta p + \left(\frac{\partial V}{\partial T}\right)_{p,t} \Delta T + \left(\frac{\partial V}{\partial t}\right)_{p,t} \Delta t$$

will always hold. For a system with energy dissipation, the second law becomes an inequality, and one is not allowed to apply the relationships based on the assumption that the law is an equality (Maxwell's thermodynamics relations) without carefully re-examining them. We shall not attempt this here, but simply note which equations were derived by taking the second law as an equality and avoid using them for irreversible processes. We assume that consistent units are used throughout. Thus the specific heat will be in ergs per moles in cgs units.

We list some useful definitions and some equations. The equations are either found in Slater (S8) or can be derived simply.

$$\beta = \frac{1}{V}\left(\frac{\partial V}{\partial T}\right)_p \tag{A-1}$$

$$\frac{1}{K} = -\frac{1}{V}\left(\frac{\partial V}{\partial p}\right)_T \tag{A-2}$$

$$\frac{1}{K_s} = -\frac{1}{V}\left(\frac{\partial V}{\partial p}\right)_S \tag{A-3}$$

$$C_v = \left(\frac{\partial U}{\partial T}\right)_V \tag{A-4}$$

$$C_p = \left(\frac{\Delta Q}{\Delta T}\right)_p \tag{A-5}$$

$$dp = -\frac{K}{V}dV + \beta K dT \qquad \text{(A-6)}$$

$$dV = -(V/K)dp + V\beta dT \qquad \text{(A-7)}$$

$$dT = (1/K\beta)dp + \frac{1}{\beta}\frac{dV}{V} \qquad \text{(A-8)}$$

$$dU = (C_p - V p\beta)dT - \left(T\beta - \frac{p}{K}\right)Vdp$$

(A-9) Using second law

$$= \frac{C_v}{K\beta}dp + \left(\frac{C_p}{V\beta} - p\right)dV \qquad \text{(A-10)}$$

$$= C_v dT + (T\beta K - p)dV \qquad \text{(A-11) Using second law}$$

$$K_s = K\frac{C_p}{C_v} \qquad \text{(A-12) Using second law}$$

$$C_p = C_v + TV\beta^2 K \qquad \text{(A-13) Using second law}$$

$$K_s = K\left(1 + \frac{TV\beta^2 K}{C_v}\right) \qquad \text{(A-14) Using second law}$$

$$\frac{C_p}{\beta K_s}\left(\frac{C_p}{C_v} - 1\right) = TV\beta \qquad \text{(A-15) Using second law}$$

REFERENCES

A1 E. A. Alexander and J. D. Lambert, Proc. Roy. Soc. (London) **179A**, 499 (1942).
A1a T. Alfrey, *Mechanical Behavior of High Polymers* (Interscience Publishers, Inc., New York, 1948).
A2 R. S. Alleman, Phys. Rev. **55**, 87 (1939).
A3 W. Altberg, Ann. Physik **11**, 405 (1903).
B1 R. Bär, Helv. Phys. Acta **10**, 332 (1937).
B2 Barrett, McNamara, and Beyer, J. Acoust. Soc. Am. **23**, 629(A) (1951).
B2a R. E. Barrett and R. T. Beyer, Phys. Rev. **84**, 1060 (1951).
B3 A. Basset, *Treatise on Hydrodynamics* (Deighton Bell and Company, 1888), Vol. II, Chapter XX.
B4 E. Bauer, Proc. Phys. Soc. (London) **62A**, 141 (1949).
B5 E. Baumgardt, Compt. rend. **202**, 203 (1936).
B6 E. Baumgardt, Compt. rend. **204**, 416 (1937).
B7 P. Bazulin, Compt. rend URSS **3**, 285 (1936).
B8 P. Bazulin, Compt. rend. URSS **14**, 273 (1937).
B9 P. Bazulin, Compt. rend. URSS **19**, 153 (1938).
B10 P. Bazulin, J. Exptl. Theor. Phys. USSR **8**, 457 (1938).
B11 P. Bazulin, J. Exptl. Theor. Phys. USSR **9**, 1147 (1939).
B12 P. Bazulin and J. M. Merson, Compt. rend. URSS **24**, 690 (1939).
B13 P. Bazulin, Compt. rend. URSS **31**, 113 (1941).
B14 P. Bazulin and M. A. Leontovich, Dokl. Akad. Nauk USSR **57**, 29 (1947).
B15 L. Bergmann, *Der Ultraschall* (third edition, reprinted by Edwards Brothers, Ann Arbor, 1942; fifth edition, S. Hirzel, Zurich, 1949).
B16 P. B. Bergmann, J. Acoust. Soc. Am. **17**, 329 (1946).
B17 R. T. Beyer and M. C. Smith, J. Acoust. Soc. Am. **18**, 424 (1946).
B18 R. T. Beyer, Am. J. Phys. **18**, 25 (1950).
B19 R. T. Beyer, J. Chem. Phys. **19**, 788 (1951).
B20 P. Biquard, Compt. rend. **196**, 257 (1933).
B21 P. Biquard, Compt. rend. **202**, 117 (1936).
B22 P. Biquard, Ann. phys., Paris **6**, 195 (1936).
B23 H. Born, Z. Physik **120**, 383 (1943).
B24 M. Born and H. S. Green, *A General Kinetic Theory of Liquids* (Cambridge University Press, London, 1949).
B25 E. Borovik, J. Exptl. Theor. Phys. USSR **18**, 48 (1948).
B26 D. G. Bourgin, Nature **122**, 133 (1928).
B27 D. G. Bourgin, Phil. Mag. **7**, 821 (1929).
B28 D. G. Bourgin, Phys. Rev. **34**, 521 (1929).
B29 D. G. Bourgin, Phys. Rev. **42**, 721 (1932).
B30 D. G. Bourgin, J. Acoust. Soc. Am. **4**, 108 (1932).
B31 D. G. Bourgin, J. Acoust. Soc. Am. **5**, 57 (1934).
B32 D. G. Bourgin, Phys. Rev. **50**, 355 (1936).
B33 R. W. Boyle and G. B. Taylor, Trans. Roy. Soc. Can. Sec. III **20**, 245 (1926).
B34 L. Brillouin, Rev. Acoust. **5**, 99 (1936).
B35 British Internal Technical Report No. 51, March, 1942.
B36 J. M. Burgers, *First Report on Viscosity and Plasticity* (Committee for the Study of Viscosity of the Academy of Sciences at Amsterdam, 1935), pp. 5–72.
B37 C. J. Burton, J. Acoust. Soc. Am. **20**, 186 (1948).
B38 W. Buss. Ann. Physik **33**, 143 (1938).
C1 S. Chapman, Trans. Roy. Soc. (London) **A217**, 115 (1918).
C2 S. Chapman and T. G. Cowling, *The Mathematical Theory of Non-Uniform Gases* (Cambridge University Press, London, 1939).
C3 Claeys, Errera, and Sack, Compt. rend. **202**, 1493 (1936).
C4 Claeys, Errera, and Sack, Trans. Faraday Soc. **33**, 136 (1937).
D1 G. Damkohler, Z. Elektrochem. **48**, 62, 116 (1942).
D2 P. Debye and F. Sears, Proc. Nat. Acad. Sci. **18**, 409 (1932).
D2a S. R. DeGroot, *Thermodynamics of Irreversible Processes* (Interscience Publishers, Inc., New York, 1951).
D3 P. A. M. Dirac, *Quantum Mechanics* (Oxford University Press, London, 1934), second edition.
D3a A. K. Dutta, Bose Research Inst. Calcutta, Trans. **12**, 115 (1936–1937).
D4 A. K. Dutta and B. B. Ghosh, Trans. Bose Res. Inst. **13**, 31 (1939).
E1 C. Eckart, Phys. Rev. **58**, 269 (1940).
E2 C. Eckart, *The Attenuation of Sound in the Sea*, NDRC Report U-236; Project NS-140 UCDWR, July 6, 1944.
E3 C. Eckart, Phys. Rev. **73**, 68 (1948).
E3a C. Eckart (unpublished lectures at University of California).
E4 A. Einstein, Sitz. Berl. Akad. p. 380 (1920).
E5 P. S. Epstein, *Textbook of Thermodynamics* (John Wiley & Sons, Inc., New York, 1937).
E6 A. Eucken and R. Becker, Z. Physik. Chem. **27B**, 219 (1934).
E7 A. Eucken and S. Aybar, Z. Physik. Chem. **46B**, 195 (1940).
E8 F. A. Everest and H. T. O'Neil, *Attenuation of Underwater Sound*, NDRC; C4-sr30-494, UCDWR, July 6, 1944.
E9 C. F. Eyring, J. Acoust. Soc. Am. **1**, 217 (1930).
F1 R. H. Fowler and E. A. Guggenheim, *Statistical Thermodynamics* (Cambridge University Press, London, 1939).
F2 F. E. Fox and G. D. Rock, J. Acoust. Soc. Am. **12**, 505 (1941).
F3 F. E. Fox and G. D. Rock, Phys. Rev. **70**, 68 (1946).
F4 J. Frenkel and J. Obraztsov, J. Phys. USSR **3**, 131 (1940); J. Exptl. Theor. Phys. USSR **9**, 1081 (1939).
F5 J. Frenkel, *A Kinetic Theory of Liquids* (The Clarendon Press, Oxford, 1946).
F6 E. F. Fricke, J. Acoust. Soc. Am. **12**, 245 (1940).
G1 J. K. Galt, J. Chem. Phys. **16**, 505 (1948).
G2 B. B. Ghosh, Indian J. Phys. **24**, 1 (1950).
G2a Glasstone, Laidler, and Eyring, *The Theory of Rate Processes* (McGraw-Hill Book Company, Inc., New York, 1941).
G2b H. Grad, Commun. Pure Appl. Math. **2**, 331 (1949).
G2c H. Grad, Paper at the 1951 Pittsburgh Meeting of the Am. Phys. Soc.
G3 M. Greenspan, Phys. Rev. **75**, 197 (1949).
G4 M. Greenspan, J. Acoust. Soc. Am. **22**, 568 (1950).
G5 E. C. Gregg, Jr., Rev. Sci. Instr. **12**, 149 (1941).
G6 E. Grossmann, Physik. Z. **35**, 83 (1934).
G7 A. Grobe, Physik. Z. **38**, 333 (1938).
G8 E. A. Guggenheim, *Thermodynamics* (Interscience Publishers, Inc., New York, 1949).
G9 S. B. Gurevich, Compt. rend. URSS **55**, 17 (1947).
H1 Leonard Hall, Phys. Rev. **73**, 775 (1948).
H2 O. Halpern, Phys. Rev. **55**, 881 (1939).
H3 G. Hammann, Ann. Physik **32**, 593 (1938).
H4 G. P. Harnwell, Am. J. Phys. **16**, 127 (1948).

H5 G. K. Hartmann and A. B, Focke, Phys. Rev. **57**, 221 (1940).
H6 P. S. H. Henry, Proc. Cambridge Phil. Soc. **28**, 249 (1932).
H7 G. Hertz and H. Mende, Z. Physik **114**, 354 (1939).
H8 G. Herzberg, *Molecular Spectra and Molecular Structure I* (D. Van Nostrand Company, Inc., New York, 1950), 2nd edition.
H9 G. Herzberg, *Molecular Spectra and Molecular Structure II* (D. Van Nostrand Company, Inc., New York, 1945).
H10 K. F. Herzfeld and F. O. Rice, Phys. Rev. **31**, 691 (1928).
H11 K. F. Herzfeld, Phys. Rev. **53**, 899 (1938).
H12 K. F. Herzfeld, J. Acoust. Soc. Am. **13**, 33 (1941).
H13 Hoff Lu, J. Acoust. Soc. Am. **23**, 12 (1951).
H14 E. Hsu, J. Acoust. Soc. Am. **17**, 127 (1945).
H15 J. C. Hubbard, Phys. Rev. **38**, 1011 (1931).
H16 J. C. Hubbard, Phys. Rev. **41**, 523 (1932).
H17 J. L. Hunter, J. Acoust. Soc. Am. **13**, 36 (1941).
H18 J. L. Hunter and F. E. Fox, J. Acoust. Soc. Am. **22**, 238 (1950).
H19 J. L. Hunter, J. Acoust. Soc. Am. **22**, 243 (1950).
I1 International Critical Tables (McGraw-Hill Book Company, Inc., New York, 1926).
J1 Sir James Jeans, *An Introduction to the Kinetic Theory of Gases* (Cambridge University Press, London, 1940).
J2 Sir James Jeans, *The Dynamical Theory of Gases* (Cambridge University Press, London, 1904), first edition.
J3 G. Joos, *Theoretical Physics* (Blackie and Son Limited, London, 1934).
K1 H. H. Keller, Physik. Z. **41**, 386 (1940).
K2 L. V. King, Can. J. Research **11**, 135, 484 (1934).
K3 G. Kirchhoff, Pogg. Ann. **134**, 177 (1868).
K4 J. G. Kirkwood, J. Chem. Phys. **14**, 180 (1946).
K5 J. G. Kirkwood, J. Chem. Phys. **15**, 72 (1947).
K6 Kirkwood, Buff, and Green, J. Chem. Phys. **17**, 988 (1949).
K7 C. Kittel, J. Chem. Phys. **14**, 614 (1946).
K8 C. Kittel, Phys. Soc. Rep. Proc. in Phys. **11**, 205 (1948).
K9 H. O. Kneser and V. O. Knudsen, Ann. Physik **21**, 682 (1935).
K10 H. O. Kneser and O. Gauler, Physik. Z. **37**, 677 (1936).
K11 H. O. Kneser, Ergeb. exakt. Naturwiss. **22**, 121 (1949).
K12 H. Knötzel and L. Knötzel, Ann. Physik **2**, 393 (1948).
K13 V. O. Knudsen, J. Acoust. Soc. Am. **5**, 64 (1933).
K14 V. O. Knudsen and E. Fricke, J. Acoust. Soc. Am. **12**, 255 (1940); Erratum, **12**, 449 (1941).
K15 A. G. Korn, thesis (Brown University, 1948).
L1 H. Lamb, *Hydrodynamics* (Cambridge University Press, London, 1932), sixth edition.
L2 J. Lamb and J. M. M. Pinkerton, Proc. Roy. Soc. (London) **A199**, 114 (1949).
L2a J. Lamb and J. H. Andreae, Nature **167**, 898 (1951).
L3 Landau and E. Teller, Physik. Z. Sowjetunion **10**, 34 (1936).
L4 Landolt-Börnstein, *Physikalische-Chemische Tabellen* (Verlag. Julius Springer, Berlin, 1923–1935).
L5 R. W. Leonard, J. Acoust. Soc. Am. **12**, 241 (1940).
L6 R. W. Leonard, J. Acoust. Soc. Am. **18**, 252 (1946).
L7 R. W. Leonard, Technical Report No. 1, UCLA June 1, 1950.
L8 M. Leontovich, J. Exptl. Theor. Phys. USSR **8**, 40 (1938).
L9 L. N. Liebermann and D. A. Wilson, WHOI Report No. 3, October, 1946.
L10 L. N. Liebermann, J. Acoust. Soc. Am. **20**, 868 (1948).
L11 L. N. Liebermann, Phys. Rev. **75**, 1415 (1949).
L12 L. N. Liebermann, Phys. Rev. **76**, 1520 (1949).
L13 A. Lindberg, Physik. Z. **41**, 457 (1940).
L14 R. B. Lindsay, Am. J. Phys. **16**, 371 (1948).
L14a R. B. Lindsay, J. Acoust. Soc. Am. **23**, 628(A) (1951).
L15 T. Litovitz, J. Acoust. Soc. Am. **23**, 75 (1951).
L16 L. B. Loeb, *The Kinetic Theory of Gases* (2nd edition, McGraw-Hill Book Company, Inc., New York, 1934).
L17 R. Lucas, Compt. rend. **203**, 459 (1936).
L18 D. G. C. Luck, Phys. Rev. **40**, 440 (1932).
M1 L. I. Mandelstam and M. A. Leontovich, J. Exptl. Theor. Phys. USSR **7**, 438 (1937).
M2 J. J. Markham, J. Acoust. Soc. Am. **22**, 628 (1950).
M2a' J. J. Markham, J. Acoust. Soc. Am. **23**, 144 (1951).
M3 J. J. Markham, Brussels Conference on Ultrasonics, 1951.
M4 K. Matta and E. G. Richardson, J. Acoust. Soc. Am. **23**, 58 (1951).
M5 J. C. Maxwell, Trans. Roy. Soc. (London) **157**, 49 (1867).
M6 J. Meixner, Ann. Physik **43**, 470 (1943).

M7 I. G. Mikhailov and S. B. Gurevich, J. Exptl. Theor. Phys. USSR **19**, 193 (1949).
M8 C. J. Moen, J. Acoust. Soc. Am. **23**, 62 (1951).
M9 P. M. Morse, *Vibration and Sound* (McGraw-Hill Book Company, Inc., New York, 1936).
M10 C. E. Mulders, Appl. Sci. Res. **B1**, 341 (1950).
N1 NDRC Summary Technical Reports, Div. 6, Vol. 8, Part I, *Physics of Sound in the Sea*.
N1a W. L. Nyborg, Chicago Meeting, Acoustical Society of America, October, 1951.
O1 H. Oberst, Akust. Z. **2**, 76 (1937).
O1a Office of Naval Research London, Technical Report ONRL–102–50, October 9, 1950).
O2 N. Otpushchennikov. J. Exptl. Theor. Phys. USSR **9**, 229 (1939).
O3 Te-Tchao Ouang, Compt. rend. **222**, 1215 (1946).
P1 S. Parthasarathy, Current Sci. (India) **6**, 501 (1938).
P1a J. G. Parker, Ph.D. thesis (Brown University, 1951).
P2 J. R. Pellam and J. K. Galt, J. Chem. Phys. **14**, 608 (1946).
P3 J. R. Pellam and J. Galt, Technical Report No. 4 NDRC Div. 14, June 10, 1946.
P4 J. R. Pellam and C. F. Squire, M.I.T. Technical Report 44, August 1, 1947; Phys. Rev. **72**, 1245 (1947).
P5 W. H. Pielemeier, Phys. Rev. **34**, 1184 (1929).
P6 J. M. M. Pinkerton, Nature **160**, 128 (1947).
P7 J. M. M. Pinkerton, Proc. Phys. Soc. (London) **62B**, 129 (1949).
P8 J. M. M. Pinkerton, Proc. Phys. Soc. (London) **62B**, 286 (1949).
P9 E. J. Pumper, J. Phys. USSR **1**, 411 (1939).
Q1 B. J. Quinn, J. Acoust. Soc. Am. **18**, 185 (1946).
R1 R. A. Rapuano, Phys. Rev. **72**, 78 (1947).
R2 Lord Rayleigh, *Theory of Sound* (Dover Publications, New York, 1945), second edition.
R3 J. E. Rhodes, Phys. Rev. **70**, 932 (1946).
R4 W. T. Richards, Revs. Modern Phys. **11**, 36 (1939).
R5 E. G. Richardson, Proc. Phys. Soc. (London) **52**, 480 (1940).
R6 P. Rieckmann, Physik. Z. **40**, 582 (1939).
R7 Ringo, Fitzgerald, and Hurdle, Phys. Rev. **72**, 87 (1947).
R8 Y. Rocard, Ann. phys. **8**, 5 (1927).
R9 Y. Rocard, Act. Sci. Ind. No. 222 (1935).
R10 V. Rojansky, *Introduction to Quantum Mechanics* (Prentice-Hall, Inc., New York, 1938).
R11 N. S. Rudenko and L. W. Schubnikov, Physik. Z. Sowjetunion **6**, 470 (1934).
R12 W. Rüfer, Ann. Physik **41**, 301 (1942).
S1 Z. Sakadi, Proc. Phys. Math. Soc. Japan **23**, 208 (1941).
S1a H. L. Saxton, J. Chem. Phys. **6**, 30 (1938).
S2 K. Schafer, Z. Physik. Chem. **46B**, 212 (1940).
S2a N. Schmidtmüller, Akust. Z. **3**, 387 (1936).
S3 E. Schreuer and K. Osterhammel, Naturwiss. **29**, 44 (1941).
S4 D. Sette, Phys. Rev. **78**, 476 (1950).
S5 D. Sette, Supplement to Nuovo cimento **6**, 1 (1949).
S6 D. Sette, Nuovo cimento **7**, 55 (1950).
S6a D. Sette, J. Chem. Phys. **18**, 1592 (1950).
S7 E. Skudrzyk, Acta. Phys. Austr. **2**, 148 (1948).
S8 J. C. Slater, *Chemical Physics* (McGraw-Hill Book Company, Inc., New York, 1939).
S9 M. C. Smith and R. T. Beyer, J. Acoust. Soc. Am. **20**, 608 (1948).
S10 Smith, Barrett, and Beyer, J. Acoust. Soc. Am. **23**, 71 (1951).
S11 C. Sörensen, Ann. Physik **27**, 70 (1936).
S12 I. S. Sokolnikoff and E. S. Sokolnikoff, *Higher Mathematics for Engineers and Physicists* (McGraw-Hill Book Company, Inc., New York, 1941), second edition.
S13 E. B. Stephenson, *Absorption Coefficients of Sound in Sea Water*, Report S-1466, NRL, August 12, 1938.
S14 E. S. Stewart, Phys. Rev. **69**, 632 (1946).
S15 G. W. Stewart and R. B. Lindsay, *Acoustics* (D. Van Nostrand Company, Inc., New York, 1930).
S16 J. L. Stewart, Rev. Sci. Instr. **17**, 59 (1946).
S17 G. G. Stokes, Trans. Cambridge Phil. Soc. **8**, 287 (1845).
S18 G. G. Stokes, Phil. Mag. **1**, 305 (1851).
T1 C. E. Teeter, Jr., J. Acoust. Soc. Am. **18**, 488 (1946).
T1a K. Tamm and G. Kurtze, Nature **168**, 346 (1951).
T1b W. J. Thaler, J. Acoust. Soc. Am. **23**, 627 (1951).
T2 Thiessen, Leslie; and Simpson, Can. J. Research **26A**, 306 (1948).

T2a J. H. C. Thompson, Trans. Roy. Soc. (London) **231**, 339 (1933).
T3 L. Tisza, Phys. Rev. **61**, 531 (1942).
T4 H. Tsien and R. Schamberg, J. Acoust. Soc. Am. **18**, 334 (1946).
V1 A. van Itterbeek and P. Mariens, Physica **4**, 609 (1937).
V2 A. van Itterbeek and P. Mariens, Physica **5**, 153 (1938).
V3 A. van Itterbeek and L. Thys, Physica **5**, 298 (1938).
V4 A. van Itterbeek and L. Thys, Physica **5**, 640 (1938).
V5 A. van Itterbeek and P. Mariens, Physica **7**, 938 (1940).
V6 A. van Itterbeek and O. van Paemel, Physica **8**, 133 (1941).
V7 A. van Itterbeek and P. Slootmakers, Physica **15**, 897 (1949).
V8 G. S. Verma, J. Chem. Phys. **18**, 1352 (1950).
W1 C. S. Wang Chang and G. E. Uhlenbeck, *On the Transport Phenomena in Rarefied Gases*, Applied Physics Laboratory, The Johns Hopkins University, Report CM-443, UMH-3-F (1948).
W2 C. S. Wang Chang, *On the Dispersion of Sound in Helium*, Applied Physics Laboratory, The Johns Hopkins University, Report CM-467, UMH-3-F (1948).
W3 P. J. Westervelt, J. Acoust. Soc. Am. **22**, 319 (1950).
W4 G. W. Wheland, *Advanced Organic Chemistry* (John Wiley and Sons, Inc., New York, 1949).
W5 G. W. Willard, J. Acoust. Soc. Am. **12**, 438 (1941).
W6 A. O. Williams, Jr., and L. A. Labaw, J. Acoust. Soc. Am. **16**, 231 (1945).
W7 F. H. Willis, J. Acoust. Soc. Am. **19**, 242 (1947).
W8 O. B. Wilson, Jr. and R. W. Leonard, J. Acoust. Soc. Am. **23**, 624(A) (1951).
Z1 I. F. Zartman, J. Acoust. Soc. Am. **21**, 171 (1949).
Z2 M. W. Zemansky, *Heat and Thermodynamics* (McGraw-Hill Book Company, Inc., New York, 1943).
Z3 A. J. Zmuda, J. Acoust. Soc. Am. **23**, 472 (1951).

Editor's Comments on Paper 27

27 Bömmel and Dransfeld: *Excitation of Very-High-Frequency Sound in Quartz*

As interest in the properties of ultrasonic radiation and its many applications increased during the post–World War II period, the problem of producing very-high-frequency radiation (that is, well above 1 MHz) at high intensity became more pressing. The standard piezoelectric crystal technique demands thinner and thinner crystals as the desired frequency increases, unless harmonics of the fundamental are employed, and with the latter there is a definite limitation on the intensity that can be produced. This paper represents an early attempt to extend the frequency range above the gigahertz figure (10^9 Hz) by exciting a quartz rod in the electrical field of a cavity resonator. This provided a powerful tool in the further use of ultrasonic radiation.

Both authors are German-born physicists who resided in the United States and worked at the Bell Telephone Laboratories at the time the work reported here was done. They have since returned to Germany.

EXCITATION OF VERY-HIGH-FREQUENCY SOUND IN QUARTZ

H. E. Bömmel and K. Dransfeld

Bell Telephone Laboratories,
Murray Hill, New Jersey
(Received September 5, 1958)

Recently Baranskii[1] reported the generation of longitudinal ultrasonic waves along the x axis of a 1.5-cm-thick quartz plate in the frequency range from 10^8 to 2×10^9 cps, using standard optical diffraction methods for detection.[2]

We have performed a series of further experiments at 10^9 to 2.5×10^9 cps as part of a program to extend research with ultrasonic waves into the microwave frequency range:

(1) A 9-mm quartz cube with optically flat surfaces was placed in the electric field of a coaxial cavity which had two opposite slots for optical observation. By orienting either the x or y axis of the cube parallel to the field, longitudinal or transverse waves could be generated, distinguished by their different diffraction angles for light. Figure 1 shows an example for longitudinal waves.

FIG. 1. Photograph of longitudinal sound beam in quartz cube. Frequency ν = 1000 Mc/sec. In the picture on the right, the HF Power was turned off.

(2) A rectangular rod $4 \times 1 \times 0.5$ cm, with the x axis along its length, was placed with one end in the electric field of the cavity, parallel to the x axis [see Fig. 2(a)]. In this way it was possible to observe the traveling sound waves in the sections l_1 inside the cavity as well as l_2 outside. From the intensity of the diffracted light at different points of the rod, the sound attenuation at 1000 Mc/sec was estimated to be of the order of 2 to 4 db/cm.

(3) The direction of sound propagation depends on the angle between the end face f and the x axis [see Fig. 2(b)]. An angle of 10° caused, under otherwise identical conditions, a change in the direction of propagation of about 25° to

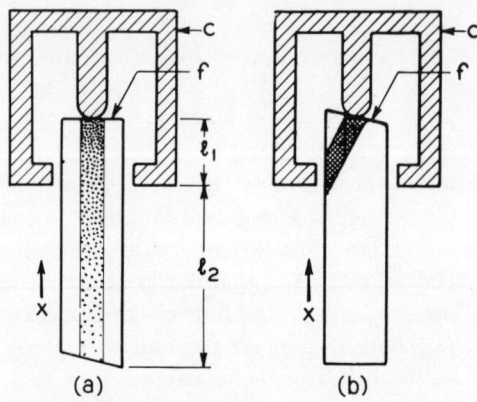

FIG. 2. Sound propagation in rectangular quartz rod extending out of coaxial cavity c. (a) End face perpendicular to x axis. (b) End face under angle of 10° to x axis.

30°, as can be explained by the elastic properties of quartz.

(4) A cylindrical x-cut quartz rod 2.5 cm long and 0.3 cm in diameter, with optically flat and parallel end faces, was placed between two identical cavities, one of which served as a transmitter and the other as a receiver. This arrangement showed the reconversion of acoustical into electromagnetic energy: at 1500 and 2.5 Mc/sec, pulses of 1 microsecond duration fed into the transmitter cavity were received about 5 microseconds later in the second cavity, corresponding to the acoustic delay in the rod. The ratio of electrical input to output power was slightly greater than 10^7. There was no comparable electrical leakage between both cavities.

Our observations, in particular the one mentioned under 3, suggest that even for homogeneous electric fields these ultrasonic waves are excited at the surface of the quartz crystals. This is understandable, because it is only here that under the influence of the uniform piezoelectric stress a displacement can be initiated. This displacement then propagates as a traveling wave into the interior.

If one assumes that most of the electric field of the cavity is concentrated in a volume V of the quartz rod, an estimate shows that the acoustic power F_1 leaving the surface is

$$F_1 = [k^2(\lambda q/V)Q]P_i, \qquad (1)$$

where k = electromechanical coupling constant (for quartz about 10^{-1}), λ = acoustic wavelength, q = cross section of the rod, Q = quality factor of the cavity, and P_i = power input. This holds for critical coupling of the input lead into the cavity. The order of magnitude of F_1 agrees with a qualitative estimate from the observed intensity of light diffracted by the sound waves.

Relation (1) can also be shown to be reciprocal, i.e., the conversion at the receiving cavity from acoustic into electromagnetic energy takes place with the same efficiency:

$$P_{\text{out}} = [k^2(\lambda q/V)Q]F_2, \qquad (2)$$

where P_{out} = output of the receiving cavity, F_2 = incident sound energy flux. Therefore, neglecting sound absorption, the output from the second cavity should be smaller than the power input into the first cavity by a factor $[k^2(\lambda q/V)Q]^2$. Allowing for sound absorption, this agrees with the results reported under 4.

---

[1]K. N. Baranskii, Doklady Akad. Nauk S.S.S.R. 114, 517 (1957) [translation: Soviet Phys. Doklady 2, 237 (1957)].

[2]See for example L. Bergmann, Ultrasonics and Its Scientific and Technical Applications (G. Bell and Sons, Ltd., London, 1938).

Editor's Comments on Papers 28 and 29

28 Debye and Sears: *On the Scattering of Light by Supersonic Waves*

29 Brillouin: *Diffusion of Light by a Transparent Homogeneous Body*

"On the Scattering of Light by Supersonic Waves" presents a good example of research stimulated by knowledge of work done in another, but closely related, field. Brillouin's studies of light scattering by pretersonic radiation in homogeneous media (Paper 29, this volume) suggested to Debye the search for light scattering by artificially produced sound radiation of lower than pretersonic frequency. Sears performed the necessary experiment and the results are set forth in their paper.

The reader is reminded that at the time this paper appeared the term "supersonic" was used to denote sound waves of frequency above the audible limit. This term was later adopted by aerodynamical scientists and engineers to denote speeds in air in excess of the velocity of sound. The term "ultrasonic" was then chosen to designate high-frequency (inaudible) sound.

Peter Debye (1884–1966) was a theoretical physical chemist. Born in the Netherlands, he held professorships at a number of European universities before coming to the United States in 1940, where he served as a professor of chemistry at Cornell University from 1940 until his retirement in 1950. He is well known for his pioneering research in the theory of specific heats, electrolytes, X-ray scattering, and the properties of high polymers.

Francis W. Sears (1898–   ) is an American physicist, educated at the Massachusetts Institute of Technology, where he was also a professor of physics from 1941 to 1955. From 1955 until his retirement in 1964, he was a professor of natural philosophy at Dartmouth College. It was at M.I.T. that the Debye–Sears effect, described in the paper presented here, was discovered. In a recent letter to the editor of this volume, Sears sheds an interesting historical sidelight on the discovery. Debye was in the United States to give lectures on X-ray diffraction. He had the idea that just as one gets Bragg reflection when X-rays are incident on the surface of a crystal, one might expect to get something analogous to Bragg reflection when a beam of visible light is reflected from the surface of a liquid that is irradiated by a beam of high-frequency sound waves normally incident on the surface. At Debye's suggestion the experiment was tried by Sears, unsuccessfully; the expected result was not observed. Sears did observe, however, that when the light was incident at a glancing angle so that some of it actually moved through the liquid nearly parallel to the surface, it showed a light diffraction pattern similar to that produced by the standard diffraction grating. When Sears showed this to Debye, further, more elaborate experimental and theoretical work led to the article presented. This was a clear case of serendipity, i.e., the turning up of a result not originally sought.

Exploration of the phenomenon of the scattering of light in passing through an isotropic medium goes back to the experimental research of John Tyndall (1868) and the theoretical work of Lord Rayleigh on the blue color of the sky (1871). Tyndall believed that the observed scattering was due to material particles in the medium. Rayleigh showed that light could be scattered by the molecules of a gas as well. The problem took another turn when M. Smoluchowski endeavored, in 1908, to explain the opalescence of substances at the critical state by assuming that light is scattered by rather large fluctuations in density taking place under such conditions. The problem was examined in greater detail by Einstein in 1912. He represented the density fluctuations by a summation of harmonic oscillation terms of many frequencies, but in assigning energy to each oscillation he used the classical equipartition law. This was rather strange since Einstein was familiar with, and indeed had had much to do with the development of, quantum theory, in which the classical equipartition law is replaced by Planck's quantized expression. Brillouin noted this point and in this paper, presented in English translation, generalized Einstein's treatment to its quantum theory equivalent. At the same time, he put more emphasis on the assumption that the density fluctuations are due to the superposition of high-frequency sound waves moving in all directions through the scattering medium. Hence his explanation of light scattering by a material medium is reduced to a problem of the scattering of light waves by sound waves associated with the thermal motions of the constituent molecules of the medium. This started a train of experimental and theoretical studies in an attempt to learn more about these "hypersonic," or as they are now called, "pretersonic," waves in all material media. They can be studied only through their effect in the scattering of light. Brillouin elaborated his theory in a longer paper in 1922, although in trying to extend it to the scattering of X-rays he made an inappropriate extrapolation.

The extent to which this early work by Brillouin has influenced more recent work in pretersonics is clearly shown in a special issue of the *Journal of the Acoustical Society of America* containing papers delivered in Brillouin's honor at the meeting of the Society in San Diego in 1969 (*J. Acoust. Soc. Amer.*, **49**, 994, 1971).

Brillouin's two early papers actually had an even more extensive influence on the development of acoustics than that already stated. They stimulated Debye to examine the scattering of light by an artificially produced ultrasonic beam in a liquid. This opened up the whole subject of the exploration of sound fields by light radiation. [See the paper on the Debye–Sears effect (Paper 28, this volume) and the associated editorial introduction.]

Leon Brillouin (1889–1969) was a French-born theoretical physicist, who spent the latter part of his professional career in the United States. He did important work in many phases of modern physics, including quantum theory, quantum statistics, solid-state physics, thermodynamics, and information theory.

PROCEEDINGS
OF THE
NATIONAL ACADEMY OF SCIENCES

ON THE SCATTERING OF LIGHT BY SUPERSONIC WAVES

By P. Debye and F. W. Sears

Department of Physics, Massachusetts Institute of Technology

Communicated April 23, 1932

1. *Introduction.*—In a paper published in 1922 Leon Brillouin[1] treated the problem of light scattering. In accordance with the fact that for low temperatures Einstein's theory of specific heat has to be abandoned for Debye's theory, Brillouin attributes the thermal density fluctuations in the body, which, in his theory, as in a previous theory of Einstein's,[2] are responsible for the scattering to a superposition of sound waves. He tries to apply his theoretical results to the explanation of x-ray scattering. We know now that this application is far from correct, as for such short waves the electronic density changes due to the atomic or molecular structure are much more important than the thermal fluctuations. For light waves, however, with a wave-length much longer than molecular distances, Brillouin's analysis leads to some remarkable results. They can be stated in the following manner. Suppose the primary light travels in figure 1 in a direction characterized by a vector $\vec{S_0}$ of length unity in this direction. Let it be assumed that of the scattered light a part is observed traveling in another direction characterized by a unit vector $\vec{S}$. Then firstly, of the sound waves of all possible directions, only those are important for the scattering which are traveling in or opposite to the direction of the vector $\vec{s} = \vec{S} - \vec{S_0}$. This can also be expressed by saying that the planes of the sound waves have to be situated such that the scattered light can be considered as optically reflected by these planes. But there is a second limitation. Of all the sound waves of direction $\pm \vec{s}$, only those of a definite wave-length Λ are effective. This wave-length is $\Lambda = \lambda/s$, if λ is the wave-length of the light and s is the length of the vector $\vec{s}$, which is $2 \sin \Theta/2$, calling Θ the angle between the primary and the secondary ray. This last condition can be expressed by saying that the consecutive planes of maximum density in the sound wave must be

separated by a distance Λ such that the well-known relation of Bragg holds. In this case the light rays reflected by the consecutive planes will have path differences of one wave-length (of light) each and therefore the reflections will be strong. So we are left with only two sound waves, traveling with the velocity of sound q, one in the direction $+\vec{s}$ and the other in the direction $-\vec{s}$. The frequency of the reflected light, according to Doppler's principle, will be changed, and instead of the primary frequency v, we should find in the scattered light the two frequencies:

$$v = v_0 \left[1 \pm 2n\ q/c\ \sin\ \theta/2\right]$$

where c is the velocity of light in vacuo and n is the index of refraction of the medium.

Gross[3] reports that with an echelon he has been able to photograph several new components of a spectral line created by the scattering process.

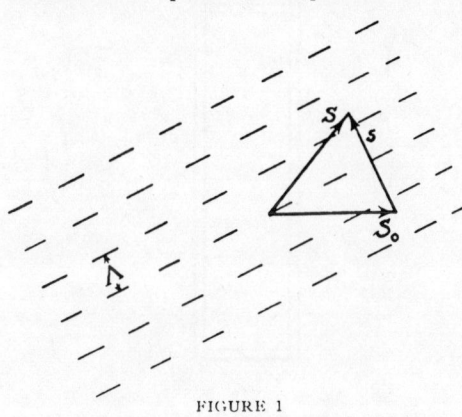

FIGURE 1

Recently, Meyer and Ramon[4] have published actual photographs of the effect. In contrast with Gross's results they were able to obtain two components only, one at either side of the primary line, which is in accordance with the theory. The shift is approximately 0.06 A. U. and checks satisfactorily with the value calculated from the known velocities of sound in the liquids used. Besides the two components, a central line of unchanged frequency v_0 appears which may partly be due to unavoidable traces of dust in the liquid, and partly to thermal fluctuations of the index of refraction as a result of molecular rotation, an effect which was not considered in the derivation of Brillouin's theory.

2. *Experiments.*—Brillouin[1] mentions the possibility of experimental verification of his calculations, making use of elastic waves set up in a liquid by a quartz crystal driven by a high-frequency oscillator, and it occurred to us that it would be interesting to try the scattering of light by these "artificial" sound waves. The experiment was set up in the following way. In a trough of rectangular cross-section, figure 2, filled with a liquid (benzene, carbon tetrachloride, etc.) a quartz crystal Q was immersed. The leads to the silvered faces of this crystal could be connected with a radio-frequency oscillator. Vibrations set up in the crystal

in this way excite supersonic waves traveling in the liquid in the direction QE. Perpendicular to this direction a parallel beam of light from a slit S and a lens L_1 passes through the liquid. The parallel beams, scattered by the illuminated part of the liquid are focussed by a telescope lens L_2. As soon as the crystal is connected to the oscillator, spectra of different orders (like the spectra of an ordinary grating), appear at the left and right of the central image of the slit. The number of orders which can be observed depends on the intensity of the vibrations. Their spacing depends only on the frequency and increases if the frequency of the oscillator is increased.

As in ordinary grating spectra, the blue is the least deviated, and the red the most. With a mercury arc source, the different mercury lines can be observed in every order. Under favorable conditions as to the intensity of the vibration more than 10 spectra to the right and to the left have been obtained. In order to see if the supersonic waves are markedly transmitted by a solid, a glass block of cross-section equal to the trough was immersed in the liquid. Light passed through the liquid in front of and behind the block showed the spectra, but through the block itself produced only the central slit image. We may safely expect, however, that using a

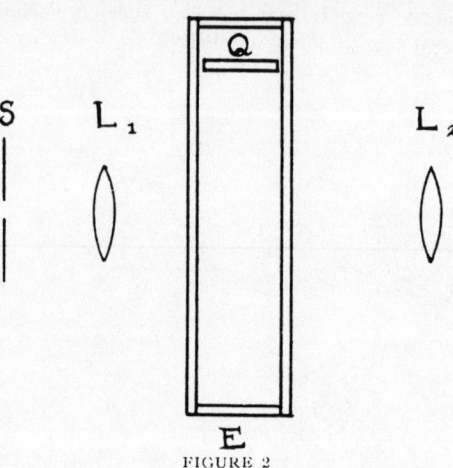

FIGURE 2

higher intensity the effect will also be seen in the solid. No marked decrease in the number of orders visible could be observed in passing the light through portions of the liquid at different distances from the crystal with liquids like benzene or carbon tetrachloride. With glycerine, however, this effect is very pronounced, no doubt due to its high viscosity.

Fixing the attention on one of the spectra, preferably of higher order, one can observe that it attains its maximum intensity if the trough is turned through a small angle such that the primary rays are no longer parallel to the planes of the supersonic waves. Different settings are required to obtain highest intensities in different orders. If the trough is turned continuously in one direction, starting from a position which gave the highest intensity to one of the orders, the intensity decreases steadily, goes through zero, increases to a value much smaller than the first maximum, decreases to zero a second time and goes up and down again passing through a still smaller maximum. The same series of events occurs in turning the trough in the other direction. To make these observations it is necessary to watch the pattern carefully. One thing, however, which can be seen at first glance is the fact that the number of orders visible on the right and the left of the central image is different except for the case in which the primary rays are passing exactly parallel to the planes of the supersonic waves. Turning the trough continuously changes the number of orders to the right or the left in the same way as has been described for the intensity.

The whole effect is rather brilliant and can easily be projected on a screen so as to make it visible to an audience. Figure 3 shows a photograph of the effect obtained in passing monochromatic light of wavelength $\lambda = 5461$ A. U. through toluene, the frequency of the supersonic waves being 5.7×10^6.

3. *Interpretation.*—As a tentative explanation it was first thought that the primary rays passing through the liquid in sheets would, as a consequence of the periodic density variations due to the supersonic waves, acquire phase differences of periodical nature. The light emerging from the grating created in the liquid would show interference patterns of the same general kind as observed with an ordinary grating. The fact that the supersonic waves are not standing waves should make no difference.

FIGURE 3

This explanation, however, fails to explain the observed intensity distribution in the various orders, as can be shown by a more detailed analysis, and a theory has been developed based on the assumption of a volume scattering, in which every volume of the liquid contributes to the total scattering in accordance with Maxwell's equations. In this way, it seems at first, following Brillouin's theory, that one would expect only one reflection for a definite angle of incidence of the primary light and one other reflection on the other side for the same angle taken with negative sign. Moreover, light passing parallel to the planes of the supersonic wave should show no effect at all. These results are evidently at variance with the observations. Taking into account, however, that the dimensions of the illuminated volume of the liquid are finite it can easily be shown that in our case Bragg's reflection angle is not sharply defined and that reflection should occur over a rather appreciable angular range. If l is the length of the path of light in the liquid, Λ the wave-length of the supersonic waves and λ the wave-length of the light, then two quantities are of importance; namely, the quotients l/Λ and Λ/λ. Only if

l/Λ is large compared to Λ/λ does a sharp definition of Bragg's angle exist. Working with a frequency of 10^7 cycles, Λ is about 0.1 mm., l is of the order 10 mm. and λ is about 0.5×10^{-3} mm. In this case, therefore, $l/\Lambda = 100$ and $\Lambda/\lambda = 200$, the quotient of these two quantities is 1/2 and cannot be considered as large. A detailed analysis shows that in such a case reflection will occur over a range of angles left and right of the critical angle which follows from Bragg's relation. Moreover the intensity variations predicted by the theory in varying the angle continuously are just the same as described in relating the rather peculiar experimental results on this point.

We are, however, still left with another difficulty. The theory predicts only the first order spectrum to the right and the first order to the left. But in the theory so far it has been assumed that the variations of the index of refraction are of purely sinusoidal character. If they are not, then we can consider the disturbance as a superposition of variations of frequencies v, $2v$, $3v$, etc., with the corresponding wave-lengths Λ, $\Lambda/2$, $\Lambda/3$, etc., provided a marked dispersion of the velocity of sound in the frequency region considered does not exist. A departure from sinusoidal character therefore accounts for the existence of the higher order spectra. This departure may be due to the non-sinusoidal cha acter of the crystal vibrations, although higher harmonics may be produced by the scattering itself, if the intensity of the supersonic waves is high enough. This is in agreement with the fact that the higher orders fade out if the intensity of the vibrations of the crystal is decreased.

Adopting the theory, the measurement of the angles θ of the different orders ρ with respect to the central image provides us with a measure of the wave-length of the supersonic wave in terms of the wave-length of the light which has been used. In fact, as in an ordinary grating

$$\sin \theta \rho = \rho \lambda / \Lambda.$$

The frequency can easily be determined with an ordinary wavemeter and so we get a very simple method for the determination of the velocity of sound.

The following table shows some preliminary measurements performed with a very simple spectrometer, together with values calculated from the density and the adiabatic compressibility.

	VELOCITY CALCULATED M./SEC.	FREQUENCY	VELOCITY OBSERVED M./SEC.	FREQUENCY	VELOCITY OBSERVED M./SEC.
Toluene	1290	1.7×10^6	1330	16.5×10^6	1310
Carbon tetrachloride	920	1.7×10^6	940	16.5×10^6	930

Up to now no indication of a change in velocity with frequency has been observed. If it exists, a more careful measurement of the angles for different orders should show it, as the measurement of say 10 orders simultaneously existing provides us with a frequency range of v to $10\,v$.

[1] L. Brillouin, *Ann. phys.*, **17**, 88 (1922).
[2] A. Einstein, *Ann. Phys.*, **33**, 1275 (1910).
[3] Gross, *Zeit. Phys.*, **63**, 685 (1930); *Naturwiss.*, **18**, 718 (1930); *Nature*, **126**, 201, 400, 603 (1930).
[4] Meyer and Ramon, *Phys. Zeitschr.*, **33**, 270 (1932).

29

Diffusion of Light by a Transparent Homogeneous Body

LEON BRILLOUIN

Translated expressly for this Benchmark volume by R. Bruce Lindsay from "Diffusion de la lumière par un corps transparent homogène," Compt. Rend. Acad. Sci., Paris, 158, 1331–1334 (1914).

When a ray of light traverses an isotropic body, part of the light is diffused in all directions. This phenomenon is visible in the blue of the sky[1] and in opalescence at the critical state.[2] I wish to complete here the theory given by Einstein,[3] in order to put it in agreement with the research of Debye on specific heats.[4]

A homogeneous body at a fixed temperature is in a state of continual vibration. We can resolve this vibration into the characteristic vibrations of the body. To each characteristic vibration Debye attributes an average energy that depends on its period in accordance with Planck's law. Among these vibrations the longitudinal waves produce fluctuations in density capable of diffusing light.

Let a body be contained in a cubical enclosure ($0 < x < L$, $0 < y < L$, $0 < z < L$). The fluctuations Δ in density at a given moment can be expressed in the form

$$\Delta = \sum_\rho \sum_\sigma \sum_\tau B_{\rho\sigma\tau} \cos\frac{\pi\rho x}{L} \cos\frac{\pi\sigma y}{L} \cos\frac{\pi\tau z}{L}$$

Each term groups a series of condensations distributed regularly in space with a wavelength

$$\Lambda = \frac{2L}{\sqrt{\rho^2 + \sigma^2 + \tau^2}} \qquad (1)$$

This term has a period (in time) $\theta = \Lambda/U$, where U is the velocity of propagation of the sound waves in the medium. The $B_{\rho\sigma\tau}$, constantly variable, have a zero average value but a mean-square that is different from zero. The term corresponding to $\rho\sigma\tau$ represents a

---

[1] Lord Rayleigh, *Scientific Papers*, Dover, New York, 1932.
[2] M. Smoluchowski, *Ann. Physik*, **25**, 205 (1908).
[3] A. Einstein, *Ann. Physik*, **33**, 1275 (1910).
[4] P. Debye, *Ann. Physik*, **39**, 789 (1912).

certain potential energy that is easily calculated from the elastic constants of the body. This potential energy[5] is proportional to B^2. This is the potential energy associated with the degree of freedom of the frequency $f = 1/\theta$. It is, on the average, equal to one half the total energy of the degree of freedom, that is, half of

$$\phi(f) = \frac{hf}{e^{hf/kT} - 1} \tag{2}$$

Einstein, on the contrary, retaining the theory of the equipartitions of energy, attributed to each degree of freedom the energy $kT/2$.

We also ought to take account of the fact that the coefficients B are limited in number. The total number of these parameters is equal to three times the number of the molecules in the volume. This sets a limit to the wavelengths of the characteristic vibration Λ_{lim}, which Debye evaluates from

$$\Lambda_{\text{lim}}^3 = \frac{4\pi}{9} \frac{\mathrm{V}}{\mathcal{N}} \tag{3}$$

where V = molecular volume and $\mathcal{N}$ = Avogadro's number.

Let us now suppose that a ray of light with direction cosines α, β, and γ falls on a small volume $0 < x < l$, $0 < y < l$, $0 < z < l$ of our medium. Suppose that the incident wave is represented by

$$\mathbf{E} = \mathbf{A} \cos 2\pi n \left(t - \frac{\alpha x + \beta y + \gamma z}{V} \right) \tag{4}$$

where V is the velocity of light in the medium. We observe the diffused or scattered light which travels along the x axis a distance D. In carrying out the method of Einstein[6] we are led to the calculation of the integral

$$\iiint \frac{\frac{Uh}{\pi l}\sqrt{x^2 + y^2 + z^2}}{e^{Uh\sqrt{x^2 + y^2 + z^2}/\pi lkT}} \frac{\sin^2(x - l\lambda/2)\sin^2(y - l\mu/2)\sin^2(z - l\nu/2)}{(x - l\lambda/2)^2 (y - l\mu/2)^2 (z - l\nu/2)^2} \, dx\, dy\, dz \tag{5}$$

(where U is the velocity of sound in the medium). This integral is taken over the interior of the sphere with its center at 0 and having radius $R = \pi l/\Lambda_{\text{lim}}$. We use the notation

$$x = \frac{\pi l}{2L}\rho, \qquad y = \frac{\pi l}{2L}\sigma, \qquad z = \frac{\pi l}{2L}\tau$$

$$\lambda = \frac{2\pi n}{V}(1 - \alpha) \qquad \mu = \frac{2\pi n}{V}\beta \qquad \nu = \frac{2\pi n}{V}\gamma$$

The above integral is easily calculated. Because of the presence of the second factor we see that the most important part of the integral will be confined to the vicinity of the point $x_c = l\lambda/2$, $y_c = l\mu/2$, $z_c = l\nu/2$, which I shall call the critical point. One can then take outside the first factor in the sum and integrate the second from $-\infty$ to $+\infty$, getting π^3 as a

---

[5] Einstein, *loc. cit.*, p. 1284, formula (6).
[6] Einstein, *loc. cit.*, pp. 1286–1293.

result. The following remark justifies this approximation. If we integrate the second factor in the neighborhood of the critical point through a sphere of radius m such that the integral equals π^3 to within 1 part in 10,000, we are able to verify that in the interior of this sphere the first factor will remain constant to within 1 part in 10,000.

If we replace the various quantities by their values, we finally find for the average intensity of the l_y component of the scattered light,

$$\overline{e_y^2} = \phi\left(\frac{U}{\lambda}\sqrt{2(1-\alpha)}\right) \frac{v_0 \, (\partial \varepsilon/\partial v)_0^2}{(\partial^2 \psi/\partial v2)_0} \left(\frac{2\pi}{\lambda}\right)^4 \frac{\Phi}{(4\pi D)^2} \frac{\Lambda_0^2}{2} \tag{6}$$

Here $U =$ velocity of sound, $v =$ specific volume, $\varepsilon =$ dielectric constant for the wavelength λ of the light employed, $\Phi =$ volume of the light scatterer, $\psi(v) =$ work required to compress isothermally 1 gram of medium from specific volume v_0 to specific volume v, and $\alpha =$ cosine of the angle between the incident light ray and the scattered light ray. The function $\phi(f)$, which in our formula replaces the coefficient kT used by Einstein, has the following meaning: for $0 < f < U/\Lambda_{\text{lim}}$, the term $\phi(f)$ is given by formula (2). For $f > U/\Lambda_{\text{lim}}$, $\phi(f) = 0$. A formula analogous to (6) holds for the component e_z.

If the incident light is natural (i.e., unpolarized), we find for the ratio of the intensities of the scattered to the incident light,

$$\frac{I_{\text{dif.}}}{I_{\text{inc.}}} = \phi\left(\frac{U}{\lambda}\sqrt{2(1-\alpha)}\right)\left(\frac{1+\alpha^2}{2}\right) v \frac{(\partial \varepsilon/\partial v)^2}{\partial^2 \psi/\partial v^2} \left(\frac{2\pi}{\lambda}\right)^4 \frac{\Phi}{(4\pi D)^2} \tag{7}$$

For wavelengths in the visible spectrum, the function ϕ remains equal to kT to 1 part in 100, and again we find Einstein's result. For ultraviolet light ϕ is equal to kT for forward scattering but diminishes as the angle increases and can thus vary by as much as 1 part in 10 for backscattering. Thus there is more light scattered in the direction of propagation of the ray than in the opposite direction. For wavelengths of the order of 10^{-8} cm (X-rays), the function ϕ, which is again equal to kT for forward scattering, decreases rapidly as the scattering angle increases, and after a certain limiting scattering angle it vanishes. The scattered energy will then be confined in a core around the incident ray. The experimental results of Friedrich[7] would be readily explained, it seems to me, by the superposition of curves (for the decrease of intensity as a function of scattering angle) corresponding to the different wavelengths of the incident X-rays.

---

[7] Friedrich, *Phys. Z.*, **14**, 317 (1913).

Editor's Comments on Papers 30 Through 34

30 **Tisza:** *On the Thermal Superconductivity of Liquid Helium II and the Bose–Einstein Statistics*

31 **Landau:** *The Theory of Superfluidity of Helium II*

32 **Atkins:** *Third and Fourth Sound in Liquid Helium II*

33 **Rudnick and Shapiro:** *Fourth Sound in Helium II*

34 **Shapiro and Rudnick:** *Experimental Determination of the Fourth Sound Velocity in Helium II*

The peculiar properties of liquid helium at temperatures close to absolute zero (below the λ point) began to attract attention in the mid 1930s. The most striking of these is the complete loss of viscosity below the λ point, leading to the assumption of the existence of a superfluid form of helium, usually referred to as helium II. Laszlo Tisza was one of the first to attempt an explanation of this astonishing behavior. In the short paper presented here he developed in some detail London's earlier suggestion that below the λ point, liquid helium behaves as an ideal gas obeying Bose–Einstein statistics, that is, it is a quantum fluid. In this way he produced an explanation of the thermal superconductivity of liquid helium. He also showed that, in addition to ordinary sound propagated at the normal velocity for a compressional wave in a fluid, liquid helium should also propagate another kind of sound, strictly a temperature wave but also accompanied by excess pressure propagation, with a velocity less than that of ordinary sound. This is what has come to be called *second sound*, although this terminology was not introduced by Tisza.

Tisza's point of view with regard to liquid helium II and its properties was adversely criticized by L. Landau, who presented his own view of the matter a few years after Tisza's work appeared. Landau's paper is presented in English translation following that of Tisza. Although Landau probably had the better of the argument over the more appropriate theoretical model of liquid helium, Tisza certainly deserves the credit for having first predicted the existence of second sound.

Laszlo Tisza (1907–) is a Hungarian-born physicist whose education was obtained in Germany and Hungary and who acquired professional experience in France and Russia. He has been a resident in the United States since 1941 and is a professor of physics at the Massachusetts Institute of Technology. He is noted for research in statistical thermodynamics and the theory of solids.

We have presented Tisza's pioneer paper on second sound in liquid helium (Paper 30). Landau disagreed with Tisza's way of looking at the properties of liquid helium and developed another point of view, now the accepted one. This is presented in the paper " The

Editor's Comments on Papers 30 Through 34

Theory of Superfluidity of Helium II." From the complete paper we have chosen to reprint, in English translation by D. ter Haar, the Introduction and Sections 1, 2, 7, and 8. It is in Section 2 that Landau introduces for the first time the concept of phonon in connection with liquid helium, a concept previously restricted to lattice vibrations in solids. In Section 7 he develops the fundamental hydrodynamical equations for Helium II, and in Section 9 he applies these to the propagation of sound in this medium. He then derives the analytical expression for the velocity of second sound.

Lev Davidovich Landau (1908–1968) was a Russian physicist who became a professor at the Moscow State University and a member of the Academy of Sciences of the USSR. He worked in many fields and made noteworthy contributions to solid-state physics, liquid state (especially liquid helium) superconductivity, nuclear physics, and quantum mechanical field theory. He received the Nobel Prize in physics in 1962.

Atkins in his article, "Third and Fourth Sound in Liquid Helium II," predicted, on the basis of rather plausible theoretical considerations, the existence of third and fourth sound in liquid helium II. His discussion ties in well with the earlier theoretical work of Landau and with the later experimental verification by Rudnick and his collaborators.

Kenneth R. Atkins (1920–) is a British-born physicist whose professional career since 1951 has been spent in Canada and the United States. A professor of physics at the University of Pennsylvania since 1954, he is particularly well known for his research in solid-state and low-temperature physics.

As has already been emphasized in the previous papers by Tisza, Landau, and Atkins, the subject of sound propagation in liquid helium II has, for some time, been an engrossing one in physical acoustics. Among the more recent research in this field that by Rudnick and his collaborators deserves special attention. We, therefore, reproduce two papers from this school. The first is "Fourth Sound in Helium II" by I. Rudnick and K. A. Shapiro. Fourth sound is a peculiar compressional wave in superfluid helium passing through a porous medium. Although its existence was also predicted by Pellam in 1948, the experiments reported in this paper were the first verification.

The second paper, "Experimental Determination of the Fourth Sound Velocity in Helium II" by Shapiro and Rudnick, pursues further the problem of fourth sound.

Isadore Rudnick (1917–), an American physicist who has spent the larger part of his professional career as a professor at the University of California in Los Angeles, is noted for his work in atmospheric acoustics, acoustical cavitation, and ultrasonics, as well as his investigations of liquid helium.

Kenneth A. Shapiro, one of Rudnick's graduate students, has recently been a member of the Technical Staff of T.R.W. Systems, Redondo Beach, California.

30

On the Thermal Superconductivity of Liquid Helium II and the Bose–Einstein Statistics

LASZLO TISZA

Translated expressly for this Benchmark volume by R. Bruce Lindsay from "Sur la supraconductibilité thermique de l'hélium II liquide et la statistique de Bose–Einstein," Compt. Rend. Acad. Sci., Paris, 207, 1035–1037 (1938).

F. London has interpreted the λ point of liquid helium as the point of condensation of a Bose–Einstein gas.[1] The hydrodynamical properties of such a system permit one to understand the paradoxical kinetic phenomena observed in helium II. We present first a brief exposition of London's idea[2] in a somewhat generalized form, and then a summary of our results on thermal superconductivity.[3]

In the neighborhood of absolute zero, helium is in a condensed state, with a certain practically constant density ρ_0. The state of one atom in the field of the rest can be characterized by a Bloch wave vector **k**. Because of the isotropy of the liquid the corresponding energy will be of the form $\varepsilon = \varepsilon_0 + \varepsilon'(k^2)$, and the atom in the state **k** will carry out translational motion with velocity

$$v = \frac{1}{\hbar}\frac{\partial \varepsilon}{\partial \mathbf{k}} = \frac{2}{\hbar}\frac{\partial \varepsilon'}{\partial k^2}\mathbf{k}$$

The atoms will be distributed on the energy levels ε according to Bose–Einstein statistics. To density ρ_0 there corresponds a temperature of condensation T_0 (of the λ point); for $T < T_0$ only a portion of the atoms with density ρ' will be distributed over the excited states, and the remainder with density $\rho'' = \rho_0 - \rho'$ will be condensed into the state ε_0 with $\mathbf{k} = \mathbf{v} = 0$. ρ' depends on the temperature alone. From the specific heat anomaly we deduce that $\rho' \cong \rho_0 (T/T_0)^5$. Even though our system is homogeneous in ordinary space, we can speak of two phases in the momentum space (phases I and II, or even atoms of kinds I and II).

Atoms of type I, having velocity of translation, exert a pressure p'; determined by a

---

[1] *Nature*, **141**, 643 (1938).

[2] *Proc. Roy. Soc. (London)*, **A153**, 576 (1936); *Phys. Rev.* (in press).

[3] See, for example, Rollin, *Physica*, **3**, 557 (1935); also J. F. Allen, R. E. Peierls, and Uddin, *Nature*, **150**, 62 (1937); and W. H. Keesom, A. P. Keesom, and B. F. Saris, *Physica*, **5**, 281 (1938).

unique parameter, either T or ρ', and compensated by van der Waals forces. If these were suppressed, p' would reduce to the pressure of an ideal gas. We shall prefer to give to p' a definition analogous to that of osmotic pressure. We shall demonstrate in a following note that very fine capillary tubes are semipermeable for helium II in that they permit phase II to pass through and refuse passage to phase I. The pressure p' can then be defined and measured with the help of the work $p'\,dV$ necessary for the adiabatic compression $-dV$ of phase I to the state of total constant density ρ_0. At the time of adiabatic compression there is, to a first approximation, no transition between the two phases. The transition of two atoms from phase I to phase II is accompanied by the release of the heat $W \sim \mathcal{N}kT$.

Nonhomogeneities of temperature also always involve nonhomogeneities of $\rho'(T)$ and $p'(T)$. If the system is adiabatically isolated, these quantities tend to vanish in stimulating a current of phase I in the direction of $-\mathrm{grad}\,T$. But since ρ_0 is constant, this current must be compensated if there is not to be a macroscopic motion by a current of phase II in the opposite direction such that the total current $\rho'u' + \rho''u'' = 0$, where u' and u'' are the average velocities. The two currents coupled in this way are described to a first approximation (neglecting irreversible effects of the second order) by a single system independent of the equations of hydrodynamics.

For phase I we then have

$$\frac{d\rho'}{dt} + \rho'\,\mathrm{div}\,u' = 0 \tag{1}$$

$$p'\frac{du'}{dt} = -\frac{\rho''}{\rho_0}\,\mathrm{grad}\,p' \tag{2}$$

The general discussion of these nonlinear equations provides the same difficulties one encounters in classical hydrodynamics. But if the currents stimulated by the nonhomogeneities of T are weak, they become equivalent to the propagation equation

$$\nabla^2 T = \frac{1}{v^2}\frac{\partial^2 T}{\partial t^2} \tag{3}$$

The same type of equation applies to p' and ρ'. The quantity v is the velocity of temperature wave s,

$$v = \sqrt{\frac{\rho''}{\rho_0}\frac{dp'}{d\rho'}} \sim \sqrt{\frac{kT}{m}\left(1 - \frac{T}{T_0}\right)^5} \tag{4}$$

Equations (3) and (4) do not apply to the usual arrangement of two reservoirs at temperatures T_1 and T_2, respectively, joined by a tube of cross section a full of helium II. The current I has a source of $\rho'u's$ atoms per second in the hot reservoir and a source $-\rho'u's$ atoms per second in the cold reservoir. The quantity of heat $W_1 \sim \rho'u'skT_1$ is produced per second by the hot reservoir, and $W_2 \sim \rho'u'skT_2$ is released to the cold reservoir. [The decrease in temperature is produced by an adiabatic dilatation of $p'(T_1)$ to $p'(T_2)$.] The energy $W_1 - W_2$ serves, at first, to accelerate the currents. After the stationary regime has been attained, the energy is dissipated by effects of the second order. The system is analogous to a thermal machine utilized for the transmission of heat. The calculation of the superconductivity would demand the determination of u' as a function of $T_1 - T_2$. But this is not possible because of the turbulence resulting from the fact that the heat current grows more slowly than $T_1 - T_2$. But the qualitative properties and the order of magnitude of the phenomenon correspond perfectly to the thermomechanical effect we have just described. The quantitative verification of the theory should be possible with the help of Eqs. (3) and (4).

The theory will not hold for $T < 0.8$ or $0.9°$, since in this region the Debye specific heat is more important than the anomalous effect considered here.[4] In very narrow capillary tubes, viscosity plays an important role.

---

[4] See N. Kurti and F. Simon, *Nature*, **142**, 207 (1938).

1
The Theory of Superfluidity of Helium II†

The quantisation of an arbitrary system of interacting particles (a liquid) is performed by means of introducing the operators of the density and of the velocity of the liquid: the commutation rules between these operators are determined (§ 1). From the results of this quantisation the general character of the distribution of the energy levels in the spectrum of a quantum liquid is determined (§ 2). The temperature dependence of the heat capacity of helium II is investigated (§ 3). It is shown that at absolute zero a quantum liquid can possess the property of superfluidity (§ 4). At non-zero temperatures it is found that two motions—a superfluid and a normal—can simultaneously exist in helium II. This can be described by means of the conception of the superfluid and normal parts of the liquid; the λ-point in helium II is connected with the disappearance of the " superfluid " part of the liquid (§ 5). The experiments made to measure the heat conductivity and viscosity of helium II are interpreted; the thermomechanical effects in helium II are considered (§ 6). A system of hydrodynamic equations is advanced describing the macroscopic motion of helium II (§ 7). By means of these equations the propagation of sound is investigated and it is shown that two velocities of sound must exist in helium II (§ 8).

LIQUID helium is known to possess a number of peculiar properties at temperatures lower than the λ-point. Of these properties the most important one is superfluidity discovered by P. L. Kapitza[1]—the lack of viscosity during the flow of helium through a thin capillary or slit.

All these properties, including the fact that helium exists as a liquid right down to absolute zero, obviously cannot be explained by the classical theory and are connected with quantum phenomena.

L. Tisza[2] suggested that helium II should be considered as a degenerate ideal Bose gas. He suggested that the atoms found in the normal state (a state of zero energy) move through the liquid without

---

†*J. Phys. USSR*, **5**, 71, 1941.

friction. This point of view, however, cannot be considered as satisfactory. Apart from the fact that liquid helium has nothing to do with an ideal gas, atoms in the normal state would not behave as "superfluid". On the contrary, nothing could prevent atoms in a normal state from colliding with excited atoms, i.e., when moving through the liquid they would experience a friction and there would be no superfluidity at all. In this way the explanation advanced by Tisza not only has no foundation in his suggestions but is in direct contradiction to them.

1. The Quantisation of the Motion of Liquids

An arbitrary system of interacting particles (a liquid) can be described in classical theory by means of the density ρ and the flow of mass j, which are determined in the following manner. Let R be the radius-vector of an arbitrary point in space and r_α—the radius-vector of a particle with a mass m_α. Then ρ is determined as

$$\rho = \sum_a m_\alpha \delta(r_\alpha - R), \tag{1.1}$$

δ being the three-dimensional δ-function and the summation is extended over all particles in the system. The volume-integral $\int \rho dV$ gives the total mass of the system. Similarly the density j of the flow of the mass is determined as

$$j = \sum_a m_\alpha v_\alpha \delta(r_\alpha - R) = \sum_a p_\alpha \delta(r_\alpha - R)$$

(v_α, p_α are the velocity and the momentum of the particle m_α).

It must be emphasised that in such a description of a liquid there is no averaging in that sense in which it is done in statistics. This description proceeds from the microscopic picture as all the particles possess (at a given moment) definite coordinates r_α and velocities v_α.

When passing over to the quantum theory ρ and j must be regarded as certain operators the form of which must be determined. For the sake of simplicity suppose that the system consists of one particle only. Then the classical density is $\rho = m\delta(r - R)$. The operator ρ must be determined in such a way that its mathematical expectation $\int \psi(r) \rho \psi(r) dV$ [$\psi(r)$ being the wave function of the particle] equals the density of the mass at the point R, i.e., $m|\psi(R)|^2$. From this it follows that the operator ρ must have the same form

$\rho = m\delta(r - R)$ and in the case of an arbitrary system of particles—correspondingly the form (1.1).

The classical density of the flow for one particle is $j = p\delta(r - R)$. It is easy to see that the corresponding operator is

$$j = \tfrac{1}{2}[p\delta(r - R) + \delta(r - R)p],$$

where p is the usual operator of momentum:

$$p = \frac{\hbar}{i}\nabla$$

(∇ denotes the differentiation with respect to r). Actually the mathematical expectation of j is

$$\int \psi(r)^* j\psi(r)dV = \frac{\hbar}{2i}\int \psi^*\nabla\delta(r - R)\psi dV + \frac{\hbar}{2i}\int \psi^*\delta(r - R)\nabla\psi dV$$

or, integrating the first term by parts:

$$\int \psi^* j\psi dV = -\frac{\hbar}{2i}\int \psi\delta(r - R)\nabla\psi^* dV + \frac{\hbar}{2i}\int \psi^*\delta(r - R)\nabla\psi dV$$

$$= \frac{\hbar}{2i}\{\psi^*(R)\nabla\psi(R) - \psi(R)\nabla\psi^*(R)\},$$

i.e., exactly what it ought to be. For an arbitrary system we have, similarly

$$j = \tfrac{1}{2}\Sigma[p_\alpha\delta(r_\alpha - R) + \delta(r_\alpha - R)p_\alpha], \tag{1.2}$$

$$p_\alpha = \frac{\hbar}{i}\nabla_\alpha.$$

We now determine the commutation rules. For the density ρ we obviously have

$$\rho_1\rho_2 - \rho_2\rho_1 = 0 \tag{1.3}$$

[ρ_1, ρ_2 denote $\rho(R_1), \rho(R_2)$, respectively].

For the sake of brevity let us consider only one term from each of the sums (1.1) and (1.2) when determining the commutation rules, as the operators corresponding to different particles commute with each other. To determine the commutation of ρ with j we write

$$j_1\rho_2 - \rho_2 j_1 = \frac{m\hbar}{2i}\{[\nabla\delta(r-R_1) + \delta(r-R_1)\nabla]\delta(r-R_2)$$
$$- \delta(r-R_2)[\nabla\delta(r-R_1) + \delta(r-R_1)\nabla]\}.$$

To simplify the expression on the right-hand side we note that the operators of the form

$$\delta(r-R_1)\nabla\delta(r-R_2)$$

can be transformed in the following way:

$$\delta(r-R_1)\nabla\delta(r-R_2) = \delta(r-R_1)(\nabla\delta(r-R_2)) + \delta(r-R_1)\delta(r-R_2)\nabla,$$

where in the first term $(\nabla\delta(r-R_2))$ denotes simply a gradient of the δ-function, i,e., ∇ is no longer an operator. Owing to the presence of the factor $\delta(r-R_1)$ in this term one can write $(\nabla\delta(R_1-R_2))$ instead of $(\nabla\delta(r-R_2))$. In this way

$$\delta(r-R_1)\nabla\delta(r-R_2) = \delta(r-R_1)(\nabla\delta(R_1-R_2)) + \delta(r-R_1)\delta(r-R_2)\nabla.$$

Similarly

$$\nabla\delta(r-R_1)\delta(r-R_2) = \delta(r-R_2)\nabla\delta(r-R_1) + \delta(r-R_1)(\nabla\delta(R_1-R_2)).$$

The result is

$$j_1\rho_2 - \rho_2 j_1 = \frac{\hbar}{i} m\delta(r-R_1)\nabla\delta(R_1-R_2)$$

or for an arbitrary system

$$j_1\rho_2 - \rho_2 j_1 = \frac{\hbar}{i}\rho_1\nabla\delta(R_1-R_2). \quad (1.4)$$

(It makes no difference whether we write ρ_1 or ρ_2 on the right-hand side in view of the presence of the δ-function of $R_1 - R_2$.)

In a similar way the commutation rules between the components of the vector j with each other can be obtained. The calculation in this case is longer and we will not enter into it here.

We introduce the operator v of the velocity of the liquid according to

$$j = \tfrac{1}{2}(\rho v + v\rho), \quad (1.5)$$

$$v = \tfrac{1}{2}\left(\frac{1}{\rho}j + j\frac{1}{\rho}\right). \quad (1.6)$$

It will be more convenient to use the operator v instead of the operator of the flow j.

For the commutation rule of ρ with v we have

$$v_1\rho_2 - \rho_2 v_1 = \tfrac{1}{2}\left(\frac{1}{\rho_1}j_1 + j_1\frac{1}{\rho_1}\right)\rho_2 - \tfrac{1}{2}\rho_2\left(\frac{1}{\rho_1}v_1 + v_1\frac{1}{\rho_1}\right)$$

$$= \frac{1}{2\rho_1}(j_1\rho_2 - \rho_2 j_1) + \tfrac{1}{2}(j_1\rho_2 - \rho_2 j_1)\frac{1}{\rho_1}$$

or, on inserting (1.4):

$$v_1\rho_2 - \rho_2 v_1 = \frac{\hbar}{i}\nabla\delta(R_1 - R_2). \tag{1.7}$$

The commutation rules for the components of v are found to be

$$v_{1i}v_{2k} - v_{2k}v_{1i} = \frac{\hbar}{i}\delta(R_1 - R_2)\frac{1}{\rho_1}(\operatorname{curl} v)_{ik}, \tag{1.8}$$

where $(\operatorname{curl} v)_{ik}$ denotes the difference

$$\frac{\partial v_k}{\partial x_i} - \frac{\partial v_i}{\partial x_k}.$$

Further on we shall also need the commutation rule between ρ and curl v. By applying the operation curl (with a differentiation with respect to coordinates R_1) to both sides of the equation (1.7) we get

$$\operatorname{curl} v_1 \cdot \rho_2 - \rho_2 \cdot \operatorname{curl} v_1 = 0. \tag{1.9}$$

It is easy to see that by applying the formulae obtained to the macroscopic movement of the liquid we get, as required, the usual hydrodynamic equations written in an operational form. The energy of a unit volume of a classical liquid considered macroscopically is

$$\frac{\rho v^2}{2} + \rho\varepsilon(\rho),$$

where $\varepsilon(\rho)$ is the internal energy of a unit mass of the liquid. It is supposed that the energy ε depends only on the density ρ of the liquid; this corresponds to the macroscopic character of the argument and is connected with a statistical averaging. For a microscopic investigation this supposition is, of course, invalid.

The corresponding quantum operator is†

$$\frac{v\rho v}{2} + \rho\varepsilon(\rho).$$

The Hamiltonian H of the liquid is an integral over the volume

$$H = \int \left\{ \frac{v\rho v}{2} + \rho\varepsilon(\rho) \right\} dV. \tag{1.10}$$

For the derivative of the density ρ with respect to time one has

$$\dot{\rho} = \frac{i}{\hbar}(H\rho - \rho H).$$

We shall denote temporarily the coordinates of the point at which ρ is taken by the index 1, and the coordinates of the variable point in the region of integration in (1.10) by the index 2. Then

$$\dot{\rho}_1 = \frac{i}{\hbar}\int \{\tfrac{1}{2}[v_2\rho_2 v_2\rho_1 - \rho_1 v_2\rho_2 v_2] + [\rho_2\varepsilon(\rho_2)\rho_1 - \rho_1\rho_2\varepsilon(\rho_2)]\}dV_2$$

In view of (1.3) the second term under the sign of the integration vanishes, and the first can be written as

$$\tfrac{1}{2}[v_2\rho_2(v_2\rho_1 - \rho_1 v_2) + (v_2\rho_1 - \rho_1 v_2)\rho_2 v_2]$$

or, by introducing (1.7):

$$\frac{\hbar}{2i}\nabla\delta(R_2 - R_1)(v_2\rho_2 + \rho_2 v_2) = \frac{\hbar}{i}\nabla\delta(R_2 - R_1)j_2,$$

In this way
$\dot{\rho}_1 = \int \nabla\delta(R_2 - R_1)j_2 dV_2 = -\int \delta(R_2 - R_1)\operatorname{div} j_2 dV_2 = -\operatorname{div} j_1,$
i.e., we come to the continuity equation in operational form:

$$\frac{\partial\rho}{\partial t} + \operatorname{div}\frac{\rho v + v\rho}{2} = 0. \tag{1.11}$$

---

† The operator $\dfrac{v\,\rho\,v}{2}$ can also be written in the form

$$\frac{\rho v^2 + v^2\rho}{4}.$$

In a similar way the derivative:

$$\dot{v} = \frac{i}{\hbar}(Hv - vH),$$

can be calculated, which brings us to the equation

$$\frac{\partial v_i}{\partial t} + \tfrac{1}{2}\left(v_k \frac{\partial v_i}{\partial x_k} + \frac{\partial v_i}{\partial x_k} v_k\right) = -\frac{1}{\rho}\frac{\partial}{\partial x_i}\frac{\partial \varepsilon}{\partial \rho}, \qquad (1.12)$$

i.e., Euler's equation in an operational form ($d\varepsilon/d\rho$ is the pressure p of the liquid).

It must be again emphasised that the equations (1.11), (1.12) are less general than the commutation rules (1.3)–(1.9), which are also valid for an exact, microscopical investigation of the liquid.

2. The Energy Spectrum of a Quantum Liquid

In the classical hydrodynamics of ideal liquids it is shown that if, at a certain moment of time, the motion is potential (curl $v = 0$) in the whole volume of the liquid, it will be potential for all other moments of time (Lagrange's theorem). It appears that this classical theorem finds its analogy in quantum hydrodynamics.

According to the commutation rules (1.9), curl v always commutes with the density ρ. The components of curl v, however, do not commute, generally speaking, either with each other or with the components of velocity v [when the operation curl is applied to the equation (1.8) the right-hand side does not vanish]. Therefore, curl v does not, generally speaking, commute with the Hamiltonian, i.e., is not conserved.

An exception is the case when over the whole volume of the liquid curl $v = 0$. In this case we have zero in the right-hand side of (1.8) and curl v commutes with ρ and v and, therefore, also with the Hamiltonian.†

In this way curl v is conserved if it is zero. In other words, a quantum liquid always possesses stationary states in which curl v equals zero over the whole volume of the liquid. Such a state might

---

† Not only with the Hamiltonian (1.10), but also with all other functions containing v, ρ and their derivates of any order with respect to the co-ordinates.

be called, by analogy to classical hydrodynamics, a state of potential motion of the liquid.

Concerning these results an analogy can be made with the angular momentum M in quantum mechanics. The commutation of two components of M with each other leads to the third component of M, with the result that all the components of M commute with each other if they are all equal to zero. It is also known that there exist no states with an infinitely small angular momentum, its first non-zero eigenvalues are of the order of $\hbar$. This is a consequence of the fact that the commutation rules are inhomogeneous—their left-hand sides are quadratic in M and the right-hand sides are linear.

A similar statement can be advanced concerning curl v in quantum hydrodynamics. Namely, no states can exist in which curl v would be non-zero, but arbitrarily small over the whole volume of the liquid. In other words, between the states of the potential (curl $v = 0$) and vortex (curl $v \neq 0$) motions of a quantum liquid there is no continuous transition.

From this the principal features of the energy spectrum of a liquid directly follow. The presence of a gap between the states of the potential and vortex motions means that between the lowest energy levels of vortex and potential motions a certain finite energy interval must exist. As to the question which of these two levels lies lower, apparently both cases are logically possible. It will be shown below that we get the phenomenon of superfluidity if we suppose that the normal level of the potential motions lies lower than the normal level of votex motions. Hence we must suppose that this very case exists in liquid helium. It must be remarked, however, that, as only one quantum liquid exists, liquid helium, the question as to whether such a distribution of the levels and hence the property of superfluidity is a general property of a quantum liquid cannot be solved experimentally.

This brings us to the following picture of the distribution of the levels in the energy spectrum of liquid helium (it must be emphasised that we do not here refer to the levels for single helium atoms but to the levels corresponding to the states of the whole liquid). This spectrum is made up of two superimposed continuous spectra. One of them corresponds to the potential motions and the other—to vortex motions. The lowest level of the vortex spectrum is situated above the lowest level of the potential spectrum, this latter level being the normal unexcited state of a liquid; the energy interval between these two levels we denote by Δ.

The value of the energy gap Δ cannot be calculated exactly. Its order of magnitude is

$$\Delta \sim \frac{\hbar^2 \rho^{2/3}}{m^{5/3}} \tag{2.1}$$

(m being the mass of the helium atom and ρ—the density of the liquid). This is the only quantity of the dimension of energy which can be built up from m, ρ and $\hbar$. This gives numerically $\Delta/k \sim 1°$, i.e., Δ, as was expected, is of the order of kT_λ, T_λ being the temperature of the λ-point of helium [cf. (3.8)].

Consider an excited level which is situated not too high above the beginning of the spectrum (vortex or potential one).

Every weakly excited state can be considered as an aggregate of a number of single "elementary excitations". As far as the excited levels of the potential spectrum are concerned, the potential internal motions of the liquid are longitudinal waves, i.e., these motions are sound waves. Therefore, the corresponding elementary excitations are simply sound quanta, i.e., phonons. The energy of the phonons is known to be a linear function of their momentum p:

$$\varepsilon = cp, \tag{2.2}$$

c being the velocity of sound. Thus, at the beginning of the potential spectrum the energy is proportional to the first power of the momentum.

An "elementary excitation" of the vortex spectrum might be called a "roton".† Those special reasons which stipulate a linear dependence of ε on p for phonons do not exist for rotons. For small momenta p the energy of the roton can be simply expanded in powers of p; in view of the isotropy of the liquid the expansion of the scalar ε in powers of the vector p only contains terms with even powers, so one may write:

$$\varepsilon = \Delta + \frac{p^2}{2\mu}, \tag{2.3}$$

---

† This name was suggested by I. E. Tamm.

where μ is an " effective mass " of the roton [in (2.2) and (2.3) the energy is measured from the normal state].†

If the number of phonons and rotons (per unit volume of the liquid) is not large, their aggregate can be regarded as a mixture of two ideal gases—a phonon gas and a roton gas. It is known that the phonon gas obeys Bose statistics. As to the rotons, they too probably obey Bose statistics. It must, however, be remarked that inasmuch as the energy of a roton always contains a quantity Δ large compared with kT (at low temperatures only when the aggregate of rotons can be treated as a gas) the difference between the Bose and Fermi statistics is not essential and one can use Boltzmann's distribution for the rotons.

---

† In a recent paper A. Bijl[3] investigated the properties of the energy spectrum of a liquid and came to the conclusion that there must be an energy gap between the normal and all excited states. This result does not seem to be plausible as it would mean, in particular, the impossibility of the propagation of sound waves with small frequencies in liquids.

†† This calculation had already been made in 1940 by A. Migdal whom I wish to thank for informing me of his results.

* * * * * * *

7. Equations of the Macroscopic Hydrodynamics of Helium II

Starting from the above considerations on the microscopic mechanism of the phenomenon of superfluidity a complete system of hydrodynamic equations can be built which would describe helium II in a macroscopic (phenomenological) way.

The starting point is the fundamental circumstance that the

motion of helium II must be described not by one velocity as in ordinary hydrodynamics but by two. One of these is the "superfluid" velocity (denoted by v_s) satisfying the condition

$$\operatorname{curl} v_s = 0. \tag{7.1}$$

On the boundary of a hard surface only the normal component of v_s becomes zero and not its tangential one corresponding to the fact that the superfluid liquid is not held back by friction against the walls of the vessel. For the "normal" velocity v_n of the liquid on the boundary with a hard surface the condition $v_n = 0$ (as in ordinary viscous liquids) must be fulfilled which expresses the fact that the normal liquid is brought to a standstill owing to the friction against the walls.

It turns out that the hydrodynamic equations with the two velocities v_s and v_n can be obtained absolutely unambiguously starting from the one condition only that they should satisfy all the conservation laws. These equations for the general case of arbitrary velocities are somewhat complicated and we shall not give them here and confine ourselves to a simplified deduction of the equations applicable to the motion with not too large velocities v_s and v_n.

Let j be the macroscopic current of the mass of liquid; it is a function of both the velocities v_s and v_n. For small velocities j can be expanded in powers of v_s and v_n. In the first approximation

$$j = \rho_s v_s + \rho_n v_n. \tag{7.2}$$

The coefficients ρ_s and ρ_n are obviously those which we called the densities of the superfluid and normal "parts" of the liquid. Their sum equals the real density ρ of helium II:

$$\rho = \rho_s + \rho_n; \tag{7.3}$$

ρ_s and ρ_n are, of course, functions of temperature. Note that the current j (7.2) is at the same time the momentum density, i.e., the momentum of a unit volume of liquid. ρ and j must satisfy the continuity equation:

$$\frac{\partial \rho}{\partial t} + \operatorname{div} j = 0. \tag{7.4}$$

We shall here write the equations applicable to a motion in which the viscosity of the "normal liquid" plays no part. Then the equation for the momentum conservation is written in the form:

$$\frac{\partial j_i}{\partial t} + \frac{\partial \Pi_{ik}}{\partial x_k} = 0 \qquad (7.5)$$

(the summation is extended over the indexes which are repeated), where the tensor Π_{ik} of the momentum current equals

$$\Pi_{ik} = p\delta_{ik} + \rho_n v_i^{(n)} v_k^{(n)} + \rho_s v_i^{(s)} v_k^{(s)}, \qquad (7.6)$$

(p being the pressure). To take into account the viscosity of the normal liquid we must add to Π_{ik} the terms expressed in the ordinary way through the coefficients of viscosity and the derivatives of the velocity v_n with respect to the coordinates.

Further, the equation for the conservation of entropy takes the form:

$$\frac{\partial S\rho}{\partial t} + \text{div}\,(\rho S v_n) = 0 \qquad (7.7)$$

(S is the entropy per unit mass of helium II). The "entropy current" equals $\rho S v_n$, as the entropy is only transferred by the normal part of the liquid. If the viscosity of the normal part is taken into account, supplementary terms must be added to the right-hand side of (7.7) expressing the increase of the entropy owing to the irreversibility of the processes.

Finally, the last equation of the complete set of hydrodynamic equations we get equalising the acceleration dv_s/dt to the force acting on a unit of the "superfluid" mass. To determine this force imagine that the unit mass of liquid is displaced from the point 1 to the point 2 in such a way that the distribution of phonons and rotons is not changed. In other words, one might say that during the transfer only the "superfluid liquid" is displaced and the distribution of the normal liquid remains unchanged. The energy E of the liquid changes during such a transfer by

$$\left(\frac{\partial E}{\partial M}\right)_1 - \left(\frac{\partial E}{\partial M}\right)_2$$

(M being the mass of the liquid). Derivatives must be taken here at constant entropy (because the entropy is connected only with the normal liquid) and at a constant momentum of the motion of the normal mass of the liquid relative to the superfluid; † besides this the volume of the liquid is considered as a constant.

From the expression obtained for the change of energy it is seen that the quantity $\partial E/\partial M$ can be regarded as a "potential energy" of the superfluid liquid, so that the force acting upon it is

$$- \operatorname{grad} \frac{\partial E}{\partial M}.$$

To calculate the derivative $\partial E/\partial M$ we notice that the derivative of the energy at constant entropy and volume is equal to the derivative of the thermodynamic potential at constant pressure and temperature. The thermodynamic potential $M\Phi$ of the liquid (Φ is the potential per unit mass) can be written in the form of the sum of the thermodynamic potential $M\Phi_0(p, T)$ of the stationary liquid and the kinetic energy $P^2/2M_n$ of the relative motion of the superfluid and normal "parts":

$$M\Phi = M\Phi_0(p, T) + \frac{P^2}{2M_n},$$

P is here the momentum of the motion of the normal mass M_n relative to the superfluid. By differentiating $M\Phi$ with respect to M at constant p, T and P and remembering that the normal mass M_n is proportional (at a given p and T) to the total mass M, we get

$$\Phi_0 - \frac{P^2}{2M_n M}.$$

If we insert $P = M_n(v_n - v_s)$ and put the ratio of the densities in place of the ratio of the masses we finally find for the derivative $(\partial E/\partial M)_{S, V, P}$ the expression

$$\Phi_0 - \frac{\rho_n}{2(\rho_n + \rho_s)}(v_n - v_s)^2.$$

---

† The motion of the superfluid liquid may be considered as external conditions in which the phonons and rotons move. Therefore, the "Lagrange function" for the motion of the normal liquid does not simply depend on its velocity v_n, but on the difference of the velocities $v_n - v_s$. The conserved momentum is, therefore, a derivative of the Lagrange function with respect to $v_n - v_s$, i.e., the momentum of the relative motion.

It follows that the hydrodynamic equation for which we were looking is of the form:

$$\frac{dv_s}{dt} = \frac{\partial v_s}{\partial t} + (v \cdot \nabla)v_s = - \text{grad}\left\{\Phi - \frac{\rho_n(_n v - v_s)^2}{2(\rho_n + \rho_s)^2}\right\}$$

(the index of Φ_0 is left out). It can be written differently if we use (7.1):

$$(v_s \cdot \nabla)v_s = \text{grad}\,\frac{v_s^2}{2}.$$

In this way

$$\frac{\partial v_s}{\partial t} = - \text{grad}\left\{\Phi + \frac{v_s^2}{2} - \frac{\rho_n(v_n - v_s)^2}{2(\rho_n + \rho_s)}\right\}. \tag{7.8}$$

The equations (7.1)–(7.8) are a complete set of hydrodynamic equations for helium II.

For a stationary flow the left-hand side of (7.8) is zero; hence:

$$\frac{v_s^2}{2} - \frac{\rho_n(v_n - v_s)^2}{2(\rho_n + \rho_s)} + \Phi = \text{const.} \tag{7.9}$$

This equation together with the next one, (7.10), plays here the role of the Bernoulli equation.

Consider now the motions at which liquid may be considered incompressible. If we take the densities ρ_n and ρ_s and entropy S as constants we find from (7.4) and (7.7) that

$$\text{div } v_s = 0, \ \text{div } v_n = 0.$$

Now, for a stationary motion we have in (7.5) $\partial j/\partial t = 0$; by using

$$\frac{\partial v_k^{(s)}}{\partial x_k} \equiv \text{div } v_s = 0, \ \frac{\partial v_k^{(n)}}{\partial x_k} = 0$$

we can rewrite (7.5) in the form

$$\nabla p + \rho_n(v_n \cdot \nabla)\,v_n + \rho_s\,(v_s \cdot \nabla)\,v_s = 0.$$

Remembering that curl $v_s = 0$ we can write this equation as

$$\nabla\left(p + \rho_n\frac{v_n^2}{2} + \rho_s\frac{v_s^2}{2}\right) = \rho_n[v_n \wedge \text{curl } v_n].$$

We project this equation on the line of the current of the normal motion, i.e., on the direction of v_n. Then on the right-hand side we get zero, so that

$$p + \rho_n \frac{v_n^2}{2} + \rho_s \frac{v_s^2}{2} = \text{const.} \tag{7.10}$$

It must be emphasised that the expression (7.10) is constant for a stationary flow only along each of the lines of current of the normal motion; and the expression (7.9) is constant over the whole volume of the liquid.

If the temperature and pressure change little over the volume of the liquid, Φ can be expanded in powers of $\Delta T = T - T_0$, $\Delta p = p - p_0$; T_0, p_0 being the temperature and pressure at a certain point in the liquid:

$$\Phi = \Phi(p_0, T_0) - S\Delta T + \frac{\Delta p}{\rho}.$$

By inserting this into (7.9) we get

$$-\frac{v_s^2}{2} + \frac{\rho_n(v_s - v_n)^2}{2\rho} + S\Delta T - \frac{\nabla p}{\rho} = \text{const.}$$

By combining this equation with equation (7.10)

$$\Delta p + \frac{\rho_n v_n^2}{2} + \frac{\rho_s v_s^2}{2} = \text{const.}$$

we get

$$\Delta T + \frac{\rho_n}{\rho S}(v_n \cdot v_n - v_s) = \text{const.} \tag{7.11}$$

This relation, like (7.10), is valid along the current lines of the normal motion.

8. Propagation of Sound in Helium II

The equations obtained can be applied to the propagation of sound in helium II. The velocity of the motion in sound waves is as usual supposed to be small and the density, pressure and entropy are almost equal to their constant equilibrium values. The terms in

(7.6) and (7.8) which are quadratic with respect to the velocities can be neglected, and in (7.7) we can take the entropy ρS in the term div $(v_n S\rho)$ out of the sign of div as this term already contains the small quantity v_n. In this way the system of hydrodynamic equations for sound waves acquires the form

$$\frac{\partial \rho}{\partial t} + \text{div } j = 0; \tag{8.1}$$

$$\frac{\partial \rho S}{\partial t} + \rho S \text{ div } v_n = 0; \tag{8.2}$$

$$\frac{\partial j}{\partial t} + \nabla p = 0; \tag{8.3}$$

$$\frac{\partial v_s}{\partial t} + \nabla \Phi = 0. \tag{8.4}$$

By differentiating (8.1) with respect to time and inserting (8.3) we get:

$$\frac{\partial^2 \rho}{\partial t^2} = \nabla^2 p \tag{8.5}$$

Further, we have

$$\frac{\partial S}{\partial t} = \frac{1}{\rho}\frac{\partial \rho S}{\partial t} - \frac{S}{\rho}\frac{\partial \rho}{\partial t} = -S \text{ div } v_n + \frac{S}{\rho} \text{ div } j,$$

or

$$\frac{\partial S}{\partial t} = \frac{S\rho_s}{\rho} \text{ div } (v_s - v_n). \tag{8.6}$$

For the thermodynamic potential the relation

$$d\Phi = -S dT + V dp = -S dT + \frac{1}{\rho} dp$$

holds (V being the specific volume). Hence we have

$$\nabla p = S\rho \nabla T + \rho \nabla \Phi,$$

D

or by introducing ∇p from (8.3) and $\nabla \Phi$ from (8.4)

$$\rho_n \frac{\partial}{\partial t}(v_n - v_s) + \rho S \nabla T = 0. \tag{8.7}$$

Differentiating (8.6) with respect to time and introducing (8.7) we find:

$$\frac{\partial^2 S}{\partial t^2} = \frac{S^2 \rho_s}{\rho_n} \nabla^2 T. \tag{8.8}$$

Two equations (8.5) and (8.8) determine the propagation of sound in helium II. It is already seen from the fact that there are two equations that there must be two velocities of sound in helium II.

Write S, ρ, p, T in the form $S = S_0 + S'$, $\rho = \rho_0 + \rho'$, etc. where the quantities with a dash represent the small changes of the corresponding quantities stipulated by the sound wave and the quantities with index zero are their constant equilibrium values. Then we can write:

$$\rho' = \frac{\partial \rho}{\partial T} T' + \frac{\partial \rho}{\partial p} p', \quad S' = \frac{\partial S}{\partial T} T' + \frac{\partial S}{\partial p} p',$$

and equations (8.5) and (8.8) take the form

$$\frac{\partial \rho}{\partial p} \frac{\partial^2 p'}{\partial t^2} - \nabla^2 p' + \frac{\partial \rho}{\partial T} \frac{\partial^2 T'}{\partial t^2} = 0,$$

$$\frac{\partial S}{\partial p} \frac{\partial^2 p'}{\partial t^2} + \frac{\partial S}{\partial T} \frac{\partial^2 T'}{\partial t^2} - \frac{S^2 \rho_s}{\rho_n} \nabla^2 T' = 0.$$

We look for a solution of these equations in the form of a plane wave in which p' and T' are proportional to a factor $e^{i\omega(t-x/u)}$ (u being the velocity of sound) and then for the conditions of solubility we get the equation:

$$\begin{vmatrix} u^2 \dfrac{\partial p}{\partial \rho} - 1 & u^2 \dfrac{\partial \rho}{\partial T} \\ u^2 \dfrac{\partial S}{\partial p} & u^2 \dfrac{\partial S}{\partial T} - S^2 \dfrac{\rho_s}{\rho_n} \end{vmatrix} = 0$$

THE THEORY OF SUPERFLUIDITY OF HELIUM II

or

$$u^4\frac{\partial(\rho, S)}{\partial(p, T)} - u^2 \left(\frac{\partial S}{\partial T} + S^2 \frac{\rho_s}{\rho_n}\frac{\partial \rho}{\partial \rho}\right) + S^2 \frac{\rho_s}{\rho_n} = 0$$

[where $\partial(\rho, S)/\partial(p, T)$ denotes the Jacobian of the transformation from ρ, S to p, T]. By means of a simple transformation with the use of the thermodynamic relations this equation can be put in the form

$$u^4 - u^2 \left[\left(\frac{\partial p}{\partial \rho}\right)_S + \frac{TS^2}{C_v}\frac{\rho_s}{\rho_n}\right] + \frac{S^2 \rho_s T(\partial p/\partial \rho)_T}{\rho_s C_v} = 0 \quad (8.9)$$

(C_v being the heat capacity of a unit mass of helium II). This quadratic equation determines two velocities of sound in helium II.

If $\rho_s = 0$, i.e., at the λ-point, one of the roots of the equation (8.9) becomes zero and we get, as we ought, only one ordinary velocity of sound

$$u = \sqrt{\left(\frac{\partial p}{\partial \rho}\right)_S}.$$

Practically for all temperatures the heat capacities C_p and C_v are close to each other. According to the known thermodynamic formula in these conditions the isothermic and adiabatic compressibilities are also close to each other, i.e.,

$$\left(\frac{\partial p}{\partial \rho}\right)_S \approx \left(\frac{\partial p}{\partial \rho}\right)_T.$$

If we denote the common value of $(\partial p/\partial \rho)_T$ and $(\partial p/\partial \rho)_S$ as $\partial p/\partial \rho$ and the common value of C_p and C_v simply we get from the equation (8.9) two velocities of sound u_1 and u_2 in the form

$$u_1^2 = \frac{\partial p}{\partial \rho}, \quad u^2 = \frac{TS^2}{C}\frac{\rho_s}{\rho_n}. \quad (8.10)$$

In this way one of the velocities (u_1) is almost constant and the other (u_2) strongly depends on the temperature becoming zero at the λ-point. At a temperature 1·33° K we get a value of about 25 m/sec for u_2. At extremely low temperatures, $\rho_n^{(px)} \gg \rho_n^{(r)}$ one gets

$$u_2 = \frac{c}{\sqrt{3}} \quad (8.11)$$

In this way as the temperature tends to zero the velocity of sound tends to constant limits $u_1 = c$, $u_2 = c/\sqrt{3}$.

Third and Fourth Sound in Liquid Helium II†

K. R. ATKINS

Department of Physics, University of Pennsylvania, Philadelphia, Pennsylvania

(Received October 29, 1958)

This article discusses the possible existence of two hitherto undetected types of wave propagation in liquid helium II. Third sound is a surface wave of long wavelength on a liquid helium film during which the normal component remains stationary and the superfluid component oscillates parallel to the wall. To treat this properly it is necessary to consider temperature changes and evaporation from the surface of the film. Fourth sound may exist in narrow two-sided channels. The normal component again remains stationary and the superfluid component oscillates parallel to the wall, but the width of the channel must remain fixed and so there are oscillations in both total density and temperature.

1. INTRODUCTION

To discuss wave propagation in liquid helium II, it is necessary to write down two separate hydrodynamical equations, one for the superfluid component and the other for the normal component. In first sound the two components move in the same direction in phase, and there is a first-order oscillation of the density but only a second-order oscillation of the temperature. In second sound the two components move in opposite directions out of phase, and the temperature oscillation is then first-order while the density oscillation is only second-order. Under the special circumstances to be considered in this article, there may exist an essentially different type of oscillation, during which the normal component remains stationary and the superfluid component alone oscillates.

2. SURFACE WAVES ON A LIQUID HELIUM FILM

Surface waves on bulk liquid helium were first discussed by Atkins[1] in order to explain the variation of surface tension with temperature. The phase velocity of such a wave is

$$v_p = \left(\frac{g\lambda}{2\pi} + \frac{2\pi\sigma}{\rho\lambda}\right)^{\frac{1}{2}}, \quad (1)$$

where λ is the wavelength, g is the acceleration due to gravity, σ is the surface tension, and ρ is the density of the liquid. The surface waves which affect the surface energy and hence the surface tension have such short wavelengths that the term involving g is unimportant.

Kuper[2] has suggested that these surface waves may be relevant to the critical velocity of superflow of a liquid helium film. According to Landau,[3] when liquid helium II flows through a narrow channel the kinetic energy of flow of the superfluid component can be dissipated only by the creation at the wall of elementary excitations of energy ϵ and momentum p, and this is not possible until the velocity of flow exceeds ϵ/p. The flow is therefore frictionless up to a critical velocity v_{crit} corresponding to the creation of those excitations for which $\epsilon/p(=v_{\text{crit}})$ is a minimum. Kuper suggests that the relevant excitations are quantized surface waves, or *ripplons*. ϵ/p is then equal to the phase velocity as given by Eq. (1), and when $\lambda \sim 0.3$ cm it assumes a minimum value ~ 10 cm sec^{-1}, which is comparable with the experimental value of about 25 cm sec^{-1} for the critical velocity of the film.

However, Atkins[4] has pointed out that for a thin film the restoring force is not gravity but the forces which are responsible for the formation of the film, including the van der Waals forces of attraction between the helium atoms and the wall on which the film is formed. Also, it is important to use the formula for a surface wave on liquid of finite depth d equal to the film thickness:

$$v_p^2 = \left(\frac{f\lambda}{2\pi} + \frac{2\pi\sigma}{\rho\lambda}\right) \tanh\left(\frac{2\pi d}{\lambda}\right). \quad (2)$$

This is a minimum when λ is infinite and gives a critical velocity:

$$v_{\text{crit}}^2 = v_{p\infty}^2 = fd. \quad (3)$$

If f were due only to van der Waals forces, we would have

$$f = 3\alpha/d^4, \quad (4)$$

and the thickness d of the film at a height H above the surface of the bulk liquid would be given by

$$\alpha/d^3 = gH, \quad (5)$$

whence

$$v_{\text{crit}} = (3gH)^{\frac{1}{2}}$$
$$\sim 50 \text{ cm sec}^{-1} \quad \text{if} \quad H \sim 1 \text{ cm}. \quad (6)$$

This is again in good order of magnitude agreement with experiment. Actually there are other factors besides the van der Waals forces entering into the

---

† Supported by the National Science Foundation.
[1] K. R. Atkins, Can. J. Phys. **31**, 1165 (1953).
[2] C. G. Kuper, Physica **22**, 1291 (1956).
[3] L. D. Landau, J. Exptl. Theoret. Phys. (U.S.S.R.) **5**, 71 (1941); **11**, 91 (1947).
[4] K. R. Atkins, Physica **23**, 1143 (1957).

formation of the film,[5-7] and the exact form of f is complicated and not completely understood, although its magnitude is probably not very different from the value arising from van der Waals forces alone.

Although the above arguments may be valid at 0°K, at a finite temperature it is necessary to use the two-fluid theory of liquid helium II and to write separate hydrodynamical equations for the normal and superfluid components. Arkhipov[8] has attempted this, but has assumed, without explicitly stating his reasons, that div $\mathbf{v}_s = 0$ and div $\mathbf{v}_n = 0$. This assumption that the two components separately behave like incompressible fluids implies that there are no temperature gradients in the liquid $[(\partial/\partial t)(\rho S) + \rho S \operatorname{div} \mathbf{v}_n = 0]$. In the analysis to be presented in this article, temperature gradients play an important role.

3. THIRD SOUND IN A LIQUID HELIUM FILM

We are interested in the case of wavelengths very long compared with the thickness of the film. This is analogous to the classical case of a long wave on water in a shallow channel, and it is well known that the oscillatory motion of an element of the fluid is mainly in a direction parallel to the bottom of the channel. We shall argue that the superfluid component oscillates in this fashion with its velocity $\mathbf{v}_s$ almost parallel to the wall on which the film is formed. Because of its viscosity, the velocity v_{nx} of the normal component parallel to the wall is negligibly small compared with v_{sx}. (The z axis is perpendicular to the wall and the x axis is in the direction of propagation of the wave.) If a plane surface oscillates parallel to itself in the bulk liquid with angular frequency ω, the motion penetrates exponentially into the normal component with a characteristic "penetration depth" $(2\eta_n/\omega\rho_n)^{\frac{1}{2}}$. In the case of a film of thickness d, since the velocity of the normal component at the wall must be zero relative to the wall, it is clear that it is very difficult to make the normal component oscillate parallel to the wall if $d \ll (2\eta_n/\omega\rho_n)^{\frac{1}{2}}$. For the superfluid component we do not have to worry about viscosity and it is usually assumed that the tangential velocity at the wall can be finite. We shall in fact assume that the velocity of the superfluid component is independent of depth, although there is no direct experimental evidence for this.

To a first approximation, then, the normal component is at rest and there is a surface wave on the superfluid component, although there must be a small motion of the normal component in the z direction perpendicular to the wall to enable the rotons and phonons to distribute themselves uniformly throughout the depth of the film. At a peak of the wave an excess of superfluid has collected and the temperature at this point is lowered (see Fig. 1), while at a trough the temperature is raised, so that in addition to the pressure gradient present in the case of a classical liquid there is an additional restoring force due to the thermomechanical effect of the temperature gradient. Also, at the trough where the film is hot it will evaporate into the vapor phase, whereas at the peak the vapor will condense on to the film. There will be a flow of vapor from the troughs to the peaks, but we shall neglect the resulting small pressure gradients within the vapor and will assume that the pressure of the vapor on the surface of the film is everywhere equal to the value p_0 which it had in the absence of the surface wave.

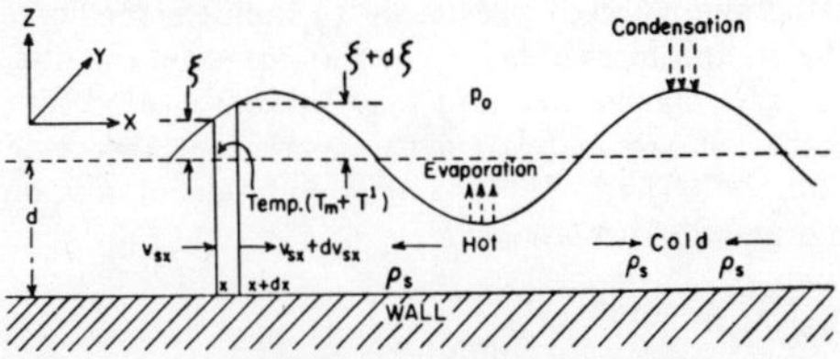

FIG. 1. Third sound in a liquid helium film.

In Fig. 1 the wave is propagating in the x direction and there is unit length of film in the y direction. At the point x the surface is raised a height ζ above its equilibrium position, the temperature is $T_m + T'$ where T_m is the mean temperature, and the velocity of the superfluid component is v_{sx}. The rate of evaporation of the film in g sec^{-1} per unit area of surface is[9]

$$\frac{dm}{dt} = \gamma \left(\frac{M}{2\pi RT}\right)^{\frac{1}{2}} \left(\frac{dp}{dT}\right)_{\text{v.p.c.}} T' = KT', \quad (7)$$

where $(dp/dT)_{\text{v.p.c.}}$ is the slope of the vapor pressure curve and γ is very close to unity. Considering a slab of film of width dx, conservation of mass gives the equation

$$\rho_s d \frac{\partial v_{sx}}{\partial x} + \rho \frac{\partial \zeta}{\partial t} + KT' = 0. \quad (8)$$

Considering the heat flow into the slab

$$\rho d C \frac{\partial T'}{\partial t} = \rho_s d \frac{\partial v_{sx}}{\partial x} ST - KLT'. \quad (9)$$

C, S, and L are the specific heat, entropy, and heat of vaporization per gram. The pressure at a point within the film has increased by

$$\delta p = \beta T' + \rho f \zeta. \quad (10)$$

$\beta = (dp/dT)_{\text{v.p.c.}}$ is the slope of the vapor pressure curve.

---

[5] K. R. Atkins, Can. J. Phys. **32**, 347 (1954).
[6] S. Franchetti, Nuovo cimento **4**, 1504 (1956); **5**, 183 (1957); **5**, 1266 (1957).
[7] K. R. Atkins, *Progress in Low-Temperature Physics*, edited by C. J. Gorter (North-Holland Publishing Company, Amsterdam, 1957), Vol. 2, p. 105.
[8] R. G. Arkhipov, J. Exptl. Theoret. Phys. (U.S.S.R.) **33**, 116 (1957) [translation: Soviet Phys. JETP **6**, 90 (1958)].

[9] Atkins, Rosenbaum, and Seki, Phys. Rev. **113**, 751 (1959).

f is the symbol used previously to indicate the force acting on unit mass of helium at the surface of the film. (The exact significance of f may be complicated, but is best discussed in relation to a particular theory of the film.[5-7]) The hydrodynamical equation of motion of the superfluid component,

$$\frac{\partial v_s}{\partial t} = -\frac{1}{\rho}\operatorname{grad}p + S\operatorname{grad}T, \quad (11)$$

becomes

$$\frac{\partial v_{sx}}{\partial t} = -f\frac{\partial \zeta}{\partial x} + \left(S - \frac{\beta}{\rho}\right)\frac{\partial T'}{\partial x}. \quad (12)$$

Second order terms have been consistently neglected.

A traveling wave solution

$$\zeta = \zeta_0 e^{i(\omega t - kx)} \quad (13)$$

plus similar expressions for T' and v_{sx}, is consistent with Eqs. (8), (9), and (12) if

$$\frac{\omega^2}{k^2} = \frac{\rho_s}{\rho}df + \frac{\rho_s}{\rho}ST\left[\left(S - \frac{\beta}{\rho}\right) - i\frac{Kf}{\rho\omega}\right] \bigg/ \left[C - i\frac{KL}{\rho\omega d}\right]. \quad (14)$$

If we ignore evaporation effects by putting $K=0$, then

$$v_p^2 = \frac{\omega^2}{k^2} = \frac{\rho_s}{\rho}\left[df + \frac{ST(S-\beta/\rho)}{C}\right]. \quad (15)$$

The first term is reminiscent of Eq. (3), but above 1°K the second term is by far the larger and the temperature gradients are consequently very important. However, if we retain the evaporation effects with $\gamma=1$ in Eq. (7), then, for frequencies small compared with 10^3 sec^{-1}, the imaginary terms in the numerator and denominator of the last term of Eq. (14) are large compared with the real terms, and so the velocity of third sound is given by

$$u_3^2 \simeq \frac{\rho_s}{\rho}df\left[1 + \frac{TS}{L}\right]. \quad (16)$$

The second term in the bracket is small and the principal difference from Eq. (3) is the factor ρ_s/ρ.

If ζ_0, T_0, and v_0 are the amplitudes of the oscillatory parts of film thickness, temperature, and superfluid velocity, then it is easily shown from Eqs. (8), (9), and (12) that

$$\frac{v_0}{u_3} = \frac{\rho}{\rho_s}\frac{\zeta_0}{d}. \quad (17)$$

u_3 has the order of magnitude of the critical velocity of the film and therefore, except very near the λ-point where ρ/ρ_s becomes large, a large fractional change in film thickness is possible without the velocity of the superfluid component approaching its critical value.

Also,

$$\frac{T_0}{T} = -i\frac{\rho dS\omega}{KL}\frac{\zeta_0}{d}. \quad (18)$$

At frequencies low enough to ensure that third sound is not appreciably attenuated, the temperature oscillation is immeasurably small and there is little hope that third sound could be detected by the type of receiver commonly used for second sound. The most hopeful method of observing third sound seems to be a direct observation of the oscillating film thickness by a suitable modification of the optical method developed by Jackson and his co-workers.[10] We are currently developing such an approach. If third sound exists, its velocity should give information about the quantity f and hence about the factors affecting the formation of the film.

4. FOURTH SOUND IN A VERY NARROW CHANNEL

In a narrow two-sided channel the normal component is again constrained to remain at rest ($\mathbf{v}_n = 0$), but, since the width of the channel remains constant, the oscillation of the superfluid component produces a first order change in density. The hydrodynamical equation of the superfluid component is

$$\frac{\partial v_{sx}}{\partial t} = -\frac{1}{\rho}\frac{\partial p}{\partial x} + S\frac{\partial T}{\partial x}. \quad (19)$$

The equation expressing conservation of mass,

$$\frac{\partial \rho}{\partial t} + \operatorname{div}(\rho_s \mathbf{v}_s + \rho_n \mathbf{v}_n) = 0, \quad (20)$$

becomes

$$\frac{\partial \rho}{\partial t} + \rho_s\frac{\partial v_{sx}}{\partial x} = 0. \quad (21)$$

Since no entropy is transported by the flow of the superfluid component,

$$\frac{\partial}{\partial t}(\rho S) + \frac{2\chi}{T}T' = 0, \quad (22)$$

where $\chi T'$ is the heat flowing per second into unit area of the wall when the liquid is at an excess temperature T'. A possible solution of Eqs. (19), (21), and (22) is a traveling wave for which

$$\frac{\omega^2}{k^2} = \frac{\rho_s}{\rho}\frac{\partial p}{\partial \rho} + \frac{\rho_n}{\rho}\left(\frac{\rho_s}{\rho_n}\frac{TS^2}{C}\right)\left(1 - i\frac{2\chi}{\rho\omega dC}\right)^{-1} \quad (23)$$

$$= \frac{\rho_s}{\rho}u_1^2 + \frac{\rho_n}{\rho}u_2^2\left(1 - i\frac{2\chi}{\rho\omega dC}\right)^{-1}, \quad (24)$$

---

[10] E. J. Burge and L. C. Jackson, Proc. Roy. Soc. (London) **A205**, 270 (1951); L. C. Jackson and D. G. Henshaw, Phil. Mag. **44**, 14 (1953); A. C. Ham and L. C. Jackson, Proc. Roy. Soc. (London) **A240**, 243 (1957).

where u_1 and u_2 are the velocities of first and second sound. At all temperatures the term in u_2^2 is small compared with the term in u_1^2 and so the attenuation per wavelength is never large. To an accuracy of better than 1%, the velocity of fourth sound is

$$u_4 \simeq \left(\frac{\rho_s}{\rho}\right)^{\frac{1}{2}} u_1. \tag{25}$$

The amplitude of the temperature oscillation is given by

$$\frac{T_0}{T} = -\frac{S}{C}\left(1 - i\frac{2\chi}{\rho\omega dC}\right)^{-1}\frac{\rho_0}{\rho}. \tag{26}$$

At high frequencies, the term involving χ can be neglected and the temperature oscillation is a first-order effect as well as the density oscillation. At low frequencies, however, there is plenty of time for the heat to escape to the wall and the fractional change in temperature is small compared with the fractional change in density. The amplitude of the oscillating part of the superfluid velocity is given by

$$v_0/u_4 = \rho_0/\rho_s. \tag{27}$$

Presumably fourth sound is strongly attenuated when v_0 exceeds the critical velocity for superflow through the channel. The critical velocity is much smaller than u_4 (except possibly in the immediate vicinity of the λ point) and so this interesting effect should occur for quite small fractional changes in density.*

ACKNOWLEDGMENTS

I am grateful to H. Flicker and H. Seki for stimulating discussions of these ideas.

---

* *Note added in proof.*—Dr. R. H. Walmsley has pointed out that, in Eqs. (23) and (26), S should be replaced by $(S-\alpha u_1^2)$, where α is the isobaric coefficient of expansion. $(1-\alpha u_1^2/S)$ is 1.6 at 2°K, 0 near 1.1°K, −12.5 at 0.8°K, and −7 near 0°K. Equation (25) therefore remains a good approximation. The temperature amplitude T_0 of Eq. (26) is somewhat modified. It is slightly increased above 1.1°K, becomes zero near 1.1°K, and changes phase with an increased magnitude below 1.1°K.

FOURTH SOUND IN He II[†]

I. Rudnick and K. A. Shapiro
Physics Department, University of California, Los Angeles, California
(Received July 25, 1962)

In addition to the well-known first and second sound in helium II, two other unattenuated modes of wave propagation in this liquid are possible and have been identified as "third sound" and "fourth sound." Both are characterized by the fact that the normal component of the fluid is locked in place and only the superfluid component oscillates in the wave propagation. Third sound, which has been recently detected,[1] is found in helium II films and depends for its existence on body forces. It is an oscillation in thickness of the film, and the temperature and pressure variations are very small. Fourth sound is a compressional wave in which the pressure and temperature variations are first-order quantities. Its wave velocity can be obtained from the linearized thermohydrodynamic equations by setting the velocity of the normal component equal to zero and is given by the expression[2,3]

$$C_4^2 = (\rho_s/\rho)C_1^2 + (\rho_n/\rho)C_2^2(1-2\alpha C_1^2/s), \quad (1)$$

where C_1, C_2, and C_4 are, respectively, the first, second, and fourth sound velocities, ρ_s and ρ_n are, respectively, the superfluid and normal fluid densities, ρ is the density of helium II, α is the isobaric coefficient of expansion, and s is the specific entropy of the fluid. In the range between 2.1°K and 1.2°K, $-2\alpha C_1^2/s$ is positive and of the order of magnitude of unity. In this temperature range, errors of approximately 1% or less are incurred by retaining only the first term in Eq. (1).

Pellam[4] in 1948 first pointed out the existence of such an unattenuated wave in a superleak[5] (a porous medium in which the normal fluid component is locked). He obtained the expression

$$C_4^2 = (\rho_s/\rho)C_1^2 + (\rho_n/\rho)C_2^2. \quad (2)$$

In addition to this mode he found a root to the wave velocity equation which was a diffusion wave and had a limiting velocity of zero when the normal fluid component was firmly locked. He identified (2) as a second-sound wave and the latter mode as a first-sound wave. Actually the wave whose root is given by Eq. (2) is, in the limit that the normal fluid component becomes unlocked, first sound, and in the same limit the diffusion wave becomes second sound. The question of whether one regards the unattenuated mode as a special case of first sound or second sound is unimportant since in fact it is neither, and Atkins[3] in a different context—namely, mode propagation in a narrow channel—appropriately gave it a unique name, fourth sound.

The purpose of this Letter is to present experimental results which verify the existence of fourth sound.

The apparatus consisted of a standing wave tube 3.16 cm long and 1 cm i.d. filled with Gelman Polypore AM-6 filter material[6] and terminated at each end by identical transducers, one of which served as a sound source and the other as a receiver. The filter material is described by the manufacturer as having an average pore size of 0.45 ± 0.02 micron and a porosity of 85%. Each transducer consists of a gold-coated film of Mylar approximately 7×10^{-4} cm thick which is stretched in a plane parallel to a flat back electrode and is separated from the electrode by a distance of the order of 10^{-3} cm. The source transducer is driven by an alternating voltage (100 volts) applied between the back electrode and the gold-coated outside surface of the film. This generates a mechanical oscillation of the Mylar film of twice the voltage frequency. The receiver transducer is polarized by a dc voltage of 200 volts, and the voltage variations produced by capacitance changes associated with diaphragm displacements are measured by a high-impedance electrical circuit. Thus the signal picked up has twice the frequency of the electric signal energizing the source and cross-talk problems are greatly reduced.

The system acts like a closed-closed tube with a fundamental frequency which ranged from 3286 cps at 1.19°K to 1874 cps at 2.1°K. Figure 1 contains the results of the measurements. The curve labelled "fourth sound" is a plot of Eq. (1). For comparison purposes a curve giving the velocity of first sound is also plotted. With a given driving voltage the received signal was a monotonically decreasing function of temperature, the sharpest drop occurring in the neighborhood of the λ point. At 2.1°K the level was 30 db below that at 1.19°K. There was absolutely no indication of a signal when the temperature

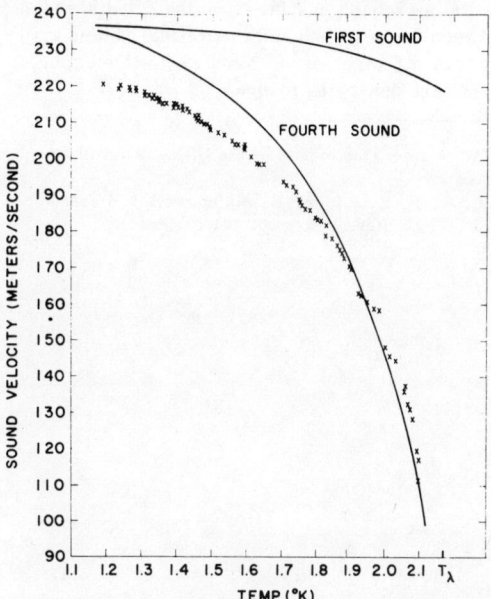

FIG. 1. Measured values of the compressional wave velocity in a half-wave resonant tube filled with superleak material. The curve labelled fourth sound is a plot of Eq. (1). The first-sound velocity is also shown.

was 0.01°K above the λ point (or indeed at any temperature above the λ point) although it was searched for at levels approximately 60 dB below the signal levels at 1.19°K. When the superleak was removed very substantial signals were observed throughout this temperature range.

The measured Q of the resonant mode decreased with increasing temperature. Its value was 82 at 1.3°K and 25 at 1.79°K. If these Q's are assumed to be due to the attenuation coefficient of fourth sound one obtains values of 0.006 cm^{-1} and 0.02 cm^{-1}, although it is much more likely that the Q's were limited because the locking of the normal fluid component was incomplete. This could be caused by residual normal fluid motion in the larger pores of the filter material, gaps between tube walls and filter material, etc. In any case the data lend support to the result that fourth sound is, in the ideal case, unattenuated.

The spirit of this investigation has been to detect the existence of fourth sound with only such accuracy as is necessary to lend credence to the results. The absolute accuracy of the measurements is probably not far from the 6% discrepancy between the observed and theoretical values at the lowest temperature. More definitive experiments are now being planned.

---

*Work supported in part by the Office of Naval Research.

[1]C. W. F. Everitt, K. R. Atkins, and A. Denenstein, Phys. Rev. Letters $\underline{8}$, 161 (1962).

[2]Walmsley in reference 3 finds a different expression for the bracketed term in Eq. (1), namely $(1 - \alpha C_1^2/s)^2$.

[3]K. R. Atkins, Phys. Rev. $\underline{113}$, 962 (1959).

[4]J. R. Pellam, Phys. Rev. $\underline{73}$, 608*(1948).

[5]The presence of the superleak introduces no added attenuation provided one can neglect irreversible heat conduction in this material or in the locked normal fluid.

[6]Gelman Instrument Company, 106 North Main Street, Chelsea, Michigan.

Copyright © 1965 by the American Physical Society
Reprinted from Phys. Rev., **137**, A1383–A1391 (1965)

Experimental Determination of the Fourth Sound Velocity in Helium II†

Kenneth A. Shapiro,*‡ and Isadore Rudnick
Physics Department, University of California, Los Angeles, California
(Received 23 September 1964)

Fourth sound is a pressure and density wave in He II, in which only the superfluid is in motion. Using packed rouge to lock the normal fluid, the existence of this wave mode has been experimentally confirmed by measuring the temperature dependence of a plane-wave resonance in a closed-closed cylindrical acoustic resonator. The measured phase velocity has a temperature dependence which agrees within 1% with that predicted theoretically. Pulse measurements using 40-kc/sec sine-wave bursts further corroborate the existence of this type of wave. The absolute value of the fourth sound velocity is affected by the coherent multiple scattering of the wave from the superleak material. The results of Twersky's and Saxon's scattering theories are compared with the data. It is found that the equation $n=(2-P)^{1/2}$, where n is the index of refraction and P is the porosity, fits the data within 3%. Values of the ratio of the superfluid to the He II density, derived from the fourth-sound data, agree within 2% with those obtained by other experimental techniques.

INTRODUCTION

THE thermohydrodynamics of He II were first developed by Tisza[1] and Landau[2] independently. Neglecting irreversible effects, these authors predicted the existence of two wave modes in the unbounded medium. In He II $\gamma-1\ll1$ where γ is the ratio of the specific heats; it is a valid approximation for many purposes to set $\gamma=1$. In this limit the density and entropy waves uncouple. The first mode—first sound—is a density (and pressure) wave with no entropy or temperature fluctuation; the normal fluid and superfluid move in phase. The second mode—second sound—is an entropy (and temperature) wave with no pressure or density fluctuations; the normal fluid and superfluid oscillate 180° out of phase with each other. These predictions have been supported by numerous experiments.[3–7]

Pellam[8] extended the analysis of wave propagation in He II to include the case in which the liquid filled a porous stationary matrix of solid material. The principal function of this matrix is to inhibit motion of the normal fluid through viscous drag. When this viscous drag is negligible, then, of course, one obtains the usual first and second sound with second-order damping. As the viscous drag increases (due to a decreased pore size) the velocities of the two modes change and attenuation of the waves increases. However, in the limit that the normal fluid becomes locked, an interesting situation occurs. The wave mode which was originally first sound suffers *no* attenuation. This mode has been called fourth sound by Atkins.[9] It has a phase velocity C_4 given by[10]

$$C_4=\{(\rho_s/\rho_0)C_1^2+(\rho_n/\rho_0)C_2^2[1-(2\beta_p C_1^2/\gamma s_0)]\}^{1/2}, \quad (1)$$

where ρ_s is the superfluid density, ρ_n is the normal fluid density, $\rho_0=\rho_s+\rho_n$, C_1 is the first-sound velocity, C_2 is the second-sound velocity, β_p is the isobaric coefficient of expansion, s_0 is the ambient entropy per unit mass (see Appendix A). There is motion only of the superfluid and there are oscillations of density, pressure, temperature and entropy per unit mass. The wave mode which was originally second sound becomes, in the limit that the normal fluid is completely immobilized, a diffusion wave with a vanishingly small phase velocity.

The superfluid can be regarded as that component of He II in the ground state and in this sense fourth sound can be regarded as a wave motion in the ground state, the normal component (i.e., the excitations) maintaining its equilibrium density.

Figure 1 shows the temperature dependence of the velocity of first, second, and fourth sound. The values for fourth sound were calculated from Eq. (1). The values of ρ_s/ρ_0 were obtained from Andronikashvilli,[11] and the values of C_1 and C_2 are from Atkins.[12]

The grains of the superleak can be considered as incompressible, immobile scattering centers, and this will cause, as Pellam has pointed out,[8] a reduction in phase velocity. The grains are in all cases very small compared to the wavelength so that the long-wavelength limit of theoretical treatments is applicable. An important complication is the fact that the density of scatterers is sufficiently high that multiple scattering effects become important. Twersky[13,14] gives the following low-frequency results for identical scatterers, randomly dis-

---

† Work supported in part by the U. S. Office of Naval Research.
* This author gratefully acknowledges the hospitality of the Physics Division of the Aspen Institute for Humanistic Studies, where part of this paper was written.
‡ National Science Foundation Postdoctoral Fellow 1964–1965.
[1] L. Tisza, J. Phys. Radium **1**, 165, 350 (1940).
[2] L. Landau, J. Phys. U.S.S.R. **5**, 71 (1941).
[3] V. P. Peshkov, J. Phys. U.S.S.R. **8**, 131 (1944).
[4] V. P. Peshkov, J. Phys. U.S.S.R. **10**, 389 (1946).
[5] V. P. Peshkov, Zh. Eksperim. i Teor. Fiz. **18**, 857, 951 (1948).
[6] V. P. Peshkov, Zh. Eksperim. i Teor. Fiz. **19**, 270 (1949).
[7] J. R. Pellam and R. B. Scott, Phys. Rev. **76**, 869 (1949).
[8] J. R. Pellam, Phys. Rev. **73**, 608 (1948).

[9] K. R. Atkins, Phys. Rev. **113**, 962 (1959).
[10] I. Rudnick and K. A. Shapiro, Phys. Rev. Letters **9**, 5, 191 (1962).
[11] E. Andronikashvilli, Zh. Eksperim. i Teor. Fiz. **18**, 424 (1948).
[12] K. R. Atkins, *Liquid Helium* (Cambridge University Press, Cambridge, England, 1959).
[13] V. Twersky, J. Math. Phys. **3**, 700, 716, 724 (1962).
[14] V. Twersky, Technical Report EDL-L25, **1964**, Sylvania Electronic Defense Laboratories, Mountain View, California (unpublished).

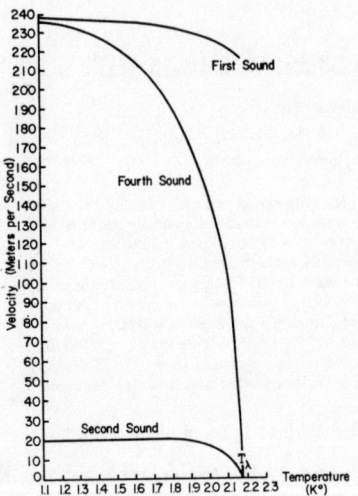

Fig. 1. Velocity of the wave modes in He II versus the absolute temperature. Fourth sound is the wave mode in which only the superfluid is in motion causing oscillations of density, pressure, temperature and entropy per unit mass. The curves for first and second sound represent the accepted experimental values. The fourth sound curve was calculated from Eq. (1).

tributed in space (n is ratio of the phase velocity in the absence of the scatterers to that in their presence, i.e., the index of refraction; and P is the fraction of volume occupied by fluid, i.e., the porosity):

Spheres:
$$n = \{(3-P)/2\}^{1/2}. \qquad (2)$$

Needles (i.e., spheroids with eccentricity approximately equal to one) whose axes are perpendicular to the direction of sound propagation but arbitrarily oriented otherwise:
$$n = \{2-P\}^{1/2}. \qquad (3)$$

These results do not hold for arbitrarily high density of scatterers.

Saxon[15] has investigated the scattering from a regular lattice of scatterers and has obtained results which are not seriously limited in this respect. For incompressible immobile spheres, in the long-wavelength limit his result is
$$n = \{1-g\}^{-1/2}, \qquad (4)$$

where g is a function of the relative geometry of the spheres and lattice, and depends on the direction of sound propagation, but is independent of the size of the spheres. For the cubic lattices, g is also independent of the direction of sound propagation (see Appendix B).

It is worthy of note that Eqs. (2)–(4) show no dispersion.

---

[15] D. Saxon (private communications).

EXPERIMENTAL TECHNIQUES

The experiment was performed as follows: solid dielectric transducers were placed at both ends of a standing wave tube, which was packed with superleak material. The composite package formed a closed-closed resonator. One transducer was electrically driven to produce pressure fluctuations in the liquid helium in the superleak, while the other was used as a receiver. The frequency of a plane-wave resonance was then determined as a function of temperature. The velocity of fourth sound was calculated from the observed frequency using the distance between the transducers and the harmonic number.

The experimental apparatus will be described under three headings: (A) Cryogenics, (B) Electronics, and (C) Acoustics.

(A) *Cryogenics*. The glass double Dewar system was of the standard type.[16] The temperature was lowered by pumping away the helium vapor with a vacuum pump which had sufficient pumping speed so that with the existing heat leaks it was possible to reach 1.1°K. The temperature was determined by measuring the vapor pressure with a Wallace and Tiernan FA 160 Gauge.

(B) *Electronics*. The electronics are shown in a block diagram in Fig. 2. The equipment consisted of a beat-frequency audio oscillator which was used to drive the source transducer. The input and output voltages were measured; the frequency was determined with a frequency counter. The output of the receiving transducer was fed into a cathode follower, amplified by 43 dB with an amplifier and a setup transformer, filtered and then displayed on one channel of a dual beam oscilloscope. The second channel of the oscilloscope was used to display the driving signal. The source transducer generated an acoustic signal whose frequency was twice that of the

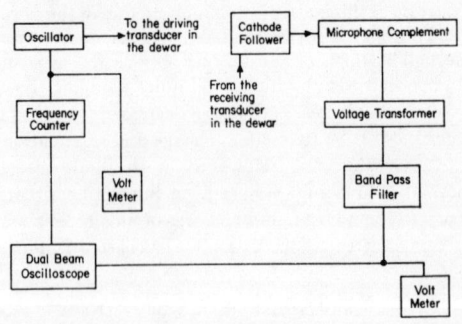

Fig. 2. Block diagram of the electronics. The driving transducer was always unpolarized and therefore the acoustic signal had twice the oscillator frequency. The input and received signals were compared on a dual beam oscilloscope and it was possible to distinguish the acoustic signal from the electromagnetic pickup by invoking the frequency doubling criterion.

---

[16] G. K. White, *Experimental Techniques in Low Temperature Physics* (Oxford University Press, Oxford, England, 1959), Chaps. 5, 4.

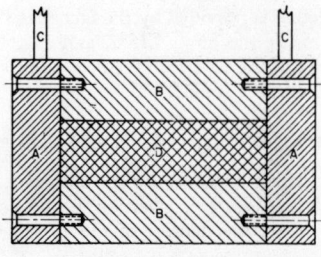

FIG. 3. Schematic of the acoustic resonator. The normal fluid was immobilized in the tiny channels of the superleak, while the superfluid was free to move. Since observations were made only on the first and second harmonics of the resonator, the velocity of fourth sound C_4, was given by $C_4 = 2fl/q$ where f is the frequency, l is the distance between the transducers and q is either one or two. Q's up to 85 were obtainable with this setup.

A. Transducers
B. Resonator Body
C. Support Tubes
D. Space For Superleak Material

driving voltage. By comparing the driving and received signals on the oscilloscope, it was possible to distinguish between acoustic signal and electromagnetic pickup since only the acoustic signal had double the drive frequency.

An Ohmite carbon resistor (27 Ω at room temperature) which was calibrated against the Wallace and Tiernan FA 160 manometer, served as an auxiliary thermometer. It was placed in close proximity to the resonator. A standard three-wire connection was used to reduce thermoelectric emf's and to couple the resistor to a Wheatstone bridge. A galvanometer, with microvolt sensitivity, was employed as a null detector for the bridge.

(C) *Acoustics*. The acoustic resonator consisted of a hollow copper cylinder with a transducer at each end, as shown in Fig. 3.

The body of the resonator had sufficient wall thickness to insure that the walls did not yield appreciably under the acoustic pressure swings, and hence did not affect the velocity of wave propagation.[17]

It is well known[18] that the resonant frequencies of a rigid cylindrical enclosure with plane rigid ends are given by

$$f = \tfrac{1}{2}C\{(q/l)^2 + (\alpha_{mn}/a)^2\}^{1/2}, \quad (5)$$

where f is the resonant frequency, C is the velocity of wave propagation, l is the distance between the ends, a is the radius of the cylinder, q is zero or a positive integer, and α_{mn} is a root of $d[J_m(\pi\alpha)]/d\alpha = 0$. $J_m(\pi\alpha)$ is a Bessel function of the first kind. The α_{mn} arise by requiring the radial velocity of the medium to be zero at the surface of the cylindrical enclosure. The plane

[17] W. P. Mason *Electromechanical Transducers and Wave Filters* (D. Van Nostrand Company, Inc., New York, 1948), Sec. 4.22.
[18] P. M. Morse, *Vibration and Sound* (McGraw-Hill Book Company, Inc., New York, 1948), 2nd ed., p. 397.

wave modes (i.e., the harmonics), characterized by $q \neq 0$, $\alpha_{mn} = \alpha_{00} = 0$, will be exclusively excited only when the driving frequency is smaller than that of the first nonsymmetric (i.e., $\alpha_{mn} = \alpha_{10} = 0.5861$) and the first radial (i.e., $\alpha_{mn} = \alpha_{01} = 1.2197$) modes. For the resonator used in this experiment ($l = 3.32$ cm and $a = 1.00$ cm) there are three harmonics below the first nonsymmetric mode and eight harmonics below the first radial mode. Since observations were made only on the first and second harmonics of the resonator, the velocity of wave propagation is given by

$$C = 2fl/q, \quad (6)$$

where q is equal to either one or two.

The interior of the cylinder was always filled with some sort of porous medium whose purpose was to immoblize the normal component of the He II.

The transducers, which bolted directly to the body of the resonator, were mechanically identical and of the solid-dielectric-condenser type.[19] Figure 4 shows a cross section of one of the transducers. The copper housing A was machined from a single block of copper. Six equally spaced countersunk clearance holes were placed through the housing so that they would mate with the tapped holes in the body of the resonator. The nylon insert B insulated the back plate C from the copper housing. The back plate formed one side of the condenser. The insert and the back plate were "press-fit" in place and the front surface was machined so as to be flush with the front of the housing. The transducer was completed by laying a piece of prestretched, $\tfrac{1}{4}$-mil-thick Mylar,[20] which was gold-coated on one side, over the body of the resonator and bolting the transducer in place. The gold-coated side of the Mylar, which completed the condenser, was in contact with the body of the resonator, and therefore electrically grounded.

The tapped hole in the top of the transducer housing was the receptacle for the brass connect D, and also was

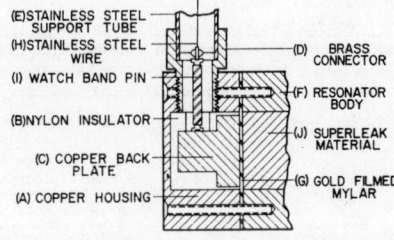

FIG. 4. Cross-section of a transducer. The $\tfrac{1}{4}$-mil Mylar film, which is gold-coated on the side in contact with the resonator body, formed the moving element of the transducer. The small mass of this Mylar diaphram caused a negligible kinetic reaction on the transducer body and consequently mechanical cross-talk was no problem.

[19] W. Kuhl, G. R. Schodder, and F. K. Schroder, Acustica 4, 5, 519 (1954).
[20] Metalized Films; Hastings & Company, Inc., 2314 Market Street, Philadelphia, Pennsylvania.

the opening for the electrical connection to the back plate.

The brass connector was soldered to the outside of the stainless steel tube E. The tubes served as a support for the resonator, and also as the outer conductor of a coaxial line. The inner conductor, which was a stretched, 3 mil diameter, stainless steel wire, was soldered to feed throughs at the top and bottom of the tube. Electrical contact between the feed-through in the bottom of the tube and the back plate was made via the watch-band pin I.

The dead capacity of the tube-transducer assembly was small enough so that the acoustic signal was reduced by less than 2.5 dB.

The resistance to ground of the coaxial line used with the pickup transducer was of the order of 10^{10} Ω. This value of the leakage resistance maintained the polarizing voltage necessary for proper operation of this transducer.

An essential feature of the transducers is the very small mass of the diaphragm. This results in a very small kinetic reaction on the transducer body and mechanical cross talk is consequently no problem. Evidence for this is provided by the absence of spurious signals of significant level and the fact that when the temperature is allowed to drift above the lambda point the residual signals are of the order of 60 dB below those found through most of the temperature range below the lambda point.

With ordinary fluids, radiation of sound out of the resonator through motion in the annuli between the transducers and the tubular body of the resonator is prevented by (1) viscous drag, and (2) the mass reactance of the fluid in this annulus. Only the latter mechanism is operative in the He II at the acoustic levels used in this experiment. This mass reactance was made large by keeping the annulus thin (a good fit between the transducers A and the resonators) and long (thick resonator body). Evidence that these measures were effective is provided by the fact that the Q of the resonant mode was 85 at 1.15°K and 25 at 1.74°K. In the absence of the superleak the tube had the predicted closed-closed tube resonances.

PRESENTATION OF THE RESULTS

The viscous wavelength for the normal fluid is given by $\lambda_n = 2\pi\{\eta_n/(\rho_n\omega)\}^{1/2}$, where η_n is the normal fluid viscosity, ρ_n is the density of the normal fluid, and $f=\omega/2\pi$ is the frequency. $\lambda_n = 301 f^{-1/2}$ μ at the lambda point, and is larger at lower temperatures. For frequencies less than 10 kc/sec the normal fluid viscous wavelength will be larger than 3.01 μ, and therefore the pore sizes in the superleak should have a maximum value which does not exceed this.

Data obtained with a Polypore superleak (average pore size 0.45±0.02 μ) have already been reported.[10] However, no multiple scattering correction was applied at that time. An empirical correction of 5.5% results in a 0.5% agreement between theory and experiment at lowest temperatures, but at 2.0°K the experimental points are 10% above the theoretical prediction. This discrepancy is probably due to the incomplete locking of the normal fluid in the larger pores of the superleak. The multiple scattering corrections are shown in Table I.

A new superleak was made using "green rouge" (Cr_2O_3) whose particles have a diameter of approximately 0.5 μ as determined from sedimentation techniques.[21] It was pounded into a hollow brass capsule. This capsule effectively filled the inside of the resonator. Its outside diameter was a mil smaller than the inside diameter of the resonator body. The ends of the capsule were terminated by thin brass plates perforated with the maximum possible number of 0.033-in. diameter holes. The walls and one end of the capsule were machined out of a solid piece of brass while the second end had its periphery threaded and screwed into the walls of the capsule.

TABLE I. The correction to the fourth sound propagation velocity due to the multiple scattering of the wave from the particles of the superleak. The theoretical results of Twersky (Refs. 13 and 14) and Saxon (Ref. 15) are included for comparison with the empirical correction. Equation (7) agrees with the empirical correction to within 3% over the entire range of porosities used.

Data	Average particle size	Porosity	Empirical correction	Equation (7)	Twersky Spheres	Twersky Needles	Hexagonal close packed z direction	Saxon Face centered cubic	Simple cubic
Ref. 10	0.45 μ	85%	5.5%	7.5%	3.75%	7.5%	0.50%	1.5%	5.5%
Green rouge, Fig. 5	0.5 μ	54.6%	23%	20.5%	11%	20.5%	3.2%	2.6%	11%
Green rouge, Fig. 6	0.5 μ	49.6%	25%	22.2%	11.8%	22.2%	3.8%	2.5%	13%
Pulse Al_2O_3	0.05 μ	80.8%	12%	9.2%	4.7%	9.2%			
Al_2O_3	0.05 μ	93.8%	1.0%	3.0%	1.5%	3.0%			
Green rouge	0.5 μ	80.0%	11.5%	9.5%	4.9%	9.5%			
Ferrero and Sacerdote (Ref. 27)	3.7 mm	39.0%	26%	27%	14.0%	27%			
	1.95 mm and 2.55 mm	38.5%	28%	27%	14.0%	27%			

[21] P. R. Day, Soil Sci. Soc. Am. Proc. 20, 167 (1956).

The capsule was weighed before and after packing and its packing density was determined to be 2.36 g/cm³; the porosity was 54.6%. The length of the actual superleak was 3.23 cm while the distance between the transducers was 3.41 cm. This left two 0.09-cm open spaces which were composed of the holes in the brass end plates and the unfilled space between the ends of the capsule and the fronts of the transducers.

The initial measurements showed that the normal fluid was not completely immobilized in the resonator because the clearance between the capsule and the resonator body was sufficiently large to allow the normal fluid component to move and therefore conduct first sound. To rectify this situation the capsule's surface was greased with vaseline and put into the resonator body. The open spaces near the transducers were filled

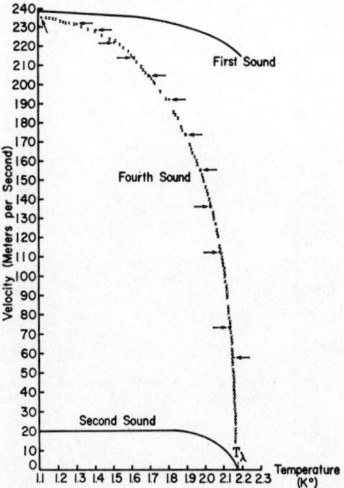

FIG. 6. The velocity of fourth sound. The x's are the experimental points for the first harmonic which varied from 2748 at 1.14°K to 186 at 2.16°K. The arrows are precise points taken from Fig. 1. The superleak was made of 0.5 μ average diameter green rouge particles and had a porosity of 49.6%.

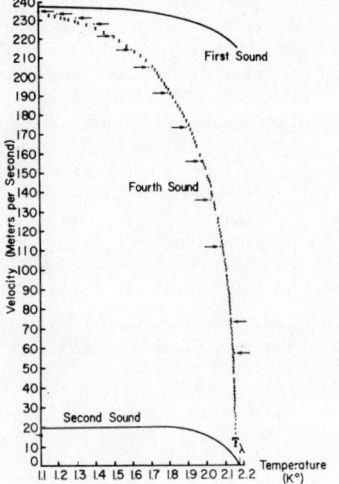

FIG. 5. The velocity of fourth sound. The x's are the experimental points for the second harmonic which varied from 5402 cps at 1.16°K to 374 cps at 2.15°K. The arrows are precise points taken from Fig. 1. The superleak was made of 0.5 μ average diameter green rouge particles and had a porosity of 54.6%.

with wafers of Gelman AM-10 Polypore[22] filter material having 0.05-μ pores. Finally, in order to more effectively seal the ends of the resonator, the transducers were seated on 0.3-mm lead washers.

The resulting measurements are shown in Fig. 5. The x's are the experimental points for the 2nd harmonic which varied from 5402 cps at 1.16°K to 374 cps at 2.15°K. The arrows are precise points taken from Fig. 1. It was necessary to apply an end correction and a multiple scattering correction to these data. The end correction was made by multiplying the length of the end plates of the brass capsule and the length of the Polypore wafers by the ratio of their porosity to that of the rouge.

This yielded an effective length of 3.52 cm between the transducers. An empirical multiple scattering correction of 23.0% results in a fit to better than 1.0% at all temperatures. Table I summarizes the scattering corrections.

In order to eliminate the necessity of an end correction, the resonator was completely filled with green rouge. A porosity of 49.6% obtained. The results for this superleak are shown in Fig. 6. This mode was the fundamental and varied from 2748 cps at 1.14°K to 186 cps at 2.16°K. The distance between the transducers was 3.43 cm. It was once again necessary to use a one-parameter fit in order to empirically correct for multiple scattering. The empirical scattering correction for this porosity was 25%. Once again Table I summarizes the multiple scattering corrections.

In Eq. (1) the dominant term in the curly brackets is the first term, $(\rho_s/\rho_0)C_1^2$. The remaining terms account for 3% of C_4 at 2.16°K, and decrease progressively as the temperature decreases being 1.7% at 2.0°K. It becomes zero at 1.1°K because of the reversal in sign of β_p at 1.17°K. It is thus clear that to the extent that C_4 and C_1 can be accurately determined such data offer a procedure alternative to the pendulum measurements of Andronikashvilli,[11] or second sound,[23] for obtaining accurate values of ρ_s/ρ_0. The experiments reported here have a high degree of internal consistency and accordingly values of ρ_s/ρ_0 were calculated from these data and are presented in Figs. 7 and 8. The arrows mark the

---

[22] Gelman Polypore Filter Material, Gelman Instrument Company, 106 North Main Street, Chelsea, Michigan (1963).

[23] D. DeKlerk, R. P. Hudson, and J. R. Pellam, Phys. Rev. **93**, 28 (1954).

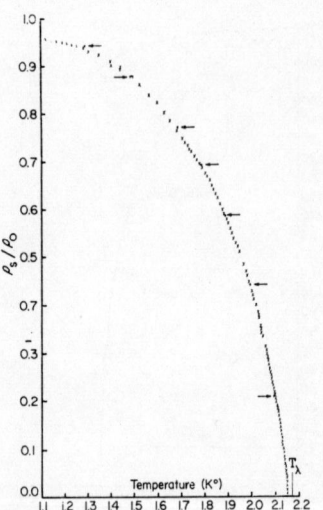

FIG. 7. The ratio of the superfluid to the total liquid density ρ_s/ρ_0, versus the absolute temperature. The x's are values calculated from the fourth sound data in Fig. 5. The arrows mark the values derived from Andronikashvilli's pendulum measurements, and are seen to be in substantial agreement with the present data.

data of Andronikashvilli[11] and are seen to be in substantial agreement with the present data.

PULSE MEASUREMENTS

The fourth sound propagation velocity has also been measured using 40 kc/sec sine wave bursts. Two Clevite PZT-4 ceramic transducers, in the form of cylinders 1-in. diameter ½-in. thick, were mounted vertically and coaxially with the opposing plane surfaces 4.20 cm apart, as described previously.[24] When a transducer was excited the lowest frequency in its natural decay was 76 kc/sec and the time constant was sufficiently long so as to interfere with the measurements. Accordingly 40-kc/sec signals were used with an appropriate filter in the receiver transducer circuit to eliminate the higher frequencies.

Superleak material (0.05 μ Linde Al_2O_3 0.05B polishing compound)[25] was pressed, using 1000 pounds per square inch, into a hollow brass cylinder whose inside diameter was 10 mil larger than the outside diameter of the PZT-4 ceramic. The resultant porosity was 80.0%. This superleak rested on the flat face of the bottom transducer; the top transducer pressed lightly on the other side of the superleak holding it in place. A large hollow brass cylinder which enclosed the entire assembly, was filled with green rouge (Cr_2O_4). The rouge filled the remaining open spaces around the transducers, and was sufficiently

---

[24] R. D. Finch, R. Kagiwada, M. Barmatz, and I. Rudnick, Phys. Rev. **134**, A1425 (1964). See Fig. 2.
[25] Linde, Production Polishing Semiconductors, Linde Company, Division of Union Carbide Corporation, Crystal Products Department, 4120 Kennedy Avenue, E. Chicago, Indiana (1961).

loosely packed so that the normal fluid was only partially locked; hence, any acoustic radiation outside the tightly packed superleak cartridge would be attenuated because of viscous damping.

The gating for the sine-wave bursts was synchronized with the 40-kc/sec signal to avoid jitter in the observed signal. Figure 9 is a photograph of an oscilloscope trace showing six well-defined and evenly spaced pulses of fourth sound at $T = 1.2°K$. Near the seventh and eighth pulses spurious signals begin to occur but fairly accurate measurements can be made on the earlier arrivals.

Measurements made in this way are in substantial agreement with the results obtained using the resonator but are not reported here because they were not made with comparable accuracy. The necessary scattering correction was found to be 12% and is listed in Table I.

The use of (1) fine particles (0.05 μ Al_2O_3) and (2) packing under high pressure (1000 lbs per sq in.) were found necessary in order to completely lock the normal fluid at all temperatures when working at 40 kc/sec. Earlier measurements made under circumstances in which these two criteria were not adequately met showed significant departures from the fourth sound velocity at higher temperatures. However, the measurements have value in that at lower temperatures one can obtain multiple scattering corrections and these are shown in Table I for Al_2O_3 with a porosity of 93.8% and green rouge with a porosity of 80%.

FURTHER DISCUSSION OF SCATTERING

An interesting byproduct of the investigation of fourth sound has been the data obtained on multiple

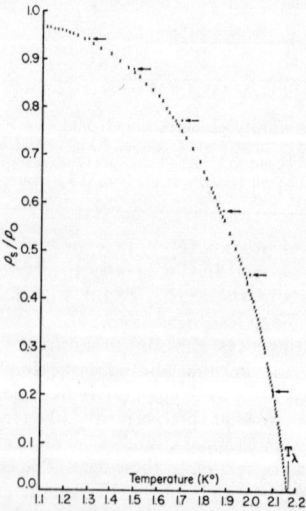

FIG. 8. The ratio of the superfluid to the total liquid density ρ_s/ρ_0, versus the absolute temperature. The x's are values calculated from the fourth sound data in Fig. 6. The arrows mark the values derived from Andronikashvilli's pendulum measurements, and are seen to be in substantial agreement with the present data.

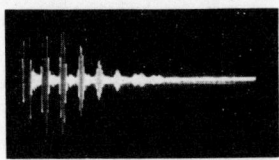

FIG. 9. Photograph of an oscilloscope trace showing 40 kc/sec fourth sound pulses at approximately 1.2°K. These signals were propagated through a superleak of 0.05 μ average diameter Al$_2$O$_3$ which had a porosity of 80.0%. Near the seventh and eighth pulse spurious signals appear, but measurements made on the earlier arrivals yield values of the fourth sound velocity in agreement with the resonator data.

scattering. Since viscous and heat conductive effects in ordinary fluids change the velocity of wave propagation through capillary porous media[26] the effect of multiple scattering on the velocity is not always apparent. Measurements of the fourth-sound velocity come very close to the situation in which viscous and thermal effects can be neglected.

The empirical scattering corrections listed in Table I can be fit to 3%, by the expression

$$n = \{2 - P\}^{1/2}, \qquad (7)$$

where n is the index of refraction and P is the porosity. Although this formula is identical with Twersky's[14] result for needles [cf. Eq. (3)] this agreement appears to be fortuitous since the particles of the superleak are not in general needles, and if they were, it would be difficult to explain why they should all have their major axis perpendicular to the direction of the original unscattered wave.[27]

CONCLUSION

The existence of fourth sound and the temperature dependence of the velocity of propagation of fourth sound are firmly established by the work described in the preceding sections.

The problem of obtaining accurate absolute values of the velocity was complicated by the multiple scattering of the wave from the particles of the superleak. The results of Twersky's[13,14] and Saxon's[15] scattering theories are compared with the data and summarized in Table I. Equation (7) is an analytic expression which yields the empirical scattering corrections to within 3% for all the cases discussed.

The values of ρ_s/ρ_0 derived from the corrected values of the experimentally determined fourth-sound velocity agree within 2% with those of other investigators.

---

[26] L. E. Kinsler and A. R. Frey, *Fundamentals of Acoustics* (John Wiley & Sons, Inc., New York, 1950), Chap. 9.

[27] M. A. Ferrero and G. G. Sacerdote, Acustica **1**, 137 (1951). Equation (7) also accounts for the lowering of the velocity of propagation of sound waves in air, when the air is trapped in the open spaces in a container filled with a random arrangement of small lead balls. The experimental velocity agrees to within 3%, with Eq. (7), if an order of magnitude correction for viscous and thermal effects is also applied.

ACKNOWLEDGMENTS

The authors wish to thank J. R. Pellam for his interest and encouragement throughout the course of this investigation, D. Saxon and V. Twersky for furnishing the results of their theoretical investigations prior to publication, and T. Holstein for illuminating discussions on multiple-scattering theory.

APPENDIX A: DERIVATION OF THE VELOCITY OF FOURTH SOUND

The linearized equations governing the reversible thermohydrodynamics of He II in unbounded space are written below in the Eulerian representation[28]

$$\partial \mathbf{v}_s/\partial t = -(\text{grad}p)/\rho + s\,\text{grad}T, \qquad (A1)$$

$$\partial \mathbf{v}_n/\partial t = -(\text{grad}p)/\rho - (\rho_s s\,\text{grad}T)/\rho_n, \qquad (A2)$$

$$\partial \rho/\partial t = -\rho_s \text{div}\mathbf{v}_s - \rho_n \text{div}\mathbf{v}_n, \qquad (A3)$$

$$\partial(\rho s)/\partial t = -\rho s\,\text{div}\mathbf{v}_n, \qquad (A4)$$

$$\rho = \rho_s + \rho_n \qquad (A5)$$

where ρ is the density of He II, $\mathbf{v}_n$ is the normal fluid velocity vector, $\mathbf{v}_s$ is the superfluid velocity vector, s is the entropy per unit mass of the He II, ρ_s is the superfluid density, ρ_n is the normal fluid density, p is the total pressure in the He II, and T is the temperature of the He II. The first equation is the force equation for the superfluid, the second is the force equation for the normal fluid, the third equation expresses conservation of mass and the fourth equation, conservation of entropy.

In order to write the equations for He II in a porous medium only Eq. (A2) has to be altered by adding $-R\mathbf{v}_n$ to the right-hand side of the equation, where R is the flow resistance for the normal fluid[29] (multiple scattering effects are neglected). Hence the porous medium acts only on the normal fluid, by impeding its flow.

Now, introduce the coordinates

$$\dot{\mathbf{r}}_1 = (\rho_n \mathbf{v}_n + \rho_s \mathbf{v}_s)/\rho \qquad (A6)$$

and

$$\dot{\mathbf{r}}_2 = \rho_s(\mathbf{v}_n - \mathbf{v}_s)/\rho, \qquad (A7)$$

where $\mathbf{v}_n$ is the normal fluid velocity vector and $\mathbf{v}_s$ is the superfluid velocity vector. $\dot{\mathbf{r}}_1$ refers to "center-of-mass" velocity of the liquid and $\dot{\mathbf{r}}_2$ to a "relative motion" with no mass flow. These are the "normal coordinates" Tisza[30] used in his description of first and second sound.

---

[28] F. London, *Superfluids* (John Wiley & Sons, Inc., New York, 1954), Vol. 2, p. 83.

[29] G. L. Pollack, California Institute of Technology thesis, p. 27, 1962 (unpublished).

[30] L. Tisza, Phys. Rev. **72**, 838 (1947).

Upon substitution of Eqs. (A6) and (A7) into Eqs. (1), (2), (3), and (4) [remembering to add $-Rv_n$ to Eq. (A2)] the linearized thermohydrodynamic equations become

$$\ddot{r}_1 - (\rho_n \ddot{r}_2/\rho_s) = -\text{grad} p/\rho + s\, \text{grad} T, \quad (A8)$$

$$\ddot{r}_1 + \ddot{r}_2 = (-\text{grad} p/\rho) - (\rho_s s\, \text{grad} T/\rho_n) - R(\dot{r}_1 + \dot{r}_2), \quad (A9)$$

$$\partial \rho/\partial t = -\rho\, \text{div}\, \dot{r}_1, \quad (A10)$$

$$\partial (\rho s)/\partial t = -\rho s\, \text{div}(\dot{r}_1 + \dot{r}_2). \quad (A11)$$

Eliminating T from Eqs. (A8) and (A9), expanding p in terms of ρ and s, and using Eqs. (A10) and (A11) to express ρ and s as functions of $\dot{r}_1$ and $\dot{r}_2$ yields

$$\dot{q}_1 = C_1^2 \{ \nabla^2 \dot{r}_1 + (T_0 s_0 \beta_p \nabla^2 \dot{r}_2)/C_p \} - \{ \rho_n R(\dot{r}_1 + \dot{r}_2)/\rho_0 \}, \quad (A12)$$

where C_1 is the first-sound velocity, T_0 is the ambient temperature, s_0 is the ambient entropy per unit mass, β_p is the isobaric expansion coefficient, C_p is the specific heat at constant pressure, ρ_0 is the ambient density, and $\dot{q}_1 = \partial \ddot{r}_1/\partial t$.

Eliminating p from Eqs. (A8) and (A9), expanding T in terms of ρ and s, and using Eqs. (A10) and (A11) to eliminate ρ and s gives

$$\dot{q}_2 = C_2^2 [\nabla^2 \dot{r}_2 + (C_1^2 \beta_p \nabla^2 \dot{r}_1)/\gamma s_0] - [\rho_s R(\dot{r}_1 + \dot{r}_2)/\rho_0], \quad (A13)$$

where C_2 is the second-sound velocity, γ is the ratio of specific heats, and $\dot{q}_2 = \partial \ddot{r}_2/\partial t$.

Seeking a plane-wave solution, put

$$\dot{r}_1 = \dot{r}_{10} \hat{e}_x \exp[j(\omega t - kx)], \quad (A14)$$

and

$$\dot{r}_2 = \dot{r}_{20} \hat{e}_x \exp[j(\omega t - kx)], \quad (A15)$$

where $\hat{e}_x$ is a unit vector in the x direction, $\dot{r}_{10}$ and $\dot{r}_{20}$ are the amplitudes of $\dot{r}_1$ and $\dot{r}_2$, ω is the angular frequency and k is the wave number. Substituting Eqs. (A14) and (A15) into Eqs. (A12) and (A13) yields

$$\dot{r}_{20} = -\dot{r}_{10} \frac{(C_1^2 C_2^2 \beta_p k^2/\gamma s_0) + (R \rho_s j\omega/\rho_0)}{(R\rho_s j\omega/\rho_0) + C_2^2 k^2 - \omega^2}, \quad (A16)$$

$$\dot{r}_{20} = -\dot{r}_{10} \frac{(R\rho_n j\omega/\rho_0) + C_1^2 k^2 - \omega^2}{(C_1^2 T_0 s_0 \beta_p k^2/C_p) + (R\rho_n j\omega/\rho_0)}. \quad (A17)$$

Equations (A16) and (A17) form a pair of linear homogeneous equations in $\dot{r}_{10}$ and $\dot{r}_{20}$. These equations have nontrivial solutions only if the determinant of the coefficients of $\dot{r}_{10}$ and $\dot{r}_{20}$ is identically zero. This condition gives the result written below:

$$C^4 - \frac{C^2}{1-(jR/\omega)} \left[(C_1^2 + C_2^2) - jR \frac{\rho_s C_1^2 + \rho_n C_2^2}{\rho_0 \omega} \right.$$
$$\left. + 2j\rho_n C_1^2 C_2^2 \frac{\beta_p R}{\rho_0 \omega \gamma s_0} \right]$$
$$+ \frac{C_1^2 C_2^2}{1-(jR/\omega)} \left(1 - \frac{\beta_p^2 T_0 C_1^2}{\gamma C_p} \right) = 0, \quad (A18)$$

where $C^2 = \omega^2/k^2$.

The two roots of Eq. (A18) for C^2 are, in general, difficult to evaluate. However, two limiting cases are of interest. When $(R/\omega) \to 0$

$$C^2 = \tfrac{1}{2}(C_1^2 + C_2^2)$$
$$\pm \tfrac{1}{2}[(C_1^2 + C_2^2)^2 - 4C_1^2 C_2^2 (1-\epsilon)]^{1/2}, \quad (A19)$$

where $\epsilon = C_1^2 \beta_p T_0/\gamma C_p$. When $\epsilon \ll 1$ this result leads to the following modes:

$$C_+^2 = C_1^2 + \epsilon [C_1^2 C_2^2/(C_1^2 - C_2^2)] \quad (A20)$$

and

$$C_-^2 = C_2^2 - \epsilon [C_1^2 C_2^2/(C_1^2 - C_2^2)], \quad (A21)$$

where C_+^2 is the root corresponding to the + sign in Eq. (A19) and C_-^2 is the root corresponding to the − sign in Eq. (A19). The results were first derived by Tisza.[1] ϵ represents a coupling term which is equal to 1.5×10^{-2} a tenth millidegree below the lambda point and is smaller at lower temperatures. Hence ϵ can be neglected, and the modes reduce to first and second sound.

When $(R/\omega) \to \infty$

$$C^4 - C^2 C_4^2 = 0, \quad (A22)$$

where C_4 is the velocity of fourth sound,

$$C_4 = \left[\frac{\rho_s C_1^2}{\rho_0} + \frac{\rho C_2^2}{\rho_0} \left(1 - \frac{2\beta_p C_1^2}{\gamma s_0} \right) \right]^{1/2}. \quad (A23)$$

Hence

$$C_+^2 = C_4^2 \quad (A24)$$

and

$$C_-^2 = 0. \quad (A25)$$

This solution also leads to two modes. The "+" mode is a wave motion which propagates with a velocity given by Eq. (A23). The "−" mode does not propagate. (This mode might be nicknamed "No Sound.")

It should be noted that the "+" mode, which is C_1^2 when $R/\omega \to 0$, goes over to C_4^2 when $R/\omega \to \infty$, and that the "−" mode, which is C_2^2 when $R/\omega \to 0$, does not propagate when $R/\omega \to \infty$.

Equation (A23) can be written as

$$C_4 = \left(\frac{\rho_s C_1^2}{\rho_0} + \frac{\rho_n C_2^2}{\rho_0}\right)^{1/2}, \quad (A26)$$

since the difference in the value of Eqs. (A23) and (A26) is 1.5% at 2.16°K and becomes smaller as the temperature is lowered. Equation (A26) is the Pellam's formula for fourth sound.

APPENDIX B: SAXON'S FORMULA FOR THE COHERENT MULTIPLE-SCATTERING CORRECTION TO THE PROPAGATION VELOCITY

The effect of coherent multiple scattering from a periodic arrangement of fixed, rigid spheres has been analyzed to a reasonable approximation by Saxon[15] using a Green's function technique. When the general formulation, which is valid for any allowable porosity, is simplified to the case of long wavelengths, the ratio of the propagation velocity in the fluid medium in the periodic lattice C, to the propagation velocity in free space C_0, is

$$C_0/C = (1-g)^{-1/2}. \quad (B1)$$

The symbol g is given by

$$g = \left[\frac{3(1-P)}{2\pi N_0}\right]^2 \sum_{\substack{s=-\infty \\ s\neq 0}}^{\infty} \frac{(\hat{e}_k \cdot \mathbf{x}_s)^2 j_1^2(2\pi|\mathbf{x}_s|)}{\mathbf{x}_s^4}, \quad (B2)$$

where P is the porosity, N_0 is the number of spheres per primitive cell, $\hat{e}_k$ is a unit vector in the direction of propagation, and $j_1(2\pi|\mathbf{x}_s|)$ is the spherical Bessel function of the first kind and first order. The vector $\mathbf{x}_s$ is expressed as

$$\mathbf{x}_s = (s_1\alpha_1 a + s_2\alpha_2 a + s_3\alpha_3 a)/K, \quad (B3)$$

where s_1, s_2, and s_3 are positive or negative integers not simultaneously zero, a is the sphere radius, and α_1, α_2, and α_3 are the primitive translation vectors of the reciprocal lattice for the lattice in question. $2a$ has been chosen as the side of the conventional unit cube.[31] The α's are expressed in terms of the sphere radius a and the periodicity of the lattice. K is a parameter which adjusts the volume of a primitive cell for the correct porosity and is given by

$$K = a\left[\frac{4\pi N_0(\alpha_1 \times \alpha_2 \cdot \alpha_3)}{3(1-P)}\right]^{1/3}. \quad (B4)$$

Hence, g depends only on the direction of propagation and on the relative geometry of the spheres and lattice.

g has the form of a symmetric second-rank tensor expressed relative to the principle axes. It can be shown[32] that the principle components of such a tensor are equal in cubic lattices. Thus, the quantity g is independent of the direction of sound propagation in such lattices.

Equation (B1) is identical to Eq. (5) of the Introduction.

It is of interest that Rayleigh[33] solved the problem of the effect on the propagation velocity of multiple scattering of a wave from a simple cubic lattice composed of rigid, fixed spheres. Rayleigh's result is

$$\frac{C_0^2}{C^2} = \frac{3-P}{2}\left[\frac{1-(0.394(1-P)^{10/3}/(3-P))}{1-(0.394(1-P)^{10/3}/2P)}\right], \quad (B5)$$

where C_0, C, and P have been defined above. This equation is valid when powers of $(1-P)$ higher than 10/3 may be neglected. The quantity in the square bracket has a maximum value of 1.025. If powers of $(1-P)^{10/3}$ are neglected this equation is identical to Twersky's formula for rigid fixed spheres [cf.: Eq. (2) of the Introduction].

Although Rayleigh's result [i.e., Eq. (B5)] does not formally resemble Saxon's formula [cf.: Eq. (B1)] when evaluated for the simple cubic lattice, numerical analyses show they agree to better than 2% for all allowed porosities.

---

[31] C. Kittel, *Introduction to Solid State Physics* (John Wiley & Sons, Inc., New York, 1956), Chap. 12; beware of errors in the formulas for the primitive translation vectors in the crystal and in the reciprocal lattice.
[32] J. F. Nye, *Physical Properties of Crystals* (Oxford University Press, Oxford, 1952), Chap. 1.
[33] Lord Rayleigh, Phil. Mag. 34, 481 (1892); *Collected Works* (Cambridge University Press, Cambridge, England), Vol. 4, p. 19.

Editor's Comments on Papers 35 and 36

35 Mason and McSkimin: *Attenuation and Scattering of High Frequency Sound Waves in Metals and Glasses*

36 Mason: *Ultrasonic Attenuation Due to Lattice–Electron Interaction in Normal Conducting Metals*

Although common observation indicates that the attenuation of sound through elastic solids, for example, a solid rod of a highly elastic material such as steel, is, in general, less than in liquids and gases at the same frequency, early experimental methods were not precise enough to provide exact values for the attenuation as a function of frequency. This shortcoming was remedied by the introduction of the pulse technique for attenuation measurements in solids by Mason and McSkimin, with results as shown in the first paper presented here, "Attenuation and Scattering of High-Frequency Sound Waves in Metals and Glasses." These investigators were able to distinguish two components of the attenuation; one proportional to the first power of the frequency, the other, due to Rayleigh scattering, proportional to the fourth power of the frequency. In this important experimental study, they laid the groundwork for subsequent measurements of attenuation in solids.

This paper also emphasizes that the attenuation of sound in solids is a very complicated affair, depending as it does on thermal conductivity, grain-boundary effects, dislocations, and interstitial atom diffusion.

At high frequencies and very low temperatures (approaching absolute zero) another interesting effect, the interaction of the free electrons in a metal with the metallic lattice, becomes evident. For a given frequency, attenuation in a metal increases as the temperature decreases until the superconducting transition is reached; at that point, the attenuation falls abruptly. This is well demonstrated in the second paper of the group, "Ultrasonic Attenuation Due to Lattice–Electron Interaction in Normal Conducting Metals." Mason also offers a rather simple explanation for the variation of attenuation in the normal conducting state, which agrees well with experimental results.

Recent work on ultrasonic attenuation in solids, with special reference to both semiconductors and superconductors, has been extensive. A good review will be found in *Physical Acoustics, Principles and Methods*, Warren P. Mason, ed., Volume IV, Part A: "Applications to Quantum and Solid State Physics" (Academic Press, New York, 1966).

Herbert J. McSkimin (1915–   ) was a colleague of Mason at the Bell Telephone Laboratories and is well known for his work on the acoustical properties of solids.

Attenuation and Scattering of High Frequency Sound Waves in Metals and Glasses

W. P. Mason and H. J. McSkimin
Bell Telephone Laboratories, Murray Hill, New Jersey

By using a pulse method, attenuation and velocity measurements have been made for aluminum and glass rods in the frequency range from 2 to 15 megacycles. The sound pulses are generated by crystals waxed to the surface of the rod. This wax joint limits the band width of the transmitted pulse and measurements are made using long pulses which approach steady state conditions. The reflected pulses show evidence of several normal modes which can be minimized by using specially shaped electrodes. Longitudinal waves show delayed pulses of smaller magnitude that are caused by the longitudinal wave breaking up into reflected longitudinal and shear waves at the boundary. This effect is small if the diameter of the rod is 20 wave-lengths or more.

The measured losses for aluminum rods show a component proportional to the frequency and another component proportional to the fourth power of the frequency. The first component is the hysteresis loss found for most solid materials. The component proportional to the fourth power of the frequency is caused by Rayleigh scattering losses which are the result of differences in the elastic constants between adjacent grains caused by changes in orientation. Calculated scattering losses agree quite well with the measured values. The fourth-power scattering law holds quite well until the grain size is equal to one-third of a wave-length. For higher frequencies the scattering loss increases more nearly with the square of the frequency. Glasses and fused quartz have a loss directly proportional to the frequency, showing that any irregularities must be of very small size

I. INTRODUCTION

DURING the last few years pulse methods and high frequency attenuation methods have been widely applied in measuring the properties of liquids[1] and in locating flaws[2] in metal castings and other solid bodies. For liquids the method consists of applying a high frequency pulse of short duration to a crystal sending out a beam of parallel rays into the liquid and picking this pulse up by another crystal accurately parallel to the first crystal. The received pulse and its series of reflections are shown on a cathode-ray oscillograph, and by measuring the relative amplitudes the attenuation between successive reflections can be measured. By changing the distance between crystals the increase in attenuation due to path length alone can be measured, and the value of the reflection coefficient can be evaluated. A measure of the velocity can be obtained by timing the difference between the sent pulse and the received pulse.

It is the purpose of the present paper to apply this method to determining the properties of solids by means of the technique of sending and receiving longitudinal and shear wave pulses.

Although no very exact mathematical solutions have been obtained for the transmission of waves in finite solids, experimentally it has been found that a good replica of a longitudinal wave is obtained through a solid rod if the diameter is many wave-lengths, the pulse traveling with the velocity for a free medium, namely,

$$v_d = \left(\frac{\lambda + 2\mu}{\rho}\right)^{\frac{1}{2}} \quad (1)$$

where λ and μ are the Lamé elastic constants of the solid and ρ the density. This main pulse is often followed by a series of pulses which are replicas of the main pulse but of smaller amplitude and delayed in time by amounts that are proportional to the radius of the rod. It has been shown that these delayed pulses are due to the incident dilational wave, which travels nearly parallel to the surface, being reflected from the surface and breaking up into a reflected dilational wave and a generated shear wave which makes a considerable angle with the periphery of the rod.

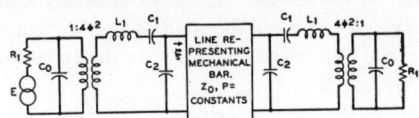

Fig. 1. Equivalent circuit representing two crystals connected to a mechanical bar.

---

[1] J. K. Galt and J. R. Pellam, J. Acous. Soc. Am. 18, 251 (1946); J. Chem. Phys. 14, 608 (1946).
[2] F. A. Firestone, J. Acous. Soc. Am. 17, 287 (1946); F. A. Firestone and J. R. Frederick, J. Acous. Soc. Am. 18, 200 (1946).

This shear wave strikes the opposite side of the rod and is partially converted back into a dilational wave which proceeds with the dilational velocity. It will be delayed by an amount which depends on the diameter of the rod and the ratio of shear and dilational velocities. If, however, the diameter of the rod is a large number of wavelengths, these accompanying pulses are small compared to the main pulse, and a measurement of the relative amplitude of the main pulse as a function of distance can be used to determine the attenuation existing in the metal. By using a shear crystal to generate a shear wave in the rod, the velocity and attenuation of shear waves can be measured. These waves are not accompanied by the phenomena of trailing pulses since the shear waves, being nearly parallel to the surface, are incident on the side walls with the angles greater than the critical angles for longitudinal waves. Hence, the properties of the solid are more easily measured with shear waves than with longitudinal.

For metals, the attenuations of both shear and longitudinal waves are proportional to the frequency for low frequencies, but for high frequencies a component of the attenuation was found which increased as the fourth power of the frequency. This was found to result from a scattering due to the finite grain size of the material, an effect similar to the scattering of sound by small particles as investigated by Rayleigh. This effect did not exist for glasses up to 10 megacycles, indicating that the grain size, if any, for glasses is very small. A theoretical formula for the scattering type of attenuation has been developed using Rayleigh's formula for scattered sound waves.

II. EXPERIMENTAL METHODS

In measuring the attenuation and velocity in a solid material, a rod of the material several feet long is used, as straight as possible and surfaced off square on the ends in a lathe so that either X-cut or Y- and rotated Y'-cuts (AT and BT) quartz crystals can be attached to the surfaces. X-cut crystals are used for longitudinal waves and Y'-cuts for shear waves. Since the rod acts as a wave guide and will conduct a wave around a very small bend, the rod does not have to be accurately straight. For attenuation and velocity measurements, the crystals can be attached to the rod by such waxes as halowax or beeswax which have a relatively high shear and longitudinal stiffness for a wax, although low compared to a crystal or metal. This small layer of a low stiffness material results in reducing the frequency range over which energy can be transferred from the crystal to the solid. This is obvious from Fig. 1 showing the equivalent circuit[3] of a crystal free on one side and connected to the high impedance rod on the other, through the small compliance C_2. If we write out the equation for the network, the particle velocity imparted to the high impedance mechanical line (the rod) becomes

$$\dot{\xi} = \frac{E_0/2\varphi}{\left[R_1 + \frac{Z_0}{4\varphi^2} + \frac{C_2}{C_1}\frac{Z_0}{4\varphi^2}\left(1 - \frac{\omega^2}{\omega R^2}\right)\right] + j\left[\omega C_2 Z_0 R_1 - \frac{1}{4\omega C_1 \varphi^2}\left(1 - \frac{\omega^2}{\omega R^2}\right)\right]}, \quad (2)$$

if we neglect C_0, which is usually tuned out with an inductance. In the equations, R_1 is the electrical resistance of the pulsing amplifier, Z_0 the mechanical characteristic impedance of the bar given by $Z_0 = (\rho v) A$ where ρ is the density and v the velocity of propagation of the bar and A the cross-sectional area, C_1 the series compliance of the crystal, C_2 the shunt compliance of the connecting layer between the crystal and the bar, $\omega_R = 2\pi f_R$ where f_R is the half-wave resonant frequency of the crystal and $4\varphi^2$ is the impedance ratio of the perfect transformer which transforms current into particle velocity, or voltage into force in a piezoelectric crystal. In terms of the crystal constants in c.g.s. units,

$$C_0 = \frac{AK}{4\pi l_t}; \quad C_1 = \frac{2}{\pi^2}\frac{l_t}{Ac}; \quad L = \frac{\rho A l_t}{2}; \quad 2\varphi = \frac{DKA}{2\pi l_t}. \quad (3)$$

[3] The equivalent circuit of a quartz crystal free on one end and driving a load on the other is derived in *Electromechanical Transducers and Wave Filters*, by W. P. Mason (D. Van Nostrand, Inc., New York), Chapter VII, p. 231.

can be set equal to zero, the energy delivered to the rod is

$$M.E. = \dot{\xi}^2 Z_0$$

$$= \frac{E^2 Z_0/4\varphi^2}{\left[R_1 + \dfrac{Z_0}{4\varphi^2}\right]^2 + \left[\dfrac{1}{4\varphi^2 \omega C_1}(1-f^2/f_R^2)\right]^2}. \quad (5)$$

Under the most favorable condition the energy delivered by the pulser will be obtained if the electrical impedance of the crystal is equal to R_1, the resistance of the pulser. Hence, the maximum input electrical energy will be

$$(E/2R_1)^2 R_1 = E^2/4R_1, \quad (6)$$

and the conversion efficiency is

$$\text{eff.} = \frac{R_1 Z_0/\varphi^2}{\left[R_1 + \dfrac{Z_0}{4\varphi^2}\right]^2 + \left[\dfrac{1}{4\varphi^2 \omega C_1}(1-f^2/f_R^2)\right]^2}. \quad (7)$$

Introducing the values above, the equations become

$$\text{eff.} = \frac{4\pi^2 (\rho v)_b v_c / D^2 K}{\left[1 + \dfrac{\pi^2 (\rho v)_b v_c}{D^2 K}\right]^2 + \left[\dfrac{\pi^3 c}{2 D^2 K}(1-f^2/f_R^2)\right]^2}$$

$$= \frac{442}{(1+110.5)^2 + [138(1-f^2/f_R^2)]^2}. \quad (8)$$

At the resonant frequency of the crystal the efficiency is about 4 percent, representing a loss of 14.6 db in going from the electrical circuit to the rod. Over a frequency range, the loss is plotted on Fig. 2. This represents quite a wide frequency range of conversion.

When the layer of wax represents a finite coupling compliance C_2, the efficiency is given in the formula

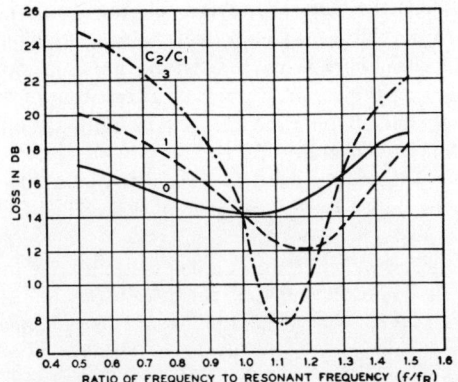

Fig. 2. Conversion loss in going from electrical circuit to mechanical bar for three ratios of wax compliance to crystal compliance.

where l_t is the thickness of the crystal, A the cross-sectional area, K the dielectric constant equal to 4.55 for X- or Y-cut quartz, c the elastic stiffness equal to 8.3×10^{11} dynes per cm^2 for X-cut quartz and 4.05×10^{11} dynes per cm^2 for Y-cut quartz, and D the piezoelectric constant relating the stress generated by the piezoelectric effect to the applied surface charge. For X-cut quartz D has a value of 13.85×10^4 in c.g.s. units and for Y-cut quartz the same value $D = 13.85 \times 10^4$. From these constants the performance of the crystal in converting electrical into mechanical power can be calculated. As an example, let us consider the conversion of electrical energy from a pulsing amplifier having a resistance R_1 equal to the static reactance of the crystal or

$$R_1 = -j/\omega_R C_0, \quad (4)$$

to mechanical energy in longitudinal form in an aluminum bar having a density equal to 2.7 and a longitudinal velocity of 6.32×10^5 cm per second. If we suppose first that the layer of adhesive connecting the crystal to the rod is so thin and so stiff that C_2, the shunt compliance,

$$\text{eff.} = \frac{4\pi^2 (\rho v)_b v_c / D^2 K}{\left[1 + \dfrac{\pi^2 (\rho v)_b v_c}{D^2 K}\left(1 + \dfrac{C_2}{C_1}\left(1 - \dfrac{f^2}{f_R^2}\right)\right)\right]^2 + \left[\dfrac{C_2}{C_1}\left(\dfrac{2}{\pi} \dfrac{(\rho v)_b v_c}{c}\right) - \dfrac{\pi^3 c}{2 D^2 K}(1-f^2/f_R^2)\right]^2}. \quad (9)$$

A plot of the conversion loss expressed in decibels i.e.,

$$db = 10 \log_{10}(P_E/P_M), \quad (10)$$

where P_E is the maximum electric input power and P_M the mechanical power in the rod, is shown plotted by Fig. 2 for three values of the

3

ratio C_2/C_1. As this ratio becomes larger the device acts as a transforming band pass filter and increases the efficiency of conversion over a narrow frequency range just above the resonant frequency of the crystal. Since this characteristic occurs both at the input and output, the transmitted band is limited to about 10 percent of the carrier frequency. This sets the minimum pulse length that can be employed, for the pulse will not build up to its full amplitude unless the length in seconds is as large as the inverse of the band width in cycles or,

$$P.L. \geqq 1/BW \text{ in cycles}, \qquad (11)$$

where the band width BW is determined by the frequency difference of the two band edges three db down from the maximum efficiency point. For a carrier of 3 megacycles, the pulse length has to be at least 3.3 microseconds in order to give a full amplitude. The Y-cut crystal has a slightly lower loss at mid-band—12.8 db—and a slightly wider band width due to the lower value of velocity for the shear wave over the longitudinal wave. However the band width for finite values of C_2/C_1 is very similar to the X-cut as shown by Fig. 2.

The experimental arrangement is shown in Fig. 3. A variable frequency oscillator is the source of the carrier frequency. This is sent through a wide band-tuned amplifier that impressed about 10 volts at 100 ohms across the crystal. The bias on the input tube of the amplifier is controlled by the pulser. Normally a high negative bias is on the grid of the amplifier tube, and the pulser puts on a positive bias of a value to overcome the negative bias and allows the amplifier to amplify for the time duration that the biasing pulse is on. The firing of the pulser is controlled by a sinusoidal wave of frequency from several hundred cycles to 5000 cycles, and this timing wave also controls the sweep circuit of the cathode-ray oscillograph. The pulser is one of conventional design and puts a square top pulse of positive voltage on the two balanced input tubes of the amplifier that are connected in a push-pull arrangement. The d.c. pulse is then balanced out in the output and does not affect the succeeding tubes. The carrier frequency, on the other hand, is inserted on the suppressor grid of one of the tubes and is not balanced out in the output. When the grid is negative, the carrier output of the tube is neutralized, and no steady state output appears in the amplifier. When the gating pulse is impressed on the input, a pulse of alternating current of controllable time duration is impressed on the sending crystal.

The receiving crystal is terminated in a 100-ohm resistance and capacity annulling coil which are connected across the input of a wide band untuned amplifier. This amplifier is terminated in a diode detector, and the rectified output is impressed across the vertical plates of the cathode-ray oscillograph. Since the horizontal sweep is controlled by the same sinusoidal wave that controls the pulses, the received pulse and any reflected pulses appear in the same position on the cathode-ray tube for successive pulses and form a picture in time of the received pulse and reflections. The position of the transmitted pulse can also be marked by establishing a slight coupling with the transmitting amplifier.

The method of measuring attenuation is as follows. The frequency of the pulse is set at the natural resonant frequency of the crystal alone, and a pulse is used that is long enough to establish steady state conditions. Since as seen from Fig. 2, this effectively results in sending a single side band, the input of the pulse may be somewhat distorted, but the steady state condition corresponds nearly to the steady state output that would result if the carrier frequency were at the exact resonant frequency of the crystal. Furthermore, the reflections obtained at this frequency are nearly perfect, since the termination impedance at the resonant frequency is effectively zero because the mass and compliance of the crystal annul each other at this frequency. If the frequency had been set at the frequency of maximum conversion, the reflection would have been far from perfect on account of the trans-

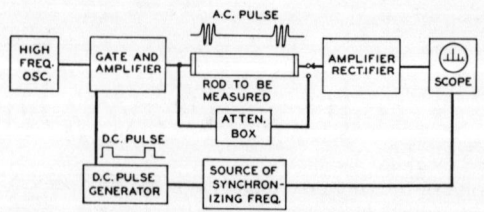

FIG. 3. Block diagram of experimental circuit.

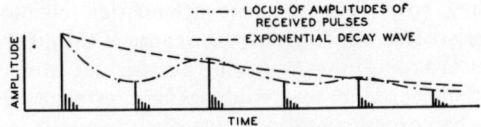

Fig. 4. Typical received pulse, time curve for a longitudinal wave.

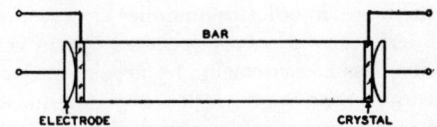

Fig. 5. "Door Knob" electrodes for producing a single normal mode for a longitudinal rod.

forming action of the glued joint. In order to evaluate the losses occurring in the wax joints, several length rods are used, for example, 1.5 inches and 1 foot. By comparing the received pulses for a given total path length, the loss per reflection can be evaluated. For example, the loss at 7.5 mc was found to be about 0.07 db per reflection for compressional waves and is small enough to be neglected. However, for shear waves, the loss put in by the wax is higher and may amount to 0.5 db per reflection at 5 mc. This has to be taken into account.

A typical series of received pulses for an aluminum rod 1 inch in diameter for a longitudinal wave is shown in Fig. 4. The frequency of this pulse was five megacycles. The trailing pulses are of rather small amplitude but become larger with respect to the main pulse for greater distances. This figure shows another feature of the successive pulses—that for a full electrode over both crystals—the successive pulses do not decrease exponentially but show an interference pattern between two or more modes which causes successive pulses to become first smaller then larger as the phases between the successive pulses cause a cancellation or addition. This effect is caused by the fact that a full coverage of the crystal generates not only the fundamental wave guide mode, which is a Bessel's function nearly vanishing at the surface, but also generates higher modes which are transmitted with a slightly higher phase velocity. This effect can be minimized by using electrodes of somewhat smaller diameter than the rod, or by shaping the back electrode of the crystal with a surface, as shown by Fig. 5, that controls an air gap between the crystal and electrode near the boundary. With this electrode, the crystal is driven much less strongly near the surface, and the wave sent out approximates the Bessel function of the first mode. As a result the secondary modes are suppressed, and the successive reflections die down nearly exponentially.

With this arrangement the reflections are exponential, and the trailing pulses are small enough to contain only a small portion of the total energy; hence fairly accurate measurements can be made for frequencies above 5 megacycles for a rod an inch in diameter. The attenuation is measured by comparing the amplitude of successive pulses and computing the attenuation by the formula,

$$\text{total attenuation in nepers} = A_x = \log_e(I_0/I_1), \quad (12)$$

where I_0 is the amplitude of the first received pulse and I_1 the amplitude of a reflected pulse which has travelled a total distance of x centimeters. By comparing successive reflected pulses the truth of the exponential law can be tested. By taking rods of different lengths the loss at each reflection can be computed and this has been found quite small for wax joints. The velocity of propagation can be measured by using a timing wave and measuring the time by which successive reflections are delayed with respect to each other. For a moderately long delay a very accurate method is to control the pulsing rate until two successive reflections are made to coincide on the scope pattern. For example, for a two-foot aluminum rod reflections were made to coincide when the pulsing rate was 5184 cycles. The velocity was then

$$v = f \times d = 5184 \times 2 \times (24 \times 2.54)$$
$$= 6.32 \times 10^5 \text{ cm/sec.} \quad (13)$$

since the time between successive reflections is the time required to travel twice the length of the rod.

For shear waves the phenomenon of delayed pulses is absent, but the phenomenon of interference between successive normal modes is present. The door knob type of electrode shown by Fig. 5 does not get rid of the upper modes but these are relatively small for a full coverage of the crystal and by measuring the amplitudes only at the high points, a true attenuation can be obtained.

Furthermore, in order to minimize errors caused by interference of waves one may obtain comparative loss measurements by keeping ratios of dimensions constant regardless of the frequency. This technique keeps the phase relations at the receiving crystal essentially the same for all frequencies

The received pulse amplitudes, expressed in relative db's, should be plotted as a function of number of wave-lengths rather than distance travelled. If the attenuation per wave-length is a constant (as it would be for elastic hysteresis alone) the curves will coincide. If this is not the case, the difference in loss per wave-length is determined by the separation of the curves. Let A_1=attenuation per wave-length at f_1 and A_2=the attenuation per wave-length at f_2. Then $A_2-A_1=\Delta A$ is the value measured experimentally. Then the losses per centimeter at f_1 and f_2 are given by

$$\frac{A_1}{\lambda_1} \text{ and } \frac{A_2}{\lambda_2}=\frac{A_1+\Delta A}{\lambda_2}.$$

The ratio of the loss per cm at f_2 to the loss per cm at f_1 is given by

$$R=(\lambda_1/\lambda_2)(1+\Delta A/A_1)$$

and hence the attenuation per unit length at any frequency f_2 can be obtained in terms of a known attenuation at f_1.

III. EXPERIMENTAL RESULTS

These experimental methods have been applied in measuring the attenuation and velocities for shear and longitudinal waves for two aluminum rods and several glass rods. The two aluminum rods were standard aluminum rods designated 17 S.-T. Although they carried the same designation they gave very different attenuation characteristics. In running down the cause of the difference, a connection between grain size and attenuation caused by scattering has been established.

The experimental data for these two rods for shear and longitudinal waves are shown plotted on Figs. 6 and 7. Data on attenuation for longitudinal waves above 5 megacycles should be reliable while that below 5 megacycles may be somewhat high because of loss of energy to the trailing pulses. The data for the shear waves should be good at all frequencies. The velocity measurements gave the same result for both rods and for all frequencies, namely,

$$v_l=6.32\times 10^5 \text{ cm/sec.};$$
$$v_s=3.13\times 10^5 \text{ cm/sec. at } 25°C. \quad (14)$$

These agree well with those calculated from the published values[4] of the elastic constants

$$\lambda=5.44\times 10^{11} \text{ dynes/cm}^2;$$
$$\mu=2.67\times 10^{11} \text{ dynes/cm}^2; \quad \rho=2.71 \quad (15)$$
$$v_l=\left(\frac{\lambda+2\mu}{\rho}\right)^{\frac{1}{2}}=6.32\times 10^5 \text{ cm/sec.};$$
$$v_s=\left(\frac{\mu}{\rho}\right)^{\frac{1}{2}}=3.14\times 10^5 \text{ cm/sec.}$$

These elastic constants agree somewhat poorly with the values[5] obtained for the aluminum

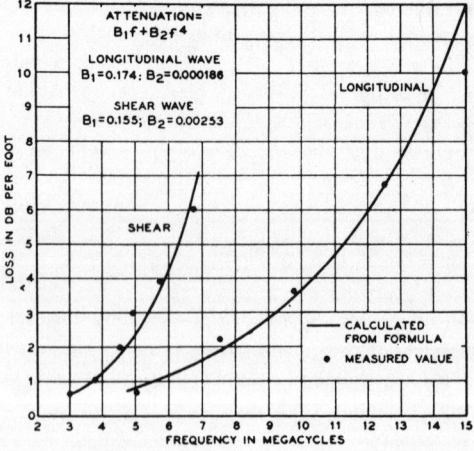

FIG. 7. Measured attenuation of longitudinal and shear waves for rod No. 2 having a grain size of 0.13 mm.

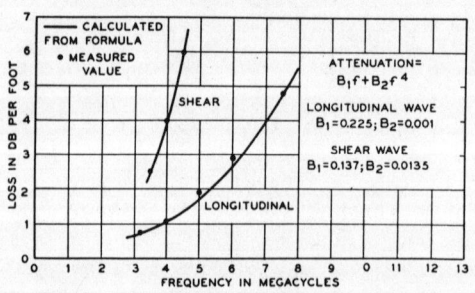

FIG. 6. Measured attenuation of longitudinal and shear waves for rod No. 1 having a grain size of 0.23 mm.

[4] Kaye and Laby, *Physical and Chemical Constants* (Longmans Green and Company, New York), p. 29.
[5] E. Goens, "Elastic constants for aluminum single crystals," Ann. d. Physik **17**, 233 (1933).

single crystal. This is a face-centered cubic which has the elastic constants

$$c_{11} = 10.76 \times 10^{11} \text{ dynes/cm}^2;$$
$$c_{12} = 6.18 \times 10^{11}; \quad c_{44} = 2.84 \times 10^{11}. \quad (16)$$

The elastic constants for any orientation are given by the transformation equations,

$$\begin{aligned}c_{11}' &= c_{11}(l_1^4 + m_1^4 + n_1^4) \\ &+ (2c_{12} + 4c_{44})(l_1^2 m_1^2 + l_1^2 n_1^2 + m_1^2 n_1^2), \\ c_{44}' &= c_{11}(l_1^2 l_2^2 + m_1^2 m_2^2 + n_1^2 n_2^2) \\ &+ 2c_{12}[l_1 l_2 m_1 m_2 + n_1 n_2 (l_1 l_2 + m_1 m_2)] \\ &+ c_{44}[(l_1 l_2 + m_1 m_2)^2 + (l_1 l_2 + n_1 n_2)^2 \\ &+ (m_1 m_2 + n_1 n_2)^2],\end{aligned} \quad (17)$$

where l_1 to n_3 are the direction cosines between the new set of axes and the old set

	x	y	z
x'	l_1	m_1	n_1
y'	l_2	m_2	n_2
z'	l_3	m_3	n_3

From these equations it can be shown that c_{11} (which should correspond to $(\lambda + 2\mu)$ for an isotropic body) can vary from 10.76×10^{11} to 11.49×10^{11} dynes/cm^2 while the shear constant can vary from 2.84×10^{11} to 2.27×10^{11}.

The average value of c_{11}' can be determined from the equation

$$\begin{aligned}c_{11}' &= c_{11} + [2(c_{12} - c_{11}) \\ &+ 4c_{44}][l_1^2 m_1^2 + l_1^2 n_1^2 + m_1^2 n_1^2], \quad (18) \\ &= 10.76 + 2.2[l_1^2 m_1^2 + l_1^2 n_1^2 + m_1^2 n_1^2].\end{aligned}$$

If we let the radius vector be in a plane through z making an angle φ from the x axis, the vector making an angle θ from z, it is readily shown that the direction cosines are

$$l_1 = \sin\theta \cos\varphi; \quad m_1 = \sin\theta \sin\varphi; \quad n_1 = \cos\theta. \quad (19)$$

Hence if all orientations are equally probable, the average space value of c_{11}' will be

$$\begin{aligned}\langle c_{11}' \rangle_{\text{Av}} &= c_{11} + [2(c_{12} - c_{11}) + 4c_{44}] \frac{\int_0^{2\pi} d\varphi \int_0^{\pi} [\sin^4\theta \sin^2\varphi \cos^2\varphi + \sin^2\theta \cos^2\theta] \sin\theta d\theta}{4\pi} \\ &= c_{11} + \frac{[2(c_{12} - c_{11}) + 4c_{44}]}{5}. \quad (20)\end{aligned}$$

For the values given above, this results in 11.20×10^{11} which is somewhat high.

The attenuation measurements of Figs. 6 and 7 show a rapidly increasing attenuation with frequency that approaches the fourth power of the frequency for high frequencies. In fact, if we express the attenuation according to the equation

$$A = B_1 f + B_2 f^4, \quad (21)$$

a good fit is obtained for both the shear and longitudinal curves for both rods. This indicates that we have a component of attenuation proportional to the frequency and another component proportional to the fourth power of the frequency. The component proportional to the frequency is the same as observed for most metals and solid materials at low frequencies[6] and indicates the presence of an elastic hysteresis. The term proportional to the fourth power is indica-

tive of a scattering of energy similar to the scattering of sound by small particles, which as Rayleigh[7] has shown, produces a scattered energy compared to the incident energy that increases as the fourth power of the frequency.

The data of Fig. 6 for rod No. 1 are fitted well at all frequencies for longitudinal waves if we take

$$B_1 = 0.225 \text{ db/ft./megacycle},$$
$$B_2 = 0.001 \text{ db/ft./megacycle}.^4 \quad (22)$$

For the theoretical values given in the next section, it is desirable to express these as nepers per cm per cycle. Since one neper = 8.68 db and one foot = 30.5 cm, these values become

$$B_1 = 0.845 \times 10^{-9} \text{ neper/cm/cycle};$$
$$B_2 = 3.74 \times 10^{-30} \text{ neper/cm/cycle}^4 \text{ (long)}. \quad (23)$$

Similarly, the shear attenuation is well repre-

---

[6] Wegel and Walther, "Internal dissipation in solids for small cyclic strains," Physics 6 (1935).

[7] Rayleigh, *Theory of Sound* (The Macmillan Company, New York, 1929), Vol. II, p. 152.

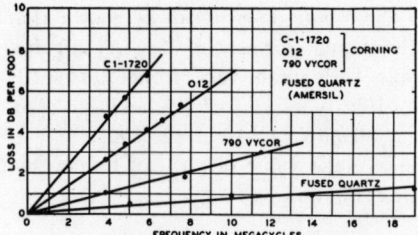

FIG. 8. Measured attenuation for shear waves of three glasses and fused quartz.

sented by the constants,

$B_1 = 0.515 \times 10^{-9}$ neper/cm/cycle;
$B_2 = 50.2 \times 10^{-30}$ neper/cm/cycle^4 (shear). (24)

For rod No. 2, the constants best fitting the curves are

$B_1 = 0.65 \times 10^{-9}$ neper/cm/cycle;
$B_2 = 0.695 \times 10^{-30}$ neper/cm/cycle^4 (long); (25)

and

$B_1 = 0.58 \times 10^{-9}$ neper/cm/cycle;
$B_2 = 9.4 \times 10^{-30}$ neper/cm/cycle^4 (shear). (26)

Since these two rods were supposed to be the same material, but gave considerably different attenuation values, some effort was spent in trying to determine the cause of the difference. Microphotographs were taken of the grain size and it was found that the average grain size[8] of rod No. 1 was 0.23±0.01 mm while that of rod No. 2 was 0.130±0.01 mm. This caused a larger scattering for rod No. 1, as discussed in the next section, and resulted in a higher attenuation.

Figure 8 shows measurements of the losses in three glass rods and a fused quartz rod of optical quality for shear wave transmission. Here any grain size irregularities must be of a very small order, and this is shown experimentally by the strict proportionality between the attenuation and the frequency. Fused quartz has the lowest loss of any material so far measured. The elastic hysteresis values for the glasses are

Material	A, neper/cm/cycle	v in cm/sec.	Q
C-1 1720 glass	4.37×10^{-9}	3.74×10^5 cm/sec.	1,970
012 glass	2.67×10^{-9}	2.80×10^5	4,200
790 Vycor	1.03×10^{-9}	3.58×10^5	8,520
Fused quartz	1.88×10^{-10}	3.76×10^5	44,500

[8] This work was done by E. E. Thomas of the Laboratories.

These values are higher than those measured for low frequencies for longitudinal waves[9] (see Table VIII) and indicate again that the elastic hysteresis is less for shear waves than for longitudinal. The 790 Vycor, which is 96 percent fused quartz, is lower than that of other glasses, and indicates the low loss associated with fused silica.

IV. CALCULATION OF ATTENUATION DUE TO SCATTERED ENERGY

A multi-crystalline rod of aluminum or other metal is made up of a number of small sized crystals that are not exactly lined up. The boundaries between these small crystals or grains can be determined by polishing and etching the surface and taking a microphotograph of the resulting etched surface. Sound scattering can occur because of a difference in density between adjacent elements of a medium or because of a difference in elasticity. It is probable that the difference in density between successive grains is negligible, but a difference in elasticity occurs since the grains are not all lined up and the elasticity depends on grain direction. Rayleigh's formula[7] for the scattering of energy of a single particle is

$$\frac{S.A.}{I.A.} = \frac{\pi T}{R \lambda^2} \left[\frac{\Delta \kappa}{\kappa} + \cos\theta \frac{\Delta \rho}{\rho} \right], \quad (29)$$

where T is the volume of the particle, λ the wavelength, κ the elasticity of the medium, $\Delta \kappa$ the difference in elasticity between the particle and the medium, ρ the density of the medium, θ the angle between the direction of observation and the direction of the incident wave, $S.A.$ the scattered amplitude, $I.A.$ the incident amplitude, and R the distance of the particle from the point of observation. For the present case we can neglect $\Delta \rho / \rho$ since the density between successive particles does not vary.

The total energy scattered from the single particle is proportional to the square of the scattered amplitude integrated over a sphere of radius R. Performing this integration, neglecting $\Delta \rho / \rho$, we find

$$S.E. = (I.A.)^2 \frac{\pi^2 T^2}{\lambda^4 R^2} \times 2\pi R^2 \left(\frac{\Delta \kappa}{\kappa} \right)^2$$
$$\times \int_0^{2\pi} \sin\theta d\theta = (I.A.)^2 \frac{4\pi^3 T^2}{\lambda^4} \left(\frac{\Delta \kappa}{\kappa} \right)^2. \quad (30)$$

[9] W. P. Mason, *Electromechanical Transducer and Wave Filter* (D. Van Nostrand Company, New York, 1942).

Now if we have a large number of grains concentrated in a volume whose cross-sectional area is A and whose length is dx, if we assume that the scattering from all the particles is random and neglect multiple scattering, the total scattered energy will be the sum of the scattered energy from each of the particles or

$$\text{Total S.E.} = (I.A.)^2 \frac{4\pi^3}{\lambda^4} \sum_{k=1}^{N} \left(T_k^2 \left(\frac{\Delta\kappa}{\kappa}\right)^2 \right)_k. \quad (31)$$

If there is no connection between the grain volume T and the inhomegeneity of the elasticity, so that the two can be summed independently, we have

$$\frac{\text{Total S.E.}}{(I.A.)^2} = \frac{4\pi^3}{\lambda^4} \sum_{k=1}^{n} T_k^2 \sum_{k=1}^{N} \frac{\left(\frac{\Delta\kappa}{\kappa}\right)^2}{N}$$

$$= \frac{4\pi^3}{\lambda^4} \sum_{k=1}^{N} T_k^2 \left\langle \left(\frac{\Delta\kappa}{\kappa}\right)^2 \right\rangle_{\text{Av}} \quad (32)$$

where $\langle(\Delta\kappa/\kappa)^2\rangle_{\text{Av}}$ is the space average of the quantity $(\Delta\kappa/\kappa)^2$. For a distribution of particle sizes that does not differ much from the average, the first summation is

$$NT^2 = VT = AdxT \quad (33)$$

where V is the volume under consideration equal to Adx. But since $A(I.A.)^2$ is proportional to the total incident energy, the ratio of the total scattered energy to the total incident energy becomes

$$\frac{T.S.E.}{T.I.E.} = \frac{4\pi^3 dxT}{\lambda^4} \left\langle \left(\frac{\Delta\kappa}{\kappa}\right)^2 \right\rangle_{\text{Av}}. \quad (34)$$

This determines an energy attenuation factor per unit length of

$$E_0 = E_I - E_S = E_I e^{-\alpha dx} = E_I(1-\alpha dx) \quad (35)$$

where E_0 is the energy of the wave out of the section, E_I the incident energy, and E_S the scattered energy which represents a total loss as far as the pick-up crystal is concerned. Hence,

$$\alpha = \frac{4\pi^3 T}{\lambda^4} \left\langle \left(\frac{\Delta\kappa}{\kappa}\right)^2 \right\rangle_{\text{Av}}. \quad (36)$$

In the measurements it was the amplitude attenuation factor that was measured, and this is half the energy attenuation factor. Then since $1/\lambda = f/v$, the constant B_2 of Eq. (21) can be written in the form

$$B_2 = \frac{2\pi^3 T}{v^4} \left\langle \left(\frac{\Delta\kappa}{\kappa}\right)^2 \right\rangle_{\text{Av}}. \quad (37)$$

If there is a range of particle sizes, the value of T tends to be larger than the average particle size obtained by counting the number in a given volume.

An approximate idea[10] of the value of the space average of $\langle(\Delta\kappa/\kappa)^2\rangle_{\text{Av}}$ can be had from the variation of c_{11}' as a function of orientation. From Eq. (17) we have

$$c_{11}' = c_{11} + [2(c_{12}-c_{11})+4c_{44}][l_1^2 m_1^2 + l_1^2 n_1^2 + m_1^2 n_1^2]$$

and

$$\langle c_{11}'\rangle_{\text{Av}} = c_{11} + \frac{[2(c_{12}-c_{11})+4c_{44}]}{5}.$$

Hence,

$$\left\langle \left[\frac{c_{11}' - \langle c_{11}'\rangle_{\text{Av}}}{\langle c_{11}'\rangle_{\text{Av}}}\right]^2 \right\rangle_{\text{Av}}$$

$$= A - B[\sin^4\theta \sin^2\varphi \cos^2\varphi + \sin^2\theta \cos^2\theta]$$
$$+ C[\sin^4\theta \cos^2\varphi \sin^2\varphi + \sin^2\theta \cos^2\theta]^2, \quad (38)$$

where

$$A = \left[\frac{2(c_{12}-c_{11})+4c_{44}}{5c_{11}+2(c_{12}-c_{11})+4c_{44}}\right]^2;$$

$$B = 10A, \quad C = 25A.$$

Integrating this equation over all orientations, we have

$$\left\langle \left(\frac{\Delta\kappa}{\kappa}\right)^2 \right\rangle_{\text{Av}} = \left\langle \left(\frac{c_{11}' - \langle c_{11}'\rangle_{\text{Av}}}{\langle c_{11}'\rangle_{\text{Av}}}\right)^2 \right\rangle_{\text{Av}}$$

$$= A\left(1 - 2 + \frac{25}{21}\right) = \frac{4}{21}A. \quad (39)$$

For the values of Eq. (16) this gives a value of

---

[10] This method for evaluating the inhomogeneity factor was suggested to the writer by Dr. C. Kittel of Massachusetts Institute of Technology. It is related to the R function of Zener, used to calculate the thermoelastic losses in solids, see Phys. Rev. **53**, 90 (1938).

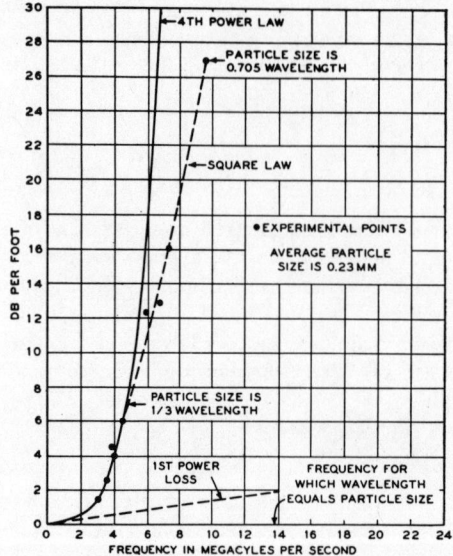

FIG. 9. Scattering losses for aluminum rod when particle size approaches the wave-length.

$\langle(\Delta\kappa/\kappa)\rangle^2_{Av} = 0.0003$. Using this value of $\langle(\Delta\kappa/\kappa)\rangle^2_{Av}$ and the particle diameter of 0.130 mm for the No. 2 rod, the theoretical value of the attenuation should be

$$B_2 = 0.134 \times 10^{-30} \text{ neper/cm/cycle}^4 \quad (40)$$

This compares with the measured value of $B_2 = 0.695 \times 10^{-30}$, which agrees as closely as could be expected considering that the particles scatter shear as well as longitudinal waves and since the shear waves are shorter they should be more efficient sources of scattering. For rod No. 1 the scattering loss should be in the ratio to rod No. 2, of

$$\left(\frac{0.23}{0.13}\right)^2 = 5.5,$$

and this agrees quite exactly with the experimental ratio of 5.4.

Since shear waves are polarized waves, the scattering formula should be somewhat different from that for longitudinal waves. The scattering formula should be the same as for light waves in which an irregularity occurs in the refractive index. This has been shown by Rayleigh[11] to be

$$\frac{S.I.}{I.I.} = \frac{N\pi T^2}{\lambda^4 R^2} \left\langle \left(\frac{\Delta\kappa}{\kappa}\right)^2 \right\rangle_{Av} (1+\cos^2\theta) \quad (41)$$

where $S.I.$ is the scattered intensity, $I.I.$ the incident intensity and θ the angle between the direction of observation and the direction of the incident ray. With this value the amplitude attenuation factor becomes

$$B_2 = \frac{8}{3} \frac{\pi^2 T}{v^4} \left\langle \left(\frac{\Delta\kappa}{\kappa}\right)^2 \right\rangle_{Av}. \quad (42)$$

The average value of $\langle(\Delta\mu'/\mu')\rangle_{Av}^2$ is considerably larger than for the longitudinal case and amounts to about 0.004. Using this value the theoretical attenuation for rod No. 2 should be

$$B_2 = 12.5 \times 10^{-30} \text{ neper/cm/cycle}^4 \quad (43)$$

compared to the measured value of $B_2 = 9.4 \times 10^{-30}$. Since shear waves are more efficient scatterers than longitudinal waves, the measurements agree much better with Rayleigh's formula for shear waves than for longitudinal. The ratio between the shear scattering losses for the two rods is again 5.5 theoretically compared to the experimental value of 5.4.

The fourth-power scattering law should hold as long as the wave-lengths are considerably larger than the grain size. If, however, the wave-length becomes comparable to the grain size, the fourth-power law no longer holds and when the grain size gets large compared to the wave-length, the scattering is independent of frequency in the case of light waves. Figure 9 shows measurements of the scattering loss for shear waves carried up to a frequency for which the grain size is nearly equal to the wave-length. This was done on rod No. 1 which had the larger size grain of 0.23-mm diameter. The fourth-power law shown by the solid line is valid up to the frequency for which the grain size is about 0.33 of the wave-length. Above this point a square law variation holds quite well as shown by the dashed line.

[11] Phil. Mag. 41, 107–120, 274–279 (1871).

Ultrasonic Attenuation Due to Lattice-Electron Interaction in Normal Conducting Metals

W. P. MASON

Bell Telephone Laboratories, Murray Hill, New Jersey
(Received November 12, 1954)

IN a recent letter to the editor,[1] Bömmel published some experimental results on the attenuation of sound waves at ultrasonic frequencies for single lead crystals, which showed that there was an increase in attenuation at very low temperatures for the normal conducting state which disappeared in the superconducting state. This attenuation difference occurred for both shear and longitudinal waves and increased in proportion to the square of the frequency. From 1.6°K to 4°K the difference was independent of the temperature and was 0.106 neper per cm for longitudinal waves of 26.65 Mc/sec and 0.061 neper per cm for shear waves of 9.5 Mc/sec. Figure 1 shows complete measurements for longitudinal waves. Bömmel has recently measured the same effect for a single tin crystal with the results shown by Fig. 2 for a longitudinal wave of 28.5 Mc/sec. A curve of similar shape is obtained for shear waves with a value of 0.036 neper per cm at 17 Mc/sec and 1.5°K for the difference between normal and superconducting states.

It is the purpose of this note to point out that a simple phenomenological concept of the interaction between the lattice vibrations and the electron gas gives values of attenuations which agree well with the measured results.

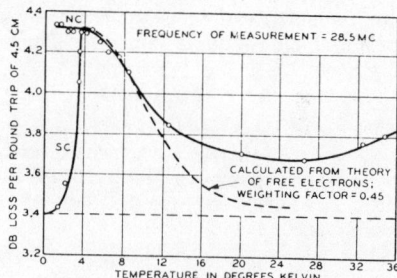

FIG. 2. Comparison of measured attenuation in tin with that calculated from free electron theory.

The concept considered is that in the normal state a lattice vibration can communicate energy to the electron gas by a viscous reaction, i.e., transfer of momentum, and is damped by the viscosity of the gas, while in the superconducting state the lattice is not able to transfer momentum to the electron gas and the damping disappears.

For the most general case the attenuations[2] caused by the energy loss due to the shear and compressional viscosities of the electron gas are for longitudinal and shear waves in the lattice

$$A_l(\text{nepers/cm}) = \frac{2\pi^2 f^2}{\rho v_l^3}\left(\frac{4}{3}\eta + \chi\right); \quad A_s = \frac{2\pi^2 f^2}{\rho v_s^3}\eta, \quad (1)$$

where ρ is the density of the crystal, f is the frequency, η the shear viscosity, χ the compressional viscosity, v_l and v_s the velocities of longitudinal and shear waves respectively. This concept accounts directly for the increase in attenuation with the square of the frequency. The velocities of the waves were measured by Bömmel and were respectively $v_l = 2.35 \times 10^5$ cm/sec, $v_s = 1.266 \times 10^5$ cm²/sec for lead and $v_l = 3.48 \times 10^5$ cm/sec, $v_s = 1.9 \times 10^5$ cm/sec for tin at 1.5°K. With these values the viscosities calculated are

Lead: $\eta = 0.787$ poise; $\chi = 0.05$ poise;
Tin: $\eta = 0.314$ poise; $\chi = 0.03$ poise. (2)

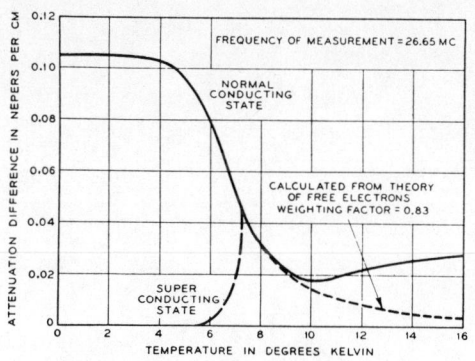

FIG. 1. Comparison of measured attenuation of a lead single crystal with that calculated from free electron theory.

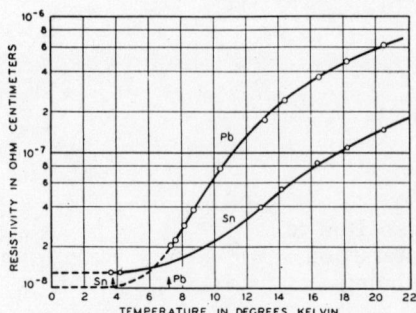

FIG. 3. Resistivity of lead and tin (along tetragonal axis) at low temperatures.

The values of compressional viscosity are small and are probably within the experimental error, nearly zero.

To see if these values are reasonable for the viscosity of an electron gas, we make use of the formula for viscosity[3]:

$$\eta = Nml\bar{v}/3, \qquad (3)$$

where N is the number of electrons per cc, m their mass, l the mean free path (in this case between electrons and lattice atoms) and $\bar{v}$ the mean velocity. These last two quantities can be evaluated from the theory of the free electron gas[4] and are

$$\bar{v}^2 = \frac{3}{5}\frac{h^2}{m^2}(3\pi^2 N)^{\frac{2}{3}}; \quad l = \frac{\sigma m \bar{v}}{Ne^2}, \qquad (4)$$

where e is the charge on the electron, $\hbar$ is Planck's constant divided by 2π, and σ is the electrical conductivity in cgs units. Introducing (4) in (3) and substituting $\sigma = 9 \times 10^{11}/R$, where R is the resistivity in ohm cm, the value of η becomes

$$\eta = 9 \times 10^{11} \hbar^2 (3\pi^2 N)^{\frac{1}{3}} / (5e^2 R). \qquad (5)$$

The value of N for monovalent metals is equal to the number of atoms, which for lead and tin are respectively 3.33×10^{22} and 3.72×10^{22} per cc. For quadrivalent metals no adequate theory exists for correlating the number of electrons with the number of atoms, but experiments[5] on tin indicate that the ratio is between 0.43 and 1.1. All the other values are known and the viscosity is determined in terms of the measured resistivity by the equations

$$\eta_l = (N_e/N_a)^{\frac{1}{3}} \times 8.6 \times 10^{-9}/R; \\ \eta_t = (N_e/N_a)^{\frac{1}{3}} \times 9.2 \times 10^{-9}/R. \qquad (6)$$

The measured resistivities[6] for lead and tin are shown in Fig. 3. Using the values of Fig. 3 and Eq. (6) to determine the viscosity, and substituting in Eq. (1) for a longitudinal wave, the calculated attenuations are shown by the dashed lines. Best agreements are obtained if the factors N_e/N_a are

$$N_e/N_a = 0.75, \text{ lead}; \quad N_e/N_a = 0.3, \text{ tin}.$$

These values account well for the shapes of the measured curves. Above 10°K additional loss occurs due to dislocation motions.

[1] H. E. Bömmel, Phys. Rev. **96**, 220 (1954).
[2] See W. P. Mason's *Piezoelectric Crystals and Their Application to Ultrasonics* (D. Van Nostrand and Company, New York, 1950), p. 478.
[3] See G. Joos, *Theoretical Physics* (Hafner Publishing Company, New York, 1950), p. 562.
[4] See C. Kittel, *Introduction to Solid State Physics* (John Wiley and Company, New York, 1953), Chap. 12.
[5] See E. H. Sondheimer, Advances in Physics **1**, No. 1, 1–43 (1952).
[6] *International Critical Tables*. The value for tin is the value at room temperature for the tetragonal axis reduced by the factors found for polycrystal tin.

Editor's Comments on Paper 37

37 Proctor and Tanttila: *Saturation of Nuclear Electric Quadrupole Energy Levels by Ultrasonic Excitation*

Experiments in microwave spectroscopy indicate that the transition frequencies between nuclear spin energy levels in certain atoms at very low temperatures are in the megahertz range. It was, therefore, tempting to believe that such transitions may be facilitated by the application of ultrasonic radiation in this range, thus providing another example of the tie between high-frequency sound radiation and atomic phenomena. The experiment described in the article "Saturation of Nuclear Electric Quadrupole Energy Levels by Ultrasonic Excitation" was an early substantiation of this presupposition. It led to further work in the same field.

W. G. Proctor (1920–), an American physicist now on the staff of Varian Associates in Palo Alto, California, is known for his work in solid-state physics.

W. H. Tanttila (1922–) is an American physicist now located at the University of Colorado. He is known for his work in low-temperature physics.

Saturation of Nuclear Electric Quadrupole Energy Levels by Ultrasonic Excitation*

W. G. Proctor and W. H. Tanttila

Department of Physics, University of Washington, Seattle, Washington

(Received April 4, 1955)

WE have observed the decrease in the population difference between the degenerate $m=\pm\frac{1}{2}$ and the $m=\pm\frac{3}{2}$ quadrupole energy levels of Cl^{35} in $NaClO_3$ following a long pulse of ultrasonic excitation at the transition frequency. The experiment was performed at the temperature of liquid nitrogen, for which the transition frequency is 30.63 Mc/sec and the thermal relaxation time 0.94 sec.

The population difference was measured by the amplitude of the transient nuclear induction signal following a short (50 μsec) pulse of radio-frequency magnetic flux at the transition frequency.[1] In our experiment, the signals were induced in a second coil, a receiver coil, perpendicular to the exciting or transmitter coil, as suggested by Dean.[2] The sodium chlorate crystal, about 1 cm³ in volume, located between the above coils, received ultrasonic energy in the (1,0,0) direction across a polished face in contact with the polished face of a long, narrow halite crystal. The latter was joined similarly to a second halite crystal, joined in turn to an X-cut quartz crystal, used as an ultrasonic transducer. Rubber-vaseline vacuum grease was used as an interface medium. The halite crystals, about 1 cm² in cross section, were each about 4 cm long, separating the quartz transducer from the $NaClO_3$ sample by about 8 cm. The transducer was excited by a second transmitter of variable frequency; current reached the transducer through a coaxial cable, of which the outer, grounded conductor was flared out at the end to enclose the quartz transducer completely and make contact with the silver coating of its outside face, thus preventing magnetic fields from originating from the transducer to a great extent.

The ultrasonic pulse was 0.3 second in duration; the rf power supplied to the transducer was about 5 watts. After a delay of about 0.03 second, the population difference was examined. The cycle was repeated at 1-second intervals. Depending upon a number of variables, it was observed that the amplitude of the transient was diminished to 20 percent or less of its equilibrium amplitude only when the transducer was excited at the transition frequency. To separate the ultrasonically induced quadrupole transitions from a spurious effect which would be obtained by dipole transitions caused by magnetic flux leaking into the sample region, the resonant transmitter and receiver coils were short-circuited by relays during the ultrasonic excitation period. Further, after a small gap ($\sim\frac{1}{2}$ mm) was introduced between the sample crystal and its neighboring halite crystal, interrupting the path of ultrasonic energy while providing a geometry and transducer loading for which one would expect almost identical leakage fluxes, no attenuation was discernable. A further possible spurious effect, due to the generation of a temperature gradient in the sample crystal, is not likely since the transducer heating should not be frequency-dependent; the transducer, driven at the third harmonic, tuned broadly when loaded.

Quantitative measurements are now under way. The experiment was performed in order to be able to measure the direct and Raman process contributions to the thermal relaxation time.[3]

We would like to thank Professor E. A. Uehling and Dr. C. H. Chang for the stimulation of their continued interest.

* This research was supported by the United States Air Force, through the Office of Scientific Research of the Air Research and Development Command.
[1] M. Bloom and R. E. Norberg, Phys. Rev. 93, 638 (1954).
[2] C. Dean, Phys. Rev. 96, 1053 (1954).
[3] J. Van Kranendonk, Physica 20, 781 (1954).

Author Citation Index

Abello, T. P., 298, 322
Airy, G. B., 139
Alexander, E. A., 402
Alfrey, T., 402
Alleman, R. S., 402
Allen, J. F., 421
Altberg, W., 402
Andreae, J. H., 403
Andree, C. A., 338
Andronikashvilli, E., 449
Arkhipov, R. G., 443
Atkins, K. R., 442, 443, 448, 449
Aybar, S., 402

Bär, H., 402
Baranskii, K. N., 407
Barmatz, M., 454
Barone, A., 195
Barrett, R. E., 402, 403
Basset, A., 402
Bauer, E., 402
Baumgardt, E., 402
Bazulin, P., 402
Becker, R., 402
Bellin, J. L. S., 208
Benioff, H., 80
Beranek, L. L., 80
Bergmann, L., 402
Bergmann, P. B., 402
Beyer, R. T., 204, 208, 402, 403
Biquard, P., 204, 402
Blokintzev, D., 209
Bloom, M., 472
Bolt, R., 210
Boltzmann, L., 303
Bömmel, H. E., 470
Bopp, F., 198, 200, 204
Borgnis, F. E., 194, 197, 198, 201, 203, 204
Born, H., 402
Born, M., 402
Borovik, E., 402
Bourgin, D. G., 303, 318, 342, 402

Boyle, R. W., 402
Boys, C. V., 171
Brillouin, L., 194, 197, 204, 402, 415
British Internal Technical Report, 402
Buff, 403
Burge, E. J., 444
Burgers, J. M., 402
Burton, 149
Burton, C. J., 402
Buss, W., 402

Cady, W. G., 280
Campbell, G. A., 85
Chapman, S., 402
Chrisler, V. L., 336
Chrzanowski, P., 80
Claeys, 402
Cook, R. K., 80
Cowling, T. G., 402
Crandall, I. B., 122, 123, 124
Curie, J., 277
Curie, O. de P., 277
Curie, P., 277

Damkohler, G., 402
Daniels, F. B., 71
Darwin, G., 126
Day, P. R., 452
Dean, C., 472
Debye, P., 311, 402, 416
De Groot, S. R., 402
DeKlerk, D., 453
Denenstein, A., 448
Dessauer, F., 80
Dirac, P. A. M., 402
Donn, W. L., 80
Dorsey, N. E., 199
Drysdale, C. F., 90
Duhem, P. M. M., 158
Dutta, A. K., 402

Earnshaw, S., 138, 139

Author Citation Index

Eckart, C., 194, 206, 402
Einstein, A., 306, 342, 402, 415, 416, 417
Epstein, P. S., 402
Errera, J., 402
Eucken, A., 299, 317, 402
Everest, F. A., 402
Everitt, C. W. F., 448
Eyring, C. F., 325, 402

Faraday, M., 175
Ferrero, M. A., 455
Finch, R. D., 454
Firestone, F. A., 459
Fitzgerald, E. R., 403
Focke, A. B., 403
Fowler, R. H., 402
Fox, F. E., 402, 403
Franchetti, S., 443
Frederick, J. R., 459
Frenkel, J., 402
Frey, A. R., 455
Fricke, E. F., 402, 403
Friedrich, W., 418

Galt, J. K., 402, 403, 459
Gauler, O., 403
Geophysical Journal, 80
Ghosh, B. B., 402
Glasstone, S., 402
Goens, E., 464
Grad, H., 402
Graetz, L., 224
Graffunder, W., 80
Green, 403
Green, H. S., 402
Greene, G., 80
Greenspan, M., 402
Gregg, E. C., Jr., 402
Grobe, A., 402
Gross, E., 415
Grossman, E., 402
Guggenheim, E. A., 402
Gurevich, S. B., 402, 403
Gutenberg, B., 80

Hahnemann, W., 231
Hall, L., 402
Hall, V. C., 114
Halpern, O., 402
Ham, A. C., 444
Hammann, G., 402
Harnwell, G. P., 402

Hartmann, G. K., 403
Heaviside, O., 108
Hecht, H., 231
Henry, P. S. H., 403
Henshaw, D. G., 444
Herschel, J., 22
Hertz, G., 198, 202, 204, 403
Herzberg, G., 403
Herzfeld, K. F., 303, 318, 324, 342, 403
Hopper, F. L., 336
Hsu, E., 403
Hubbard, J. C., 403
Hudson, R. P., 453
Hueter, T., 210
Hugoniot, H., 156
Hunter, J. L., 403
Hurdle, B. G., 403

The Institute of Radio Engineers, 223

Jackson, L. C., 444
Jeans, J., 342, 403
Jenkins, R. T., 206
Joos, G., 403, 470

Kagiwada, R., 454
Kaye, G. W. C., 464
Keesom, A. P., 421
Keesom, W. H., 421
Keller, H. H., 403
Kenkel, F., 268
Kennelly, A. E., 90, 231
King, L. V., 403
Kinsler, L. E., 455
Kirchhoff, G., 179, 298, 403
Kirkwood, J. G., 403
Kittel, C., 403, 457, 470
Kneser, H. O., 316, 318, 342, 403
Knötzel, H., 403
Knötzel, L., 403
Knudsen, V. O., 336, 341, 403
Korn, A. G., 403
Kuhl, W., 451
Kundt, A., 7
Kuper, C. G., 442
Kurowaka, K., 90
Kurti, N., 423
Kurtze, G., 403

Labaw, L. W., 210, 404
Laby, T. H., 464
Laidler, K. J., 402

Lamb, H., 90, 96, 99, 112, 127, 131, 176, 198, 202, 206, 403
Lamb, J., 403
Lambert, J. D., 402
Landau, L. D., 403, 442, 449
Landolt-Börnstein, 403
Langevin, P., 280
Lemmon, K. T., 80
Leonard, R. W., 403, 404
Leontovich, M. A., 402, 403
Leslie, 403
Leslie, J., 22
Liebermann, L. N., 403
Lindberg, A., 403
Lindsay, R. B., 114, 115, 403
Litovitz, T., 403
Loeb, L. B., 403
London, F., 421, 455
Lorentz, H. A., 303, 313
Lu, H., 403
Lucas, R., 403
Luck, D. G. C., 403

McGuinness, W. T., 80
McNamara, F. L., 402
Mallock, A., 171
Mandelstam, L. I., 403
Mariens, P., 404
Markham, J. J., 194, 403
Mason, W. P., 115, 451, 460, 466, 470
Matta, K., 403
Maxwell, J. C., 10, 403
Meixner, J., 403
Mende, H., 198, 202, 204, 403
Mendousse, J. S., 204
Merson, J. M., 402
Meyer, E., 323, 415
Meyer, O. E., 10
Mikhailov, I. G., 403
Miller, C. E., 336
Milne-Thomson, L. M., 195, 196
Moen, C. J., 403
Morse, P. M., 403, 451
Mulders, C. E., 403

Neklepajew, N., 298, 322
Nicolson, A. M., 224
Norberg, R. E., 472
Norris, R. F., 338
Nuovo, R., 195
Nyborg, W. L., 403
Nye, J. F., 457

Oberst, H., 403
Obraztsov, J., 402
Office of Naval Research London, 403
Olson, H. F., 119
O'Neil, H. T., 206, 402
Osterhammel, K., 403
Otpushchennikov, N., 403
Ouang, T.-T., 403

Parker, J. G., 403
Parthasarathy, S., 403
Peierls, R. E., 421
Pellam, J. R., 403, 448, 449, 453, 459
Perolle, M., 31
Peshkov, V. P., 449
Pielemeier, W. H., 298, 322, 403
Pierce, G. W., 277, 295, 298, 318
Pinkerton, J. M. M., 403
Poisson, S. D., 24, 136
Pollack, G. L., 455
Poynting, J. H., 204
Pumper, E. J., 403

Quinn, B. J., 403

Ramm, W., 415
Rankine, W. J. M., 150
Rapuano, R. A., 403
Reynolds, O., 190
Rhodes, J. E., 403
Rice, F. O., 303, 318, 342, 403
Rice, O. K., 318, 322, 324
Rich, D. L., 298, 322
Richards, W. T., 403
Richardson, E. G., 403
Richter, G., 204
Rieckmann, P., 403
Riemann, G. F. B., 144
Ringo, 403
Rocard, Y., 403
Rock, G. D., 402
Rojansky, V., 403
Rollin, B. V., 421
Rosenbaum, B., 443
Rudenko, N. S., 403
Rudnick, I., 449, 454
Rüfer, W., 403

Sabine, P. E., 323
Sacerdote, G. G., 455
Sack, H. S., 402
Sadowski, A., 80

Sakadi, Z., 403
Saris, B. F., 421
Saxon, D., 450
Saxter, L., 80
Saxton, H. L., 403
Schaefer, C., 204
Schafer, K., 403
Schaffhauser, J., 80
Schamberg, R., 404
Schechter, A., 318
Schmidtmüller, N., 403
Schoch, A., 204
Schodder, G. R., 451
Schreuer, E., 403
Schroder, F. K., 451
Schubnikov, L. W., 403
Scott, R. B., 449
Sears, F., 402
Seki, H., 443
Semenoff, N., 318
Sette, D., 403
Shapiro, K. A., 449
Simon, F., 423
Simpson, F. W., 403
Skudrzyk, E., 403
Slater, J. C., 403
Slootmakers, P., 404
Smaluchowski, M., 416
Smirnov, A. A., 207
Smith, M. C., 402, 403
Sokolnikoff, E. S., 403
Sokolnikoff, I. S., 403
Sondheimer, E. H., 470
Sörensen, C., 403
Squire, C. F., 403
Stefan, J., 14
Stephenson, E. B., 403
Stewart, E. S., 403
Stewart, G. W., 114, 115, 403
Stewart, J. L., 403
Stokes, C. G., 298, 299
Stokes, G. G., 10, 137, 149, 403
Strutt, J. W. (Lord Rayleigh), 89, 96, 97, 109, 112, 131, 138, 158, 175, 176, 190, 193, 204, 298, 323, 403, 416, 457, 465, 468

Tamm, K., 403
Taylor, G. B., 402
Teeter, C. E., Jr., 403
Teller, E., 403
Thaler, W. J., 403
Thiessen, 403
Thompson, J. H. C., 404
Thuras, 206
Thys, L., 404
Tisza, L., 404, 449, 455
Tsien, H., 404
Twersky, V., 449

Uddin, M. Z., 421
Uhlenbeck, G. E., 404

van Itterbeek, A., 404
Van Kranendonk, J., 472
van Paemel, O., 404
Verma, G. S., 404
Voigt, W., 224, 227
von Helmholtz, H. L. F., 7
von Lüde, K., 317

Walther, C. H., 465
Wang Chang, C. S., 404
Weaver, W., 204
Weber, H., 149
Wegel, R. L., 231, 465
Westervelt, P. J., 194, 196, 203, 204, 206, 208, 404
Wheland, G. W., 404
White, G. K., 450
Willard, G. W., 404
Williams, A. O., Jr., 210, 404
Willis, F. H., 404
Wilson, D. A., 403
Wilson, O. B., Jr., 404
Winkelmann, A., 224

Young, J. M., 80

Zartman, I. F., 404
Zemansky, M. W., 404
Zener, C. M., 321
Zmuda, A. J., 404

Subject Index

Absorption of sound
  in associated versus nonassociated liquids, 392ff.
  experimental values for gases and liquids, 384ff., 387ff.
  in gas, 336ff.
  at high frequency (ultrasound), 298ff.
  in liquids with large relaxational effects, 390ff.
  methods of measurement, 379ff.
  in real gases, 376
  in solutions, 395ff.
  due to viscosity, 375ff.
  in water, 388ff.
Acoustic beam
  absorption and reflection of, 201ff.
  interaction with surrounding medium, 200
Acoustic capacitance, 88, 90
Acoustic end-fire array, 208ff.
Acoustic filters, 83ff., 92ff.
Acoustic impedance, of orifices, tubes, and horns, 58ff.
Acoustic inertance, 88ff.
Acoustic pressure, 58
Acoustic streaming, 175ff., 181ff.
Acoustic transmission line, 84ff.
Acoustic volume, 58
Acoustic wave propagation in gas mixture, 269ff.
Adiabatic invariant
  in pendulum motion, 186
  for vibrating string, 187
Amplitude of sound waves, 35
Atmosphere
  reflection of sound in, 67ff.
  sound propagation in, 67ff.
Attenuation
  of longitudinal waves in solids, 460ff.
  of shear waves in solids, 460
  of sound
    in aluminum, 464
    in atmosphere, 69ff.
    in a fluid, due to heat conduction, 13ff.
    in a gas, due to relaxation process, 320ff.
    in glass and quartz, 466
    in helium gas, 372
    in a room, due to air absorption, 325ff.
    due to viscosity, 350, 367

Bandpass filter, 111ff.
Beam of sound, 35
Bell-in-jar experiment
  Boyle, 20
  Stokes, 30ff.
Biological effects of ultrasonic radiation, 257ff.
Bose–Einstein statistics, 421
Boundary conditions in acoustic lines, 116ff.
Bow wave of a bullet, 171
Boyle's law, 138
Branch impedance in acoustic filter, 85ff., 100ff.
Brillouin scattering of light by pretersonic radiation, 416ff.
Brillouin's stress tensor in fluids, 196
Bulk viscosity
  effect on sound absorption in a gas, 378
  of a fluid, 375

Cavity resonator for producing pretersonic radiation, 406
Characteristic impedance of a tube, 99
Chemical reactions, affected by ultrasonic radiation, 256ff.
Condensation in fluid flow, 24
Continuity, equation of, for a fluid, 7

Decay of sound intensity in a room, 325ff.
  as function of humidity, temperature, and frequency, 342ff.

Subject Index

Direction cosines of sound ray, 37
Direct method of sound absorption measurement, 381ff.
Dispersion of sound, 272
   in gases, 303ff., 310
   at high frequency (ultrasound), 298ff.
   in solids, 248ff.
   due to viscosity, 351ff., 367
Dissociated gases, sound propagation in, 268ff.

Earthquake waves, 75ff.
Eikonal, 37
   wavefronts, 52ff.
Elastic constants, Stokes relation for, in a fluid, 10
Electric circuit for producing mechanical vibrations, 280ff.
Electric tuning fork, 214
Emulsions and fogs produced by ultrasonic radiation, 253ff.
Energy
   density of, in sound wave in a fluid, 199ff.
   exchanges of, between gas molecules, 316ff.
   propagation of, in finite amplitude waves, 151ff.
Energy spectrum of a quantum liquid, 430ff.
Equation of state, for heat radiation, 351ff.
   in relation to propagation of acoustic waves, 347ff.

Filtration of sound
   in air, 83ff., 96ff.
   in solid rods, 114ff.
Finite amplitude wave, 136ff.
   change of type in, 140ff.
   discontinuity in, 142ff.
   Earnshaw's method, 139ff.
   Hugoniot's method, 152ff.
   of permanent regime, 146
   Rankine's method, 150ff.
   Riemann's method, 144ff.
   thermodynamics of, 153ff.
Fluid, equation of motion of, 8
Fourth sound
   in liquid helium II, 446ff.
   in narrow channel, liquid helium, 444

Gravity waves in water, 73
Group velocity, 190

Harmonic sound waves, 35ff.
High-frequency sound waves (ultrasonics), 240ff.
   Langevin's study of, 241
   pressure due to, 243ff.
High-pass filter, 109
Heat developed by ultrasonic waves, 247ff., 251
Heat radiation, effect on sound propagation, 351ff.
Humidity, effect on sound reverberation in a room, 323ff.

Index of refraction, 37
Infrasonic instruments, 70ff.
Infrasound, 67ff.
   in the atmosphere, 69ff.
Intensity contours, in ray acoustics, 47ff.
Intensity of sound
   calculated from ray pattern, 43ff.
   calculated from wave theory, 53ff.
Interferometer method of sound absorption measurement, 381
Internal energy states in gases, 303ff.
Intersecting plane sound waves, 206

Kirchhoff's circuit laws in acoustic transmission, 101

Lattice–electron interaction in metals, effect on ultrasonic attenuation, 469ff.
Loaded string as a transverse wave filter, 122ff.
Longitudinal vibrations of solid rods, 228ff.
Low-pass filter, 105ff.
Lumped impedance theory of acoustic filtration, 84ff.

Macroscopic hydrodynamical equation of liquid helium II, 433ff.
Macrosonics, 135ff.
Magnetic storms, relation with infrasound, 77ff.
Mass-action law, 271
Mechanical impedance, 58ff.
Microbarom, 73
Microseism, 73ff.
Molar specific heat, 304
Musical sounds, 213ff.

Nonlinear acoustic equations, 176ff.

Ocean waves, 73
Optical method, of sound absorption measurement, 380

Pass and attenuation bands
   in air acoustic filter, 89ff., 106ff.
   in solid acoustic filter, 89ff., 106ff.
Pendulum motion with varying string length, 185ff.
Phase of wave, 35
Phone, 59ff.
Phonometer, 60ff.
Phonon, 432
Piezoelectric crystals, 223ff., 277ff.
Piezoelectric resonator, 223ff.
   construction of, 236ff.
   reaction on electric circuit, 232ff.
Plane-wave propagation in a tube, 96ff.
Poulsen arc, 241
Pretersonic waves in quartz, 406
Propagation constant, in acoustic filters, 104
Pulse method of sound-absorption measurement, 382

Quantization of liquid motion, 425ff.
Quartz piezoelectric crystals, 225ff.
Quartz wind, 175

Radiation pressure
   Boltzmann's method, 191
   Brillouin's method, 195ff.
   in sound waves, 185ff., 193ff.
Range of sound ray, 42, 45
Ray
   acoustics of, validity of, 51ff.
   of sound, 34
   tracing, in sound propagation, 41ff.
Relaxation
   in a fluid, dispersion and absorption due to, 354ff.
   kinetic theory of, 361ff.
   thermodynamic theory of, 363ff.
   time, 300, 307, 317ff, 348
Resonance circle, 230ff.
Reverberation time method, of sound absorption measurement, 382
Rijke tube, 218ff.

Rochelle salt, 224
Roton, 432

Scattering
   of fourth sound, 455
   of light by sound
      Brillouin, 410ff.
      Debye and Sears, 412ff.
   in solids, effect on attenuation, 466ff.
   in solids, by lead single crystal, 469
   of sound by sound, 206ff.
Second sound in liquid helium II, 421ff., 441
Serendipity, in scientific research, 409
Shadow zone in ray acoustics, 53
Shear viscosity, 349, 375
Shock waves, 156ff.
Snell's law of refraction of sound, 40
Solid rod, longitudinal wave propagation through, 115ff.
Sonic ray plotter, 48ff.
Sound
   absorption coefficients of, as function of humidity, 333ff.
   production of, by heat, 214ff.
   propagation of
      in a gas, influence of heat conduction on, 7ff.
      in liquid helium II, 438ff.
   radiation
      from vibrating body into surrounding medium, 22ff.
      from vibrating sphere, 29ff.
   rays, for various temperature gradients, 48ff.
   waves in a tube, attenuation of, 15ff.
Spherical sound waves, in fluid, attenuation of, due to heat conduction, 14
Stationary surface wave in a solid, 130ff.
Stationary ultrasonic waves, 283ff.
Stratosphere, sound propagation in, 70
Structural relaxation, 359ff.
Superfluidity of liquid helium II, 424
Surface waves
   on liquid helium film, 442ff.
   in solids, 126ff.
   velocity of propagation of, 130

Temporal relaxation and absorption, 369ff.
Thermal relaxation, effect on sound radiation, 352ff., 359

Subject Index

Thermal superconductivity, in liquid helium II, 421
Thèvenin's theorem, 107
Third sound, in liquid helium film, 443
Tornadic storms, relation to infrasound, 78ff.
Transmission anomaly in ray acoustics, 46
Troposphere, sound propagation in, 69ff.

Ultrasonic attenuation in solids, 459ff.
Ultrasonic excitation of nuclear electric quadrupole energy levels in solids, 472
Ultrasonic interferometer, 250ff.
Ultrasonic radiation in gases, 282ff.
Ultrasonic scattering in solids, 459ff.
Ultrasonic stationary waves in tubes, 245ff.
Ultrasonic transverse waves in solids, 246ff.
Ultrasonic waves in solids, method of production, 460ff.

Velocity
  of sound
    in the atmosphere, 69ff.
    determined by light scattering, 414
    first, second, and fourth, in liquid helium II, as a function of temperature, 450ff.
    in a fluid, 9
    fourth, in liquid helium II, 449ff.
    fourth, theoretical derivation of, 455ff.
    in a gas, 304
    in helium gas, 372
  of transverse ultrasonic waves in solids, 248ff.
  of ultrasound in gases, 277ff.
Velocity potential in fluid flow, 24
Vibrational energy state of a gas, 316ff.
Viscosity effects in nonlinear acoustics, 178ff.
Vorticity in a fluid, produced by sound waves, 180ff.

Wavefront
  differential equation of, 37
  of sound wave, 34
  of spherical wave, 35